STUDENT SOLUTIONS MANUAL

KEVIN BODDEN   RANDY GALLAHER

LEWIS & CLARK COMMUNITY COLLEGE

ELEMENTARY ALGEBRA

GRAPHS &
AUTHENTIC
APPLICATIONS

JAY LEHMANN

PEARSON

Prentice Hall

Upper Saddle River, NJ 07458

Vice President and Editorial Director, Mathematics: Christine Hoag
Editor-in-Chief: Paul Murphy
Sponsoring Editor: Mary Beckwith
Print Supplement Editor: Georgina Brown
Senior Managing Editor: Linda Mihatov Behrens
Project Manager, Production: Robert Merenoff
Supplement Cover Manager: Paul Gourhan
Supplement Cover Designer: Victoria Colotta
Operations Specialist: Ilene Kahn
Senior Operations Supervisor: Diane Peirano

© 2008 Pearson Education, Inc.
Pearson Prentice Hall
Pearson Education, Inc.
Upper Saddle River, NJ 07458

Printed in the United States of America

10 9 8 7 6 5 4 3 2

ISBN-13: 978-0-13-220168-1   Standalone

ISBN-10:   0-13-220168-2   Standalone

ISBN-13: 978-0-13-220174-2   Component

ISBN-10:   0-13-220174-7   Component

Pearson Education Ltd., *London*
Pearson Education Australia Pty. Ltd., *Sydney*
Pearson Education Singapore, Pte. Ltd.
Pearson Education North Asia Ltd., *Hong Kong*
Pearson Education Canada, Inc., *Toronto*
Pearson Educación de Mexico, S.A. de C.V.
Pearson Education—Japan, *Tokyo*
Pearson Education Malaysia, Pte. Ltd.

Table of Contents

Chapter 1
Introduction to Modeling

Homework 1.1

1. 25 thousand fans attended the Coldplay rock concert.

3. 159 million Americans used cell phones in 2003.

5. 4.4 million iPods® were sold in 2004.

7. The company lost $45 thousand that year.

9. The statement $t = 9$ represents the year 2009 (9 years after 2009).

11. The statement $t = -3$ represents the year 2002 (3 years before 2005)

13. Answers may vary. One possibility is:
Let h be the height (in inches) of a person. Then h can represent the numbers 67 and 72, but h cannot represent the numbers -5 and 0.

15. Answers may vary. One possibility is:
Let p be the price (in dollars) of an audio CD. Then p can represent the numbers 12.99 and 17.99, but p cannot represent the numbers -2 and -8.

17. Answers may vary. One possibility is:
Let T be the total time (in hours) that a person works in a week. Then T can represent the numbers 15 and 40, but T cannot represent the numbers 240 and -10.

19. Answers may vary. One possibility is:
Let s be the annual salary (in thousands of dollars) of a person. Then s can represent the numbers 25 and 32, but s cannot represent the numbers -15 and -9.

21. **a.** Answers may vary. Possibilities follow:

4 inches / 6 inches

3 inches / 8 inches

1 inch / 24 inches

b. In the described situation, the symbols W and L are variables. Their values can change.

c. In the described situation, the symbol A is a constant. Its value is fixed at 24 square inches.

23. **a.** Answers may vary. Possibilities follow:

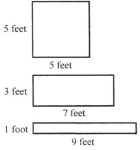

5 feet / 5 feet

3 feet / 7 feet

1 foot / 9 feet

b. In the described situation, the symbols W and L are variables. Their values can change.

c. In the described situation, the symbol P is a constant. Its value is fixed at 20 feet.

25. **a.** Answers may vary. Possibilities follow:

1 inch / 4 inches

2 inches / 5 inches

3 inches / 6 inches

b. In the described situation, the symbols W, L, and A are all variables. All of their values can change.

c. In the described situation, none of the symbols are constants. All of their values can change.

27. **a.** Answers may vary. Possibilities follow:

2 yards / 2 yards

2 yards / 3 yards

2 yards / 4 yards

b. In the described situation, the symbols *L* and *P* are variables. Their values can change.

c. In the described situation, the symbol *W* is a constant. Its value is fixed at 2 yards.

29.

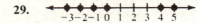

31.

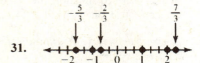

33.

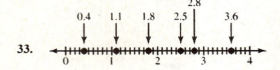

35.

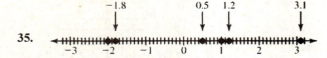

37. The counting numbers between 3 and 8 are 4, 5, 6, and 7.

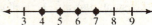

39. The integers between -2 and 2, inclusive, are $-2, -1, 0, 1,$ and 2.

41. The integers between -1 and 4, inclusive, are $-1, 0, 1, 2, 3,$ and 4.

43. The negative integers between -4 and 4 are $-3, -2,$ and -1.

45. The counting numbers in the list are 3 and 356.

47. The only negative integer in the list is -4.

49. The irrational numbers in the list are $\sqrt{7}$ and π.

51. Answers may vary. Three examples of negative integers are: $-2, -5,$ and -7.

53. Answers may vary. Three examples of negative integers less than -7 are: $-8, -9,$ and -27.

55. Answers may vary. Three examples of integers that are not counting numbers are: $-2, -5,$ and -40.

57. Answers may vary. Three examples of rational numbers between 1 and 2 are: $\dfrac{5}{4}, \dfrac{3}{2},$ and $\dfrac{7}{4}$.

59. Answers may vary. Three examples of real numbers between -3 and -2 are: $-2.1, -2.3,$ and -2.8.

61. $\dfrac{10+12+6+9+15+14}{6} = \dfrac{66}{6} = 11$

The average number of units taken per semester is 11 units.

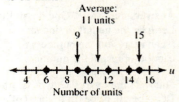

63. $\dfrac{16+14+13+11+10}{5} = \dfrac{64}{5} = 12.8$

The average percentage of disposable personal annual income spent on food is 12.8%.

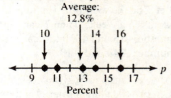

65.

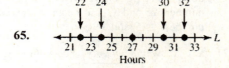

67.

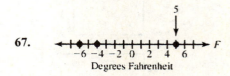

69. a. $\dfrac{0+2+9+19+62}{5} = \dfrac{92}{5} = 18.4$

The average is 18.4 low-carb ice cream products per year.

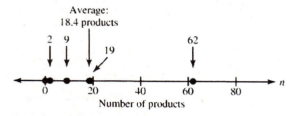

b. The number of low-carb ice cream products increased between 2000 and 2004. The number of products available went up each year.

c. The annual *increases* in the number of low-carb ice cream products increased between 2000 and 2004. The annual increases were

Years	Increase
2000 to 2001	$2 - 0 = 2$
2001 to 2002	$9 - 2 = 7$
2002 to 2003	$19 - 9 = 10$
2003 to 2004	$62 - 19 = 43$

71. a. $\dfrac{86+162+215+247}{4} = \dfrac{710}{4} = 177.5$

The average is 177.5 thousand complaints of identity theft per year.

Average:
177.5 thousand complaints

86   162   215

247

```
+--+--+--+--+--+--+--+--+-- n
0   50  100 150 200 250
```
Thousands of complaints

b. The number of complaints increased between 2001 and 2004. The number of complaints went up each year.

c. The *increases* in the number of complaints per year decreased between 2001 and 2004. The annual increases were

Years	Increase
2001 to 2002	$162 - 86 = 76$
2002 to 2003	$215 - 162 = 53$
2003 to 2004	$247 - 215 = 32$

73. a. -5

b. No. Explanations may vary.
One possibility follows: The words "below zero" make the temperature negative, so T represents negative and positive numbers.

75. a. i. $\dfrac{7+9}{2} = \dfrac{16}{2} = 8$

```
<---+----+----+----+----+--->
    6    7    8    9    10
```

ii. $\dfrac{1+5}{2} = \dfrac{6}{2} = 3$

```
<--+----+----+----+----+----+----+-->
   0    1    2    3    4    5    6
```

iii. $\dfrac{2+8}{2} = \dfrac{10}{2} = 5$

```
<--+---+---+---+---+---+---+---+-->
   1   2   3   4   5   6   7   8   9
```

b. Answers may vary. One possibility follows: The average value is perfectly centered between the two values being averaged.

c. There are infinitely many numbers between 0 and 1.

77. The types of numbers discussed in this section are: real numbers, rational number, irrational numbers, integers, and counting numbers (or natural numbers). Descriptions may vary.

Homework 1.2

1–15 odds.

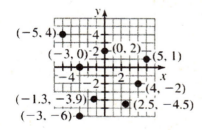

17. The x-coordinate is 2.

19. Presumably, the longer a student studies, the better his or her score will be on the quiz. So, the score s depends on the number of hours studying n. Thus, n is the independent variable and s is the dependent variable.

21. As a girl gets older, she will grow taller (at least for a while). So, the height h depends on the age a. Thus, a is the independent variable and h is the dependent variable.

23. As the number of credits in which a student enrolls increases, so will the tuition. So, the tuition t depends on the number of credits c. Thus, c is the independent variable and t is the dependent variable.

25. As the floor area of a classroom increases, the more students can comfortably fit into the classroom. So, the number of students n who fit comfortably into a classroom depends on the floor area A. Thus, A is the independent variable and n is the dependent variable.

27. The height of the baseball changes as time passes, going up at first and then coming back down. So, the height of the baseball h depends on the number of seconds t after it is hit. Thus, t is the independent variable and h is the dependent variable.

29. The average number of magazine subscriptions sold per week depends on the number of hour worked per week. So, t is the independent variable and n is the dependent variable. The ordered pair (32, 43) means that $t = 32$ and $n = 43$. A telemarketer who works 32 hours per week will sell an average of 43 magazine subscriptions per week.

31. The percentage of volunteers depends on age. So, A is the independent variable and p is the dependent variable. The ordered pair (21, 38) means that $t = 21$ and $n = 38$. This means that 38% of Americans at age 21 years say that they volunteer.

33. The amount of defense spending depends on the year. So, t is the independent variable and b is the dependent variable. The ordered pair (2, 328) means that $t = 2$ and $b = 328$. This means that $328 billion was spent on defense in $2000 + 2 = 2002$.

35. The number of travelers who booked trips online depends on the year. So, t is the independent variable and p is the dependent variable. The ordered pair $(-2, 42)$ means that $t = -2$ and $b = 42$. This means that 42 million travelers booked trips online in $2005 + (-2) = 2003$.

37.

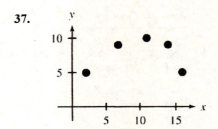

39. Point A is 4 units to the left of the origin and 3 units down. Thus, its coordinates are $(-4, -3)$.

    Point B is 5 units to the left of the origin on the x-axis. Thus, its coordinates are $(-5, 0)$.

    Point C is 2 units to the left of the origin and 4 units up. Thus, its coordinates are $(-2, 4)$.

    Point D is 1 unit to the right of the origin and 3 units up. Thus, its coordinates are $(1, 3)$.

    Point E is 2 units below the origin on the y-axis. Thus, its coordinates are $(0, -2)$.

    Point F is 5 units to the right of the origin and down 4 units. Thus, its coordinates are $(5, -4)$.

41. a.

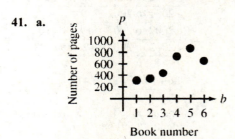

Book number

   b. The fifth book has the greatest number of pages (870 pages).

   c.

Books	Increase in pages
1 to 2	$341 - 309 = 32$
2 to 3	$435 - 341 = 94$
3 to 4	$734 - 435 = 299$
4 to 5	$870 - 734 = 136$
5 to 6	$652 - 870 = -218$

   The greatest increase in the number of pages occurs from the third to the fourth book. This can be seen in the graph because there is a larger vertical change between the third and fourth points.

43. a.

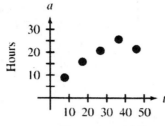

Years since 1960

b. The average time that mothers spent doing paid work was the greatest in 1995. That year, mothers spent an average of 26 hours per week doing paid work.

c. The average time that mothers spent doing paid work was the least in 1965. That year, mothers spent an average of 9 hours per week doing paid work.

45. a.

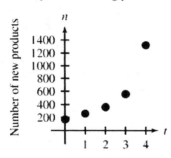

Years since 2000

b. The number of new products that contain Splenda increased over the years 2001 to 2004. The number of products is larger each successive year.

c. The annual *increases* in the number of new products that contain Splenda increased between 2000 and 2004. The annual increases were

Years	Increase in pages
2000 to 2001	$261 - 183 = 78$
2001 to 2002	$365 - 261 = 104$
2002 to 2003	$561 - 365 = 196$
2003 to 2004	$1330 - 561 = 769$

47. a.

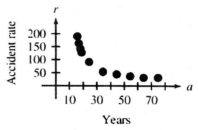

Years

b. The 60–69-year-old age group has the lowest accident rate (31.3).

c. The 16-year-old age group has the highest accident rate (190.3).

d. The greatest change in accident rates occurs between ages 16 and 17 years. The changes were

Years	Change in accident rates
16 to 17	$163.2 - 190.3 = -27.1$
17 to 18	$142.9 - 163.2 = -20.3$
18 to 19	$127.8 - 142.9 = -15.1$
19 to 24.5	$91.4 - 127.8 = -36.4$
24.5 to 34.5	$54.7 - 91.4 = -36.7$
34.5 to 44.5	$43.9 - 54.7 = -10.8$
44.5 to 54.5	$36.4 - 43.9 = -7.5$
54.5 to 64.5	$31.3 - 36.4 = -5.1$
64.5 to 75	$32.1 - 31.3 = 0.8$

We cannot be sure of this statement from the data provided because after age 19, the rates cover a larger range of ages.

e. Answers will vary.

49. a.

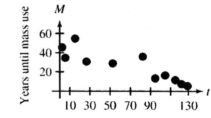

Years since 1870

b. It has taken less time for recent inventions to reach mass use. Explanations may vary.

c. No. It took longer for the microwave to reach mass use than it did for several other earlier inventions.

d. Answers will vary.

e. It took longer for the automobile to reach mass use that it did for earlier inventions of electricity and the telephone. Explanations may vary.

51. a. The average starting salary is highest for the field of computer science. That salary is approximately $53 thousand.

b. The average starting salary is lowest for the field of social science. That salary is approximately $32 thousand.

c. The average beginning salary for employees with a mathematics degree is approximately $44 thousand.

53. The ordered pairs selected and scattergrams will vary. The points will lie on the same vertical line. Explanations will vary.

55. There are three possibilities for the position of the other two vertices.

(1) If the two provided vertices $(2, 1)$ and $(2, 5)$ are on the left side of the square, then the length of the sides of the square is $5 - 1 = 4$. The coordinates of the other two vertices are: $(2 + 4, 1) = (6, 1)$ and $(2 + 4, 5) = (6, 5)$.

(2) If the two provided vertices $(2, 1)$ and $(2, 5)$ are on the right side of the square, then the length of the sides of the square is $5 - 1 = 4$. The coordinates of the other two vertices are: $(2 - 4, 1) = (-2, 1)$ and $(2 - 4, 5) = (-2, 5)$.

(3) If the two provided vertices $(2, 1)$ and $(2, 5)$ are on opposite corners of the square (that is, if they are the ends of a diagonal), then this diagonal given is vertical with length $5 - 1 = 4$. The other diagonal will be horizontal with length 4. Since the diagonals of squares bisect each other, the coordinates of the other vertices are: $\left(2 - 2, \frac{1 + 5}{2}\right) = (0, 3)$ and $\left(2 + 2, \frac{1 + 5}{2}\right) = (4, 3)$.

57. a. For points that lie in quadrant I, the x-coordinate is positive and the y-coordinate is positive.

b. For points that lie in quadrant II, the x-coordinate is negative and the y-coordinate is positive.

c. For points that lie in quadrant III, the x-coordinate is negative and the y-coordinate is negative.

d. For points that lie in quadrant IV, the x-coordinate is positive and the y-coordinate is negative.

59. Answers will vary.

Homework 1.3

1. The line contains the point $(-2, 2)$, so $y = 2$ when $x = -2$.

3. The line contains the point $(6, -2)$, so $x = 6$ when $y = -2$.

5. The line and the x-axis intersect at $(2, 0)$, so the x-intercept is $(2, 0)$.

7. The line contains the point $(-3, -2)$, so $y = -2$ when $x = -3$.

9. The line contains the point $(-6, -3)$, so $x = -6$ when $y = -3$.

11. The line and the y-axis intersect at $(0, -1)$, so the y-intercept is $(0, -1)$.

13. a–b.

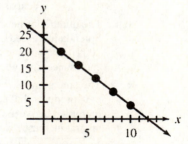

c. The line contains the point $(3, 18)$, so $y = 18$ when $x = 3$.

d. The line contains the point $(9, 6)$, so $x = 9$ when $y = 6$.

e. The line and the y-axis intersect at $(0, 24)$, so the y-intercept is $(0, 24)$.

f. The line and the x-axis intersect at $(12, 0)$, so the x-intercept is $(12, 0)$.

15. a. The line contains the point (2, 18), so $v = 18$ when $t = 2$. This means that, after 2 hours of pumping, 18 thousand gallons of water will be in the basement.

   b. The line contains the point (4.2, 5), so $t = 4.2$ when $v = 5$. This means that 5 thousand gallons will remain in the basement after 4.2 hours of pumping.

   c. The line and the v-axis intersect at (0, 30), so $v = 30$ when $t = 0$. This means that 30 thousand gallons of water were in the basement before any water was pumped out.

   d. The line and the t-axis intersect at (5, 0), so $t = 5$ when $v = 0$. This means that all of the water will pumped out of the basement after 5 hours.

17. No, a line is not a reasonable model . The data points do not lie close to one line.

19. a.

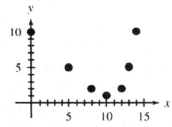

   b. No, there is not a linear relationship between x and y. The data points do not lie close to one line.

21. a.

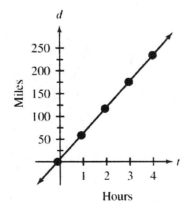

   b. The line contains the point (2.5, 150), so $d = 150$ when $t = 2.5$. This means the student has traveled 150 miles in 2.5 hours.

   c. The line contains the point (3.5, 210), so $t = 3.5$ when $d = 210$. This means that it took the student 3.5 hour to travel 210 miles.

23. a.

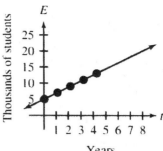

   b. The line contains the point (6, 17), so $E = 17$ when $t = 6$. We predict that the college's enrollment will be 17 thousand students when it has been open for 6 years.

   c. The line contains the point (7, 19), so $t = 7$ when $E = 19$. We predict that the college will be open for 7 years when its enrollment will reach 19 thousand.

25. a.

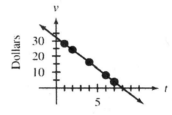

   b. The line contains the point (5, 12), so $t = 5$ when $v = 12$. This means the value of the stock was $12 in 2000 + 5 = 2005.

   c. The line and the t-axis intersect at (8, 0), so $t = 8$ when $v = 0$. This means that the stock will have no value in 2000 + 8 = 2008.

   d. The line and the v-axis intersect at (0, 32), so $v = 32$ when $t = 0$. This means that the value of the stock was $32 in 2000 + 0 = 2000.

27. a.

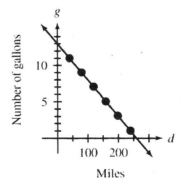

b. The line contains the point (140, 6), so $g = 6$ when $d = 140$. This means that there will still be 6 gallons of gasoline in the tank after the driver has gone 140 miles.

c. The line contains the point (220, 2), so $d = 220$ when $g = 2$. The driver has gone 220 miles when 2 gallons of gasoline remain in the tank.

d. The line and the d-axis intersect at (260, 0), so $d = 260$ when $g = 0$. This means that the gasoline tank will be empty after 260 miles of driving (if no refueling takes place).

e. The line and the g-axis intersect at (0, 13), so $g = 13$ when $d = 0$. This means that the car has a 13-gallon gasoline tank.

29. a.

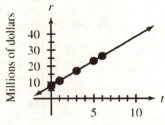

Years since 2000

b. The year 2010 corresponds to $t = 10$. The line contains the point (10, 38), so $r = 38$ when $t = 10$. We predict that the revenue of the company will be $38 million in 2010.

c. The line contains the point (2, 14), so $t = 2$ when $r = 14$. This means the revenue was $14 million in 2000 + 2 = 2002.

d. The line and the r-axis intersect at (0, 8), so $r = 8$ when $t = 0$. This means that the revenue of the company was $8 million in the year 2000.

31. a.

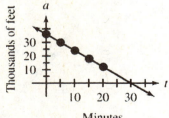

Minutes

b. The line contains the point (12, 22), so $a = 22$ when $t = 12$. This means the altitude of the plane is 22 thousand feet 12 minutes after it began its decent.

c. The line contains the point (30, 0), so $t = 30$ when $d = 0$. This means that it will take 30 minutes for the plane to reach the ground.

d. The prediction in part (c) will be an underestimate. A slower decent the last 2000 feet means it will take longer to reach the ground than predicted.

33. a.

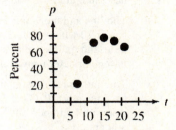

Years since 1980

b. No, there is not a linear relationship between t and p. The data points do not lie close to one line.

35. No. The 5 is the y-coordinate of ordered pair (2, 5), not the y-intercept.

37. No. The y-coordinate of an x-intercept must be 0. The x-intercept might be (2, 0), but not (0, 2).

39. No. An x-intercept is a point that corresponds to an ordered pair with two coordinates, not a single number. The x-intercept might be (5, 0), but not just 5.

41. Answers will vary. One possibility follows:

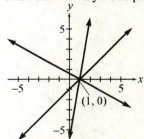

43. a. Answers will vary. One possibility follows:

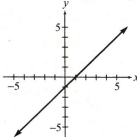

    i. The line shown contains the point $(2, 1)$, so the output is 1 when the input is 2. There is only one output.

    ii. The line shown contains the point $(4, 3)$, so the output is 3 when the input is 4. There is only one output.

    iii. The line shown contains the point $(-3, -4)$, so the output is -4 when the input is -3. There is only one output.

b. For the line shown, a single input leads to a single output.

c. For *any* nonvertical line, a single input leads to a single output.

45. Answers will vary.

Homework 1.4

Throughout this section, answers may vary.

1. a. and c.

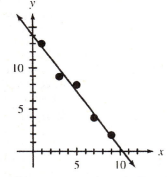

b. The variables are approximately linearly related.

d. $(8, 3.2)$

e. $(5.9, 6)$

f. $(0, 14)$

g. $(10.3, 0)$

3. a–b. First, we list the values of t and n in the table below. For example, $t = 4$ represents 1984 because 1984 is 4 years after 1980.

Years since 1980	Number of Polls
t	n
0	26
4	42
8	50
12	86
16	99
20	136

We then create the scatter plot and linear model.

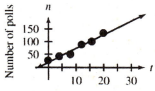

Years since 1980

c. We estimate that the line contains the point $(24, 150)$, so $t = 24$ when $n = 150$. We predict that there were 150 presidential election polls in the first seven months of $1980 + 24 = 2004$.

d. The year 2008 corresponds to $t = 28$. We estimate that the line contains the point $(28, 171)$, so $n = 171$ when $t = 28$. We predict that the year 2008 will have 171 presidential election polls in the first 7 months.

The first 7 months of 2008 consists of about 30 weeks. Now, $171 \div 30 = 5.7$, so we predict that there will be about 6 polls per week in the first 7 months of 2008.

5. a. and c. First, we list the values of t and n in the table below. For example, $t = 5$ represents 1985 because 1985 is 5 years after 1980.

Years since 1980	Number of Species
t	n
0	281
5	384
10	596
15	962
20	1244
25	1264

We then create the scatter plot and linear model.

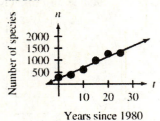

Years since 1980

b. The variables are approximately linearly related.

d. We estimate that the line contains the point (17, 1000), so $t = 17$ when $n = 1000$. We predict that there were 1000 species listed as endangered in $1980 + 17 = 1997$.

e. The year 2011 is represented by $t = 31$. We estimate that the line contains the point (31, 1620), so $n = 1620$ when $t = 31$. We predict that there will be 1620 species listed as endangered or threatened in 2011.

7. a–b. First, we list the values of t and n in the table below. For example, $t = 1$ represents 2001 because 2001 is 1 year after 2000.

Years since 2000	Bumping Rate
t	r
0	20
1	18
2	17
3	15
4	13
5	11

We then create the scatter plot and linear model.

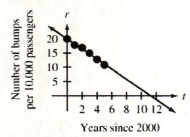

Years since 2000

c. We estimate that the line and the r-axis intersect at (0, 20), so $r = 20$ when $t = 0$. We predict that the voluntary bumping rate in 2000 was 20 bumps per 10,000 passengers.

d. We estimate that the line contains the point (9, 4), so $t = 9$ when $r = 4$. We predict that the bump rate will be 4 bumps per 10,000 passengers in 2009.

e. We estimate that the line and the t-axis intersect at (11, 0), so $t = 11$ when $r = 0$. We predict that there will be no voluntary bumping in 2011. Note: It is highly likely that model breakdown has occurred.

9. a–b.

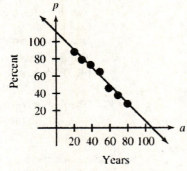

Years

c. We estimate that the line contains the point (19, 92), so $p = 92$ when $a = 19$. We predict that the 92% of Americans at age 19 go to the movies.

d. Note that half is represented by $p = 50\%$. We estimate that the line contains the point (59, 50), so $p = 50$ when $a = 59$. We predict that the half of Americans at age 59 go to the movies.

e. We estimate that the line and the a-axis intersect at (107, 0), so $a = 107$ when $p = 0$. We predict that no 107-year-old Americans go to the movies. Note: Model breakdown has occurred.

11. a–b. First, we list the values of t and n in the table below. For example, $t = 5$ represents 1990 because 1990 is 5 year2 after 1980.

Years since 1980	Average Salary (in thousands of dollars)
t	s
5	24
10	31
15	37
20	42
24	46

We then create the scatter plot and linear model.

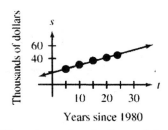

Years since 1980

c. We estimate that the line and the *s*-axis intersect at (0, 19), so $s = 19$ when $t = 0$. We predict that the average salary of public school teachers was $19 thousand in 1980.

d. The year 2010 is represented by $t = 30$. We estimate that the line contains the point (30, 53), so $s = 53$ when $t = 30$. We predict that the average salary of public school teachers will be $53 thousand in 2010.

e. We estimate that the line contains the point (27, 50), so $t = 27$ when $s = 50$. We predict the average salary of public school teachers will be $50 thousand in 1980 + 27 = 2007.

13. a–b.

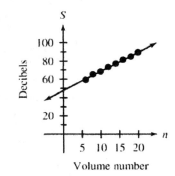

Volume number

c. We estimate that the line contains the point (19, 88), so $S = 88$ when $n = 19$. We predict that the sound level will be 88 decibels when the volume number is 19.

d. Note that the sound level of a noisy street corner is 80 decibels. We estimate that the line contains the point (15, 80), so $n = 15$ when $S = 80$. We predict that the volume number 15 is comparable to a noisy street corner.

15. a. The year 2001 is represented by $t = 6$. We estimate that the linear model contains the point (6, 99). We predict that the infection rate in 2001 is 99 infections per 1000 PCs a month.

b. From the table, the actual rate in 2001 was 113 infections per 1000 PCs a month.

c. Since the predicted rate is lower than the actual rate, the prediction is an underestimate. We can tell this from the graph since the line is under the data point. The error in the estimate is $99 - 113 = -14$ infections per 1000 PCs a month.

17. If the data point (c, p) is below the linear model, then the model will overestimate the value of p when $t = c$.

19. Answers will vary.

21. Answers may vary. One possibility follows: *Linearly related* indicates that the data follow a perfect straight line. *Approximately linearly related* indicates that the data do not follow a perfect straight line, but are close to being in a straight line.

23. Not necessarily. Model breakdown may occur for some points on the line.

Chapter 1 Review Exercises

1. The DVD revenue was $17.5 billion in 2003.

2. $t = 16$ represents the year 1995 + 16 = 2011.

3. Answers may vary. One possibility is: Let *p* be the percentage of students who are full-time students. Then *p* can represent the numbers 60 and 70, but *p* cannot represent the numbers -12 and 107.

4. a. Answers may vary. Possibilities follow:

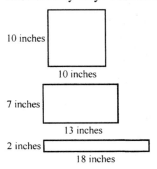

b. In the described situation, the symbols *W* and *L* are variables. Their values can change.

c. In the described situation, the symbol *P* is a constant. Its value is fixed at 40 inches.

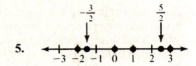

5.

6. The negative integers between −5 and 5 are −4,−3,−2, and −1.

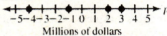

7. The numbers listed (in millions) are: 2, −4,−1, and 3.

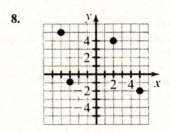

Millions of dollars

8.

9. The *y*-coordinate is −6.

10. The *x*-coordinate is −4.

11. The percentage *p* of home owners depends on age *a*. Thus, *a* is the independent variable and *p* is the dependent variable.

12. Presumably, the more education a person has, the higher his or her salary will be. So, the average salary *a* depends on the years of education *t*. Thus, *t* is the independent variable and *a* is the dependent variable.

13. The number of billionaires depends on the year. So, *t* is the independent variable and *n* is the dependent variable. The ordered pair (4, 313) means that *t* = 4 and *n* = 313. This means that there were 313 U.S. billionaires in the year 2000 + 4 = 2004.

14. The number of injuries depends on the year. So, *t* is the independent variable and *n* is the dependent variable. The ordered pair (10, 10.6) means that *t* = 10 and *n* = 10.6. This means that there were 10.6 thousand injuries on amusement park rides in the year 1990 + 10 = 2000.

15.

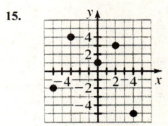

16. a.

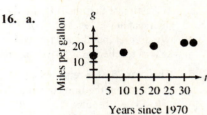

Years since 1970

b. The average gas mileage was highest in the years 2000 and 2003.

c. The average gas mileage was lowest in the year 1970.

17. a. France generates the largest percentage of its electricity by nuclear power. This percentage is about 78%.

b. The United States generates the smallest percentage of it electricity by nuclear power. This percentage is about 20%.

c. Sweden generates about 50% of its electricity by nuclear power.

18. The line contains the point (−2,−1), so $y = -1$ when $x = -2$.

19. The line contains the point (6,−5), so $y = -5$ when $x = 6$.

20. The line contains the point (4,−4), so $x = 4$ when $y = -4$.

21. The line contains the point (−6,1), so $x = -6$ when $y = 1$.

22. The line and the *y*-axis intersect at (0,−2), so the *y*-intercept is (0,−2).

23. The line and the *x*-axis intersect at (−4, 0), so the *x*-intercept is (−4, 0).

24. a–b.

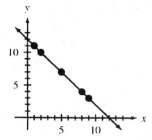

c. The line contains the point (11, 1), so $y = 1$ when $x = 11$.

d. The line contains the point (7, 5), so $x = 7$ when $y = 5$.

e. The line and the x-axis intersect at (12, 0), so the x-intercept is (12, 0).

f. The line and the y-axis intersect at (0, 12), so the y-intercept is (0, 12).

25. a.

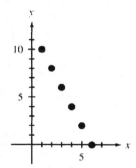

b. The variables x and y are linearly related.

26. a.

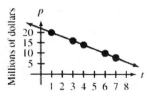

b. The year 2010 is represented by $t = 10$. The line contains the point (10, 2), so $p = 2$ when $t = 10$. We predict the profit will be $2 million in 2010.

c. The line contains the point (2, 18), so $t = 2$ when $p = 18$. We predict that profit was $18 million in the year $2000 + 2 = 2002$.

d. The line and the p-axis intersect at (0, 22), so $p = 22$ when $t = 0$. This means that the profit was $22 billion in the year 2000.

e. The line and the t-axis intersect at (11, 0), so $t = 11$ when $p = 0$. This means that the profit will be $0 in the year 2011.

27. The y-coordinate of an x-intercept of a line is 0.

28. a. and c.

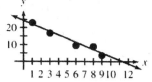

b. The variables are approximately linearly related.

d. (5, 13.5)

e. (2, 20)

f. (0, 24.3)

g. (11.2, 0)

29. a–b. First, we list the values of t and p in the table below. For example, $t = 2$ represents 1992 because 1992 is 2 years after 1990.

Years since 1990	Percent
t	p
0	12
2	13
4	14
6	17
8	18
10	20
12	22
14	25

We then create the scatter plot and linear model.

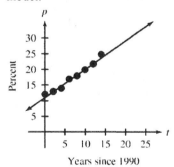

c. The year 2006 corresponds to $t = 16$. We estimate that the line contains the point (6, 26), so $p = 26$ when $t = 6$. We predict that 26% of American Adults were obese in the year 2006.

d. We estimate that the line contains the point (20, 30), so $t = 20$ when $p = 30$. We predict that 30% of American Adults will be obese in the year $1990 + 20 = 2010$.

30. a–b. First, we list the values of *t* and *n* in the table below. For example, *t* = 1 represents 1956 because 1956 is 1 year after 1955.

Years since 1955	Stolen Bases
t	*n*
1	40
2	38
3	31
4	27
5	25
6	18
7	18
8	8

We then create the scatter plot and linear model.

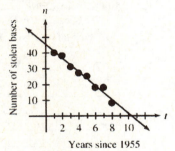

Years since 1955

c. The line and the *n*-axis intersect at (0, 45.2), so *n* = 45.2 when *t* = 0. This means that, according to the model, Mays stole about 45 bases in 1955.

d. The line and the *t*-axis intersect at (10.4, 0), so *t* = 10.4 when *v* = 0. Now 1955 + 10.4 = 1965.4. According to the model, Mays did not steal any bases in 1965.

e. Since the predicted number of stolen bases (45) is higher than the actual number of stolen bases (24), the prediction is an overestimate. Model breakdown has occurred. Explanations will vary.

f. For the year 1971, our linear model will predict a negative number of stolen bases, which is an underestimate. Model breakdown has occurred. Explanations will vary.

Chapter 1 Test

1. a. Answers may vary. Possibilities follow:

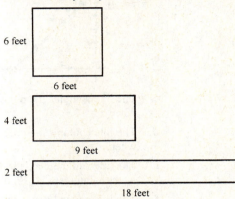

b. In the described situation, the symbols *W* and *L* are variables. Their values can change.

c. In the described situation, the symbol *A* is a constant. Its value is fixed at 36 square feet.

2. The integers between -4 and 2, inclusive, are $-4, -3, -2, -1, 0, 1,$ and 2.

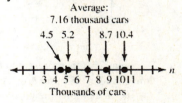

3. The numbers listed are: -5, 7, 2, and -3.

4. $\dfrac{4.5+5.2+7.0+8.7+10.4}{5} = \dfrac{35.8}{5} = 7.16$

The average number of electric cars in use per year is 7.16 thousand cars.

5. As the number of tickets increases, so will the cost. So, the cost *c* depends on the number of tickets *n*. Thus, *n* is the independent variable and *c* is the dependent variable.

6. The percentage of privately owned ATMs depends on the year. So, *t* is the independent variable and *p* is the dependent variable. The ordered pair (3, 27) means that *t* = 3 and *p* = 27. This means that 27% of ATMs were privately owned in the year $2000 + 3 = 2003$.

7. a.

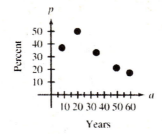

b. The point (21, 50) is the highest point. This means that Americans in the age group 18–24 are the most likely to have been without health insurance.

c. The point (59.5, 17) is the lowest point. This means that Americans in the age group 55–64 are the least likely to have been without health insurance.

8. The line contains the point $(-4, -3)$, so $y = -3$ when $x = -4$.

9. The line contains the point $(4, 1)$, so $x = 4$ when $y = 1$.

10. The line and the y-axis intersect at $(0, -1)$, so the y-intercept is $(0, -1)$.

11. The line and the x-axis intersect at $(2, 0)$, so the x-intercept is $(2, 0)$.

12. a.

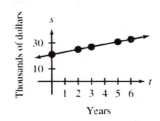

b. The line contains the point (4, 29), so $s = 29$ when $t = 4$. We predict the person's salary will be $29 thousand after she has worked 4 years at the company.

c. The line contains the point (7, 35), so $t = 7$ when $s = 35$. We predict that person's salary will be $35 thousand after she has worked 7 years at the company.

d. The line and the s-axis intersect at (0, 21), so $s = 21$ when $t = 0$. This means that, when the person was initially hired, her salary was $21 thousand.

13. Answers may vary.

14. a–b. First, we list the values of t and n in the table below. For example, $t = 4$ represents 1999 because 1999 is 4 years after 1995.

Years since 1995	Number of Massage Therapists and Body Workers (thousands)
t	n
4	24
5	39
6	41
7	55
8	68
9	81

We then create the scatter plot and linear model.

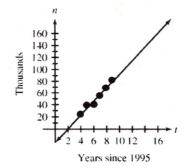

c. The year 2010 corresponds to $t = 15$. We estimate that the line contains the point (15, 145), so $n = 145$ when $t = 15$. We predict that there will be 145 thousand massage therapists and body workers in the year 2010.

d. We estimate that the line contains the point (11, 100), so $t = 11$ when $n = 100$. We predict that there were 100 thousand massage therapists and body workers in the year 1995 + 11 = 2006.

15. If the data point (c, p) is below the linear model, then the model will overestimate the value of p when $t = c$.

Chapter 2
Operations and Expressions

Homework 2.1

1. Substitute 6 for x in $x + 2$:
$(6) + 2 = 8$

3. Substitute 6 for x in $9 - x$:
$9 - (6) = 3$

5. Substitute 6 for x in $7x$:
$7(6) = 42$

7. Substitute 6 for x in $x \div 3$:
$6 \div 3 = 2$

9. Substitute 6 for x in $x + x$:
$(6) + (6) = 12$

11. Substitute 6 for x in $x \cdot x$:
$(6)(6) = 36$

13. Substitute 4 for n in $13n$:
$13(4) = 52$
So, the person spends $52 on audio CDs.

15. Substitute 440 for T in $T \div 5$:
$440 \div 5 = 88$
The student's average test score was 88.

17. a.

Number of Shares	Total Value (dollars)
1	$5 \cdot 1$
2	$5 \cdot 2$
3	$5 \cdot 3$
4	$5 \cdot 4$
n	$5n$

The expression $5n$ represents the total value of the shares.

b. Substitute 7 for n in $5n$:
$5(7) = 35$
So, the total value of 7 shares is $35.

19. a.

Tuition (dollars)	Total Cost (dollars)
400	$400 + 12$
401	$401 + 12$
402	$402 + 12$
403	$403 + 12$
t	$t + 12$

The expression $t + 12$ represents the total cost for a student.

b. Substitute 417 for t in $t + 12$:
$417 + 12 = 429$
So, if tuition is $417, the total cost will be $429.

21. a.

Number of Hours of Courses	Total Cost (dollars)
1	$87 \cdot 1$
2	$87 \cdot 2$
3	$87 \cdot 3$
4	$87 \cdot 4$
n	$87n$

The expression $87n$ represents the total cost (in dollars) of enrolling in n hours of classes.

b. Substitute 15 for n in $87n$:
$87(15) = 1305$
So, the total cost of enrolling in 15 hours of classes is $1305.

23. $x + 4$
Substitute 8 for x in $x + 4$:
$(8) + 4 = 12$

25. $x \div 2$
Substitute 8 for x in $x \div 2$:
$(8) \div 2 = 4$

27. $x - 5$
Substitute 8 for x in $x - 5$:
$(8) - 5 = 3$

29. $7x$
Substitute 8 for x in $7x$:
$7(8) = 56$

31. $16 \div x$
Substitute 8 for x in $16 \div x$:
$16 \div (8) = 2$

33. The quotient of the number and 2.

35. 7 minus the number.

37. 5 more than the number.

39. The product of 9 and the number.

41. Subtract 7 from the number.

43. The product of the number and 2.

45. Substitute 6 for x and 3 for y in the expression $x + y$:
$(6) + (3) = 9$

47. Substitute 6 for x and 3 for y in the expression $x - y$:
$(6) - (3) = 3$

49. Substitute 6 for x and 3 for y in the expression yx:
$(3)(6) = 18$

51. xy
Substitute 9 for x and 3 for y in the expression xy:
$(9)(3) = 27$

53. $x - y$
Substitute 9 for x and 3 for y in the expression $x - y$:
$(9) - (3) = 6$

55. Substitute 62 for r and 3 for t in the expression rt:
$(62)(3) = 186$
So, after 3 hours the car will have traveled 186 miles.

57. Substitute 240 for m and 12 for g in the expression $m \div g$:
$240 \div 12 = 20$
So, on a trip of 240 miles, the car averages 20 miles per gallon.

59. Substitute 315,000 for C and 485,000 for R in the expression $R - C$.
$485,000 - 315,000 = 170,000$
So, if the total revenue is \$485 thousand and the total cost is \$315 thousand, the company will have a profit of \$170 thousand.

61. a.

t	$5t$
1	$5 \cdot 1 = 5$
2	$5 \cdot 2 = 10$
3	$5 \cdot 3 = 15$
4	$5 \cdot 4 = 20$

So, the person earns \$5, \$10, \$15, and \$20 for working 1, 2, 3, and 4 hours, respectively.

b. The person makes \$5 per hour. The hourly rate is a constant while the number of hours worked is a variable. In the expression $5t$, the constant is 5 and the variable is t.

c. Answers may vary. For each additional hour worked, the total earned increases by \$5.

63. a.

t	$50t$
1	$50 \cdot 1 = 50$
2	$50 \cdot 2 = 100$
3	$50 \cdot 3 = 150$
4	$50 \cdot 4 = 200$

So, the person drives 50, 100, 150, and 200 miles after driving 1, 2, 3, and 4 hours, respectively.

b. The person is traveling at a speed of 50 miles per hour. The speed is a constant while the number of hours driving is a variable. In the expression, $50t$, the constant is 50 and the variable is t.

c. Answers may vary. For each additional hour of driving, the total distance driven increases by 50 miles.

65. Answers may vary.

67. Answers may vary.

Homework 2.2

1. The denominator of $\dfrac{3}{7}$ is 7.

3. $20 = 4 \cdot 5$
$= (2 \cdot 2) \cdot 5$
$= 2 \cdot 2 \cdot 5$

5. $36 = 4 \cdot 9$
$= (2 \cdot 2) \cdot (3 \cdot 3)$
$= 2 \cdot 2 \cdot 3 \cdot 3$

7. $45 = 9 \cdot 5$
$= (3 \cdot 3) \cdot 5$
$= 3 \cdot 3 \cdot 5$

9. $78 = 3 \cdot 26$
$= 3 \cdot (2 \cdot 13)$
$= 2 \cdot 3 \cdot 13$

11. $\dfrac{6}{8} = \dfrac{2 \cdot 3}{2 \cdot 2 \cdot 2} = \dfrac{2}{2} \cdot \dfrac{3}{2 \cdot 2} = \dfrac{3}{2 \cdot 2} = \dfrac{3}{4}$

13. $\dfrac{3}{12} = \dfrac{3 \cdot 1}{3 \cdot 2 \cdot 2} = \dfrac{3}{3} \cdot \dfrac{1}{2 \cdot 2} = \dfrac{1}{2 \cdot 2} = \dfrac{1}{4}$

15. $\dfrac{18}{30} = \dfrac{2 \cdot 3 \cdot 3}{2 \cdot 3 \cdot 5} = \dfrac{2 \cdot 3}{2 \cdot 3} \cdot \dfrac{3}{5} = \dfrac{3}{5}$

17. $\dfrac{20}{50} = \dfrac{2 \cdot 2 \cdot 5}{2 \cdot 5 \cdot 5} = \dfrac{2 \cdot 2}{2 \cdot 2} \cdot \dfrac{2}{5} = \dfrac{2}{5}$

19. $\dfrac{5}{25} = \dfrac{5 \cdot 1}{5 \cdot 5} = \dfrac{5}{5} \cdot \dfrac{1}{5} = \dfrac{1}{5}$

21. $\dfrac{20}{24} = \dfrac{2 \cdot 2 \cdot 5}{2 \cdot 2 \cdot 2 \cdot 3} = \dfrac{2 \cdot 2}{2 \cdot 2} \cdot \dfrac{5}{2 \cdot 3} = \dfrac{5}{2 \cdot 3} = \dfrac{5}{6}$

23. $\dfrac{1}{3} \cdot \dfrac{2}{5} = \dfrac{1 \cdot 2}{3 \cdot 5} = \dfrac{2}{15}$

25. $\dfrac{4}{5} \cdot \dfrac{3}{8} = \dfrac{4 \cdot 3}{5 \cdot 8} = \dfrac{2 \cdot 2 \cdot 3}{5 \cdot 2 \cdot 2 \cdot 2} = \dfrac{3}{5 \cdot 2} = \dfrac{3}{10}$

27. $\dfrac{5}{21} \cdot 7 = \dfrac{5}{21} \cdot \dfrac{7}{1} = \dfrac{5 \cdot 7}{21 \cdot 1} = \dfrac{5 \cdot 7}{3 \cdot 7} = \dfrac{5}{3}$

29. $\dfrac{5}{8} \div \dfrac{3}{4} = \dfrac{5}{8} \cdot \dfrac{4}{3} = \dfrac{5 \cdot 4}{8 \cdot 3} = \dfrac{5 \cdot 2 \cdot 2}{2 \cdot 2 \cdot 2 \cdot 3} = \dfrac{5}{2 \cdot 3} = \dfrac{5}{6}$

31. $\dfrac{8}{9} \div \dfrac{4}{3} = \dfrac{8}{9} \cdot \dfrac{3}{4} = \dfrac{8 \cdot 3}{9 \cdot 4} = \dfrac{2 \cdot 2 \cdot 2 \cdot 3}{3 \cdot 3 \cdot 2 \cdot 2} = \dfrac{2}{3}$

33. $\dfrac{2}{3} \div 5 = \dfrac{2}{3} \cdot \dfrac{1}{5} = \dfrac{2 \cdot 1}{3 \cdot 5} = \dfrac{2}{15}$

35. $\dfrac{2}{7} + \dfrac{3}{7} = \dfrac{2+3}{7} = \dfrac{5}{7}$

37. $\dfrac{5}{8} + \dfrac{1}{8} = \dfrac{5+1}{8} = \dfrac{6}{8} = \dfrac{3}{4}$

39. $\dfrac{4}{5} - \dfrac{3}{5} = \dfrac{4-3}{5} = \dfrac{1}{5}$

41. $\dfrac{11}{12} - \dfrac{7}{12} = \dfrac{11-7}{12} = \dfrac{4}{12} = \dfrac{1}{3}$

43. The LCD is 4:
$\dfrac{1}{4} + \dfrac{1}{2} = \dfrac{1}{4} + \dfrac{1}{2} \cdot \dfrac{2}{2} = \dfrac{1}{4} + \dfrac{2}{4} = \dfrac{3}{4}$

45. The LCD is 12:
$\dfrac{5}{6} + \dfrac{3}{4} = \dfrac{5}{6} \cdot \dfrac{2}{2} + \dfrac{3}{4} \cdot \dfrac{3}{3} = \dfrac{10}{12} + \dfrac{9}{12} = \dfrac{19}{12}$

47. The LCD is 3:
$4 + \dfrac{2}{3} = \dfrac{4}{1} \cdot \dfrac{3}{3} + \dfrac{2}{3} = \dfrac{12}{3} + \dfrac{2}{3} = \dfrac{14}{3}$

49. The LCD is 9:
$\dfrac{7}{9} - \dfrac{2}{3} = \dfrac{7}{9} - \dfrac{2}{3} \cdot \dfrac{3}{3} = \dfrac{7}{9} - \dfrac{6}{9} = \dfrac{1}{9}$

51. The LCD is 63:
$\dfrac{5}{9} - \dfrac{2}{7} = \dfrac{5}{9} \cdot \dfrac{7}{7} - \dfrac{2}{7} \cdot \dfrac{9}{9} = \dfrac{35}{63} - \dfrac{18}{63} = \dfrac{17}{63}$

53. The LCD is 5:
$3 - \dfrac{4}{5} = \dfrac{3}{1} \cdot \dfrac{5}{5} - \dfrac{4}{5} = \dfrac{15}{5} - \dfrac{4}{5} = \dfrac{11}{5}$

55. $\dfrac{3172}{3172} = 1$

57. $\dfrac{599}{1} = 599$

59. $\dfrac{842}{0}$ is undefined since division by 0 is not defined.

61. $\dfrac{0}{621} = 0$

63. $\dfrac{824}{631} \cdot \dfrac{631}{824} = \dfrac{824 \cdot 631}{824 \cdot 631} = 1$

65. $\dfrac{544}{293} - \dfrac{544}{293} = \dfrac{0}{293} = 0$

67. Substitute 4 for w and 12 for z in the expression $\dfrac{w}{z}$:

$$\dfrac{4}{12} = \dfrac{2 \cdot 2 \cdot 1}{2 \cdot 2 \cdot 3} = \dfrac{1}{3}$$

69. Substitute 4 for w, 3 for x, 5 for y, and 12 for z in the expression $\dfrac{x}{w} \div \dfrac{y}{z}$:

$$\dfrac{3}{4} \div \dfrac{5}{12} = \dfrac{3}{4} \cdot \dfrac{12}{5} = \dfrac{36}{20} = \dfrac{9}{5}$$

71. Substitute 4 for w, 3 for x, 5 for y, and 12 for z in the expression $\dfrac{x}{w} - \dfrac{y}{z}$: $\dfrac{3}{4} - \dfrac{5}{12}$

The LCD is 12:

$$\dfrac{3}{4} - \dfrac{5}{12} = \dfrac{3}{4} \cdot \dfrac{3}{3} - \dfrac{5}{12} = \dfrac{9}{12} - \dfrac{5}{12} = \dfrac{4}{12} = \dfrac{1}{3}$$

73. $\dfrac{19}{97} \cdot \dfrac{65}{74} \approx 0.17$

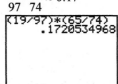

75. $\dfrac{684}{795} \div \dfrac{24}{37} \approx 1.33$

77. $\dfrac{89}{102} - \dfrac{59}{133} \approx 0.43$

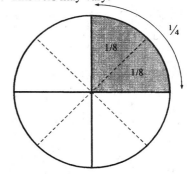

79. Answers may vary.

81. The area of a rectangle is given by the expression LW. Substitute $\frac{2}{5}$ for L and $\frac{1}{4}$ for W in the expression.

$$\dfrac{2}{5} \cdot \dfrac{1}{4} = \dfrac{2}{20} = \dfrac{1}{10}$$

The area of the plot is $\frac{1}{10}$ square mile.

83. Let t = the fraction of points from tests and f = the fraction of points from the final exam. The fraction of points from homework and quizzes is given by the expression $1 - t - f$. Substitute $\frac{1}{2}$ for t and $\frac{1}{4}$ for f in the expression.

$$1 - t - f = 1 - \dfrac{1}{2} - \dfrac{1}{4}$$
$$= \dfrac{1}{1} \cdot \dfrac{4}{4} - \dfrac{1}{2} \cdot \dfrac{2}{2} - \dfrac{1}{4}$$
$$= \dfrac{4}{4} - \dfrac{2}{4} - \dfrac{1}{4}$$
$$= \dfrac{4 - 2 - 1}{4}$$
$$= \dfrac{1}{4}$$

So, $\frac{1}{4}$ of the total points are from homework assignments and quizzes.

85. The quotient of the number and 3.

87.

Number of People	Cost per Person (dollars per person)
2	$\frac{19}{2}$
3	$\frac{19}{3}$
4	$\frac{19}{4}$
5	$\frac{19}{5}$
n	$\frac{19}{n}$

So, if n friends share the pizza, each will pay $\frac{19}{n}$ dollars.

89. a. i. $\dfrac{5}{6} \cdot \dfrac{2}{3} = \dfrac{5 \cdot 2}{6 \cdot 3} = \dfrac{10}{18} = \dfrac{5}{9}$

ii. $\dfrac{5}{6} \div \dfrac{2}{3} = \dfrac{5}{6} \cdot \dfrac{3}{2} = \dfrac{5 \cdot 3}{6 \cdot 2} = \dfrac{15}{12} = \dfrac{5}{4}$

iii. The LCD is 6:
$$\dfrac{5}{6} + \dfrac{2}{3} = \dfrac{5}{6} + \dfrac{2}{3} \cdot \dfrac{2}{2} = \dfrac{5}{6} + \dfrac{4}{6} = \dfrac{9}{6} = \dfrac{3}{2}$$

iv. The LCD is 6:
$$\dfrac{5}{6} - \dfrac{2}{3} = \dfrac{5}{6} - \dfrac{2}{3} \cdot \dfrac{2}{2} = \dfrac{5}{6} - \dfrac{4}{6} = \dfrac{1}{6}$$

b. Answers may vary.

91. Answers may vary. For multiplication, the student did not need to get a common denominator. The problem could be worked by simply multiplying across.
$$\dfrac{1}{2} \cdot \dfrac{1}{3} = \dfrac{1 \cdot 1}{2 \cdot 3} = \dfrac{1}{6}$$

93. Answers may vary. The student should have only multiplied the numerator by 2. Rewrite 2 as $\frac{2}{1}$ and then multiply across.
$$2 \cdot \dfrac{3}{5} = \dfrac{2}{1} \cdot \dfrac{3}{5} = \dfrac{2 \cdot 3}{1 \cdot 5} = \dfrac{6}{5}$$

95. a., b. $\frac{1}{4} = \frac{8}{32}$, $\frac{3}{16} = \frac{6}{32}$, $\frac{1}{8} = \frac{4}{32}$

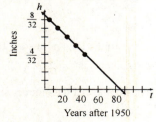

c. 2005 corresponds to $t = 55$. From the graph, the height when $t = 55$ is $h = \frac{3}{32}$ inch.

d. From the graph, the height will be $\frac{1}{16}$ inch when $t = 65$. That is, in 2015.

e. The t-intercept is the point where the graph crosses the t-axis. From the graph, the t-intercept is $(85, 0)$. There will be no grass on the putting greens in 2035 (when $t = 85$). This prediction is very unlikely.

97. a. i. $-1 + (-5) = -6$

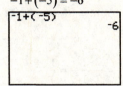

ii. $-6 + (-2) = -8$

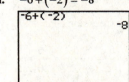

iii. $-3 + (-4) = -7$

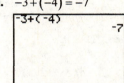

b. The results are all negative.

c. $-4 + (-5) = -9$

d. Answers may vary. To add two negative numbers, add the numbers without the negatives, then make the resulting value negative.

Homework 2.3

1. $-(-4) = 4$

3. $-(-(-7)) = -(7) = -7$

5. $|3| = 3$ because 3 is a distance of 3 units from 0 on a number line.

7. $|-8| = 8$ because -8 is a distance of 8 units from 0 on a number line.

9. $-|4| = -(4) = -4$

11. $-|-7| = -(7) = -7$

13. The numbers have different signs so subtract the smaller absolute value from the larger.
$|-7| - |2| = 7 - 2 = 5$
Since $|-7|$ is greater than $|2|$, the sum is negative.
$2 + (-7) = -5$

15. The numbers have the same sign so add the absolute values.
$|-1| + |-4| = 1 + 4 = 5$
The numbers are negative, so the sum is negative.
$-1 + (-4) = -5$

17. The numbers have different signs so subtract the smaller absolute value from the larger.
$|7| - |-5| = 7 - 5 = 2$
Since $|7|$ is greater than $|-5|$, the sum is positive.
$7 + (-5) = 2$

19. The numbers have different signs so subtract the smaller absolute value from the larger.
$|-8| - |5| = 8 - 5 = 3$
Since $|-8|$ is greater than $|5|$, the sum is negative.
$-8 + 5 = -3$

21. The numbers have the same sign so add the absolute values.
$|-7| + |-3| = 7 + 3 = 10$
The numbers are negative, so the sum is negative.
$-7 + (-3) = -10$

23. The numbers have different signs so subtract the smaller absolute value from the larger.
$|-7| - |4| = 7 - 4 = 3$
Since $|-7|$ is greater than $|4|$, the sum is negative.
$4 + (-7) = -3$

25. $1 + (-1) = 0$ because the numbers are opposites and the sum of opposites is 0.

27. $-4 + 4 = 0$ because the numbers are opposites and the sum of opposites is 0.

29. The numbers have different signs so subtract the smaller absolute value from the larger.
$|-25| - |12| = 25 - 12 = 13$
Since $|-25|$ is greater than $|12|$, the sum is negative.
$12 + (-25) = -13$

31. The numbers have different signs so subtract the smaller absolute value from the larger.
$|-39| - |17| = 39 - 17 = 22$
Since $|-39|$ is greater than $|17|$, the sum is negative.
$-39 + 17 = -22$

33. The numbers have the same sign so add the absolute values.
$|-246| + |-899| = 246 + 899 = 1145$
The numbers are negative, so the sum is negative.
$-246 + (-899) = -1145$

35. $25,371 + (-25,371) = 0$ because the numbers are opposites and the sum of opposites is 0.

37. The numbers have the same sign so add the absolute values.
$|-4.1| + |-2.6| = 4.1 + 2.6 = 6.7$
The numbers are negative, so the sum is negative.
$-4.1 + (-2.6) = -6.7$

39. The numbers have different signs so subtract the smaller absolute value from the larger.

$$\left|-5\right| - \left|0.2\right| = 5 - 0.2 = 4.8$$

Since $\left|-5\right|$ is greater than $\left|0.2\right|$, the sum is negative.

$$-5 + 0.2 = -4.8$$

41. The numbers have different signs so subtract the smaller absolute value from the larger.

$$\left|-99.9\right| - \left|2.6\right| = 99.9 - 2.6 = 97.3$$

Since $\left|-99.9\right|$ is greater than $\left|2.6\right|$, the sum is negative.

$$2.6 + \left(-99.9\right) = -97.3$$

43. The numbers have different signs so subtract the smaller absolute value from the larger.

$$\left|\frac{5}{7}\right| - \left|-\frac{3}{7}\right| = \frac{5}{7} - \frac{3}{7} = \frac{5-3}{7} = \frac{2}{7}$$

Since $\left|\frac{5}{7}\right|$ is greater than $\left|-\frac{3}{7}\right|$, the sum is positive.

$$\frac{5}{7} + \left(-\frac{3}{7}\right) = \frac{2}{7}$$

45. The numbers have different signs so subtract the smaller absolute value from the larger.

$$\left|-\frac{5}{8}\right| - \left|\frac{3}{8}\right| = \frac{5}{8} - \frac{3}{8} = \frac{5-3}{8} = \frac{2}{8} = \frac{1}{4}$$

Since $\left|-\frac{5}{8}\right|$ is greater than $\left|\frac{3}{8}\right|$, the sum is negative.

$$-\frac{5}{8} + \frac{3}{8} = -\frac{1}{4}$$

47. The numbers have the same sign so add the absolute values.

$$\left|-\frac{1}{4}\right| + \left|-\frac{1}{2}\right| = \frac{1}{4} + \frac{1}{2} = \frac{1}{4} + \frac{1}{2} \cdot \frac{2}{2} = \frac{1}{4} + \frac{2}{4} = \frac{3}{4}$$

The numbers are negative, so the sum is negative.

$$-\frac{1}{4} + \left(-\frac{1}{2}\right) = -\frac{3}{4}$$

49. The numbers have different signs so subtract the smaller absolute value from the larger.

$$\left|\frac{5}{6}\right| - \left|-\frac{1}{4}\right| = \frac{5}{6} - \frac{1}{4} = \frac{5}{6} \cdot \frac{2}{2} - \frac{1}{4} \cdot \frac{3}{3} = \frac{10}{12} - \frac{3}{12} = \frac{7}{12}$$

Since $\left|\frac{5}{6}\right|$ is greater than $\left|-\frac{1}{4}\right|$, the sum is

positive.

$$\frac{5}{6} + \left(-\frac{1}{4}\right) = \frac{7}{12}$$

51. $-325.89 + 6547.29 = 6221.4$

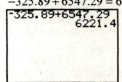

53. $-17,835.69 + \left(-79,735.45\right) = -97,571.14$

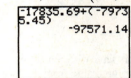

55. $-\dfrac{34}{983} + \left(-\dfrac{19}{251}\right) \approx -0.11$

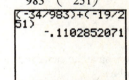

57. Substitute -4 for a and 3 for b in the expression $a + b$ and then find the sum:

$$\left(-4\right) + \left(3\right) = -4 + 3 = -1$$

59. Substitute -4 for a and -2 for c in the expression $a + c$ and then find the sum:

$$\left(-4\right) + \left(-2\right) = -6$$

61. $x + 2$

Substitute -6 for x in the expression and then find the sum:

$$\left(-6\right) + 2 = -4$$

63. $-4 + x$

Substitute -6 for x in the expression and then find the sum:

$$-4 + \left(-6\right) = -10$$

65. The balance is $-75 + 250$ dollars.

The numbers have different signs so subtract the smaller absolute value from the larger.

$$\left|250\right| - \left|-75\right| = 250 - 75 = 175$$

Since $\left|250\right|$ is greater than $\left|-75\right|$, the sum is positive.

$$-75 + 250 = 175$$

So, the balance is $175.

67. We can find the final balance be finding the balance after each transaction.

Transaction	Balance
Transfer	$-89.00 + 300.00 = 211.00$
State Farm	$211.00 - 91.22 = 119.78$
MCI	$119.78 - 44.26 = 75.52$
Paycheck	$75.52 + 870 = 945.52$

So, the final balance is $945.52.

69. The new balance is $-5471 + 2600$.
The numbers have different signs so subtract the smaller absolute value from the larger.
$|-5471| - |2600| = 5471 - 2600 = 2871$
Since $|-5471|$ is greater than $|2600|$, the sum is negative.
$-5471 + 2600 = -2871$
So, the new balance is -2871 dollars.

71. The balance after sending the check is
$-3496 + 2500 = -996$.

The balance after buying the camera is
$-996 + (-629) = -1625$.

The balance after buying the film is
$-1625 + (-8) = -1633$.

So, the final balance is -1633 dollars.

73. The current temperature is $-5 + 9$.
The numbers have different signs so subtract the smaller absolute value from the larger.
$|9| - |-5| = 9 - 5 = 4$
Since $|9|$ is greater than $|-5|$, the sum is positive.
$-5 + 9 = 4$
So, the current temperature is $4°$ F.

75. a.

Weight before Diet (pounds)	Weight after Diet (pounds)
160	$160 + (-20)$
165	$165 + (-20)$
170	$170 + (-20)$
175	$175 + (-20)$
B	$B + (-20)$

From the last row of the table, we see that the expression $B + (-20)$ represents the person's current weight (in pounds).

b. Evaluate $B + (-20)$ for $B = 169$.
$169 + (-20) = 149$
The person's current weight is 149 pounds.

77. a.

Deposit (dollars)	New Balance (dollars)
50	$-80 + 50$
100	$-80 + 100$
150	$-80 + 150$
200	$-80 + 200$
d	$-80 + d$

From the last row of the table, we see that the expression $-80 + d$ represents the new balance (in dollars).

b. Evaluate $-80 + d$ for $d = 125$.
$-80 + 125 = 45$
The new balance is $45.

79. If a is negative and b is negative, the sum $a + b$ will also be negative.

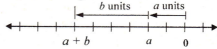

81. If $a + b = 0$ then the numbers are opposites, or the numbers are both 0.

83. a. Substitute -3 for a in $-a$:
$-(-3) = 3$

b. Substitute -4 for a in $-a$:
$-(-4) = 4$

c. Substitute -6 for a in $-a$:
$-(-6) = 6$

d. The student is not correct. The expression $-a$ represents the opposite of a. If a is positive, the result will be negative. However, if a is negative, the result will be positive, and if a is 0, the result is 0.

Homework 2.4

1. $6 - 8 = 6 + (-8) = -2$

3. $-1 - 5 = -1 + (-5) = -6$

5. $2 - (-7) = 2 + 7 = 9$

7. $-3 - (-2) = -3 + 2 = -1$

9. $4 - 7 = 4 + (-7) = -3$

11. $4 - (-7) = 4 + 7 = 11$

13. $-3 - 3 = -3 + (-3) = -6$

15. $-54 - 25 = -54 + (-25) = -79$

17. $381 - (-39) = 381 + 39 = 420$

19. $2.5 - 7.9 = 2.5 + (-7.9) = -5.4$

21. $-6.5 - 4.8 = -6.5 + (-4.8) = -11.3$

23. $3.8 - (-1.9) = 3.8 + 1.9 = 5.7$

25. $13.6 - (-2.38) = 13.6 + 2.38 = 15.98$

27. $-\dfrac{1}{3} - \dfrac{2}{3} = -\dfrac{1}{3} + \left(-\dfrac{2}{3}\right) = -\dfrac{3}{3} = -1$

29. $-\dfrac{1}{8} - \left(-\dfrac{5}{8}\right) = -\dfrac{1}{8} + \dfrac{5}{8} = \dfrac{4}{8} = \dfrac{1}{2}$

31. $\dfrac{1}{2} - \left(-\dfrac{1}{4}\right) = \dfrac{1}{2} + \dfrac{1}{4} = \dfrac{1}{2} \cdot \dfrac{2}{2} + \dfrac{1}{4} = \dfrac{2}{4} + \dfrac{1}{4} = \dfrac{3}{4}$

33. $-\dfrac{1}{6} - \dfrac{3}{8} = -\dfrac{1}{6} + \left(-\dfrac{3}{8}\right)$

$$= -\dfrac{1}{6} \cdot \dfrac{4}{4} + \left(-\dfrac{3}{8} \cdot \dfrac{3}{3}\right)$$

$$= -\dfrac{4}{24} + \left(-\dfrac{9}{24}\right)$$

$$= -\dfrac{13}{24}$$

35. $-5 + 7 = 2$

37. $-6 - (-4) = -6 + 4 = -2$

39. $\dfrac{3}{8} - \dfrac{5}{8} = \dfrac{3}{8} + \left(-\dfrac{5}{8}\right) = -\dfrac{2}{8} = -\dfrac{1}{4}$

41. $-4.9 - (-2.2) = -4.9 + 2.2 = -2.7$

43. $-2 + (-5) = -7$

45. $10 - 12 = 10 + (-12) = -2$

47. $-234.913 - 2893.26 \approx -3128.17$

```
-234.913-2893.26
         -3128.173
```

49. $29{,}643.52 - (-83{,}284.39) = 112{,}927.91$

```
29643.52-(-83284
.39)
         112927.91
```

51. $-\dfrac{17}{89} - \dfrac{51}{67} \approx -0.95$

```
(-17/89)-(51/67)
       -.9522052658
```

53. $7 - 19 = 7 + (-19) = -12$

So, the current temperature is $-12°\text{F}$.

55. $7 - (-4) = 7 + 4 = 11$

The change in temperature is $11°\text{F}$.

57. **a.** $-4 - (8) = -4 + (-8) = -12$

The change in temperature is $-12°\text{F}$.

b. To estimate the change in temperature over the past hour, we divide the change over two hours by 2.

$$\dfrac{-12}{2} = -6$$

The estimated change in temperature over the past hour is $-6°\text{F}$.

c. Answers may vary. The change in temperature is affected by the time of day in addition to the weather conditions. Thus, temperature change need not be uniform.

59. $20,320 - (-282) = 20,320 + 282 = 20,602$

The change in elevation is 20,602 feet.

61. a.

Year	Percent who voted	Change in percentage points
1980	59.2	–
1984	59.9	$59.9 - 59.2 = 0.7$
1988	57.4	$57.4 - 59.9 = -2.5$
1992	61.9	$61.9 - 57.4 = 4.5$
1996	54.2	$54.2 - 61.9 = -7.7$
2000	54.7	$54.7 - 54.2 = 0.5$
2004	60.7	$60.7 - 54.7 = 6.0$

b. The greatest increase in percent turnout was 6.0% between 2000 and 2004.

c. The absolute value of the change in percent turnout between 1992 and 1996 is larger than the others. The large decrease in percent turnout suggests that not many of those 11 million people voted.

63. a.

Years	Changes in Retail Sales (billions of $)	Retail Sales (billions of $)
1996 – 1997	0.0	2.6
1997 – 1998	–1.1	$2.6 + (-1.1) = 1.5$
1998 – 1999	–0.1	$1.5 + (-0.1) = 1.4$
1999 – 2000	–0.2	$1.4 + (-0.2) = 1.2$
2000 – 2001	0.0	$1.2 + 0.0 = 1.2$
2001 – 2002	0.3	$1.2 + 0.3 = 1.5$
2002 – 2003	0.6	$1.5 + 0.6 = 2.1$

So, the sales in 2003 were $2.1 billion.

b. Increasing sales are indicated by positive changes. Thus, retail sales were increasing during the period 2001 – 2003.

c. Decreasing sales are indicated by negative changes. Thus, retail sales were decreasing during the period 1997 – 2000.

65. a.

Score on the Second Exam (points)	Change in Score (points)
80	$80 - 87$
85	$85 - 87$
90	$90 - 87$
95	$95 - 87$
p	$p - 87$

From the last row of the table, we see that the expression $p - 87$ represents the change in score (in points) from the first exam to the second exam.

b. Evaluate $p - 87$ for $p = 81$:

$81 - 87 = 81 + (-87) = -6$

So, if a student has a score of 81 on the second exam, their score decreased 6 points from their first exam.

67. a.

Change in Enrollment	Current Enrollment
100	$100 + 24,500$
200	$200 + 24,500$
300	$300 + 24,500$
400	$400 + 24,500$
c	$c + 24,500$

From the last row of the table, we see that the expression $c + 24,500$ represents the current enrollment.

b. Evaluate $c + 24,500$ for $c = -700$:

$-700 + 24,500 = 23,800$

So, if the change in enrollment is -700, the current enrollment would be 23,800 due to the decrease in enrollment of 700 students in the past year.

69. The student should have changed the subtraction to adding the opposite instead of subtracting the opposite.

$7 - (-5) = 7 + 5 = 12$

71. a. i. $b - a = 5 - 3 = 5 + (-3) = 2$

ii. $9 - 1 = 9 + (-1) = 8$

iii. $7 - 2 = 7 + (-2) = 5$

b. Answers may vary. Since the quantity increased, the final number is larger than the beginning number. When finding the change in quantity, we subtract the beginning number from the final number. Since the final number is bigger, the result will be positive.

73. Evaluate $a+b$ for $a=-5$ and $b=2$:
$$(-5)+(2)=-5+2=-3$$

75. Evaluate $a-b$ for $a=-5$ and $b=2$:
$$(-5)-(2)=-5+(-2)=-7$$

77. Evaluate $b-c$ for $b=2$ and $c=-7$:
$$(2)-(-7)=2+7=9$$

79. $-3-x$
Evaluate the expression for $x=-5$:
$$-3-(-5)=-3+5=2$$

81. $x-8$
Evaluate the expression for $x=-5$:
$$(-5)-8=-5+(-8)=-13$$

83. $x-(-2)$
Evaluate the expression for $x=-5$:
$$(-5)-(-2)=-5+2=-3$$

85. a. i. $-2(5)=-10$

ii. $-4(6)=-24$

iii. $-7(9)=-63$

b. Answers may vary. The results are all negative.

c. $-3(7)=-21$

d. Answers may vary. To multiply two numbers with different signs, multiply their absolute values and make the result negative.

87. a. $(8)-(5)=8+(-5)=3$

b. $(5)-(8)=5+(-8)=-3$

c. The results have the same absolute value, but different signs.

d. $a-b$ for $a=-2$ and $b=4$:
$$(-2)-(4)=-2+(-4)=-6$$
$b-a$ for $a=-2$ and $b=4$:
$$(4)-(-2)=4+2=6$$
The results have the same absolute value, but different signs.

e. Answers may vary. The results should have the same absolute value, but different signs.

f. The results for $a-b$ and $b-a$ will have the same absolute value, but different signs.

Homework 2.5

1. $63\%=63.0\%=0.63$

3. $9\%=9.0\%=0.09$

5. $0.08=8\%$

7. $7.3\%=0.073$

9. $0.052=5.2\%$

11. $0.35(8)=2.8$; so, 35% of \$8 is \$2.80.

13. $0.05(2500)=125$; so, 5% of 2500 students is 125 students.

15. $0.025(7000)=175$; so, 2.5% of 7000 cars is 175 cars.

17. Since the numbers have different signs, the product is negative: $-2(6)=-12$

19. Since the numbers have the same sign, the product is positive: $-3(-6)=18$

21. Since the numbers have different signs, the product is negative: $1(-1)=-1$

23. Since the numbers have different signs, the quotient is negative: $-40 \div 5 = -8$

25. Since the numbers have different signs, the quotient is negative: $25 \div (-5) = -5$

27. Since the numbers have the same sign, the quotient is positive: $-56 \div (-7) = 8$

29. Since the numbers have the same sign, the product is positive: $-15(-37) = 555$

31. Since the numbers have different signs, the quotient is negative: $936 \div (-24) = -39$

33. Since the numbers have the same sign, the product is positive: $-0.2(-0.4) = 0.08$

35. Since the numbers have different signs, the product is negative: $2.5(-0.39) = -0.975$

37. Since the numbers have different signs, the quotient is negative: $-0.06 \div 0.2 = -0.3$

39. Since the numbers have different signs, the quotient is negative: $\dfrac{36}{-4} = 36 \div (-4) = -9$

41. Since the numbers have the same sign, the quotient is positive: $\dfrac{-32}{-8} = -32 \div (-8) = 4$

43. Since the numbers have different signs, the product is negative: $\dfrac{1}{2}\left(-\dfrac{1}{5}\right) = -\dfrac{1}{10}$

45. Since the numbers have the same sign, the product is positive: $\left(-\dfrac{4}{9}\right)\left(-\dfrac{3}{20}\right) = \dfrac{12}{180} = \dfrac{1}{15}$

47. Since the numbers have different signs, the quotient is negative:
$$-\dfrac{3}{4} \div \dfrac{7}{6} = -\dfrac{3}{4} \cdot \dfrac{6}{7} = -\dfrac{18}{28} = -\dfrac{9}{14}$$

49. Since the numbers have the same sign, the quotient is positive:
$$-\dfrac{24}{35} \div \left(-\dfrac{16}{25}\right) = \dfrac{24}{35} \cdot \dfrac{25}{16} = \dfrac{600}{560} = \dfrac{15}{14}$$

51. $6 + (-9) = -3$

53. $-39 \div (-3) = 13$

55. $4 - (-2) = 4 + 2 = 6$

57. $10(-10) = -100$

59.
$$-\dfrac{3}{4} + \dfrac{1}{2} = -\dfrac{3}{4} + \dfrac{1}{2} \cdot \dfrac{2}{2}$$
$$= -\dfrac{3}{4} + \dfrac{2}{4}$$
$$= \dfrac{-3 + 2}{4}$$
$$= -\dfrac{1}{4}$$

61. $\left(-\dfrac{10}{7}\right)\left(-\dfrac{14}{15}\right) = \dfrac{140}{105} = \dfrac{4}{3}$

63.
$$\dfrac{3}{4} - \dfrac{5}{3} = \dfrac{3}{4} \cdot \dfrac{3}{3} - \dfrac{5}{3} \cdot \dfrac{4}{4}$$
$$= \dfrac{9}{12} - \dfrac{20}{12}$$
$$= \dfrac{9 - 20}{12}$$
$$= \dfrac{-11}{12}$$
$$= -\dfrac{11}{12}$$

65. $-\dfrac{3}{8} \div \dfrac{5}{6} = -\dfrac{3}{8} \cdot \dfrac{6}{5} = -\dfrac{18}{40} = -\dfrac{9}{20}$

67. $\dfrac{-16}{20} = -\dfrac{2 \cdot 2 \cdot 2 \cdot 2}{2 \cdot 2 \cdot 5} = -\dfrac{2 \cdot 2}{5} = -\dfrac{4}{5}$

69. $\dfrac{-18}{-24} = \dfrac{2 \cdot 3 \cdot 3}{2 \cdot 2 \cdot 2 \cdot 3} = \dfrac{3}{2 \cdot 2} = \dfrac{3}{4}$

71. $\dfrac{3}{-4} + \dfrac{1}{4} = \dfrac{-3}{4} + \dfrac{1}{4} = \dfrac{-3 + 1}{4} = \dfrac{-2}{4} = -\dfrac{1}{2}$

73. $\dfrac{4}{7} - \left(\dfrac{3}{-7}\right) = \dfrac{4}{7} + \dfrac{3}{7} = \dfrac{4 + 3}{7} = \dfrac{7}{7} = 1$

75. $\dfrac{5}{6} + \dfrac{7}{-8} = \dfrac{5}{6} + \dfrac{-7}{8}$

$= \dfrac{5}{6} \cdot \dfrac{4}{4} + \dfrac{-7}{8} \cdot \dfrac{3}{3}$

$= \dfrac{20}{24} + \dfrac{-21}{24}$

$= \dfrac{20 + (-21)}{24}$

$= -\dfrac{1}{24}$

77. $-26.87(-381.572) \approx 10{,}252.84$

```
-26.87(-381.572)
      10252.83964
```

79. $222.045 \div (-32.76) \approx -6.78$

```
222.045/(-32.76)
     -6.777930403
```

81. $-\dfrac{11}{18}\left(-\dfrac{15}{19}\right) \approx 0.48$

```
(-11/18)(-15/19)
      .4824561404
```

83. $-\dfrac{59}{13} \div \dfrac{27}{48} \approx -8.07$

```
(-59/13)/(27/48)
     -8.068376068
```

85. Evaluate ab for $a = -6$ and $b = 4$:
$(-6)(4) = -24$

87. Evaluate $\dfrac{a}{b}$ for $a = -6$ and $b = 4$:

$\dfrac{-6}{4} = -\dfrac{3}{2}$

89. Evaluate $-ac$ for $a = -6$ and $c = -8$:
$-(-6)(-8) = -(48) = -48$

91. Evaluate $-\dfrac{b}{c}$ for $b = 4$ and $c = -8$:

$-\dfrac{4}{(-8)} = \dfrac{4}{8} = \dfrac{1}{2}$

93. $\dfrac{w}{2}$

Evaluate the expression for $w = -8$:

$\dfrac{(-8)}{2} = -4$

95. $w(-5)$

Evaluate the expression for $w = -8$:

$(-8)(-5) = 40$

97. $\dfrac{6}{8} = \dfrac{3}{4}$

99. $\dfrac{1776 \text{ ft}}{790 \text{ ft}} = \dfrac{888}{395} \approx \dfrac{2.25}{1}$

So, the Freedom Tower would be 2.25 times as tall as the John Hancock Tower.

101. $\dfrac{313 \text{ billionaires}}{228 \text{ billionaires}} \approx \dfrac{1.37}{1}$

So, there were 1.37 times as many billionaires in the U.S. in 2004 as in 2002.

103. a. $\dfrac{4 \text{ red bell peppers}}{5 \text{ black olives}} = \dfrac{0.8 \text{ red bell pepper}}{1 \text{ black olives}}$

For each black olive used, 0.8 red bell pepper is required.

b. $\dfrac{5 \text{ black olives}}{4 \text{ red bell peppers}} = \dfrac{1.25 \text{ black olives}}{1 \text{ red bell pepper}}$

For each red bell pepper used, 1.25 black olives would be required.

105. a. $\dfrac{37{,}682.49}{2951.0} \approx \dfrac{12.77}{1}$

The FTE enrollment is 12.77 times as large at Texas A&M University than at St. Olaf College.

b. $\dfrac{982.4}{335.1} \approx \dfrac{2.93}{1}$

The FTE faculty at the University of Massachusetts – Amherst is 2.93 times as large as that at Butler University.

c. Butler: $\dfrac{4168.0}{335.1} \approx \dfrac{12.44}{1}$

St. Olaf: $\dfrac{2951.0}{248.7} \approx \dfrac{11.87}{1}$

Stonehill College: $\dfrac{2351.0}{178.0} \approx \dfrac{13.21}{1}$

UMass – Amherst: $\dfrac{17,016.2}{982.4} \approx \dfrac{17.32}{1}$

Texas A&M: $\dfrac{37,682.49}{1785.9} = \dfrac{21.1}{1}$

d. Texas A&M University has the largest of FTE enrollment to FTE faculty, while St. Olaf College has the smallest.

e. Answers may vary. The individual is not correct. Although Stonehill College has the smallest FTE enrollment, it also has the smallest FTE faculty.

107. a. $\dfrac{-4360 \text{ dollars}}{-1825 \text{ dollars}} = \dfrac{872}{365} \approx \dfrac{2.39}{1}$

b. For each $1 she pays towards her MasterCard account, she should pay about $2.39 towards her Discover account.

109. $0.15(3720) = 558$
$-3720 + 558 = -3162$
The new balance would be -3162 dollars.

111. $12.3(2.40) = 29.52$
$0 - 29.52 = -29.52$
The new balance is -29.52 dollars.

113. a. $-2 + (-4) = -6$

b. $-2(-4) = 8$

c. The second statement is clearer because it indicates an operation as well. From parts (a) and (b), we see that just having two negative numbers is not enough to guarantee that the result is positive.

d. Answers may vary.

115. $\dfrac{a}{b} = \dfrac{-a}{-b}$; $\dfrac{-a}{b} = \dfrac{a}{-b} = -\dfrac{a}{b} = -\dfrac{-a}{-b}$

117. Answers may vary.
$4(-5) = (-5) + (-5) + (-5) + (-5)$

119. If ab is negative, we can say that the two numbers have different signs. We cannot say which is positive and which is negative, but we do know they have different signs.

121. If $ab = 0$ then at least one of the numbers must be 0. That is, either $a = 0$, $b = 0$, or both.

123. a. $(8 \div 2) \cdot 4 = 4 \cdot 4 = 16$

b. $8 \div (2 \cdot 4) = 8 \div 8 = 1$

c. The results are different, so the order of the operations makes a difference.

Homework 2.6

1. $4^3 = 4 \cdot 4 \cdot 4 = 16 \cdot 4 = 64$

3. $2^5 = 2 \cdot 2 \cdot 2 \cdot 2 \cdot 2$
$= 4 \cdot 2 \cdot 2 \cdot 2$
$= 8 \cdot 2 \cdot 2$
$= 16 \cdot 2$
$= 32$

5. $-8^2 = -(8 \cdot 8) = -64$

7. $(-8)^2 = (-8)(-8) = 64$

9. $\left(\dfrac{6}{7}\right)^2 = \left(\dfrac{6}{7}\right)\left(\dfrac{6}{7}\right) = \dfrac{36}{49}$

11. $3 \cdot (5 - 1) = 3 \cdot 4 = 12$

13. $(2 - 5)(9 - 3) = (-3)(6) = -18$

15. $4 - (3 - 8) + 1 = 4 - (-5) + 1$
$= 4 + 5 + 1$
$= 9 + 1$
$= 10$

17. $\dfrac{1-10}{1+2} = \dfrac{-9}{3} = -3$

19. $\dfrac{2-(-3)}{5-7} = \dfrac{2+3}{5-7} = \dfrac{5}{-2} = -\dfrac{5}{2}$

21. $\dfrac{4-(-6)}{-7-8} = \dfrac{4+6}{-7-8} = \dfrac{10}{-15} = -\dfrac{2}{3}$

23. $6+8\div 2 = 6+4 = 10$

25. $-5-4\cdot 3 = -5-12 = -17$

27. $20\div(-2)\cdot 5 = -10\cdot 5 = -50$

29. $-9-4+3 = -13+3 = -10$

31. $3(5-1)-4(-2) = 3(4)-4(-2)$
$\qquad\qquad\qquad = 12-(-8)$
$\qquad\qquad\qquad = 12+8$
$\qquad\qquad\qquad = 20$

33. $15\div 3-(2-7)(2) = 15\div 3-(-5)(2)$
$\qquad\qquad\qquad\quad = 5-(-10)$
$\qquad\qquad\qquad\quad = 5+10$
$\qquad\qquad\qquad\quad = 15$

35. $\dfrac{7}{8}-\dfrac{3}{4}\cdot\dfrac{1}{2} = \dfrac{7}{8}-\dfrac{3}{8} = \dfrac{7-3}{8} = \dfrac{4}{8} = \dfrac{1}{2}$

37. $2+5^2 = 2+(5\cdot 5) = 2+25 = 27$

39. $-3(4)^2 = -3(4\cdot 4) = -3(16) = -48$

41. $\dfrac{2^4}{4^2} = \dfrac{2\cdot 2\cdot 2\cdot 2}{4\cdot 4} = \dfrac{16}{16} = 1$

43. $4^3-3^4 = (4\cdot 4\cdot 4)-(3\cdot 3\cdot 3\cdot 3)$
$\qquad\qquad = 64-81$
$\qquad\qquad = -17$

45. $45\div 3^2 = 45\div(3\cdot 3)$
$\qquad\qquad = 45\div 9$
$\qquad\qquad = 5$

47. $(-1)^2-(-1)^3 = (-1)(-1)-(-1)(-1)(-1)$
$\qquad\qquad\qquad = 1-(-1)$
$\qquad\qquad\qquad = 1+1$
$\qquad\qquad\qquad = 2$

49. $-5(3)^2+4 = -5(3\cdot 3)+4$
$\qquad\qquad\quad = -5(9)+4$
$\qquad\qquad\quad = -45+4$
$\qquad\qquad\quad = -41$

51. $-4(-1)^2-2(-1)+5 = -4\big[(-1)(-1)\big]-2(-1)+5$
$\qquad\qquad\qquad\qquad = -4(1)-2(-1)+5$
$\qquad\qquad\qquad\qquad = -4-(-2)+5$
$\qquad\qquad\qquad\qquad = -4+2+5$
$\qquad\qquad\qquad\qquad = -2+5$
$\qquad\qquad\qquad\qquad = 3$

53. $\dfrac{9-6^2}{12+3^2} = \dfrac{9-(6\cdot 6)}{12+(3\cdot 3)} = \dfrac{9-36}{12+9} = \dfrac{-27}{21} = -\dfrac{9}{7}$

55. $8-(9-5)^2-1 = 8-(4)^2-1$
$\qquad\qquad\qquad = 8-(4\cdot 4)-1$
$\qquad\qquad\qquad = 8-16-1$
$\qquad\qquad\qquad = -8-1$
$\qquad\qquad\qquad = -9$

57. $8^2+2(4-8)^2\div(-2) = 8^2+2(-4)^2\div(-2)$
$\qquad\qquad\qquad\qquad = (8\cdot 8)+2\big[(-4)(-4)\big]\div(-2)$
$\qquad\qquad\qquad\qquad = 64+2(16)\div(-2)$
$\qquad\qquad\qquad\qquad = 64+32\div(-2)$
$\qquad\qquad\qquad\qquad = 64+(-16)$
$\qquad\qquad\qquad\qquad = 48$

59. $13.28-35.2(17.9)+9.43\div 2.75 \approx -613.37$

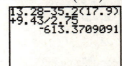

61. $5.82 - 3.16^3 \div 4.29 \approx -1.54$

```
5.82-3.16^3/4.29
        -1.535360373
```

63. $\dfrac{(25.36)(-3.42) - 17.89}{33.26 + 45.32} \approx -1.33$

```
((25.36)(-3.42)-
17.89)/(33.26+45
.32)
        -1.331397302
```

65. Evaluate $a + bc$ for $a = -2$, $b = -4$, and $c = 3$:

$(-2) + (-4)(3) = (-2) + (-12) = -14$

67. Evaluate $ac - b \div a$ for $a = -2$, $b = -4$, and $c = 3$:

$(-2)(3) - (-4) \div (-2) = (-6) - (2)$
$\qquad\qquad\qquad\qquad\quad = -8$

69. Evaluate $a^2 - c^2$ for $a = -2$, $b = -4$, and $c = 3$:

$(-2)^2 - (3)^2 = (-2)(-2) - (3)(3)$
$\qquad\qquad\quad = 4 - 9$
$\qquad\qquad\quad = -5$

71. Evaluate $b^2 - 4ac$ for $a = -2$, $b = -4$, and $c = 3$:

$(-4)^2 - 4(-2)(3) = (-4)(-4) - 4(-2)(3)$
$\qquad\qquad\qquad\quad = 16 - (-24)$
$\qquad\qquad\qquad\quad = 16 + 24$
$\qquad\qquad\qquad\quad = 40$

73. Evaluate $\dfrac{-b - c^2}{2a}$ for $a = -2$, $b = -4$, and $c = 3$:

$\dfrac{-(-4) - (3)^2}{2(-2)} = \dfrac{-(-4) - (3)(3)}{2(-2)}$
$\qquad\qquad\qquad = \dfrac{4 - 9}{-4}$
$\qquad\qquad\qquad = \dfrac{-5}{-4}$
$\qquad\qquad\qquad = \dfrac{5}{4}$

75. Substitute $a = 3$, $b = -10$, $c = -6$, and $d = 1$ in the expression $\dfrac{a - b}{c - d}$:

$\dfrac{(3) - (-10)}{(-6) - (1)} = \dfrac{3 + 10}{(-6) + (-1)} = \dfrac{13}{-7} = -\dfrac{13}{7}$

77. Substitute $a = -3$, $b = 7$, $c = 1$, and $d = -3$ in the expression $\dfrac{a - b}{c - d}$:

$\dfrac{(-3) - (7)}{(1) - (-3)} = \dfrac{-3 + (-7)}{1 + 3} = \dfrac{-10}{4} = -\dfrac{5}{2}$

79. Substitute $a = -8$, $b = -2$, $c = -15$, and $d = -5$ in the expression $\dfrac{a - b}{c - d}$:

$\dfrac{(-8) - (-2)}{(-15) - (-5)} = \dfrac{-8 + 2}{-15 + 5} = \dfrac{-6}{-10} = \dfrac{3}{5}$

81. Evaluate $-3x^2$ for $x = -3$:

$-3(-3)^2 = -3(-3)(-3) = -27$

83. Evaluate $-x^2 + x$ for $x = -3$:

$-(-3)^2 + (-3) = -(-3)(-3) + (-3)$
$\qquad\qquad\qquad = -9 + (-3)$
$\qquad\qquad\qquad = -12$

85. Evaluate $2x^2 - 3x + 5$ for $x = -3$:

$2(-3)^2 - 3(-3) + 5 = 2(-3)(-3) - 3(-3) + 5$
$\qquad\qquad\qquad\qquad = 18 - (-9) + 5$
$\qquad\qquad\qquad\qquad = 18 + 9 + 5$
$\qquad\qquad\qquad\qquad = 27 + 5$
$\qquad\qquad\qquad\qquad = 32$

87. a.

Years Since 1975	Congressional pay (thousands of dollars)
0	$4 \cdot 0 + 44.6$
1	$4 \cdot 1 + 44.6$
2	$4 \cdot 2 + 44.6$
3	$4 \cdot 3 + 44.6$
4	$4 \cdot 4 + 44.6$
t	$4t + 44.6$

From the last row of the table, we see that the expression $4t + 44.6$ represents the congressional pay (in thousands of dollars) t years after 1975.

b. Substitute 35 for t in $4t + 44.6$:

$$4(35) + 44.6 = 140 + 44.6 = 184.6$$

So, in 2010 (35 years after 1975) the congressional pay will be about \$184.6 thousand.

89. a.

Years Since 1980	Population (thousands)
0	$-2 \cdot 0 + 145$
1	$-2 \cdot 1 + 145$
2	$-2 \cdot 2 + 145$
3	$-2 \cdot 3 + 145$
4	$-2 \cdot 4 + 145$
t	$-2t + 145$

From the last row of the table, we see that the expression $-2t + 145$ represents the population of Gary (in thousands) t years after 1980.

b. Substitute 31 for t in $-2t + 145$:

$$-2(31) + 145 = -62 + 145 = 83$$

So, the population of Gary will be about 83 thousand in 2011 (31 years after 1980).

91. $5 + (-6)x$

Evaluate the expression for $x = -4$:

$$5 + (-6)(-4) = 5 + 24 = 29$$

93. $\dfrac{x}{-2} - 3$

Evaluate the expression for $x = -4$:

$$\dfrac{-4}{-2} - 3 = 2 - 3 = -1$$

95. Substitute 2 for s in the expression s^3 :

$$(2)^3 = 2 \cdot 2 \cdot 2 = 8$$

So, the volume is 8 cubic feet.

97. In the first line, the student multiplied the 2 and 3 in the first term before doing the exponentiation. Should be:

$$2(3)^2 + 2(3) + 1 = 2(9) + 2(3) + 1 = 18 + 6 + 1 = 25$$

99. The student is incorrect. In the expression -4^2, only the 4 is being squared. That is,

$$-4^2 = -\left(4^2\right) = -(4 \cdot 4) = -16 .$$

101. a. $(11 - 3) \div (1 - 5) = (8) \div (-4) = -2$

b. $11 - 3 \div 1 - 5 = 11 - 3 - 5$
$$= 8 - 5$$
$$= 3$$

c. The student did not group the numerator and denominator together as required. They should enter the expression in the manner displayed in part (a).

```
(11-3)/(1-5)
               -2
```

103. a. $\big((2)(3)\big)(4) = (6)(4) = 24$

b. $(2)\big((3)(4)\big) = (2)(12) = 24$

c. The results are the same.

d. $\big((4)(-2)\big)(5) = (-8)(5) = -40$
$$(4)\big((-2)(5)\big) = (4)(-10) = -40$$
The results are the same.

e. Answers may vary. In each case, the results should be the same.

f. Yes; this is the Associative Property of Multiplication.

g. Answers may vary. For a series of multiplications, the way in which the multiplications are grouped does not matter.

105. a.

x	x^2
-2	$(-2)^2 = (-2)(-2) = 4$
-1	$(-1)^2 = (-1)(-1) = 1$
0	$(0)^2 = (0)(0) = 0$
1	$(1)^2 = (1)(1) = 1$
2	$(2)^2 = (2)(2) = 4$

b. nonnegative (note that 0 is neither positive nor negative)

c. For any real number x, the value of x^2 is always nonnegative.

Chapter 2 Review Exercises

1. $8 + (-2) = 6$

2. $(-5) + (-7) = -12$

3. $6 - 9 = 6 + (-9) = -3$

4. $8 - (-2) = 8 + 2 = 10$

5. Since the numbers have different signs, the product will be negative: $8(-2) = -16$

6. Since the numbers have different signs, the quotient will be negative: $8 \div (-2) = -4$

7. $-24 \div (10 - 2) = -24 \div (8) = -3$

8. $(2 - 6)(5 - 8) = (-4)(-3) = 12$

9. $\dfrac{7 - 2}{2 - 7} = \dfrac{5}{-5} = -\dfrac{5}{5} = -1$

10. $\dfrac{2 - 8}{3 - (-1)} = \dfrac{2 - 8}{3 + 1} = \dfrac{-6}{4} = -\dfrac{6}{4} = -\dfrac{3}{2}$

11. $\dfrac{3 - 5(-6)}{-2 - 1} = \dfrac{3 - (-30)}{-2 - 1}$
$= \dfrac{3 + 30}{-2 + (-1)}$
$= \dfrac{33}{-3}$
$= -\dfrac{33}{3}$
$= -11$

12. $3(-5) + 2 = -15 + 2 = -13$

13. $-4 + 2(-6) = -4 + (-12) = -16$

14. $2 - 12 \div 2 = 2 - 6 = -4$

15. $4(-6) \div (-3) = -24 \div (-3) = 8$

16. $8 \div (-2) \cdot 5 = (-4) \cdot 5 = -20$

17. $2(4 - 7) - (8 - 2) = 2(-3) - (6)$
$= (-6) - (6)$
$= (-6) + (-6)$
$= -12$

18. $-2(3 - 6) + 18 \div (-9) = -2(-3) + 18 \div (-9)$
$= 6 + 18 \div (-9)$
$= 6 + (-2)$
$= 4$

19. $-14 \div (-7) - 3(1 - 5) = -14 \div (-7) - 3(-4)$
$= 2 - 3(-4)$
$= 2 - (-12)$
$= 2 + 12$
$= 14$

20. Since the number have the same sign, the product will be positive: $-0.3(-0.2) = 0.06$

21. $4.2 - (-6.7) = 4.2 + 6.7 = 10.9$

22. $\dfrac{4}{9}\left(-\dfrac{3}{10}\right) = -\dfrac{12}{90} = -\dfrac{2}{15}$

23. $\left(-\dfrac{8}{15}\right) \div \left(-\dfrac{16}{25}\right) = \left(-\dfrac{8}{15}\right)\left(-\dfrac{25}{16}\right)$
$= \dfrac{8}{15} \cdot \dfrac{25}{16}$
$= \dfrac{200}{240}$
$= \dfrac{5}{6}$

24. $\dfrac{5}{9} - \left(-\dfrac{2}{9}\right) = \dfrac{5}{9} + \dfrac{2}{9} = \dfrac{5 + 2}{9} = \dfrac{7}{9}$

25. $-\dfrac{5}{6} + \dfrac{7}{8} = -\dfrac{5}{6} \cdot \dfrac{4}{4} + \dfrac{7}{8} \cdot \dfrac{3}{3}$
$= \dfrac{-20}{24} + \dfrac{21}{24}$
$= \dfrac{-20 + 21}{24}$
$= \dfrac{1}{24}$

26. $\dfrac{-5}{2} - \dfrac{7}{-3} = \dfrac{-5}{2} + \dfrac{7}{3}$

$\qquad = \dfrac{-5}{2} \cdot \dfrac{3}{3} + \dfrac{7}{3} \cdot \dfrac{2}{2}$

$\qquad = \dfrac{-15}{6} + \dfrac{14}{6}$

$\qquad = \dfrac{-15+14}{6}$

$\qquad = -\dfrac{1}{6}$

27. $(-8)^2 = (-8)(-8) = 64$

28. $-8^2 = -(8 \cdot 8) = -64$

29. $2^4 = 2 \cdot 2 \cdot 2 \cdot 2 = 16$

30. $\left(\dfrac{3}{4}\right)^3 = \dfrac{3}{4} \cdot \dfrac{3}{4} \cdot \dfrac{3}{4} = \dfrac{3 \cdot 3 \cdot 3}{4 \cdot 4 \cdot 4} = \dfrac{27}{64}$

31. $-6(3)^2 = -6(3 \cdot 3) = -6(9) = -54$

32. $24 \div 2^3 = 24 \div (2 \cdot 2 \cdot 2)$

$\qquad = 24 \div 8$

$\qquad = 3$

33. $(-2)^3 - 4(-2) = (-2)(-2)(-2) - 4(-2)$

$\qquad = -8 - 4(-2)$

$\qquad = -8 - (-8)$

$\qquad = -8 + 8$

$\qquad = 0$

34. $\dfrac{2^3}{3 + 3^2} = \dfrac{2 \cdot 2 \cdot 2}{3 + (3 \cdot 3)} = \dfrac{8}{3+9} = \dfrac{8}{12} = \dfrac{2}{3}$

35. $\dfrac{17 - (-3)^2}{5 - 4^2} = \dfrac{17 - (-3)(-3)}{5 - (4 \cdot 4)}$

$\qquad = \dfrac{17 - 9}{5 - 16}$

$\qquad = \dfrac{8}{-11}$

$\qquad = -\dfrac{8}{11}$

36. $-3(2)^2 - 4(2) + 1 = -3(2 \cdot 2) - 4(2) + 1$

$\qquad = -3(4) - 4(2) + 1$

$\qquad = -12 - 8 + 1$

$\qquad = -20 + 1$

$\qquad = -19$

37. $24 \div (3-5)^3 = 24 \div (-2)^3$

$\qquad = 24 \div \left[(-2)(-2)(-2)\right]$

$\qquad = 24 \div (-8)$

$\qquad = -3$

38. $7^2 - 3(2-5)^2 \div (-3) = 7^2 - 3(-3)^2 \div (-3)$

$\qquad = (7 \cdot 7) - 3\left[(-3)(-3)\right] \div (-3)$

$\qquad = 49 - 27 \div (-3)$

$\qquad = 49 - (-9)$

$\qquad = 49 + 9$

$\qquad = 58$

39. $\dfrac{-18}{-24} = \dfrac{18}{24} = \dfrac{2 \cdot 3 \cdot 3}{2 \cdot 2 \cdot 2 \cdot 3} = \dfrac{3}{2 \cdot 2} = \dfrac{3}{4}$

40. $\dfrac{-28}{35} = -\dfrac{28}{35} = -\dfrac{2 \cdot 2 \cdot 7}{5 \cdot 7} = -\dfrac{2 \cdot 2}{5} = -\dfrac{4}{5}$

41. $-5.7 + 2.3^4 \div (-9.4) \approx -8.68$

```
-5.7+2.3^4/(-9.4
)
         -8.677031915
```

42. $\dfrac{3.5(17.4) - 97.6}{54.2 \div 8.4 - 65.3} \approx 0.62$

```
(3.5*17.4-97.6)/
(54.2/8.4-65.3)
       .6236446027
```

43. Substitute $\dfrac{1}{4}$ for W and $\dfrac{5}{6}$ for L in the expression $2L + 2W$:

$$2\left(\dfrac{5}{6}\right) + 2\left(\dfrac{1}{4}\right) = \dfrac{2}{1} \cdot \dfrac{5}{6} + \dfrac{2}{1} \cdot \dfrac{1}{4}$$

$$= \dfrac{10}{6} + \dfrac{2}{4}$$

$$= \dfrac{10}{6} + \dfrac{1}{2}$$

$$= \dfrac{10}{6} + \dfrac{1}{2} \cdot \dfrac{3}{3}$$

$$= \dfrac{10}{6} + \dfrac{3}{6}$$

$$= \dfrac{13}{6}$$

So, the perimeter of the rectangle is $\dfrac{13}{6} = 2\dfrac{1}{6}$ yards.

44. $-4789 + 800 - (102.99 + 3.50)$

$= -4789 + 800 - 106.49$

$= -3989 - 106.49$

$= -4095.49$

The student now owes the credit card company $4095.49.

45. $27,800 - 32,500 = -4700$

The plane had a change in altitude of -4700 feet.

46. a. $-8 - 4 = -12$

The change in temperature is $-12°\,\mathrm{F}$.

b. Divide the change for the past three hours by 3 to estimate the change over 1 hour.

$$\dfrac{-12}{3} = -4$$

The estimated change of the past hour is $-4°\,\mathrm{F}$.

c. Answers may vary. Temperature need not change uniformly.

47. a. $50 - 24 = 26$

The change in private contributions to Democratic conventions from 1996 to 2000 was $26 million.

b. $3 - 8 = -5$

The change in the private contributions to Republican conventions from 1984 to 1988 was -5 million dollars.

c. The greatest change in private contributions to Democratic conventions occurred between 1996 and 2000. The change was $26 million [from part (a)].

d. $64 - 22 = 42$

The greatest change in private contributions to Republican conventions occurred between 2000 and 2004. The change was $42 million.

48. $\dfrac{145.5 \text{ million ringtones}}{63.6 \text{ million ringtones}} \approx \dfrac{2.29}{1}$

The number of ringtones sold in 2004 was about 2.29 times the number sold in 2003.

49. $75\% = 75.0\% = 0.75$

50. $2.9\% = 0.029$

51. $0.87(43) = 37.41$

So, 87% of $43 is $37.41.

52. $0.08(925) = 74$

So, 8% of 925 students is 74 students.

53. $-5493 + 0.2(5493) = -5493 + 1098.6$

$$= -4394.4$$

The new balance is -4394.4 dollars.

54. Substitute 2 for a and -4 for c in the expression $ac + c \div a$:

$$(2)(-4) + (-4) \div (2) = -8 + (-4) \div (2)$$

$$= -8 + (-2)$$

$$= -10$$

55. Substitute 2 for a, -5 for b, and -4 for c in the expression $b^2 - 4ac$:

$$(-5)^2 - 4(2)(-4) = (-5)(-5) - 4(2)(-4)$$

$$= 25 - (-32)$$

$$= 25 + 32$$

$$= 57$$

56. Substitute 2 for a, -5 for b, and -4 for c in the expression $a(b-c)$:

$$(2)\big((-5)-(-4)\big) = 2(-5+4)$$
$$= 2(-1)$$
$$= -2$$

57. Substitute 2 for a, -5 for b, and -4 for c in the expression $\dfrac{-b-c^2}{2a}$:

$$\frac{-(-5)-(-4)^2}{2(2)} = \frac{-(-5)-(-4)(-4)}{2(2)}$$
$$= \frac{5-16}{4}$$
$$= \frac{-11}{4}$$
$$= -\frac{11}{4}$$

58. Substitute -4 for c in the expression $2c^2 - 5c + 3$:

$$2(-4)^2 - 5(-4) + 3 = 2(-4)(-4) - 5(-4) + 3$$
$$= 32 - (-20) + 3$$
$$= 32 + 20 + 3$$
$$= 52 + 3$$
$$= 55$$

59. Substitute 2 for a, -5 for b, -4 for c, and 10 for d in the expression $\dfrac{a-b}{c-d}$:

$$\frac{(2)-(-5)}{(-4)-(10)} = \frac{2+5}{-4+(-10)} = \frac{7}{-14} = -\frac{7}{14} = -\frac{1}{2}$$

60. $x+5$

Evaluate the expression for $x = -3$:

$$(-3)+5 = 2$$

61. $-7-x$

Evaluate the expression for $x = -3$:

$$-7-(-3) = -7+3 = -4$$

62. $2-x(4)$

Evaluate the expression for $x = -3$:

$$2-(-3)(4) = 2-(-12)$$
$$= 2+12$$
$$= 14$$

63. $1 + \dfrac{-24}{x}$

Evaluate the expression for $x = -3$:

$$1 + \frac{-24}{-3} = 1+8 = 9$$

64. Substitute 650 for T and 13 for n.

$$\frac{650}{13} = 50$$

Each player must pay $50 for the team to join the softball league.

65. a.

Time (hours)	Volume of Water (cubic feet)
0	$-50 \cdot 0 + 400$
1	$-50 \cdot 1 + 400$
2	$-50 \cdot 2 + 400$
3	$-50 \cdot 3 + 400$
4	$-50 \cdot 4 + 400$
t	$-50t + 400$

From the last row of the table, we see that the expression $-50t + 400$ represents the volume of water (in cubic feet) remaining in the basement after water has been pumped out for t hours.

b. Substitute 7 for t in $-50t + 400$:

$$-50(7) + 400 = -350 + 400 = 50$$

After 7 hours of pumping, there will be 50 cubic feet of water remaining in the basement.

Chapter 2 Test

1. $-8-5 = -8+(-5) = -13$

2. Since the two numbers have the same sign, the product will be positive: $-7(-9) = 63$

3. $-3+9 \div (-3) = -3 + (-3) = -6$

4. $(4-2)(3-7) = (2)(-4) = -8$

5. $\dfrac{4-7}{-1-5} = \dfrac{-3}{-6} = \dfrac{1}{2}$

6. $5-(2-10)\div(-4) = 5-(-8)\div(-4)$
$$= 5-2$$
$$= 3$$

7. $-20\div 5-(2-9)(-3) = -20\div 5-(-7)(-3)$
$$= -4-21$$
$$= -25$$

8. Since the two numbers have different signs, the product will be negative: $0.4(-0.2) = -0.08$

9. $-\dfrac{27}{10}\div\dfrac{18}{75} = -\dfrac{27}{10}\cdot\dfrac{75}{18}$
$$= -\dfrac{3\cdot 3\cdot 3\cdot 5\cdot 5}{2\cdot 5\cdot 2\cdot 3\cdot 3} = -\dfrac{3\cdot 3\cdot 5}{2\cdot 2} = -\dfrac{45}{4}$$

10. $-\dfrac{3}{10}+\dfrac{5}{8} = -\dfrac{3}{10}\cdot\dfrac{4}{4}+\dfrac{5}{8}\cdot\dfrac{5}{5}$
$$= \dfrac{-12}{40}+\dfrac{25}{40}$$
$$= \dfrac{-12+25}{40}$$
$$= \dfrac{13}{40}$$

11. $3^4 = 3\cdot 3\cdot 3\cdot 3 = 81$

12. $-4^2 = -(4\cdot 4) = -16$

13. $7+2^3-3^2 = 7+(2\cdot 2\cdot 2)-(3\cdot 3)$
$$= 7+8-9$$
$$= 15-9$$
$$= 6$$

14. $1-(3-7)^2+10\div(-5) = 1-(-4)^2+10\div(-5)$
$$= 1-(-4)(-4)+10\div(-5)$$
$$= 1-16+(-2)$$
$$= -15+(-2)$$
$$= -17$$

15. $\dfrac{84}{-16} = -\dfrac{84}{16} = -\dfrac{2\cdot 2\cdot 3\cdot 7}{2\cdot 2\cdot 2\cdot 2} = -\dfrac{3\cdot 7}{2\cdot 2} = -\dfrac{21}{4}$

16. $5-9 = 5+(-9) = -4$
The current temperature is -4°F.

17. a. $6.5-5.8 = 0.7$
The change in the tax audit rate from 2001 to 2003 was 0.7 audit per 1000 tax returns.

b. $5.8-9.0 = -3.2$
The change in the tax audit rate from 1999 to 2001 was -3.2 audits per 1000 tax returns.

c. Answers may vary. The table seems to indicate the IRS increases the audit rate during weaker economic times.

18. $\dfrac{19.82}{9.14}\approx\dfrac{2.17}{1}$
The average ticket price in 2004 was about 2.17 times the average price in 1991.

19. Substitute -6 for a, -2 for b, and 5 for c in the expression $ac-\dfrac{a}{b}$:
$$(-6)(5)-\dfrac{(-6)}{(-2)} = -30-\dfrac{(-6)}{(-2)}$$
$$= -30-3$$
$$= -30+(-3)$$
$$= -33$$

20. Substitute -6 for a, -2 for b, 5 for c, and -1 for d in the expression $\dfrac{a-b}{c-d}$:
$$\dfrac{(-6)-(-2)}{(5)-(-1)} = \dfrac{-6+2}{5+1} = \dfrac{-4}{6} = -\dfrac{4}{6} = -\dfrac{2}{3}$$

21. Substitute -6 for a, -2 for b, and 5 for c in the expression $a+b^3+c^2$:
$$(-6)+(-2)^3+(5)^2 = (-6)+(-2)(-2)(-2)+(5\cdot 5)$$
$$= -6+(-8)+25$$
$$= -14+25$$
$$= 11$$

22. Substitute -6 for a, -2 for b, and 5 for c in the expression b^2-4ac:
$$(-2)^2-4(-6)(5) = (-2)(-2)-4(-6)(5)$$
$$= 4-(-120)$$
$$= 4+120$$
$$= 124$$

23. $2x - 3x$

Evaluate the expression for $x = -5$:

$2(-5) - 3(-5) = -10 - (-15)$

$\qquad\qquad\qquad = -10 + 15$

$\qquad\qquad\qquad = 5$

24. $\dfrac{-10}{x} - 6$

Evaluate the expression for $x = -5$:

$\dfrac{-10}{-5} - 6 = 2 - 6 = -4$

25. a.

Years since 1999	Number of Books Published
0	$7 \cdot 0 + 25$
1	$7 \cdot 1 + 25$
2	$7 \cdot 2 + 25$
3	$7 \cdot 3 + 25$
4	$7 \cdot 4 + 25$
t	$7t + 25$

From the last row of the table, we see that the expression $7t + 25$ represents the number of books on obesity published in the year that is t years after 1999.

b. Substitute 11 for t in $7t + 25$:

$7(11) + 25 = 77 + 25 = 102$

So, in 2010 (11 years after 1999) there will be 102 books on obesity published.

Cumulative Review of Chapters 1 – 2

1. a. Answers may vary. Some possibilities:

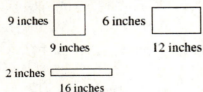

9 inches [   ]   6 inches [   ]

     9 inches   12 inches

2 inches [   ]

     16 inches

b. W and L are variables because their values are not fixed.

c. P is a constant because the perimeter is fixed at 36 inches.

2.

3.

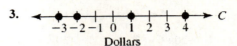

Dollars

4. The x-coordinate is -5.

5. Independent variable: t

Dependent variable: V

6. a.

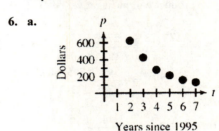

Years since 1995

b. The average price was the most in 1997.

c. The average price was the least in 2002.

d. The average price decreased the most between 1997 and 1998. The change in price was $425 - 625 = -200$ dollars.

e. The average price decreased the least between 2001 and 2002. The change in price was $120 - 150 = -30$ dollars.

7. The input $x = -4$ leads to the output $y = -3$, so $y = -3$ when $x = -4$.

8. The output $y = 1$ originates from the input $x = 4$, so $y = 1$ when $x = 4$.

9. The line and the y-axis intersect at the point $(0, -1)$ so the y-intercept is $(0, -1)$.

10. The line and the x-axis intersect at the point $(2, 0)$, so the x-intercept is $(2, 0)$.

11. a.

Months

b. The input $t = 4$ leads to the output $B = 12$, so the balance is $12 thousand after 4 months.

c. The output $B = 6$ originates from the input $t = 7$, so the balance will be $6 thousand after 7 months.

d. The line and the B-axis intersect at the point $(0, 20)$, so the B-intercept is $(0, 20)$. The original balance was $20 thousand when she was laid off.

e. The line and the t-axis intersect at the point $(10, 0)$, so the t-intercept is $(10, 0)$. The checking account will be depleted after 10 months.

12. a., b.

Years since 1995

c. The line and the c-axis intersect at the point $(0, 22)$, so the c-intercept is $(0, 22)$. The average monthly spending on cable TV per household in 1995 was $22.

d. The output $c = 43$ originates roughly from the input $t = 13$, so the average monthly spending will be $43 in 2008.

e. For 2010, the input is $t = 15$. The input $t = 15$ leads roughly to the output $c = 46$, so the average monthly spending will be $46 in 2010.

13. $\dfrac{3(-8)+15}{2-7(2)} = \dfrac{-24+15}{2-14} = \dfrac{-9}{-12} = \dfrac{3}{4}$

14. $-4(3) + 6 - 20 \div (-10) = -12 + 6 - 20 \div (-10)$
$$= -12 + 6 - (-2)$$
$$= -12 + 6 + 2$$
$$= -6 + 2$$
$$= -4$$

15. $\left(-\dfrac{14}{15}\right) \div \left(-\dfrac{35}{27}\right) = \left(-\dfrac{14}{15}\right)\left(-\dfrac{27}{35}\right)$
$$= \dfrac{14}{15} \cdot \dfrac{27}{35}$$
$$= \dfrac{2 \cdot 7 \cdot 3 \cdot 3 \cdot 3}{3 \cdot 5 \cdot 5 \cdot 7}$$
$$= \dfrac{2 \cdot 3 \cdot 3}{5 \cdot 5}$$
$$= \dfrac{18}{25}$$

16. $\dfrac{3}{8} - \dfrac{5}{6} = \dfrac{3}{8} \cdot \dfrac{3}{3} - \dfrac{5}{6} \cdot \dfrac{4}{4}$
$$= \dfrac{9}{24} - \dfrac{20}{24}$$
$$= \dfrac{9 - 20}{24}$$
$$= -\dfrac{11}{24}$$

17. $4 - (7-9)^4 + 20 \div (-4)$
$$= 4 - (-2)^4 + 20 \div (-4)$$
$$= 4 - (-2)(-2)(-2)(-2) + 20 \div (-4)$$
$$= 4 - 16 + 20 \div (-4)$$
$$= 4 - 16 + (-5)$$
$$= -12 + (-5)$$
$$= -17$$

18. $\dfrac{5 - 3^2}{4^2 + 2} = \dfrac{5 - (3 \cdot 3)}{(4 \cdot 4) + 2} = \dfrac{5 - 9}{16 + 2} = \dfrac{-4}{18} = -\dfrac{4}{18} = -\dfrac{2}{9}$

19. $-3 - 5 = -8$
The change in temperature is $-8°\text{F}$.

20. $-2692 + 850 - 23 = -1842 - 23 = -1865$

The student will now owe the credit card company $1865.

21. Substitute 1 for a, -4 for b, -3 for c, and 7 for d in the expression $\dfrac{a-b}{c-d}$:

$$\frac{(1)-(-4)}{(-3)-(7)} = \frac{1+4}{-3+(-7)} = \frac{5}{-10} = -\frac{5}{10} = -\frac{1}{2}$$

22. Substitute 2 for a, -3 for b, and -5 for c in the expression $b^2 - 4ac$:

$$(-3)^2 - 4(2)(-5) = (-3)(-3) - 4(2)(-5)$$
$$= 9 - (-40)$$
$$= 9 + 40$$
$$= 49$$

23. $x - \dfrac{(-12)}{x}$

Evaluate the expression for $x = -4$:

$$(-4) - \frac{(-12)}{(-4)} = (-4) - 3$$
$$= (-4) + (-3)$$
$$= -7$$

24. $-2x + 7$

Evaluate the expression for $x = -4$:

$$-2(-4) + 7 = 8 + 7 = 15$$

25. Evaluate $\dfrac{100(v - 42)}{42}$ for $v = 45$:

$$\frac{100(45 - 42)}{42} = \frac{100(3)}{42} = \frac{300}{42} = \frac{50}{7} \approx 7.14$$

A stock value today of $45 represents about a 7.14% growth of the investment.

26. a.

Years since 2000	Sales (thousands of cameras)
0	$4 \cdot 0 + 15$
1	$4 \cdot 1 + 15$
2	$4 \cdot 2 + 15$
3	$4 \cdot 3 + 15$
4	$4 \cdot 4 + 15$
t	$4t + 15$

From the last row of the table, we see that the expression $4t + 15$ represents the sales (in thousands of cameras) in the year that is t years since 2000.

b. Evaluate $4t + 15$ for $t = 11$:

$$4(11) + 15 = 44 + 15 = 59$$

In 2011 (11 years after 2000), the camera company will sell 59 thousand cameras.

Chapter 3
Using the Slope to Graph Linear Equations

Homework 3.1

1. Check $(-3, -10)$: $-10 \overset{?}{=} 2(-3) - 4$

$-10 \overset{?}{=} -6 - 4$

$-10 \overset{?}{=} -10$ True

So, $(-3, -10)$ is a solution of $y = 2x - 4$.

Check $(1, -3)$: $-3 \overset{?}{=} 2(1) - 4$

$-3 \overset{?}{=} 2 - 4$

$-3 \overset{?}{=} -2$ False

So, $(1, -3)$ is a not solution of $y = 2x - 4$.

Check $(2, 0)$: $0 \overset{?}{=} 2(2) - 4$

$0 \overset{?}{=} 4 - 4$

$0 \overset{?}{=} 0$ True

So, $(2, 0)$ is a solution of $y = 2x - 4$.

3. Check $(-1, 4)$: $4 \overset{?}{=} -3(-1) + 7$

$4 \overset{?}{=} 3 + 7$

$4 \overset{?}{=} 10$ False

So, $(-1, 4)$ is not a solution of $y = -3x + 7$.

Check $(0, 7)$: $7 \overset{?}{=} -3(0) + 7$

$7 \overset{?}{=} 0 + 7$

$7 \overset{?}{=} 7$ True

So, $(0, 7)$ is a solution of $y = -3x + 7$.

Check $(4, -5)$: $-5 \overset{?}{=} -3(4) + 7$

$-5 \overset{?}{=} -12 + 7$

$-5 \overset{?}{=} -5$ True

So, $(4, -5)$ is a solution of $y = -3x + 7$.

5. $y = x + 2$

To find the y-intercept, let $x = 0$ and solve for y. From the table that follows, we see that the y-intercept is $(0, 2)$. We also find two other solutions to the equation.

x	y
0	$(0) + 2 = 2$
1	$(1) + 2 = 3$
2	$(2) + 2 = 4$

We plot points $(0, 2)$, $(1, 3)$, and $(2, 4)$ and sketch the line through them.

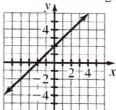

7. $y = x - 4$

To find the y-intercept, let $x = 0$ and solve for y. From the table that follows, we see that the y-intercept is $(0, -4)$. We also find two other solutions to the equation.

x	y
0	$(0) - 4 = -4$
1	$(1) - 4 = -3$
2	$(2) - 4 = -2$

We plot points $(0, -4)$, $(1, -3)$, and $(2, -2)$ and sketch the line through them.

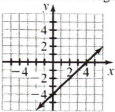

9. $y = 2x$

To find the y-intercept, let $x = 0$ and solve for y. From the table that follows, we see that the y-intercept is $(0, 0)$. We also find two other solutions to the equation.

x	y
0	$2(0) = 0$
1	$2(1) = 2$
2	$2(2) = 4$

We plot points $(0, 0)$, $(1, 2)$, and $(2, 4)$ and sketch the line through them.

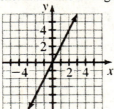

11. $y = -3x$

To find the y-intercept, let $x = 0$ and solve for y. From the table that follows, we see that the y-intercept is $(0, 0)$. We also find two other solutions to the equation.

x	y
-1	$-3(-1) = 3$
0	$-3(0) = 0$
1	$-3(1) = -3$

We plot points $(-1, 3)$, $(0, 0)$, and $(1, -3)$ and sketch the line through them.

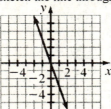

13. $y = x$

To find the y-intercept, let $x = 0$ and solve for y. From the table that follows, we see that the y-intercept is $(0, 0)$. We also find two other solutions to the equation.

x	y
0	0
1	1
2	2

We plot points $(0, 0)$, $(1, 1)$, and $(2, 2)$ and sketch the line through them.

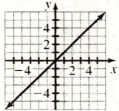

15. $y = \dfrac{1}{3}x$

To find the y-intercept, let $x = 0$ and solve for y. From the table that follows, we see that the y-intercept is $(0, 0)$. We also find two other solutions to the equation.

x	y
-3	$\dfrac{1}{3}(-3) = -1$
0	$\dfrac{1}{3}(0) = 0$
3	$\dfrac{1}{3}(3) = 1$

We plot points $(-3, -1)$, $(0, 0)$, and $(3, 1)$ and sketch the line through them.

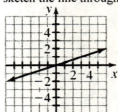

17. $y = -\dfrac{5}{3}x$

To find the y-intercept, let $x = 0$ and solve for y. From the table that follows, we see that the y-intercept is $(0, 0)$. We also find two other solutions to the equation.

x	y
-3	$-\dfrac{5}{3}(-3) = 5$
0	$-\dfrac{5}{3}(0) = 0$
3	$-\dfrac{5}{3}(3) = -5$

We plot points $(-3, 5)$, $(0, 0)$, and $(3, -5)$ and sketch the line through them.

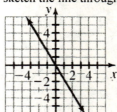

19. $y = 2x + 1$

To find the y-intercept, let $x = 0$ and solve for y. From the table that follows, we see that the y-intercept is (0, 1). We also find two other solutions to the equation.

x	y
0	$2(0) + 1 = 1$
1	$2(1) + 1 = 3$
2	$2(2) + 1 = 5$

We plot points (0, 1), (1, 3), and (2, 5) and sketch the line through them.

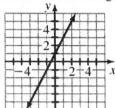

21. $y = 5x - 3$

To find the y-intercept, let $x = 0$ and solve for y. From the table that follows, we see that the y-intercept is $(0, -3)$. We also find two other solutions to the equation.

x	y
0	$5(0) - 3 = -3$
1	$5(1) - 3 = 2$
2	$5(2) - 3 = 7$

We plot points $(0, -3)$, (1, 2), and (2, 7) and sketch the line through them.

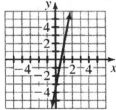

23. $y = -3x + 5$

To find the y-intercept, let $x = 0$ and solve for y. From the table that follows, we see that the y-intercept is (0, 5). We also find two other solutions to the equation.

x	y
0	$-3(0) + 5 = 5$
1	$-3(1) + 5 = 2$
2	$-3(2) + 5 = -1$

We plot points (0, 5), (1, 2), and $(2, -1)$ and sketch the line through them.

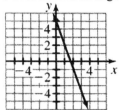

25. $y = -2x - 3$

To find the y-intercept, let $x = 0$ and solve for y. From the table that follows, we see that the y-intercept is $(0, -3)$. We also find two other solutions to the equation.

x	y
-1	$-2(-1) - 3 = -1$
0	$-2(0) - 3 = -3$
1	$-2(1) - 3 = -5$

We plot points $(-1, -1)$, $(0, -3)$, and $(1, -5)$ and sketch the line through them.

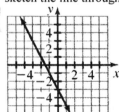

27. $y = \frac{1}{2}x - 3$

To find the y-intercept, let $x = 0$ and solve for y. From the table that follows, we see that the y-intercept is $(0, -3)$. We also find two other solutions to the equation.

x	y
0	$\frac{1}{2}(0) - 3 = -3$
2	$\frac{1}{2}(2) - 3 = -2$
4	$\frac{1}{2}(4) - 3 = -1$

We plot points $(0, -3), (2, -2)$, and $(4, -1)$ and sketch the line through them.

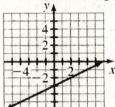

29. $y = -\frac{2}{3}x + 1$

To find the y-intercept, let $x = 0$ and solve for y. From the table that follows, we see that the y-intercept is $(0, 1)$. We also find two other solutions to the equation.

x	y
-3	$-\frac{2}{3}(-3) + 1 = 3$
0	$-\frac{2}{3}(0) + 1 = 1$
3	$-\frac{2}{3}(3) + 1 = -1$

We plot points $(-3, 3), (0, 1)$, and $(3, -1)$ and sketch the line through them.

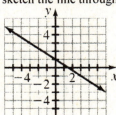

31. a. Answers may vary. One possibility follows:

x	y
0	$2(0) - 3 = -3$
1	$2(1) - 3 = -1$
2	$2(2) - 3 = 1$

b. We plot points $(0, -3), (1, -1)$, and $(2, 1)$ from the table in part (a) and sketch the line through them.

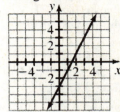

c. For each solution, the y-coordinate is 3 less than twice the x-coordinate.

33. a. $y = 3x + 1$

i. For the input $x = 2$, the output is $y = 3(2) + 1 = 7$. There is one output.

ii. For the input $x = 4$, the output is $y = 3(4) + 1 = 13$. There is one output.

iii. For the input $x = -2$, the output is $y = 3(-2) + 1 = -5$. There is one output.

b. For $y = 3x + 1$, only one output will originate from any single input. Explanations may vary. One possibility follows: This is true because the equation is a nonvertical line.

c. Answers will vary. In each case, there is one output.

d. One output will originate from any single input. Explanations may vary. One possibility follows: This is true because the equation is a nonvertical line.

e. For any equation of the form $y = mx + b$, one output will originate from any single input. Explanations may vary. One possibility follows: This is true because the equation will always be a nonvertical line.

35. a. i. $y = 3x$

We find three solutions to the equation.

x	y
-1	$3(-1) = -3$
0	$3(0) = 0$
1	$3(1) = 3$

We plot points $(-1, -3)$, $(0, 0)$, and $(1, 3)$ and sketch the line through them.

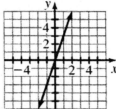

From the graph, we see that the x-intercept is $(0, 0)$ and the y-intercept is also $(0, 0)$.

ii. $y = -2x$

We find three solutions to the equation.

x	y
-1	$-2(-1) = 2$
0	$-2(0) = 0$
1	$-2(1) = -2$

We plot points $(-1, 2)$, $(0, 0)$, and $(1, -2)$ and sketch the line through them.

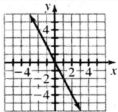

From the graph, we see that the x-intercept is $(0, 0)$ and the y-intercept is also $(0, 0)$.

iii. $y = \dfrac{2}{5}x$

We find three solutions to the equation.

x	y
-5	$\dfrac{2}{5}(-5) = -2$
0	$\dfrac{2}{5}(0) = 0$
5	$\dfrac{2}{5}(5) = 2$

We plot points $(-5, -2)$, $(0, 0)$, and $(5, 2)$ and sketch the line through them.

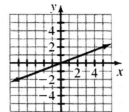

From the graph, we see that the x-intercept is $(0, 0)$ and the y-intercept is also $(0, 0)$.

b. For an equation of the form $y = mx$, $m \neq 0$, the x-intercept will be $(0, 0)$ and the y-intercept will also be $(0, 0)$.

37. Answers may vary. One possibility follows:

x	y
-4	0
-2	1
0	2
2	3
4	4

39. The line contains the point $(-4, 3)$, so $y = 3$ when $x = -4$.

41. The line contains the point $(2, 0)$, so $y = 0$ when $x = 2$.

43. The line contains the point $(4, -1)$, so $x = 4$ when $y = -1$.

45. The line contains the point $(-2, 2)$, so $x = -2$ when $y = 2$.

47. The line contains points C, D, and E, so they represent ordered pairs that satisfy the equation.

49. Answers may vary. One possibility follows: Let $x = -1$, then $y = (-1) + 2 = 1$. The point $(-1, 1)$ is a solution of $y = x + 2$ that lies in Quadrant II. There are infinitely many solutions to this equation that lie in Quadrant II.

51. Notice that in each case, we add 3 to the value of the x-coordinate and obtain the value of the y-coordinate. That is, $3 = 0 + 3$, $4 = 1 + 3$, $5 = 2 + 3$, $6 = 3 + 3$, and $7 = 4 + 3$. The equation of the line that contains the points listed is $y = x + 3$.

53. Notice that for each ordered pair, the *x*-coordinate and *y*-coordinate have the same value. The equation of the line that contains the points listed is $y = x$.

55. a. Answers may vary. One possibility follows:

x	y
−2	−6
−1	−3
0	0
1	3
2	6

b. Notice that in each case the value of the *y*-coordinate is three times the value of the *x*-coordinate. The equation of the line is $y = 3x$.

57. $x + y = 5$

Notice that the points (0, 5), (1, 4), and (2, 3) are all solutions to this equation since each pair adds to 5 (that is, $0 + 5 = 5$, $1 + 4 = 5$, and $2 + 3 = 5$).

We plot points and sketch the line through them.

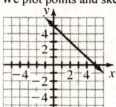

59 – 61. Answers may vary.

Homework 3.2

1. a.

Drink Cost (dollars)	Total Cost (dollars)
d	*T*
2	$2 + 3$
3	$3 + 3$
4	$4 + 3$
5	$5 + 3$
d	$d + 3$

So, the equation is $T = d + 3$.

b. $\underset{\text{dollars}}{T} = \underset{\text{dollars}}{d} + \underset{\text{dollars}}{3}$

So, the units on both sides of the equation are dollars, suggesting the equation is correct.

c. In the following table, we substitute values for *d* in the equation $T = d + 3$ to find the corresponding values for *T*. Then, we plot the points and sketch a line that contains the points.

d	T
2	$(2) + 3 = 5$
3	$(3) + 3 = 6$
4	$(4) + 3 = 7$
5	$(5) + 3 = 8$

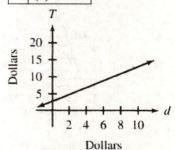

d. To find the *T*-intercept, let $d = 0$ and solve for *T*: $T = (0) + 3 = 3$. The *T*-intercept is (0, 3). This means that, if a person does not buy any drinks, then the total cost is $3.

e. The line contains the point (10, 13), so $T = 13$ when $d = 10$ as illustrated.

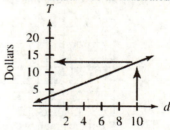

Thus, the total cost will be $13 if $10 is spent on drinks.

3. a.

Number of Credits	Total Cost (in dollars)
c	*T*
3	$56 \cdot 3$
6	$56 \cdot 6$
9	$56 \cdot 9$
12	$56 \cdot 12$
c	$56 \cdot c$

So, the equation is $T = 56c$.

b. $\underset{\text{dollars}}{T} = \underset{\substack{\text{dollars}\\ \text{credit}}}{56} \cdot \underset{\text{credits}}{c}$

We use the fact that $\dfrac{\text{credits}}{\text{credits}} = 1$ to simplify the right-hand side of the equation:

$\dfrac{\text{dollars}}{\text{credit}} \cdot \text{credit} = \text{dollars}$.

So, the units on both sides of the equation are dollars, suggesting the equation is correct.

c. In the following table, we substitute values for c in the equation $T = 56c$ to find the corresponding values for T. Then, we plot the points and sketch a line that contains the points.

c	T
3	$56 \cdot 3 = 168$
6	$56 \cdot 6 = 336$
9	$56 \cdot 9 = 504$
12	$56 \cdot 12 = 672$

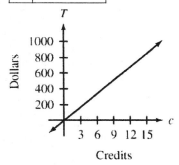

d. The line contains the point $(15, 840)$, so $T = 840$ when $c = 15$ as illustrated.

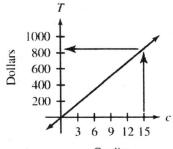

Thus, the total cost of tuition for 15 credits is $840.

5. a.

Time at Company (years)	Salary (thousands of dollars)
t	s
0	$3 \cdot 0 + 24$
1	$3 \cdot 1 + 24$
2	$3 \cdot 2 + 24$
3	$3 \cdot 3 + 24$
4	$3 \cdot 4 + 24$
t	$3 \cdot t + 24$

So, the equation is $s = 3t + 24$.

b. $\underset{\text{thous. dollars}}{s} = \underset{\substack{\text{thous. dollars}\\ \text{year}}}{3} \cdot \underset{\text{years}}{t} + \underset{\text{thous. dollars}}{24}$

We use the fact that $\dfrac{\text{years}}{\text{years}} = 1$ to simplify the right-hand side of the equation:

$\dfrac{\text{thousand dollars}}{\text{year}} \cdot \text{years} + \text{thousand dollars}$.

$= \text{thousand dollars} + \text{thousand dollars}$

So, the units on both sides of the equation are thousands of dollars, suggesting the equation is correct.

c. In the following table, we substitute values for t in the equation $s = 3t + 24$ to find the corresponding values for s. Then, we plot the points and sketch a line that contains the points.

t	s
0	$3 \cdot 0 + 24 = 24$
1	$3 \cdot 1 + 24 = 27$
2	$3 \cdot 2 + 24 = 30$
3	$3 \cdot 3 + 24 = 33$
4	$3 \cdot 4 + 24 = 36$

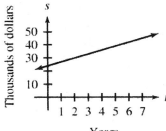

d. From the table and graph, we see the s-intercept is $(0, 24)$. This indicates that the starting salary is $24 thousand.

e. The line contains the point (6, 43), so $s = 43$ when $t = 6$ as illustrated.

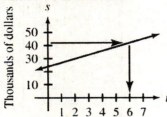

Years

Thus, the person's salary will be $43 thousand after 6 years.

7. a.

Time (weeks)	Value (dollars)
t	v
0	$65 - 5 \cdot 0$
1	$65 - 5 \cdot 1$
2	$65 - 5 \cdot 2$
3	$65 - 5 \cdot 3$
4	$65 - 5 \cdot 4$
t	$65 - 5 \cdot t$

So, the equation is $v = 65 - 5t$.

b.
$$\underset{\text{dollars}}{v} = \underset{\text{dollars}}{65} - \underset{\frac{\text{dollars}}{\text{week}}}{5} \cdot \underset{\text{weeks}}{t}$$

We use the fact that $\dfrac{\text{weeks}}{\text{weeks}} = 1$ to simplify the right-hand side of the equation:

$$\text{dollars} - \frac{\text{dollars}}{\text{week}} \cdot \text{weeks} = \text{dollars} - \text{dollars}.$$

So, the units on both sides of the equation are dollars, suggesting the equation is correct.

c. In the following table, we substitute values for t in the equation $v = 65 - 5t$ to find the corresponding values for v. Then, we plot the points and sketch a line that contains the points.

t	v
0	$65 - 5 \cdot 0 = 65$
1	$65 - 5 \cdot 1 = 60$
2	$65 - 5 \cdot 2 = 55$
3	$65 - 5 \cdot 3 = 50$
4	$65 - 5 \cdot 4 = 45$

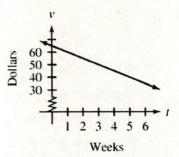

Weeks

d. The line contains the point (6, 35), so $v = 35$ when $t = 6$ as illustrated.

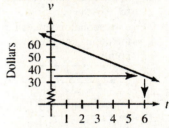

Weeks

Thus, 6 weeks after the bad publicity is released, the stock will be worth $35.

9. a. The actual baking time a is 5 minutes less than the suggested baking time r. So, the equation is $a = r - 5$.

b.
$$\underset{\text{minutes}}{a} = \underset{\text{minutes}}{r} - \underset{\text{minutes}}{5}$$

So, the units on both sides of the equation are minutes, suggesting the equation is correct.

c. In the following table, we substitute values for r in the equation $a = r - 5$ to find the corresponding values for a. Then, we plot the points and sketch a line that contains the points.

t	v
10	$10 - 5 = 5$
20	$20 - 5 = 15$
30	$30 - 5 = 25$

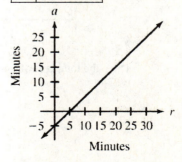

Minutes

d. The line contains the point (28, 23), so $r = 28$ when $a = 23$ as illustrated.

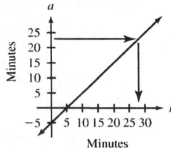

Minutes

So, if the actual baking time is 23 minutes, then the suggested baking time is 28 minutes.

11. a. The distance traveled d is 60 times the number of hours t. Thus, the equation is $d = 60t$.

b.
$$\underset{\text{miles}}{d} = \underset{\substack{\text{miles} \\ \text{hour}}}{60} \cdot \underset{\text{hours}}{t}$$

We use the fact that $\dfrac{\text{hours}}{\text{hours}} = 1$ to simplify the right-hand side of the equation:

$$\dfrac{\text{miles}}{\text{hour}} \cdot \text{hours} = \text{miles}.$$

So, the units on both sides of the equation are miles, suggesting the equation is correct.

c. In the following table, we substitute values for r in the equation $a = r - 5$ to find the corresponding values for a. Then, we plot the points and sketch a line that contains the points.

t	d
0	$60 \cdot 0 = 0$
1	$60 \cdot 1 = 60$
2	$60 \cdot 2 = 120$

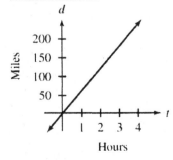

Hours

d. From the table and graph, we see the s-intercept is (0, 0). This indicates that the person will not travel any distance in 0 hours.

e. The line contains the point (2.5, 150), so $t = 2.5$ when $d = 150$ as illustrated.

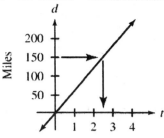

Hours

So, it will take the person 2.5 hours to travel 150 miles.

13. a. The student will pay $26u$ for tuition, plus $31 in fees ($30 for parking and $1 for student representation). Thus, the total one-semester cost T for u units of classes will be $T = 26u + 31$.

b.
$$\underset{\text{dollars}}{T} = \underset{\substack{\text{dollars} \\ \text{unit}}}{26} \cdot \underset{\text{units}}{u} + \underset{\text{dollars}}{31}$$

We use the fact that $\dfrac{\text{units}}{\text{units}} = 1$ to simplify the right-hand side of the equation:

$$\dfrac{\text{dollars}}{\text{unit}} \cdot \text{units} + \text{dollars} = \text{dollars} + \text{dollars}.$$

So, the units on both sides of the equation are dollars, suggesting the equation is correct.

c. If $u = 15$, then $T = 26(15) + 31 = 421$. Fifteen units of classes will cost a total of $421.

15. a. If the person drives for t hours, then he or she will use $2t$ gallons of gasoline. Since the tank begins with 11 gallons of gas, the amount of gasoline g left in the tank is $g = 11 - 2t$.

b.
$$\underset{\text{gallons}}{g} = \underset{\text{gallons}}{11} - \underset{\substack{\text{gallons} \\ \text{hour}}}{2} \cdot \underset{\text{hours}}{t}$$

We use the fact that $\dfrac{\text{hours}}{\text{hours}} = 1$ to simplify the right-hand side of the equation:

$$\text{gallons} - \dfrac{\text{gallons}}{\text{hour}} \cdot \text{hours} = \text{gallons} - \text{gallons}.$$

So, the units on both sides of the equation are gallons, suggesting the equation is correct.

c. In the following table, we substitute values for t in the equation $g = 11 - 2t$ to find the corresponding values for g. Then, we plot the points and sketch a line that contains the points.

t	d
0	$11 - 2 \cdot 0 = 11$
2	$11 - 2 \cdot 2 = 7$
4	$11 - 4 \cdot 2 = 3$

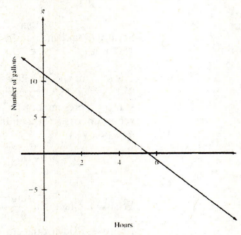

d. The line contains the point (5, 1), so $t = 5$ when $g = 1$ as shown in the illustration that follows. So, the person can drive for 5 hours before refueling when 1 gallon of gasoline is left in the tank.

e. The line contains the point $(8, -5)$, so $g = -5$ when $t = 8$ as shown in the illustration that follows. This implies that -5 gallons of gasoline would be in the tank, which does not make sense. Model breakdown has occurred.

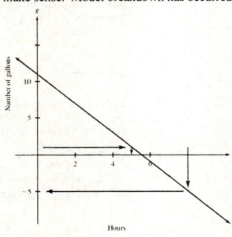

17. a. Answers may vary. One possible table follows:

Years Since 2000 t	Sales (millions of pounds) s
0	$4.3 + 2.5 \cdot 0 = 4.3$
1	$4.3 + 2.5 \cdot 1 = 6.8$
2	$4.3 + 2.5 \cdot 2 = 9.3$
3	$4.3 + 2.5 \cdot 3 = 11.8$
4	$4.3 + 2.5 \cdot 4 = 14.3$

b. The equation is $s = 4.3 + 2.5t$.

c. We plot the points found in part (a) and sketch a line that contains the points.

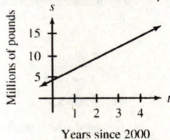

Years since 2000

19. Since 2000, the person owned a constant 45 CDs. So, the equation of the model is $n = 45$. To graph the equation, we list some corresponding values of t and n in the following table. Then we plot the points and sketch a line that contains the points.

Years Since 2000 t	Number of CDs owned n
0	45
1	45
2	45
3	45
4	45

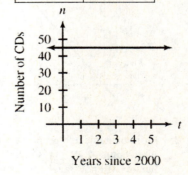

Years since 2000

21. The graph of $x = 3$ is a vertical line. Note that x must be 3, but y can have any value. Some solutions of the equation are listed in the table that follows. We plot the corresponding points and sketch the line through them.

x	y
3	−1
3	0
3	1

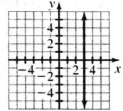

23. The graph of $y = 1$ is a horizontal line. Note that y must be 1, but x can have any value. Some solutions of the equation are listed in the table that follows. We plot the corresponding points and sketch the line through them.

x	y
−1	1
0	1
1	1

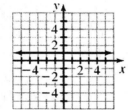

25. The graph of $y = -2$ is a horizontal line. Note that y must be −2, but x can have any value. Some solutions of the equation are listed in the table that follows. We plot the corresponding points and sketch the line through them.

x	y
−1	−2
0	−2
1	−2

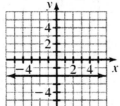

27. The graph of $x = -1$ is a vertical line. Note that x must be −1, but y can have any value. Some solutions of the equation are listed in the table that follows. We plot the corresponding points and sketch the line through them.

x	y
−1	−1
−1	0
−1	1

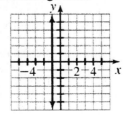

29. The graph of $x = 0$ is a vertical line. Note that x must be 0, but y can have any value. Some solutions of the equation are listed in the table that follows. We plot the corresponding points and sketch the line through them. Note the graph coincides with the y-axis.

x	y
0	−1
0	0
0	1

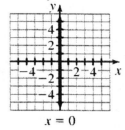

$x = 0$

31. To graph $y = x - 2$, we find three solutions to the equation using the table that follows. We then plot the corresponding points and sketch the line through them.

x	y
−1	$(-1) - 2 = -3$
0	$(0) - 2 = -2$
1	$(1) - 2 = -1$

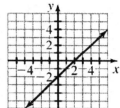

33. The graph of $y = 2$ is a horizontal line. Note that y must be 2, but x can have any value. Some solutions of the equation are listed in the table that follows. We plot the corresponding points and sketch the line through them.

x	y
−1	2
0	2
1	2

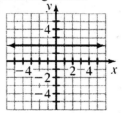

35. To graph $y = -3x + 1$, we find three solutions to the equation using the table that follows. We then plot the corresponding points and sketch the line through them.

x	y
−1	$-3(-1) + 1 = 4$
0	$-3(0) + 1 = 1$
1	$-3(1) + 1 = -2$

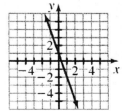

37. To graph $y = \dfrac{3}{5}x$, we find three solutions to the equation using the table that follows. We plot the corresponding points and sketch the line through them.

x	y
-5	$\dfrac{3}{5}(-5) = -3$
0	$\dfrac{3}{5}(0) = 0$
5	$\dfrac{3}{5}(5) = 3$

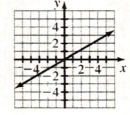

39. To graph $y = -\dfrac{5}{3}x + 1$, we find three solutions to the equation using the table that follows. We plot the corresponding points and sketch the line through them.

x	y
-3	$-\dfrac{5}{3}(-3) + 1 = 6$
0	$-\dfrac{5}{3}(0) + 1 = 1$
3	$-\dfrac{5}{3}(3) + 1 = -4$

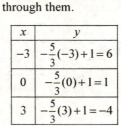
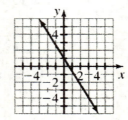

41. To graph $y = 4x - 3$, we find three solutions to the equation using the table that follows. We then plot the corresponding points and sketch the line through them.

x	y
0	$4(0) - 3 = -3$
1	$4(1) - 3 = 1$
2	$4(2) - 3 = 5$

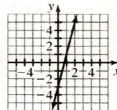

43. The graph of $x = -4$ is a vertical line. Note that x must be -4, but y can have any value. Some solutions of the equation are listed in the table that follows. We plot the corresponding points and sketch the line through them.

x	y
-4	-1
-4	0
-4	1

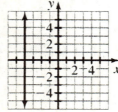

45. The graph is of a vertical line. Each point on the line has an x-coordinate -3, so the equation of the line is $x = -3$.

47. **a.** **i.** The change in the x-coordinate is $5 - 1 = 4$.

    ii. The change in the y-coordinate is $2 - 2 = 0$.

  b. **i.** The change in the x-coordinate is $1 - 5 = -4$.

    ii. The change in the y-coordinate is $2 - 2 = 0$.

49. **a.** **i.** The change in the x-coordinate is $2 - (-3) = 2 + 3 = 5$.

    ii. The change in the y-coordinate is $4 - 1 = 3$.

  b. **i.** The change in the x-coordinate is $-3 - 2 = -5$.

    ii. The change in the y-coordinate is $1 - 4 = -3$.

51. **a.** We let $Y_1 = x$, $Y_2 = 2x$, and $Y_3 = 3x$, and graph using a standard viewing window.

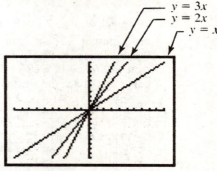

  b. The order of steepness, from least to greatest steepness is: $y = x$, $y = 2x$, $y = 3x$.

  c. Answers may vary. One possibility follows: For equations of the form $y = mx$, the steepness of the line increases as m increases.

  d. Answers may vary. One possibility follows: The line $y = 4x$ will be steeper than $y = x$, $y = 2x$, and $y = 3x$.

53. Answers will vary.

Homework 3.3

1. Slope of road $A = \dfrac{210 \text{ feet}}{3500 \text{ feet}} = 0.06$

Slope of road $B = \dfrac{275 \text{ feet}}{5000 \text{ feet}} = 0.055$

Thus, road A is steeper since it has a larger slope.

3. Slope of ski run $A = \dfrac{80 \text{ yards}}{400 \text{ yards}} = 0.2$

Slope of ski run $B = \dfrac{90 \text{ yards}}{600 \text{ yards}} = 0.15$

Thus, ski run A is steeper since it has a larger slope.

5.

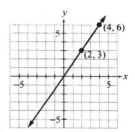

The run is $4 - 2 = 2$. The rise is $6 - 3 = 3$.

The slope is $m = \dfrac{\text{rise}}{\text{run}} = \dfrac{3}{2}$.

7.

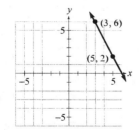

The run is $5 - 3 = 2$. The rise is $2 - 6 = -4$.

The slope is $m = \dfrac{\text{rise}}{\text{run}} = \dfrac{-4}{2} = -2$.

9.

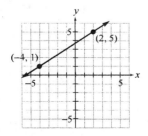

The run is $2 - (-4) = 2 + 4 = 6$. The rise is $5 - 1 = 4$. The slope is $m = \dfrac{\text{rise}}{\text{run}} = \dfrac{4}{6} = \dfrac{2}{3}$.

11.

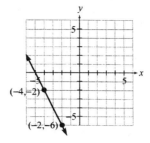

The run is $-2 - (-4) = -2 + 4 = 2$.

The rise is $-6 - (-2) = -6 + 2 = -4$.

The slope is $m = \dfrac{\text{rise}}{\text{run}} = \dfrac{-4}{2} = -2$.

13. Using the slope formula with $(x_1, y_1) = (1, 5)$ and $(x_2, y_2) = (3, 9)$, the slope is

$m = \dfrac{y_2 - y_1}{x_2 - x_1} = \dfrac{9 - 5}{3 - 1} = \dfrac{4}{2} = 2$.

The slope is positive, so the line is increasing.

15. Using the slope formula with $(x_1, y_1) = (3, 10)$ and $(x_2, y_2) = (5, 2)$, the slope is

$m = \dfrac{y_2 - y_1}{x_2 - x_1} = \dfrac{2 - 10}{5 - 3} = \dfrac{-8}{2} = -4$.

The slope is negative, so the line is decreasing.

17. Using the slope formula with $(x_1, y_1) = (2, 1)$ and $(x_2, y_2) = (8, 4)$, the slope is

$m = \dfrac{y_2 - y_1}{x_2 - x_1} = \dfrac{4 - 1}{8 - 2} = \dfrac{3}{6} = \dfrac{1}{2}$.

The slope is positive, so the line is increasing.

19. Using the slope formula with $(x_1, y_1) = (2, 5)$ and $(x_2, y_2) = (8, 3)$, the slope is

$m = \dfrac{y_2 - y_1}{x_2 - x_1} = \dfrac{3 - 5}{8 - 2} = \dfrac{-2}{6} = -\dfrac{1}{3}$.

The slope is negative, so the line is decreasing.

21. Using the slope formula with $(x_1, y_1) = (-2, 4)$ and $(x_2, y_2) = (3, -1)$, the slope is
$$m = \frac{y_2 - y_1}{x_2 - x_1} = \frac{-1-4}{3-(-2)} = \frac{-1-4}{3+2} = \frac{-5}{5} = -1.$$
The slope is negative, so the line is decreasing.

23. Using the slope formula with $(x_1, y_1) = (5, -2)$ and $(x_2, y_2) = (9, -4)$, the slope is
$$m = \frac{y_2 - y_1}{x_2 - x_1} = \frac{-4-(-2)}{9-5} = \frac{-4+2}{9-5} = \frac{-2}{4} = -\frac{1}{2}.$$
The slope is negative, so the line is decreasing.

25. Using the slope formula with $(x_1, y_1) = (-7, -1)$ and $(x_2, y_2) = (-2, 9)$, the slope is
$$m = \frac{y_2 - y_1}{x_2 - x_1} = \frac{9-(-1)}{-2-(-7)} = \frac{9+1}{-2+7} = \frac{10}{5} = 2.$$
The slope is positive, so the line is increasing.

27. Using the slope formula with $(x_1, y_1) = (-6, -9)$ and $(x_2, y_2) = (-2, -3)$, the slope is
$$m = \frac{y_2 - y_1}{x_2 - x_1} = \frac{-3-(-9)}{-2-(-6)} = \frac{-3+9}{-2+6} = \frac{6}{4} = \frac{3}{2}.$$
The slope is positive, so the line is increasing.

29. Using the slope formula with $(x_1, y_1) = (6, -1)$ and $(x_2, y_2) = (-4, 7)$, the slope is
$$m = \frac{y_2 - y_1}{x_2 - x_1} = \frac{7-(-1)}{-4-6} = \frac{7+1}{-4-6} = \frac{8}{-10} = -\frac{4}{5}.$$
The slope is negative, so the line is decreasing.

31. Using the slope formula with $(x_1, y_1) = (-2, -11)$ and $(x_2, y_2) = (7, -5)$, the slope is
$$m = \frac{y_2 - y_1}{x_2 - x_1} = \frac{-5-(-11)}{7-(-2)} = \frac{-5+11}{7+2} = \frac{6}{9} = \frac{2}{3}.$$
The slope is positive, so the line is increasing.

33. Using the slope formula with $(x_1, y_1) = (0, 0)$ and $(x_2, y_2) = (4, -2)$, the slope is
$$m = \frac{y_2 - y_1}{x_2 - x_1} = \frac{-2-0}{4-0} = \frac{-2}{4} = -\frac{1}{2}.$$
The slope is negative, so the line is decreasing.

35. Using the slope formula with $(x_1, y_1) = (3, 5)$ and $(x_2, y_2) = (7, 5)$, the slope is
$$m = \frac{y_2 - y_1}{x_2 - x_1} = \frac{5-5}{7-3} = \frac{0}{4} = 0.$$
So, the line is horizontal.

37. Using the slope formula with $(x_1, y_1) = (-3, -1)$ and $(x_2, y_2) = (-3, -2)$, the slope is
$$m = \frac{y_2 - y_1}{x_2 - x_1} = \frac{-2-(-1)}{-3-(-3)} = \frac{-2+1}{-3+3} = \frac{-1}{0}, \text{ which is}$$
undefined. So the line is vertical.

39. Using the slope formula with $(x_1, y_1) = (-3.2, 5.1)$ and $(x_2, y_2) = (-2.8, 1.4)$, the slope is
$$m = \frac{y_2 - y_1}{x_2 - x_1} = \frac{1.4-5.1}{-2.8-(-3.2)} = \frac{-3.7}{0.4} = -9.25.$$
The slope is negative, so the line is decreasing.

41. Using the slope formula with $(x_1, y_1) = (4.9, -2.7)$ and $(x_2, y_2) = (6.3, -1.1)$, the slope is
$$m = \frac{y_2 - y_1}{x_2 - x_1} = \frac{-1.1-(-2.7)}{6.3-4.9} = \frac{1.6}{1.4} \approx 1.14.$$
The slope is positive, so the line is increasing.

43. Using the slope formula with $(x_1, y_1) = (-4.97, -3.25)$ and $(x_2, y_2) = (-9.64, -2.27)$, the slope is
$$m = \frac{y_2 - y_1}{x_2 - x_1} = \frac{-2.27-(-3.25)}{-9.64-(-4.97)} = \frac{0.98}{-4.67} \approx -0.21.$$
The slope is negative, so the line is decreasing.

45. Using the slope formula with $(x_1, y_1) = (-2.45, -6.71)$ and $(x_2, y_2) = (4.88, -1.53)$, the slope is
$$m = \frac{y_2 - y_1}{x_2 - x_1} = \frac{-1.53-(-6.71)}{4.88-(-2.45)} = \frac{5.18}{7.33} \approx 0.71.$$
The slope is positive, so the line is increasing.

47. The line contains the points $(0, -1)$ and $(3, 1)$, so the slope is $m = \dfrac{1-(-1)}{3-0} = \dfrac{2}{3}$.

49. The line contains the points $(1, 4)$ and $(2, 1)$, so the slope is $m = \dfrac{1-4}{2-1} = \dfrac{-3}{1} = -3$.

51. a. The line slants downward from left to right, so the slope is negative.

b. The line slants upward from left to right, so the slope is positive.

c. The line is vertical, so the slope is undefined.

d. The line is horizontal, so the slope is zero.

53 – 59. Answers will vary.

61. Descriptions of errors may vary. One possibility follows: The student has calculated with the run over the rise rather than rise over run. The correct slope is $m = \dfrac{7-3}{4-1} = \dfrac{4}{3}$.

63. Descriptions of errors may vary. One possibility follows: The student has subtracted incorrectly for both the run and the rise. The correct slope is $m = \dfrac{8-(-5)}{3-(-1)} = \dfrac{8+5}{3+1} = \dfrac{13}{4}$.

65. Sketches may vary. One possibility follows: The graphs shown are of the lines $y = 2x$ and $y = 3x$.

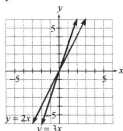

Yes, the steeper line has the greater slope.

67. a. Sketches may vary. One possibility follows: The graphs shown are of the lines $y = -2x$ and $y = -3x$.

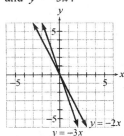

b. No. $-3 < -2$, but the line with slope -3 is steeper than the line with slope -2.

c. $|-3| = 3$; $|-2| = 2$

Yes. $|-3| > |-2|$. The steeper line has the slope with the greater absolute value.

d. Answers may vary. One possibility follows: A line having the slope with a greater absolute value will be steeper than a line with a lesser absolute value.

69. Answers may vary. One possibility follows: The slope of the line containing $(2, 1)$ and $(3, 4)$ is $m = \dfrac{4-1}{3-2} = \dfrac{3}{1} = 3$. From the point $(3, 4)$, if we run 1 and rise 3, we will find another point on the line: $(3 + 1, 4 + 3) = (4, 7)$. Now, from the point $(4, 7)$, if we run 1 and rise 3, we will find another point on the line: $(4 + 1, 7 + 3) = (5, 10)$. Now, from the point $(5, 10)$, if we run 1 and rise 3, we will find another point on the line: $(5 + 1, 10 + 3) = (6, 13)$. In summary, the points $(4, 7)$, $(5, 10)$, and $(6, 13)$ are on the line.

71. Answers may vary. One possibility follows: For a vertical line, the horizontal change (the run) is 0. So when calculating the slope, we divide by 0, which is undefined.

73. Answers may vary. One possibility follows: The slope of an increasing line is positive because the horizontal change (the run) and the vertical change (the rise) have the same sign.

75. No, Ladder A is not necessarily steeper than Ladder B. Explanations may vary. One example follows: The slope of each ladder depends not just on how high it reaches up the building (the rise), but it also depends on the length of each ladder and how far the base of each ladder is set from the wall (the run). The illustrations below show a situation where Ladder A is steeper than Ladder B, where Ladder B is steeper than Ladder A, and were Ladders A and B are equally steep. In all three cases, Ladder A reaches a higher point on the building than Ladder B.

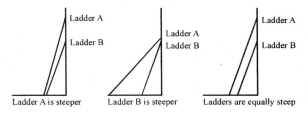

55

Homework 3.4

1. We first plot the *y*-intercept (0, 1). The slope is $\frac{2}{3}$, so the run is 3 and the rise is 2. From (0, 1), we count 3 units to the right and 2 units up, where we plot the point (3, 3). We then sketch the line that contains these two points.

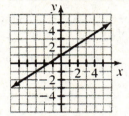

3. We first plot the *y*-intercept (0, 4). The slope is $-\frac{5}{2} = \frac{-5}{2}$, so the run is 2 and the rise is -5. From (0, 4), we count 2 units to the right and 5 units down, where we plot the point $(2, -1)$. We then sketch the line that contains these two points.

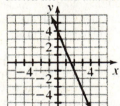

5. We first plot the *y*-intercept (0, 0). The slope is $-\frac{3}{2} = \frac{-3}{2}$, so the run is 2 and the rise is -3. From (0, 0), we count 2 units to the right and 3 units down, where we plot the point $(2, -3)$. We then sketch the line that contains these two points.

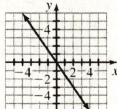

7. We first plot the *y*-intercept (0, 1). The slope is $2 = \frac{2}{1}$, so the run is 1 and the rise is 2. From (0, 1), we count 1 unit to the right and 2 units up, where we plot the point (1, 3). We then sketch the line that contains these two points.

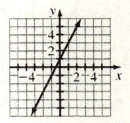

9. We first plot the *y*-intercept $(0, -2)$. The slope is $-3 = \frac{-3}{1}$, so the run is 1 and the rise is -3. From $(0, -2)$, we count 1 unit to the right and 3 units down, where we plot the point $(1, -5)$. We then sketch the line that contains these two points.

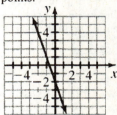

11. We first plot the *y*-intercept (0, 3). The slope is $-1 = \frac{-1}{1}$, so the run is 1 and the rise is -1. From (0, 3), we count 1 unit to the right and 1 unit down, where we plot the point (1, 2). We then sketch the line that contains these two points.

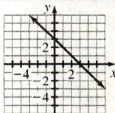

13. We first plot the point $(4, -5)$. The slope is 0, so the line is horizontal. We sketch the horizontal line that contains $(4, -5)$.

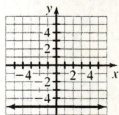

15. We first plot the point $(2,-1)$. The slope is undefined, so the line is vertical. We sketch the vertical line that contains $(2,-1)$.

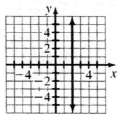

17. The equation $y = \frac{2}{3}x - 1$ is in slope-intercept form, so the slope is $m = \frac{2}{3}$ and the y-intercept is $(0,-1)$. We first plot $(0,-1)$. From this point we move 3 units to the right and 2 units up, where we plot the point $(3, 1)$. We then sketch the line that contains these two points.

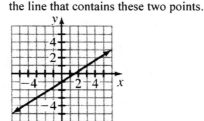

19. The equation $y = -\frac{1}{3}x + 4$ is in slope-intercept form, so the slope is $m = -\frac{1}{3}$ and the y-intercept is $(0, 4)$. We first plot $(0, 4)$. From this point we move 3 units to the right and 1 unit down, where we plot the point $(3, 3)$. We then sketch the line that contains these two points.

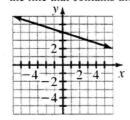

21. The equation $y = \frac{4}{3}x + 2$ is in slope-intercept form, so the slope is $m = \frac{4}{3}$ and the y-intercept is $(0, 2)$. We first plot $(0, 2)$. From this point we move 3 units to the right and 4 units up, where we plot the point $(3, 6)$. We then sketch the line that contains these two points.

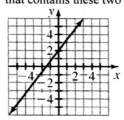

23. The equation $y = -\frac{4}{5}x - 1$ is in slope-intercept form, so the slope is $m = -\frac{4}{5}$ and the y-intercept is $(0,-1)$. We first plot $(0,-1)$. From this point we move 5 units to the right and 4 units down, where we plot the point $(5,-5)$. We then sketch the line that contains these two points.

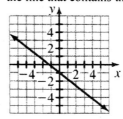

25. The equation $y = \frac{1}{2}x$ is in slope-intercept form, so the slope is $m = \frac{1}{2}$ and the y-intercept is $(0, 0)$. We first plot $(0, 0)$. From this point we move 2 units to the right and 1 unit up, where we plot the point $(2, 1)$. We then sketch the line that contains these two points.

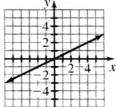

27. The equation $y = -\frac{5}{3}x$ is in slope-intercept form, so the slope is $m = -\frac{5}{3}$ and the y-intercept is $(0, 0)$. We first plot $(0, 0)$. From this point we move 3 units to the right and 5 units down, where we plot the point $(3, -5)$. We then sketch the line that contains these two points.

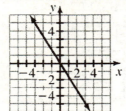

29. The equation $y = 4x - 2$ is in slope-intercept form, so the slope is $m = 4 = \frac{4}{1}$ and the y-intercept is $(0, -2)$. We first plot $(0, -2)$. From this point we move 1 unit to the right and 4 units up, where we plot the point $(1, 2)$. We then sketch the line that contains these two points.

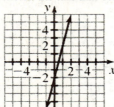

Points chosen for verification may vary.

31. The equation $y = -2x + 4$ is in slope-intercept form, so the slope is $m = -2 = \frac{-2}{1}$ and the y-intercept is $(0, 4)$. We first plot $(0, 4)$. From this point we move 1 unit to the right and 2 units down, where we plot the point $(1, 2)$. We then sketch the line that contains these two points.

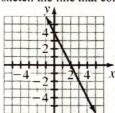

Points chosen for verification may vary.

33. The equation $y = -4x - 1$ is in slope-intercept form, so the slope is $m = -4 = \frac{-4}{1}$ and the y-intercept is $(0, -1)$. We first plot $(0, -1)$. From this point we move 1 unit to the right and 4 units down, where we plot the point $(1, -5)$. We then sketch the line that contains these two points.

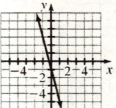

Points chosen for verification may vary.

35. The equation $y = x + 1$ is in slope-intercept form, so the slope is $m = 1 = \frac{1}{1}$ and the y-intercept is $(0, 1)$. We first plot $(0, 1)$. From this point we move 1 unit to the right and 1 unit up, where we plot the point $(1, 2)$. We then sketch the line that contains these two points.

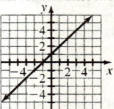

Points chosen for verification may vary.

37. The equation $y = -x + 3$ is in slope-intercept form, so the slope is $m = -1 = \frac{-1}{1}$ and the y-intercept is $(0, 3)$. We first plot $(0, 3)$. From this point we move 1 unit to the right and 1 unit down, where we plot the point $(1, 2)$. We then sketch the line that contains these two points.

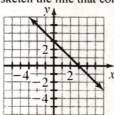

Points chosen for verification may vary.

39. The equation $y = -3x$ is in slope-intercept form, so the slope is $m = -3 = \dfrac{-3}{1}$ and the y-intercept is $(0, 0)$. We first plot $(0, 0)$. From this point we move 1 unit to the right and 3 units down, where we plot the point $(1, -3)$. We then sketch the line that contains these two points.

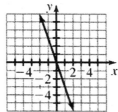

Points chosen for verification may vary.

41. The equation $y = x$ is in slope-intercept form, so the slope is $m = 1 = \dfrac{1}{1}$ and the y-intercept is $(0, 0)$. We first plot $(0, 0)$. From this point we move 1 unit to the right and 1 unit up, where we plot the point $(1, 1)$. We then sketch the line that contains these two points.

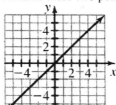

Points chosen for verification may vary.

43. The equation $y = -3$ is in slope-intercept form, so the slope is $m = 0$ and the y-intercept is $(0, -3)$. Since the slope is 0, the line is horizontal. We sketch the horizontal line with y-intercept $(0, -3)$.

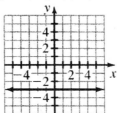

Points chosen for verification may vary.

45. The equation $y = 0$ is in slope-intercept form, so the slope is $m = 0$ and the y-intercept is $(0, 0)$. Since the slope is 0, the line is horizontal. We sketch the horizontal line with y-intercept $(0, 0)$. Note the graph coincides with the x-axis.

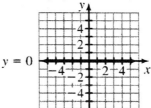

Points chosen for verification may vary.

47. a. The equation $p = 0.8t + 30.3$ is in slope-intercept form, so the slope is $m = 0.8 = \dfrac{4}{5}$ and the p-intercept is $(0, 30.3)$. We first plot $(0, 30.3)$. From this point we move 5 units to the right and 4 units up, where we plot the point $(5, 34.3)$. We then sketch the line that contains these two points.

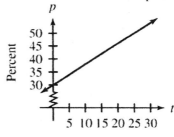

Years since 1985

b. Note that half is 50%. The line contains the point $(25, 50)$, so $t = 25$ when $p = 50$ as illustrated.

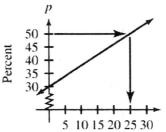

Years since 1985

Thus, half of the degrees in medicine will be earned by women in the year $1985 + 25 = 2010$.

49. a. The equation $n = -8.6t + 199.4$ is in slope-intercept form, so the slope is

$m = -8.6 = \dfrac{-8.6}{1}$ and the n-intercept is

$(0, 199.4)$. We first plot $(0, 199.4)$. From this point we move 1 unit to the right and 8.6 units down, where we plot the point $(1, 190.8)$. We then sketch the line that contains these two points.

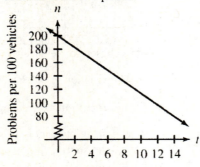

b. The line approximately contains the point $(14, 81)$, so $t = 14$ when $n = 81$ as illustrated.

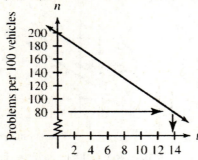

Thus, the average number of problems for all vehicles (of any brand) will reach 81 per 100 cars in the year $1995 + 14 = 2009$.

51. a. The line increases from left to right, so the m is positive. It crosses the y-axis below the x-axis, so b is negative.

b. The line is horizontal, so the m is zero. It crosses the y-axis below the x-axis, so b is negative.

c. The line decreases from left to right, so the m is negative. It crosses the y-axis above the x-axis, so b is positive.

d. The line increases from left to right, so the m is positive. It crosses the y-axis above the x-axis, so b is positive.

53. Specific graphs may vary. Since m is positive, the line must increase from left to right. Since b is positive, the line must cross the y-axis above the x-axis. One possible graph follows:

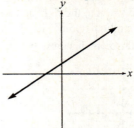

55. Specific graphs may vary. Since m is negative, the line must decrease from left to right. Since b is negative, the line must cross the y-axis below the x-axis. One possible graph follows:

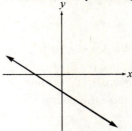

57. Specific graphs may vary. Since $m = 0$, the line must be horizontal. Since b is negative, the line must cross the y-axis below the x-axis. One possible graph follows:

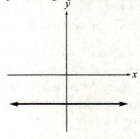

59. We substitute 3 for m and -4 for b in the equation $y = mx + b$ to obtain

$y = 3x + (-4)$
$y = 3x - 4$

61. We substitute $-\dfrac{6}{5}$ for m and 3 for b in the

equation $y = mx + b$ to obtain $y = -\dfrac{6}{5}x + 3$.

63. We substitute $-\dfrac{2}{7}$ for m and 0 for b in the

equation $y = mx + b$ to obtain $y = -\dfrac{2}{7}x + 0$ or

$y = -\dfrac{2}{7}x$.

65. The y-intercept of the line is $(0, 3)$. The run is 1 when the rise is 2, so the slope of the line is

$m = \dfrac{2}{1} = 2$. By substituting 2 for m and 3 for b

in the equation $y = mx + b$, we have $y = 2x + 3$.

67. The slope of $y = \dfrac{2}{5}x + 1$ is $m_1 = \dfrac{2}{5}$. The slope of

$y = -\dfrac{5}{2}x - 3$ is $m_2 = -\dfrac{5}{2}$. Since the slope $-\dfrac{5}{2}$

is the opposite of the reciprocal of the slope $\dfrac{2}{5}$,

the lines are perpendicular.

69. The slope of $y = 3x - 1$ is $m_1 = 3$. The slope of
$y = -3x + 2$ is $m_2 = -3$. Since the slopes are not
equal, the lines are not parallel. Since the slope
-3 is not the opposite of the reciprocal of the slope
3, the lines are not perpendicular. That is, the lines
are neither parallel nor perpendicular.

71. The slope of $y = -4x + 2$ is $m_1 = -4$. The slope
of $y = -4x + 3$ is $m_2 = -4$. Since the slopes are
equal, the lines are parallel.

73. The slope of $y = \dfrac{2}{3}x - 1$ is $m_1 = \dfrac{2}{3}$. The slope of

$y = \dfrac{3}{2}x + 3$ is $m_2 = \dfrac{3}{2}$. Since the slopes are not

equal, the lines are not parallel. Since the slope

$\dfrac{3}{2}$ is not the opposite of the reciprocal of the

slope $\dfrac{2}{3}$, the lines are not perpendicular. That is,

the lines are neither parallel nor perpendicular.

75. The slope of $y = 2$ is $m_1 = 0$. The slope of
$y = -4$ is $m_2 = 0$. Since the slopes are equal,
the lines are parallel. Note that both lines are
horizontal, confirming that they are parallel.

77. The graph of the line $x = 0$ is vertical. The
graph of the line $y = 0$ is horizontal. Thus, the
two lines are perpendicular.

79. No, the lines sketched are not parallel.
Explanations may vary. One possibility follows:
The lines do not have equal slopes. The red line

has slope $-\dfrac{3}{11}$, while the blue line has slope $-\dfrac{3}{10}$.

81. No. The slope is 2, not $2x$.

83. The equation $y = 2x - 1$ is in slope-intercept

form, so the slope is $m = 2 = \dfrac{2}{1}$ and the y-

intercept is $(0, -1)$. We first plot $(0, -1)$. From
this point we move 1 unit to the right and 2 units
up, where we plot the point $(1, 1)$. We then
sketch the line that contains these two points.

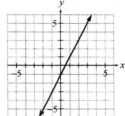

Points chosen for verification may vary. Two
possible points follow:

Check $(-1, -3)$ Check $(2, 3)$

$-3 \overset{?}{=} 2(-1) - 1$ $3 \overset{?}{=} 2(2) - 1$

$-3 \overset{?}{=} -2 - 1$ $3 \overset{?}{=} 4 - 1$

$-3 \overset{?}{=} -3$ True $3 \overset{?}{=} 3$ True

85. The solutions of $y = \dfrac{1}{2}x + 2$ can be described by

the equation, by a table, by a graph, or by words.

a. The equation $y = \dfrac{1}{2}x + 2$ is in slope-

intercept form, so the slope is $m = \dfrac{1}{2}$ and

the y-intercept is $(0, 2)$. We first plot
$(0, 2)$. From this point we move 2 units to
the right and 1 unit up, where we plot the
point $(2, 3)$. We then sketch the line that
contains these two points.

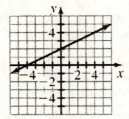

b. Answers may vary. One possibility follows:

x	y
-2	$\frac{1}{2}(-2)+2=1$
0	$\frac{1}{2}(0)+2=2$
2	$\frac{1}{2}(2)+2=3$

c. Answers may vary. One possibility follows: For each solution, the y-coordinate is two more than half the x-coordinate.

87. Answers may vary. One possibility follows: The graph provided contains the points $(-1,-2)$ and $(1, 4)$. In verifying these solutions, however, we obtain the following:

Check $(-1,-2)$

$-2 \overset{?}{=} -3(-1)+1$

$-2 \overset{?}{=} 3+1$

$-2 \overset{?}{=} 4$ False

Check $(1, 4)$

$4 \overset{?}{=} -3(1)+1$

$4 \overset{?}{=} -3+1$

$4 \overset{?}{=} -2$ False

Neither point is a solution to the equation $y=-3x-1$, so the graph is incorrect.

To obtain the correct graph, we recognize that the equation $y=-3x+1$ is in slope-intercept form, so the slope is $m=-3=\dfrac{-3}{1}$ and the y-intercept is $(0,1)$. We first plot $(0,1)$. From this point we move 1 unit to the right and 3 units down, where we plot the point $(1,-2)$. We then sketch the line that contains these two points.

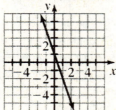

89. The line contains the point $(-3,-2)$, so $y=-2$ when $x=-3$.

91. The line contains the point $(3, 0)$, so $x=3$ when $y=0$.

93. The line contains the points $(-3,-2)$ and $(3, 0)$, so the slope is $m=\dfrac{0-(-1)}{3-0}=\dfrac{1}{3}$.

95. a. We first plot $(-1, 1)$. From this point we move 1 unit to the right and 2 units up, where we plot the point $(0, 3)$. We then sketch the line that contains these two points.

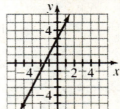

b. The slope of the line is $m=2$ and the y-intercept is $(0, 3)$. By substituting 2 for m and 3 for b in the equation $y=mx+b$, we have $y=2x+3$.

97. a. All three lines are of the form $y=b$, so in each case $m=0$. That is, the slope of each line is zero.

b. The slope of the graph of any equation of the form $y=k$, where k is a constant, is zero.

99. a. k is greater than m. We can tell because the line $y=kx+c$ is steeper than the line $y=mx+b$, so the slope of $y=kx+c$ is greater.

b. b is greater than c. We can tell because the line $y=mx+b$ crosses the y-axis above the line $y=kx+c$, so the y-coordinate of the y-intercept of $y=mx+b$ is greater.

Homework 3.5

1. $\dfrac{\text{change in salary}}{\text{change in time}} = \dfrac{\$12,400}{8 \text{ years}} = \dfrac{\$1550}{1 \text{ year}}$

The average rate of change is $1550 per year.

3. $\dfrac{\text{change in altitude}}{\text{change in time}} = \dfrac{-24,750 \text{ feet}}{15 \text{ minutes}} = \dfrac{-1650}{1 \text{ minute}}$

The average rate of change is -1650 feet per minute.

5. $\dfrac{\text{change in number of employers}}{\text{change in time}}$

$= \dfrac{228 \text{ employers} - 96 \text{ employers}}{\text{year } 2004 - \text{ year } 1999}$

$= \dfrac{132 \text{ employers}}{5 \text{ years}}$

$= \dfrac{26.4 \text{ employers}}{1 \text{ year}}$

The average rate of change is 26.4 employers per year.

7. $\dfrac{\text{change in percent}}{\text{change in time}} = \dfrac{30\% - 13\%}{\text{year } 2004 - \text{ year } 1999}$

$= \dfrac{17\%}{5 \text{ years}}$

$= \dfrac{3.4\%}{1 \text{ year}}$

The average rate of change is 3.4% per year.

9. $\dfrac{\text{change in number of Steller's sea lions}}{\text{change in time}}$

$= \dfrac{25 \text{ thousand sea lions} - 53 \text{ thousand sea lions}}{\text{year } 2000 - \text{ year } 1989}$

$= \dfrac{-28 \text{ thousand sea lions}}{11 \text{ years}}$

$\approx \dfrac{-2.55 \text{ thousand sea lions}}{1 \text{ year}}$

The average rate of change is approximately -2.55 thousand Steller's sea lions per year.

11. $\dfrac{\text{change in amount of oil exported}}{\text{change in time}}$

$= \dfrac{455 \text{ million barrels} - 800 \text{ million barrels}}{\text{year } 2002 - \text{ year } 1999}$

$= \dfrac{-345 \text{ thousand sea lions}}{3 \text{ years}}$

$= \dfrac{-155 \text{ million barrels}}{1 \text{ year}}$

The average rate of change is -155 million barrels per year.

13. $\dfrac{\text{change in income}}{\text{change in family size}} = \dfrac{\$44,775 - \$29,925}{7 \text{ people} - 4 \text{ people}}$

$= \dfrac{\$14,850}{3 \text{ people}}$

$= \dfrac{\$4950}{1 \text{ person}}$

The average rate of change is $4950 per person.

15. a. Yes, the relationship is linear since the rate of change is constant. The slope is 4. It means that the revenue is increasing by $4 million per year.

b. i. We know the slope is 4. Since t is the number of years since 2005, the r-intercept is (0, 3). Substituting 4 for m and 3 for b in the equation $r = mt + b$, we obtain $r = 4t + 3$.

ii. Answers may vary. One possibility follows:

Years Since 2005 t	Revenue (millions of dollars) s
0	$4(0) + 3 = 3$
1	$4(1) + 3 = 7$
2	$4(2) + 3 = 11$
3	$4(3) + 3 = 15$
4	$4(4) + 3 = 19$

iii. Plot the points from part (ii) and sketch the line that contains these points.

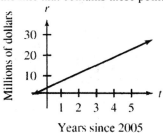

Years since 2005

17. a. Yes, the relationship is approximately linear since the rate of change is approximately constant. The slope is -21.9. It means that the number of drive-in movie sites in the United States is decreasing by about 21.9 sites per year.

b. Since t is the number of years since 2004, the n-intercept is (0, 402). It means that, in the year 2004, there were 402 drive-in movie sites in the United States.

c. Substituting -21.9 for m and 402 for b in the equation $n = mt + b$, we obtain $n = -21.9t + 402$.

d. The year 2011 is $t = 7$ years after 2004.
$n = -21.9(7) + 402 = 248.7$
The model predicts that there will be about 249 drive-in movie sites in the U.S. in 2011.

e. The year 2024 is $t = 20$ years after 2004.
$n = -21.9(20) + 402 = -36$
The model predicts that there will be -36 drive-in movie sites in the U.S. in 2024. Model breakdown has occurred.

19. a. The slope is -650. It means that the balance in the savings account declines by $650 per month.

b. Since t is the number of months since September 1, the B-intercept is (0, 4700). It means that, on September 1, the balance in the savings account was $4700.

c. Substituting -650 for m and 4700 for b in the equation $B = mt + b$, we obtain $B = -650t + 4700$.

d. $\underset{\text{dollars}}{B} = \underset{\substack{\text{dollars} \\ \text{month}}}{-650} \cdot \underset{\text{months}}{t} + \underset{\text{dollars}}{4700}$

We use the fact that $\dfrac{\text{months}}{\text{months}} = 1$ to simplify the right-hand side of the equation:
$\dfrac{\text{dollars}}{\text{month}} \cdot \text{months} + \text{dollars} = \text{dollars} + \text{dollars}$.
So, the units on both sides of the equation are dollars, suggesting the equation is correct.

e. Substituting $t = 6$ in the equation, we obtain $B = -650(6) + 4700 = 800$.
The model predicts that the savings account balance on March 1 will be $800.

21. a. The slope is 385. It means that the tuition increases by $385 per credit.

b. The tuition is $385 per credit, plus a $10 fee. Thus, the equation is $T = 385c + 10$.

c. $\underset{\text{dollars}}{T} = \underset{\substack{\text{dollars} \\ \text{credit}}}{385} \cdot \underset{\text{credits}}{c} + \underset{\text{dollars}}{10}$

We use the fact that $\dfrac{\text{credits}}{\text{credits}} = 1$ to simplify the right-hand side of the equation:
$\dfrac{\text{dollars}}{\text{credit}} \cdot \text{credits} + \text{dollars} = \text{dollars} + \text{dollars}$.
So, the units on both sides of the equation are dollars, suggesting the equation is correct.

d. Substituting $c = 9$ in the equation, we obtain $T = 385(9) + 10 = 3475$.
The tuition for 9 credits is $3475.

23. a. The slope is -0.02. It means that the car uses 0.02 gallons of gasoline per mile.

b. Since d is the number of miles driven, the G-intercept is (0, 11.9). It means that there were 11.9 gallons of gasoline in the tank at the start of the trip (when the tank is full).

c. Substituting -0.02 for m and 11.9 for b in the equation $G = md + b$, we obtain $G = -0.02d + 11.9$.

d. $\underset{\text{gallons}}{G} = \underset{\substack{\text{gallons} \\ \text{mile}}}{-0.02} \cdot \underset{\text{miles}}{d} + \underset{\text{gallons}}{10}$

We use the fact that $\dfrac{\text{miles}}{\text{miles}} = 1$ to simplify the right-hand side of the equation:
$\dfrac{\text{gallons}}{\text{mile}} \cdot \text{miles} + \text{gallons} = \text{gallons} + \text{gallons}$.
So, the units on both sides of the equation are gallons, suggesting the equation is correct.

e. Substituting $d = 525$ in the equation, we obtain $G = -0.02(525) + 11.9 = 1.4$. This means that 1.4 gallons remain in the tank. Thus, it will take $11.9 - 1.4 = 10.5$ gallons of gasoline to fill up the tank.

25. a. The slope is -3.65. It means that the number of refineries is decreasing by about 3.65 refineries per year.

b. Since t is the number of years since 2000, the n-intercept is (0, 157.31). It means that there were about 157 refineries in 2000.

c. The year 2011 is $t = 11$ years after 2000. Substituting $t = 11$ in the equation, we obtain $n = -3.65(11) + 157.31 = 117.16$. The model predicts that there will be about 117 refineries in the year 2011.

27. a.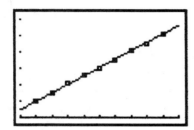

Yes, the line comes close to the data points.

b. The slope is 20.7. It means that the number of Internet users is increasing by 20.7 million users per year.

c.

Years	Rates of change
1996 to 1997	$60 - 39 = 21$
1997 to 1998	$84 - 60 = 24$
1998 to 1999	$105 - 84 = 21$
1990 to 2000	$122 - 105 = 17$
2000 to 2001	$143 - 122 = 21$
2001 to 2002	$166 - 143 = 23$
2002 to 2003	$183 - 166 = 17$
2003 to 2004	$207 - 183 = 24$

All of the rates of change are in millions of Internet users per year. Comparisons may vary.

d. The d-intercept is (0, 19.6). It means there were 19.6 million Internet users in 1995.

e. The year 2010 is represented by $t = 15$. Substituting 15 for t in the equation gives $n = 20.7(15) + 19.6 = 330.1$. The model predicts that there will be 330.1 million Internet users in 2010. This is more than the predicted U.S. population, so it appears that model breakdown has occurred.

29. a.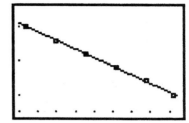

Yes, the line comes close to the data points.

b. Substituting 400 for s in the equation gives $G = -0.00254(400) + 4.58 \approx 3.56$. The model predicts that the students qualifying core GPA is 3.56. Since the actual qualifying core GPA is 3.55, the prediction is an overestimate of 0.01.

c. The slope is -0.00254. It means that the qualifying core GPA decreases by 0.00254 for an increase of 1 point on the SAT.

d. The G-intercept is (0, 4.58). It means the qualifying core GPA is 4.58 for a SAT score of 0. Since the highest GPA possible is 4.0 and the lowest possible SAT score is 400, model breakdown has occurred.

31. a. Yes, the car is traveling at a constant speed. This can be seen by the linear relationship between time and distance, showing a constant slope.

b. The line contains the points (0, 0) and (2, 120), so the slope is
$$m = \frac{120 \text{ miles} - 0 \text{ miles}}{2 \text{ hours} - 0 \text{ hours}} = \frac{120 \text{ miles}}{2 \text{ hours}} = 60$$
miles per hour.

33. a. Yes, the volume of gasoline that remains n the tank is decreasing at a constant rate. This can be seen by the linear relationship between time and volume, showing a constant slope.

b. The line contains the points (0, 12) and (3, 9), so the slope is
$$m = \frac{9 \text{ gallons} - 12 \text{ gallons}}{3 \text{ hours} - 1 \text{ hours}}$$
$$= \frac{-3 \text{ gallons}}{2 \text{ hours}}$$
$$= -1.5 \text{ gallons per hour}$$

35. Yes, t and c are linearly related. The slope is 30. It means that the total charge increases by \$30 per hour.

37. Equation 1 could be linear because, for each unit increase in x, y increases by a constant 2 units.
Equation 2 is not linear because the change in y is not constant.
Equation 3 could be linear because, for each unit increase in x, y decreases by a constant 2 units.
Equation 4 is not linear because the change in y is not constant.

39. For Equation 1, for each unit increase in x, y increases by 5 units.

x	y
0	3
1	8
2	$8+5=13$
3	$13+5=18$
4	$18+5=23$

For Equation 2, for each unit increase in x, y decreases by 7 units.

x	y
0	99
1	92
2	$92-7=85$
3	$85-7=78$
4	$78-7=71$

For Equation 3, $m = \dfrac{12-16}{23-21} = \dfrac{-4}{2} = -2$. For each unit increase in x, y decreases by 2 units.

x	y
21	16
22	$16-2=14$
23	12
24	$12-2=10$
25	$10-2=8$

For Equation 4, $m = \dfrac{29-23}{47-45} = \dfrac{6}{2} = 3$. For each unit increase in x, y increases by 3 units.

x	y
43	$20-3=17$
44	$23-3=20$
45	23
46	$29-3=26$
47	29

41. For Set 1, the slope is $m = \dfrac{7-5}{1-0} = \dfrac{2}{1} = 2$, and the y-intercept is $(0, 5)$. Substituting 2 for m and 5 for b in the equation $y = mx + b$, we obtain $y = 2x + 5$.

For Set 2, the slope is $m = \dfrac{17-20}{1-0} = \dfrac{-3}{1} = -3$, and the y-intercept is $(0, 20)$. Substituting -3 for m and 20 for b in the equation $y = mx + b$, we obtain $y = -3x + 20$.

For Set 3, the slope is $m = \dfrac{29-21}{1-0} = \dfrac{8}{1} = 8$, and the y-intercept is $(0, 21)$. Substituting 8 for m and 21 for b in the equation $y = mx + b$, we obtain $y = 8x + 21$.

For Set 4, the slope is $m = \dfrac{4-9}{1-0} = \dfrac{-5}{1} = -5$, and the y-intercept is $(0, 9)$. Substituting -5 for m and 9 for b in the equation $y = mx + b$, we obtain $y = -5x + 9$.

43. Since the slope is 7, if x is increased by 1, y will increase by 7.

45. Answers may vary. One possible table follows:

x	y
0	$-6(0)+40=40$
1	$-6(1)+40=34$
2	$-6(2)+40=28$
3	$-6(3)+40=22$
4	$-6(4)+40=16$

Note $40-6=34$, $34-6=28$, $28-6=22$, and $22-6=16$. This shows that, if x is increased by 1, then y decreases by 6.

47. **a.** The slope 3 is the number multiplied times x.

b. If the run is 1, then the rise is 3.

c. As the value of x increases by 1, the value of y increases by 3.

d. Answers will vary.

49. Answers may vary.

Chapter 3 Review Exercises

1. Check $(-3, 9)$: $9 \overset{?}{=} -2(-3)+3$

$9 \overset{?}{=} 6+3$

$9 \overset{?}{=} 9$ True

So, $(-3, 9)$ is a solution of $y = -2x+3$.

Check $(1, 2)$: $2 \overset{?}{=} -2(1)+3$

$2 \overset{?}{=} -2+3$

$2 \overset{?}{=} 1$ False

So, $(1, 2)$ is a not solution of $y = -2x+3$.

Check $(4, -5)$: $-5 \overset{?}{=} -2(4)+3$

$-5 \overset{?}{=} -8+3$

$-5 \overset{?}{=} -5$ True

So, $(4, -5)$ is a solution of $y = -2x+3$.

2. The line contains the point $(2, -1)$, so $y = -1$ when $x = 2$.

3. The line contains the point $(-2, -3)$, so $y = -3$ when $x = -2$.

4. The line contains the point $(0, -2)$, so $y = -2$ when $x = 0$.

5. The line contains the point $(-2, -3)$, so $x = -2$ when $y = -3$.

6. The line contains the point $(-4, -4)$, so $x = -4$ when $y = -4$.

7. The line contains the point $(4, 0)$, so $x = 4$ when $y = 0$.

8. Slope of plane $A = \dfrac{6500 \text{ feet}}{12,700 \text{ feet}} \approx 0.512$

Slope of plane $B = \dfrac{7400 \text{ feet}}{15,600 \text{ feet}} \approx 0.474$

Thus, plane A is climbing at a greater incline since it has a larger slope.

9. Using the slope formula with $(x_1, y_1) = (-3, 1)$ and $(x_2, y_2) = (2, 11)$, the slope is
$m = \dfrac{y_2 - y_1}{x_2 - x_1} = \dfrac{11-1}{2-(-3)} = \dfrac{11-1}{2+3} = \dfrac{10}{5} = 2$.
The slope is positive, so the line is increasing.

10. Using the slope formula with $(x_1, y_1) = (-2, -4)$ and $(x_2, y_2) = (1, -7)$, the slope is
$m = \dfrac{y_2 - y_1}{x_2 - x_1} = \dfrac{-7-(-4)}{1-(-2)} = \dfrac{-7+4}{1+2} = \dfrac{-3}{3} = -1$.
The slope is negative, so the line is decreasing.

11. Using the slope formula with $(x_1, y_1) = (4, -3)$ and $(x_2, y_2) = (8, -1)$, the slope is
$m = \dfrac{y_2 - y_1}{x_2 - x_1} = \dfrac{-1-(-3)}{8-4} = \dfrac{-1+3}{8-4} = \dfrac{2}{4} = \dfrac{1}{2}$.
The slope is positive, so the line is increasing.

12. Using the slope formula with $(x_1, y_1) = (-6, 0)$ and $(x_2, y_2) = (0, -3)$, the slope is
$m = \dfrac{y_2 - y_1}{x_2 - x_1} = \dfrac{-3-0}{0-(-6)} = \dfrac{-3-0}{0+6} = \dfrac{-3}{6} = -\dfrac{1}{2}$.
The slope is negative, so the line is decreasing.

13. Using the slope formula with $(x_1, y_1) = (-5, 5)$ and $(x_2, y_2) = (2, -2)$, the slope is
$m = \dfrac{y_2 - y_1}{x_2 - x_1} = \dfrac{-2-5}{2-(-5)} = \dfrac{-2-5}{2+5} = \dfrac{-7}{7} = -1$.
The slope is negative, so the line is decreasing.

14. Using the slope formula with $(x_1, y_1) = (-10, -3)$ and $(x_2, y_2) = (-4, -5)$, the slope is
$m = \dfrac{y_2 - y_1}{x_2 - x_1} = \dfrac{-5-(-3)}{-4-(-10)} = \dfrac{-5+3}{-4+10} = \dfrac{-2}{6} = -\dfrac{1}{3}$.
The slope is negative, so the line is decreasing.

15. Using the slope formula with $(x_1, y_1) = (-5, 2)$ and $(x_2, y_2) = (3, -7)$, the slope is
$m = \dfrac{y_2 - y_1}{x_2 - x_1} = \dfrac{-7-2}{3-(-5)} = \dfrac{-7-2}{3+5} = \dfrac{-9}{8} = -\dfrac{9}{8}$.
The slope is negative, so the line is decreasing.

16. Using the slope formula with $(x_1, y_1) = (-4, -1)$ and $(x_2, y_2) = (2, -5)$, the slope is

$$m = \frac{y_2 - y_1}{x_2 - x_1} = \frac{-5 - (-1)}{2 - (-4)} = \frac{-5 + 1}{2 + 4} = \frac{-4}{6} = -\frac{2}{3}.$$

The slope is negative, so the line is decreasing.

17. Using the slope formula with $(x_1, y_1) = (-4, 7)$ and $(x_2, y_2) = (-4, -3)$, the slope is

$$m = \frac{y_2 - y_1}{x_2 - x_1} = \frac{-3 - 7}{-4 - (-4)} = \frac{-3 - 7}{-4 + 4} = \frac{-10}{0}, \text{ which}$$

is undefined. So the line is vertical.

18. Using the slope formula with $(x_1, y_1) = (-5, 2)$ and $(x_2, y_2) = (-1, 2)$, the slope is

$$m = \frac{y_2 - y_1}{x_2 - x_1} = \frac{2 - 2}{-1 - (-5)} = \frac{2 - 2}{-1 + 5} = \frac{0}{4} = 0.$$

So, the line is horizontal.

19. Using the slope formula with $(x_1, y_1) = (5.4, 7.9)$ and $(x_2, y_2) = (8.3, -2.6)$, the slope is

$$m = \frac{y_2 - y_1}{x_2 - x_1} = \frac{-2.6 - 7.9}{8.3 - 5.4} = \frac{-10.5}{2.9} \approx -3.62.$$

The slope is negative, so the line is decreasing.

20. Using the slope formula with $(x_1, y_1) = (-8.74, -2.38)$ and $(x_2, y_2) = (-1.16, 4.77)$, the slope is

$$m = \frac{y_2 - y_1}{x_2 - x_1} = \frac{4.77 - (-2.38)}{-1.16 - (-8.74)} = \frac{7.15}{7.58} \approx 0.94.$$

The slope is positive, so the line is increasing.

21. Answers may vary. One possible solution follows:

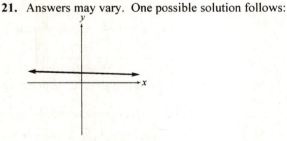

22. We first plot the y-intercept $(0, -4)$. The slope is $3 = \frac{3}{1}$, so the run is 1 and the rise is 3. From $(0, -4)$, we count 1 unit to the right and 3 units up, where we plot the point $(1, -1)$. We then sketch the line that contains these two points.

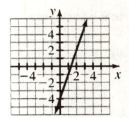

23. We first plot the y-intercept $(0, 1)$. The slope is $\frac{4}{3}$, so the run is 3 and the rise is 4. From $(0, 1)$, we count 3 units to the right and 4 units up, where we plot the point $(3, 5)$. We then sketch the line that contains these two points.

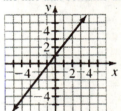

24. We first plot the point $(2, -3)$. The slope is 0, so the line is horizontal. We sketch the horizontal line that contains $(2, -3)$.

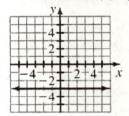

25. The equation $y = \frac{3}{4}x - 1$ is in slope-intercept form, so the slope is $m = \frac{3}{4}$ and the y-intercept is $(0, -1)$. We first plot $(0, -1)$. From this point we move 4 units to the right and 3 units up, where we plot the point $(4, 2)$. We then sketch the line that contains these two points.

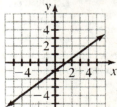

26. The equation $y = -\dfrac{1}{2}x + 3$ is in slope-intercept form, so the slope is $m = -\dfrac{1}{2}$ and the y-intercept is $(0, 3)$. We first plot $(0, 3)$. From this point we move 2 units to the right and 1 unit down, where we plot the point $(2, 2)$. We then sketch the line that contains these two points.

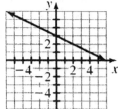

27. The equation $y = -\dfrac{2}{5}x - 1$ is in slope-intercept form, so the slope is $m = -\dfrac{2}{5}$ and the y-intercept is $(0, -1)$. We first plot $(0, -1)$. From this point we move 5 units to the right and 2 units down, where we plot the point $(5, -3)$. We then sketch the line that contains these two points.

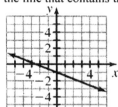

28. The equation $y = \dfrac{2}{3}x$ is in slope-intercept form, so the slope is $m = \dfrac{2}{3}$ and the y-intercept is $(0, 0)$. We first plot $(0, 0)$. From this point we move 3 units to the right and 2 units up, where we plot the point $(3, 2)$. We then sketch the line that contains these two points.

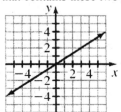

29. The equation $y = -4x$ is in slope-intercept form, so the slope is $m = -4 = \dfrac{-4}{1}$ and the y-intercept is $(0, 0)$. We first plot $(0, 0)$. From this point we move 1 unit to the right and 4 units down, where we plot the point $(1, -4)$. We then sketch the line that contains these two points.

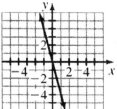

30. The equation $y = 2x - 4$ is in slope-intercept form, so the slope is $m = 2 = \dfrac{2}{1}$ and the y-intercept is $(0, -4)$. We first plot $(0, -4)$. From this point we move 1 unit to the right and 2 units up, where we plot the point $(1, -2)$. We then sketch the line that contains these two points.

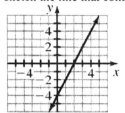

31. The equation $y = -3x + 1$ is in slope-intercept form, so the slope is $m = -3 = \dfrac{-3}{1}$ and the y-intercept is $(0, 1)$. We first plot $(0, 1)$. From this point we move 1 unit to the right and 3 units down, where we plot the point $(1, -2)$. We then sketch the line that contains these two points.

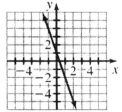

32. The equation $y = x + 2$ is in slope-intercept form, so the slope is $m = 1 = \dfrac{1}{1}$ and the y-intercept is $(0, 2)$. We first plot $(0, 2)$. From this point we move 1 unit to the right and 1 unit up, where we plot the point $(1, 3)$. We then sketch the line that contains these two points.

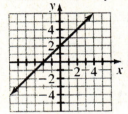

33. The equation $y = -5$ is in slope-intercept form, so the slope is $m = 0$ and the y-intercept is $(0, -5)$. Since the slope is 0, the line is horizontal. We sketch the horizontal line with y-intercept $(0, -5)$.

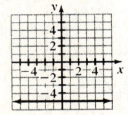

34. The graph of $x = -3$ is a vertical line. Note that x must be -3, but y can have any value. Some solutions of the equation are listed in the table that follows. We plot the corresponding points and sketch the line through them.

x	y
-3	-1
-3	0
-3	1

35. The equation $y = 2$ is in slope-intercept form, so the slope is $m = 0$ and the y-intercept is $(0, 2)$. Since the slope is 0, the line is horizontal. We sketch the horizontal line with y-intercept $(0, 2)$.

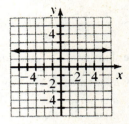

36. The solutions of $y = -2x + 1$ can be described by the equation, by a table, by a graph, or by words.

   a. Answers may vary. One possibility follows:

x	y
-1	$-2(-1) + 1 = 3$
0	$-2(0) + 1 = 1$
1	$-2(1) + 1 = -1$

   b. The equation $y = -2x + 1$ is in slope-intercept form, so the slope is $m = -2$ and the y-intercept is $(0, 1)$. We first plot $(0, 1)$. From this point we move 1 unit to the right and 2 units down, where we plot the point $(1, -1)$. We then sketch the line that contains these two points.

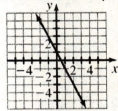

   c. Answers may vary. One possibility follows: For each solution, the y-coordinate is one more than -2 times the x-coordinate.

37. a.

Time (months)	Balance (thousands of dollars)
t	B
0	$19 - 3 \cdot 0$
1	$19 - 3 \cdot 1$
2	$19 - 3 \cdot 2$
3	$19 - 3 \cdot 3$
t	$19 - 3 \cdot t$

So, the equation is $B = 19 - 3t$.

   b. In the following table, we substitute values for t in the equation $B = 19 - 3t$ to find the corresponding values for B. Then, we plot the points and sketch a line that contains the points.

t	v
0	$19 - 3 \cdot 0 = 19$
1	$19 - 3 \cdot 1 = 16$
2	$19 - 3 \cdot 2 = 13$
3	$19 - 3 \cdot 3 = 10$

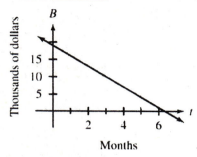

Months

c. From the table and graph, we see the B-intercept is $(0, 19)$. This indicates that, when the person first lost his job, the balance was $19 thousand.

d. The line contains the point $(5, 4)$, so $B = 4$ when $t = 5$ as illustrated.

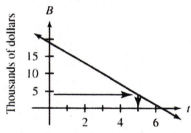

Months

Thus, if the balance is $4 thousand, then the person has been unemployed for 5 months.

38 The graph is of a vertical line. Since each point on the line has an x-coordinate of 5, the equation of the line is $x = 5$.

39. The y-intercept of the line is $(0, 4)$. The run is 3 when the rise is -2, so the slope of the line is $m = -\dfrac{2}{3}$. By substituting $-\dfrac{2}{3}$ for m and 4 for b in the equation $y = mx + b$, we have $y = -\dfrac{2}{3}x + 4$.

40. The slope of $y = 3x - 2$ is $m_1 = 3$. The slope of $y = \dfrac{1}{3}x + 6$ is $m_2 = \dfrac{1}{3}$. Since the slopes are not equal, the lines are not parallel. Since the slope

$\dfrac{1}{3}$ is not the opposite of the reciprocal of the slope 3, the lines are not perpendicular. Thus, the lines are neither parallel nor perpendicular.

41. The slope of $y = \dfrac{4}{7}x + 1$ is $m_1 = \dfrac{4}{7}$. The slope of $y = -\dfrac{7}{4}x - 5$ is $m_2 = -\dfrac{7}{4}$. Since the slope $-\dfrac{7}{4}$ is the opposite of the reciprocal of the slope $\dfrac{4}{7}$, the lines are perpendicular.

42. The graph of the line $x = -2$ is vertical. The graph of the line $y = 5$ is horizontal. Thus, the two lines are perpendicular.

43. The graphs of the lines $x = -4$ and $x = 1$ are both vertical. Thus, the two lines are parallel.

44. $\dfrac{\text{change in temperature}}{\text{change in time}} = \dfrac{-6°F}{4 \text{ hours}} = \dfrac{-1.5°F}{1 \text{ hour}}$

The average rate of change is $-1.5°F$ per hour.

45. $\dfrac{\text{change in sales}}{\text{change in time}} = \dfrac{\$1.2 \text{ billion} - \$0.9 \text{ billion}}{\text{year } 2004 - \text{ year } 2000}$

$= \dfrac{\$0.3 \text{ billion}}{4 \text{ years}}$

$= \dfrac{\$0.075 \text{ billion}}{1 \text{ year}}$

$= \dfrac{\$75 \text{ million}}{1 \text{ year}}$

The average rate of change is $0.075 billion per year, or $75 million per year.

46. a. The total charge is $2 per mile, plus a $2.50 flat fee. Thus, the equation is $c = 2d + 2.5$.

b. $\underset{\text{dollars}}{c} = \underset{\frac{\text{dollars}}{\text{mile}}}{2} \cdot \underset{\text{miles}}{d} + \underset{\text{dollars}}{2.5}$

We use the fact that $\dfrac{\text{miles}}{\text{miles}} = 1$ to simplify the right-hand side of the equation:

$\dfrac{\text{dollars}}{\text{mile}} \cdot \text{miles} + \text{dollars} = \text{dollars} + \text{dollars}$.

So, the units on both sides of the equation are dollars, suggesting the equation is correct.

c. Substituting $d = 17$ in the equation, we obtain $c = 2(17) + 2.5 = 36.5$.
The total cost of a 17-mile trip will be $36.50.

47. a. The slope is -4. It means that the person loses 4 pounds per month.

b. Since t is the number of months since the person began the weight-loss program, the w-intercept is (0, 195). Substituting -4 for m and 195 for b in the equation $w = mt + b$, we obtain $w = -4t + 195$.

c. Substituting $t = 6$ in the equation, we obtain $w = -4(6) + 195 = 171$. The person's goal was 171 pounds.

48. a. The slope is 17.75. It means that the average monthly cost increases by $17.75 per year.

b. Since t is the number of years since the 2001, the c-intercept is (0, 516). Substituting 17.75 for m and 516 for b in the equation $c = mt + b$, we obtain $c = 17.75t + 516$.

c. The year 2011 is represented by $t = 10$. Substituting $t = 10$ in the equation, we obtain $c = 17.75(10) + 516 = 693.5$. We predict that the average monthly day-care cost in 2011 will be $693.50.

49. Yes, C and n are linearly related. The slope is 130. It means that the cost is $130 per calculator.

50. Equation 1 could be linear because, for each unit increase in x, y decreases by a constant 3 units.
Equation 2 is not linear because the change in y is not constant.
Equation 3 could be linear because, y is constant at 4.
Equation 1 could be linear because, for each unit increase in x, y increases by a constant 5 units.

51. For Equation 1, for each unit increase in x, y decreases by 9 units.

x	y
0	50
1	41
2	$41 - 9 = 32$
3	$32 - 9 = 23$
4	$23 - 9 = 14$

For Equation 2, for each unit increase in x, y increases by 4 units.

x	y
0	12
1	16
2	$16 + 4 = 20$
3	$20 + 4 = 24$
4	$24 + 4 = 28$

For Equation 3, $m = \dfrac{19 - 25}{64 - 61} = \dfrac{-6}{3} = -2$. For each unit increase in x, y decreases by 2 units.

x	y
61	25
62	$25 - 2 = 23$
63	$23 - 2 = 21$
64	19
65	$19 - 2 = 17$

For Equation 4, $m = \dfrac{8 - (-4)}{30 - 26} = \dfrac{12}{4} = 3$. For each unit increase in x, y increases by 3 units.

x	y
26	-4
27	$-4 + 3 = -1$
28	$-1 + 3 = 2$
29	$2 + 3 = 5$
30	8

52. The slope of $y = -6x + 39$ is -6. This means that, when the value of x increases by 1, the value of y decreases by 6.

Chapter 3 Test

1. The line contains the point $(-3, 3)$, so $y = 3$ when $x = -3$.

2. The line contains the point $(3, -1)$, so $x = 3$ when $y = -1$.

3. The line intersects the y-axis at the point (0, 1), so the y-intercept is (0, 1).

4. The line intersects the x-axis at approximately the point (1.5, 0), so the x-intercept is (1.5, 0).

5. Slope of ski run $A = \dfrac{115 \text{ yards}}{580 \text{ yards}} \approx 0.198$

   Slope of ski run $B = \dfrac{150 \text{ yards}}{675 \text{ yards}} \approx 0.222$

   Thus, ski run B is steeper since it has a larger slope.

6. Using the slope formula with $(x_1, y_1) = (3, -8)$ and $(x_2, y_2) = (5, -2)$, the slope is
   $$m = \frac{y_2 - y_1}{x_2 - x_1} = \frac{-2 - (-8)}{5 - 3} = \frac{-2 + 8}{5 - 3} = \frac{6}{2} = 3.$$
   The slope is positive, so the line is increasing.

7. Using the slope formula with $(x_1, y_1) = (-4, -1)$ and $(x_2, y_2) = (2, -4)$, the slope is
   $$m = \frac{y_2 - y_1}{x_2 - x_1} = \frac{-4 - (-1)}{2 - (-4)} = \frac{-4 + 1}{2 + 4} = \frac{-3}{6} = -\frac{1}{2}.$$
   The slope is negative, so the line is decreasing.

8. Using the slope formula with $(x_1, y_1) = (-5, 4)$ and $(x_2, y_2) = (1, 4)$, the slope is
   $$m = \frac{y_2 - y_1}{x_2 - x_1} = \frac{4 - 4}{-1 - (-5)} = \frac{4 - 4}{-1 + 5} = \frac{0}{4} = 0.$$
   So, the line is horizontal.

9. Using the slope formula with $(x_1, y_1) = (-2, -7)$ and $(x_2, y_2) = (-2, 3)$, the slope is
   $$m = \frac{y_2 - y_1}{x_2 - x_1} = \frac{3 - (-7)}{-2 - (-2)} = \frac{3 + 7}{-2 + 2} = \frac{10}{0}, \text{ which is}$$
   undefined. So the line is vertical.

10. Using the slope formula with $(x_1, y_1) = (-5.99, -3.27)$ and $(x_2, y_2) = (2.83, 8.12)$, the slope is
    $$m = \frac{y_2 - y_1}{x_2 - x_1} = \frac{8.12 - (-3.27)}{2.83 - (-5.99)} = \frac{11.39}{8.82} \approx 1.29.$$
    The slope is positive, so the line is increasing.

11. We first plot the y-intercept $(0, -3)$. The slope is $\dfrac{2}{5}$, so the run is 5 and the rise is 2. From $(0, -3)$, we count 5 units to the right and 2 units up, where we plot the point $(5, -1)$. We then sketch the line that contains these two points.

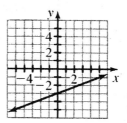

12. The equation $y = -\dfrac{3}{2}x + 2$ is in slope-intercept form, so the slope is $m = -\dfrac{3}{2}$ and the y-intercept is $(0, 2)$. We first plot $(0, 2)$. From this point we move 2 units to the right and 3 units down, where we plot the point $(2, -1)$. We then sketch the line that contains these two points.

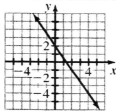

13. The equation $y = \dfrac{5}{6}x$ is in slope-intercept form, so the slope is $m = \dfrac{5}{6}$ and the y-intercept is $(0, 0)$. We first plot $(0, 0)$. From this point we move 6 units to the right and 5 units up, where we plot the point $(6, 5)$. We then sketch the line that contains these two points.

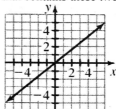

14. The equation $y = 3x - 4$ is in slope-intercept form, so the slope is $m = 3 = \dfrac{3}{1}$ and the y-intercept is $(0, -4)$. We first plot $(0, -4)$. From this point we move 1 unit to the right and 3 units up, where we plot the point $(1, -1)$. We then sketch the line that contains these two points.

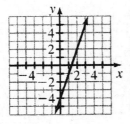

15. The equation $y = 2$ is in slope-intercept form, so the slope is $m = 0$ and the y-intercept is $(0, 2)$. Since the slope is 0, the line is horizontal. We sketch the horizontal line with y-intercept $(0, 2)$.

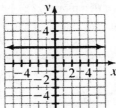

16. The equation $y = -2x + 3$ is in slope-intercept form, so the slope is $m = -2 = \dfrac{-2}{1}$ and the y-intercept is $(0, 3)$. We first plot $(0, 3)$. From this point we move 1 unit to the right and 2 units down, where we plot the point $(1, 1)$. We then sketch the line that contains these two points.

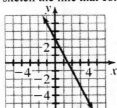

17. The y-intercept of the line is $(0, 1)$. The run is 2 when the rise is 1, so the slope of the line is $m = \dfrac{1}{2}$. By substituting $\dfrac{1}{2}$ for m and 1 for b in the equation $y = mx + b$, we have $y = \dfrac{1}{2}x + 1$.

18. a.

Time (years)	Car's value (thousands of dollars)
t	v
0	$17 - 2 \cdot 0$
1	$17 - 2 \cdot 1$
2	$17 - 2 \cdot 2$
3	$17 - 2 \cdot 3$
t	$17 - 2 \cdot t$

So, the equation is $v = 17 - 2t$, or $v = -2t + 17$.

b. In the following table, we substitute values for t in the equation $v = -2t + 17$ to find the corresponding values for v. Then, we plot the points and sketch a line that contains the points.

t	v
0	$17 - 2 \cdot 0 = 17$
1	$17 - 2 \cdot 1 = 15$
2	$17 - 2 \cdot 2 = 13$
3	$17 - 2 \cdot 3 = 11$

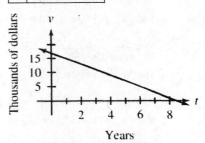

c. From the table and graph, we see the v-intercept is $(0, 17)$. This indicates that, the used car is currently worth \$17 thousand.

d. The line contains the point $(6, 5)$, so $v = 5$ when $t = 6$ as illustrated.

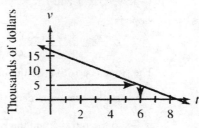

Thus, the value of the car will be \$5 thousand 6 years from now.

19. a. The line increases from left to right, so the *m* is positive. It crosses the *y*-axis above the *x*-axis, so *b* is positive.

    b. The line is horizontal, so the *m* is zero. It crosses the *y*-axis below the *x*-axis, so *b* is negative.

    c. The line decreases from left to right, so the *m* is negative. It crosses the *y*-axis above the *x*-axis, so *b* is positive.

    d. The line decreases from left to right, so the *m* is negative. It crosses the *y*-axis below the *x*-axis, so *b* is negative.

20. The slope of $y = \frac{2}{5}x + 3$ is $m_1 = \frac{2}{5}$. The slope of $y = \frac{5}{2}x - 7$ is $m_2 = \frac{5}{2}$. Since the slopes are not equal, the lines are not parallel. Since the slope $\frac{5}{2}$ is not the opposite of the reciprocal of the slope $\frac{2}{5}$, the lines are not perpendicular. Thus, the lines are neither parallel nor perpendicular.

21. The slope of $y = -3x + 8$ is $m_1 = -3$. The slope of $y = -3x - 1$ is $m_2 = -3$. Since the slopes are equal, the lines are parallel.

22. $\dfrac{\text{change in balance}}{\text{change in time}} = \dfrac{-\$3150}{6 \text{ months}} = \dfrac{-\$525}{1 \text{ month}}$
The average rate of change is $-\$525$ per month.

23. $\dfrac{\text{change in number of doses}}{\text{change in time}}$
$= \dfrac{100 \text{ million doses} - 77.9 \text{ million doses}}{\text{year } 2004 - \text{ year } 2000}$
$= \dfrac{22.1 \text{ million doses}}{4 \text{ years}}$
$\approx \dfrac{5.53 \text{ million doses}}{1 \text{ year}}$
The average rate of change is approximately 5.53 million doses per year.

24. a. The slope is 0.17. It means that total sales are increasing by $0.17 billion per year.

    b. Since *t* is the number of years since the 2003, the *s*-intercept is (0, 1.6). Substituting 0.17 for *m* and 1.6 for *b* in the equation $s = mt + b$, we obtain $s = 0.17t + 1.6$.

c. The year 2010 is represented by $t = 7$ since 2010 is 7 years after 2003. Substituting $t = 7$ in the equation, we obtain $c = 0.17(7) + 1.6 = 2.79$. We predict that total sales in 2010 will be $2.79 billion.

25. a.

Yes, the line comes close to the data points.

    b. The slope is 0.24. It means that the cooking time for a turkey increases by 0.24 hour per pound (or 14.4 minutes per pound).

    c. The *T*-intercept is (0, 1.64). It means the cooking time for a 0-pound turkey is 1.64 hours. Model breakdown has occurred.

    e. Substituting 19 for *w* in the equation gives $T = 0.24(19) + 1.64 = 6.2$. The model predicts that the cooking time for a 19-pound turkey will be 6.2 hours.

26. a. Yes, the person is running at a constant speed. This can be seen by the linear relationship between minutes and miles, showing a constant slope.

    b. The line contains the points (0, 0) and (30, 4), so the slope is
$$m = \dfrac{4 \text{ miles} - 0 \text{ miles}}{30 \text{ minutes} - 0 \text{ minutes}}$$
$$= \dfrac{4 \text{ miles}}{30 \text{ minutes}}$$
$$= \dfrac{4 \text{ miles}}{0.5 \text{ hour}}$$
$$= \dfrac{8 \text{ miles}}{1 \text{ hour}}$$
The person is running at a speed of 8 miles per hour.

27. For Set 1, the slope is $m = \dfrac{22 - 25}{1 - 0} = \dfrac{-3}{1} = -3$, and the *y*-intercept is (0, 25). Substituting -3 for *m* and 25 for *b* in the equation $y = mx + b$, we obtain $y = -3x + 25$.

For Set 2, the slope is $m = \dfrac{6-2}{1-0} = \dfrac{4}{1} = 4$, and the y-intercept is (0, 2). Substituting 4 for m and 2 for b in the equation $y = mx + b$, we obtain $y = 4x + 2$.

For Set 3, the slope is $m = \dfrac{7-12}{1-0} = \dfrac{-5}{1} = -5$, and the y-intercept is (0, 12). Substituting -5 for m and 12 for b in the equation $y = mx + b$, we obtain $y = -5x + 12$.

For Set 4, the slope is $m = \dfrac{53-47}{1-0} = \dfrac{6}{1} = 6$, and the y-intercept is (0, 47). Substituting 6 for m and 47 for b in the equation $y = mx + b$, we obtain $y = 6x + 47$.

28. The slope of $y = 3x - 8$ is 3. This means that, when the value of x increases by 1, the value of y decreases by 3.

Chapter 4
Simplifying Expressions and Solving Equations

Homework 4.1

1. $2(5x) = (2\cdot5)x = 10x$

3. $-4(-9x) = -(4(-9))x = -(-36)x = 36x$

5. $\frac{1}{2}(-8x) = \left(\frac{1}{2}\cdot(-8)\right)x = -4x$

7. $7\left(\frac{x}{4}\right) = \left(7\cdot\frac{1}{4}\right)x = \frac{7}{4}x = \frac{7x}{4}$

9. $3(x+9) = 3\cdot x + 3\cdot 9$
$\quad = 3x + 27$

11. $(x-5)2 = x\cdot2 - 5\cdot2$
$\quad = 2x - 10$

13. $-2(t+5) = -2\cdot t + (-2)\cdot5$
$\quad = -2t - 10$

15. $-5(6-2x) = -5\cdot6 - (-5)(2x)$
$\quad = -30 + 10x$

17. $(4x+7)(-6) = 4x(-6) + 7(-6)$
$\quad = -24x - 42$

19. $2(3x-5y) = 2\cdot3x - 2\cdot5y$
$\quad = 6x - 10y$

21. $-5(4x+3y-8) = -5(4x) + (-5)(3y) - (-5)(8)$
$\quad = -20x - 15y + 40$

23. $3 + 2(x+1) = 3 + 2\cdot x + 2\cdot1$
$\quad = 3 + 2x + 2$
$\quad = 2x + 3 + 2$
$\quad = 2x + 5$

25. $-0.3(x+0.2) = (-0.3)x + (-0.3)(0.2)$
$\quad = -0.3x - 0.06$

27. $4(a+3) + 7 = 4\cdot a + 4\cdot3 + 7$
$\quad = 4a + 12 + 7$
$\quad = 4a + 19$

29. $-3(4x-2) + 3 = -3(4x) - (-3)(2) + 3$
$\quad = -12x + 6 + 3$
$\quad = -12x + 9$

31. $4 - 3(3a-5) = 4 - 3(3a) - (-3)(5)$
$\quad = 4 - 9a + 15$
$\quad = -9a + 4 + 15$
$\quad = -9a + 19$

33. $-3.7 + 4.2(2.5x-8.3)$
$\quad = -3.7 + 4.2(2.5x) - 4.2(8.3)$
$\quad = -3.7 + 10.5x - 34.86$
$\quad = 10.5x - 3.7 - 34.86$
$\quad = 10.5x - 38.56$

35. $-(t+2) = -1(t+2)$
$\quad = -t + (-1)(2)$
$\quad = -t - 2$

37. $-(8x-9y) = -1(8x-9y)$
$\quad = -1(8x) - (-1)(9y)$
$\quad = -8x + 9y$

39. $-(5x+8y-1) = -1(5x+8y-1)$
$\quad = -1(5x) + (-1)(8y) - (-1)(1)$
$\quad = -5x - 8y + 1$

41. $-(3x+2) + 5 = -1(3x+2) + 5$
$\quad = -1(3x) + (-1)(2) + 5$
$\quad = -3x - 2 + 5$
$\quad = -3x + 3$

43. $8 - (x+3) = 8 - 1(x+3)$
$\quad = 8 - 1x + (-1)(3)$
$\quad = 8 - x - 3$
$\quad = -x + 8 - 3$
$\quad = -x + 5$

45.
$$-2-(2t-5)=-2-1(2t-5)$$
$$=-2-1(2t)-(-1)(5)$$
$$=-2-2t+5$$
$$=-2t-2+5$$
$$=-2t+3$$

47.
$$\frac{1}{2}(4x+8)=\frac{1}{2}(4x)+\frac{1}{2}(8)$$
$$=2x+4$$

49. $2(4x)=(2\cdot4)x=8x$

Evaluate $2(4x)$ for $x=2$:
$$2(4(2))=2(8)=16$$

Evaluate $8x$ for $x=2$:
$$8(2)=16$$
The results are the same. Additional choices of x will vary.

51.
$$5(x-7)=5x-5(7)$$
$$=5x-35$$
Evaluate $5(x-7)$ for $x=2$:
$$5(2-7)=5(-5)=-25$$
Evaluate $5x-35$ for $x=2$:
$$5(2)-35=10-35=-25$$
The results are the same. Additional choices of x will vary.

53.
$$5-3(x+4)=5-3x+(-3)(4)$$
$$=5-3x-12$$
$$=-3x+5-12$$
$$=-3x-7$$

55.
$$3(x+2)=3x+3(2)$$
$$=3x+6$$

57.
$$-4(2x-5)=-4(2x)-(-4)(5)$$
$$=-8x+20$$

59.
$$5-2(x-4)=5-2x-(-2)(4)$$
$$=5-2x+8$$
$$=-2x+5+8$$
$$=-2x+13$$

61.
$$-2+7(2x+1)=-2+7(2x)+7(1)$$
$$=-2+14x+7$$
$$=14x-2+7$$
$$=14x+5$$

63. a.
$$2(x-3)+4=2x-2(3)+4$$
$$=2x-6+4$$
$$=2x-2$$

b. Evaluate $2(x-3)+4$ for $x=5$:
$$2(5-3)+4=2(2)+4=4+4=8$$

c. Evaluate $2x-2$ for $x=5$:
$$2(5)-2=10-2=8$$

d. The results are the same. The simplification might be correct.

65. Evaluate $3(x+4)$ for $x=2$:
$$3(2+4)=3(6)=18$$
Evaluate $3x+4$ for $x=2$:
$$3(2)+4=6+4=10$$
The results are not the same; the simplification is incorrect.

67. a. Evaluate $a(bc)$ for $a=2$, $b=3$, and $c=4$:
$$2(3\cdot4)=2(12)=24$$

Evaluate $(ab)(ac)$ for $a(bc)$ for $a=2$, $b=3$, and $c=4$:
$$(2\cdot3)(3\cdot4)=(6)(12)=72$$

b. The results are not the same, so the expressions are not equivalent. The simplification is incorrect.

c. $a(bc)=(ab)c$

69. $y = 2(x-3)$

$\quad = 2x - 2(3)$

$\quad = 2x - 6$

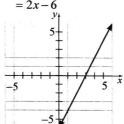

71. $y = -3(x+1)$

$\quad = -3x + (-3)(1)$

$\quad = -3x - 3$

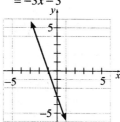

73. $y = -2(3x)$

$\quad = (-2 \cdot 3)x$

$\quad = -6x$

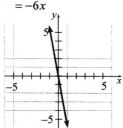

75. Answers may vary.

77. Answers may vary. Some possibilities:
$2(x-3)$, $2(x-2)-2$, $2(x+1)-8$

79. Evaluate $a-(b-c)=(a-b)-c$ for $a=7$,
$b=5$, and $c=1$:

$7-(5-1)=(7-5)-1$

$\qquad 7-4=2-1$

$\qquad\quad 3=1 \quad$ false

The result is a contradiction, so the statement
$a-(b-c)=(a-b)-c$ is false. There is no
associative law for subtraction.

81. Step 1: commutative law for addition
Step 2: associative law for addition
Step 3: commutative law for addition
Step 4: associative law for addition

83. Step 1: commutative law for multiplication
Step 2: distributive law
Step 3: commutative law for multiplication

85. Step 1: $\quad -a = -1 \cdot a$
Step 2: associative law for multiplication
Step 3: the product of two real numbers with
different signs is negative.
Step 4: $\quad -1 \cdot a = -a$
Step 5: $\quad -(-a) = a$

Homework 4.2

1. $2x + 5x = (2+5)x$

$\qquad\quad = 7x$

3. $9x - 4x = (9-4)x$

$\qquad\quad = 5x$

5. $-8w - 5w = (-8-5)w$

$\qquad\qquad\ = -13w$

7. $-t + 5t = -1t + 5t$

$\qquad\quad = (-1+5)t$

$\qquad\quad = 4t$

9. $6.6x - 7.1x = (6.6 - 7.1)x$

$\qquad\qquad\quad = -0.5x$

11. $\dfrac{2}{3}x + \dfrac{5}{3}x = \left(\dfrac{2}{3} + \dfrac{5}{3}\right)x$

$\qquad\qquad\ = \left(\dfrac{2+5}{3}\right)x$

$\qquad\qquad\ = \dfrac{7}{3}x$

13. $2 + 4x - 5 - 7x = 4x - 7x + 2 - 5$

$\qquad\qquad\qquad\ = -3x - 3$

15. $-3p + 2 + p - 9 = -3p + p + 2 - 9$

$\qquad\qquad\qquad\quad = -2p - 7$

17. $3y+5x-2y-2x+1=5x-2x+3y-2y+1$
$\qquad\qquad\qquad\qquad\quad =3x+y+1$

19. $-4.6x+3.9y+2.1-5.3x-2.8y$
$\qquad =-4.6x-5.3x+3.9y-2.8y+2.1$
$\qquad =-9.9x+1.1y+2.1$

21. $-3(a-5)+2a=-3a+15+2a$
$\qquad\qquad\qquad\quad =-3a+2a+15$
$\qquad\qquad\qquad\quad =-a+15$

23. $5.2(8.3x+4.9)-2.4=43.16x+25.48-2.4$
$\qquad\qquad\qquad\qquad\quad =43.16x+23.08$

25. $4(3a-2b)-5a=12a-8b-5a$
$\qquad\qquad\qquad\quad =12a-5a-8b$
$\qquad\qquad\qquad\quad =7a-8b$

27. $8-2(x+3)+x=8-2x-6+x$
$\qquad\qquad\qquad\quad =-2x+x+8-6$
$\qquad\qquad\qquad\quad =-x+2$

29. $6x-(4x-3y)-5y=6x-4x+3y-5y$
$\qquad\qquad\qquad\qquad =2x-2y$

31. $2t-3(5t+2)+1=2t-15t-6+1$
$\qquad\qquad\qquad\qquad =-13t-5$

33. $6-2(x+3y)+2y=6-2x-6y+2y$
$\qquad\qquad\qquad\qquad =-2x-6y+2y+6$
$\qquad\qquad\qquad\qquad =-2x-4y+6$

35. $-3(x-2)-5(x+4)=-3x+6-5x-20$
$\qquad\qquad\qquad\qquad\quad =-3x-5x+6-20$
$\qquad\qquad\qquad\qquad\quad =-8x-14$

37. $6(2x-3y)-4(9x+5y)=12x-18y-36x-20y$
$\qquad\qquad\qquad\qquad\qquad =12x-36x-18y-20y$
$\qquad\qquad\qquad\qquad\qquad =-24x-38y$

39. $-(x-1)-(1-x)=-1x+1-1+x$
$\qquad\qquad\qquad\qquad =-x+x+1-1$
$\qquad\qquad\qquad\qquad =0x+0$
$\qquad\qquad\qquad\qquad =0$

41. $2x-5y-3(2x-4y+7)=2x-5y-6x+12y-21$
$\qquad\qquad\qquad\qquad\qquad =2x-6x-5y+12y-21$
$\qquad\qquad\qquad\qquad\qquad =-4x+7y-21$

43. $5(2x-4y)-(3x-7y+2)$
$\qquad =10x-20y-3x+7y-2$
$\qquad =10x-3x-20y+7y-2$
$\qquad =7x-13y-2$

45. $\frac{2}{7}(a+1)-\frac{4}{7}(a-1)=\frac{2}{7}a+\frac{2}{7}-\frac{4}{7}a+\frac{4}{7}$
$\qquad\qquad\qquad\qquad =\frac{2}{7}a-\frac{4}{7}a+\frac{2}{7}+\frac{4}{7}$
$\qquad\qquad\qquad\qquad =-\frac{2}{7}a+\frac{6}{7}$

47. $5x-\frac{1}{2}(4x+6)=5x-\frac{1}{2}(4x)-\frac{1}{2}(6)$
$\qquad\qquad\qquad\qquad =5x-2x-3$
$\qquad\qquad\qquad\qquad =3x-3$

49. $x+5x=1x+5x$
$\qquad\qquad =(1+5)x$
$\qquad\qquad =6x$

51. $4(x-2)=4x-4(2)$
$\qquad\qquad =4x-8$

53. $x+3(x-7)=x+3x-21$
$\qquad\qquad\qquad =4x-21$

55. $2x-4(x+6)=2x-4x-24$
$\qquad\qquad\qquad =-2x-24$

57. Twice the number plus 6 times the number. (answers may vary)
$2x+6x=(2+6)x$
$\qquad\quad =8x$

59. 7 times the difference of the number and 5. (answers may vary)
$7(x-5)=7x-7(5)$
$\qquad\quad =7x-35$

61. The number, plus 5 times the sum of the number and 1. (answers may vary)

$$x + 5(x+1) = x + 5x + 5$$
$$= 1x + 5x + 5$$
$$= 6x + 5$$

63. Twice the number, minus 3 times the difference of the number and 9. (answers may vary)

$$2x - 3(x-9) = 2x - 3x + 27$$
$$= -x + 27$$

65.
$$(3x-7) + (5x+2) = 3x - 7 + 5x + 2$$
$$= 3x + 5x - 7 + 2$$
$$= 8x - 5$$

67.
$$(4x+8) - (7x-1) = 4x + 8 - 7x + 1$$
$$= 4x - 7x + 8 + 1$$
$$= -3x + 9$$

69.
$$-2x + 5 - 3 + 7x = -2x + 7x + 5 - 3$$
$$= 5x + 2$$

Evaluate for $x = 4$:

$$-2(4) + 5 - 3 + 7(4) = 5(4) + 2$$
$$-8 + 5 - 3 + 28 = 20 + 2$$
$$-3 - 3 + 28 = 22$$
$$-6 + 28 = 22$$
$$22 = 22$$

The results are the same. Similar results will be obtained using other values for x.

71.
$$4(x+2) - (x-3) = 4x + 8 - 1x + 3$$
$$= 4x - 1x + 8 + 3$$
$$= 3x + 11$$

Evaluate for $x = 4$:

$$4(4+2) - (4-3) = 3(4) + 11$$
$$4(6) - 1 = 12 + 11$$
$$24 - 1 = 12 + 11$$
$$23 = 23$$

The results are the same. Similar results will be obtained using other values for x.

73. a.
$$3(x+4) + 5x = 3x + 12 + 5x$$
$$= 3x + 5x + 12$$
$$= 8x + 12$$

b. Evaluate $3(x+4) + 5x$ for $x = 2$:

$$3(2+4) + 5(2) = 3(6) + 5(2)$$
$$= 18 + 10$$
$$= 28$$

c. Evaluate $8x + 12$ for $x = 2$:

$$8(2) + 12 = 16 + 12 = 28$$

d. The results are the same. The simplification may be correct.

75. $-2(x-3) = -2x + 6 = 2(3-x) = -3(x-2) + x$

77.

Expression	Evaluate for $x = 2$
$4(x-3) + 5x - 1$	$4(2-3) + 5(2) - 1 = 4(-1) + 5(2) - 1$
	$= -4 + 10 - 1$
	$= 5$
$4x - 3 + 5x - 1$	$4(2) - 3 + 5(2) - 1 = 8 - 3 + 10 - 1$
	$= 14$
$4x + 5x - 3 - 1$	$4(2) + 5(2) - 3 - 1 = 8 + 10 - 3 - 1$
	$= 14$
$9x - 4$	$9(2) - 4 = 18 - 4$
	$= 14$

The mistake occurred in the second step.

Correct simplification:

$$4(x-3) + 5x - 1 = 4x - 12 + 5x - 1$$
$$= 4x + 5x - 12 - 1$$
$$= 9x - 13$$

79. Answers may vary.

81. $y = 3x - 5x$
$= (3-5)x$
$= -2x$

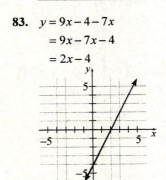

83. $y = 9x - 4 - 7x$
$= 9x - 7x - 4$
$= 2x - 4$

85. $y = 4(2x-1) - 5x$
$= 8x - 4 - 5x$
$= 8x - 5x - 4$
$= 3x - 4$

87. Answers may vary.

Homework 4.3

1. $3x + 1 = 7$
$3(2) + 1 \overset{?}{=} 7$
$6 + 1 \overset{?}{=} 7$
$7 \overset{?}{=} 7$ true
So, 2 is a solution to the equation.

3. $5(2x - 1) = 0$
$5(2(2) - 1) \overset{?}{=} 0$
$5(4 - 1) \overset{?}{=} 0$
$5(3) \overset{?}{=} 0$
$15 \overset{?}{=} 0$ false
So, 2 is not a solution to the equation.

5. $12 - x = 2(4x - 3)$
$12 - (2) \overset{?}{=} 2(4(2) - 3)$
$10 \overset{?}{=} 2(8 - 3)$
$10 \overset{?}{=} 2(5)$
$10 \overset{?}{=} 10$ true
So, 2 is a solution to the equation.

7. $x - 3 = 2$
$x - 3 + 3 = 2 + 3$
$x = 5$
Check: $x - 3 = 2$
$5 - 3 \overset{?}{=} 2$
$2 \overset{?}{=} 2$ true
So, the solution is 5.

9. $x + 6 = -8$
$x + 6 - 6 = -8 - 6$
$x = -14$
Check: $x + 6 = -8$
$-14 + 6 \overset{?}{=} -8$
$-8 \overset{?}{=} -8$ true
So, the solution is -14.

11. $t - 9 = 15$
$t - 9 + 9 = 15 + 9$
$t = 24$
Check: $t - 9 = 15$
$24 - 9 \overset{?}{=} 15$
$15 \overset{?}{=} 15$ true
So, the solution is 24.

13. $x + 11 = -17$

$x + 11 - 11 = -17 - 11$

$x = -28$

Check: $x + 11 = -17$

$-28 + 11 \overset{?}{=} -17$

$-17 \overset{?}{=} -17$ true

So, the solution is -28.

15. $-5 = x - 2$

$-5 + 2 = x - 2 + 2$

$-3 = x$

Check: $-5 = x - 2$

$-5 \overset{?}{=} -3 - 2$

$-5 \overset{?}{=} -5$ true

So, the solution is -3.

17. $x - 3 = 0$

$x - 3 + 3 = 0 + 3$

$x = 3$

Check: $x - 3 = 0$

$3 - 3 \overset{?}{=} 0$

$0 \overset{?}{=} 0$ true

So, the solution is 3.

19. $6r = 18$

$\dfrac{6r}{6} = \dfrac{18}{6}$

$r = 3$

Check: $6r = 18$

$6(3) \overset{?}{=} 18$

$18 \overset{?}{=} 18$ true

So, the solution is 3.

21. $-3x = 12$

$\dfrac{-3x}{-3} = \dfrac{12}{-3}$

$x = -4$

Check: $-3x = 12$

$-3(-4) \overset{?}{=} 12$

$12 \overset{?}{=} 12$ true

So, the solution is -4.

23. $15 = 3x$

$\dfrac{15}{3} = \dfrac{3x}{3}$

$5 = x$

Check: $15 = 3x$

$15 \overset{?}{=} 3(5)$

$15 \overset{?}{=} 15$ true

So, the solution is 5.

25. $6x = 8$

$\dfrac{6x}{6} = \dfrac{8}{6}$

$x = \dfrac{8}{6}$

$x = \dfrac{4}{3}$

Check: $6x = 8$

$6\left(\dfrac{4}{3}\right) \overset{?}{=} 8$

$\dfrac{24}{3} \overset{?}{=} 8$

$8 \overset{?}{=} 8$ true

So, the solution is $\dfrac{4}{3}$.

27. $-10x = -12$

$$\frac{-10x}{-10} = \frac{-12}{-10}$$

$$x = \frac{12}{10}$$

$$x = \frac{6}{5}$$

Check: $\quad -10x = -12$

$$-10\left(\frac{6}{5}\right) \overset{?}{=} -12$$

$$\frac{-60}{5} \overset{?}{=} -12$$

$$-12 \overset{?}{=} -12 \text{ true}$$

So, the solution is $\dfrac{6}{5}$.

29. $-2x = 0$

$$\frac{-2x}{-2} = \frac{0}{-2}$$

$$x = 0$$

Check: $\quad -2x = 0$

$$-2(0) \overset{?}{=} 0$$

$$0 \overset{?}{=} 0 \text{ true}$$

So, the solution is 0.

31. $\quad \dfrac{1}{3}t = 5$

$$3\left(\frac{1}{3}t\right) = 3(5)$$

$$t = 15$$

Check: $\quad \dfrac{1}{3}t = 5$

$$\frac{1}{3}(15) \overset{?}{=} 5$$

$$5 \overset{?}{=} 5 \text{ true}$$

So, the solution is 15.

33. $\qquad -\dfrac{2}{7}x = -3$

$$\left(-\frac{7}{2}\right)\left(-\frac{2}{7}x\right) = \left(-\frac{7}{2}\right)(-3)$$

$$x = \frac{21}{2}$$

Check: $\quad -\dfrac{2}{7}x = -3$

$$-\frac{2}{7}\left(\frac{21}{2}\right) \overset{?}{=} -3$$

$$-\frac{42}{14} \overset{?}{=} -3$$

$$-3 \overset{?}{=} -3 \text{ true}$$

So, the solution is $\dfrac{21}{2}$.

35. $\qquad -9 = \dfrac{3x}{4}$

$$\left(\frac{4}{3}\right)(-9) = \left(\frac{4}{3}\right)\left(\frac{3x}{4}\right)$$

$$\frac{-36}{3} = x$$

$$-12 = x$$

Check: $-9 = \dfrac{3x}{4}$

$$-9 \overset{?}{=} \frac{3(-12)}{4}$$

$$-9 \overset{?}{=} \frac{-36}{4}$$

$$-9 \overset{?}{=} -9 \text{ true}$$

So, the solution is -12.

37. $\quad \dfrac{2}{5}p = -\dfrac{4}{3}$

$$\frac{5}{2} \cdot \frac{2}{5}p = \frac{5}{2}\left(-\frac{4}{3}\right)$$

$$p = -\frac{20}{6}$$

$$p = -\frac{10}{3}$$

Check: $\dfrac{2}{5}p = -\dfrac{4}{3}$

$\dfrac{2}{5}\left(-\dfrac{10}{3}\right) \overset{?}{=} -\dfrac{4}{3}$

$-\dfrac{20}{15} \overset{?}{=} -\dfrac{4}{3}$

$-\dfrac{4}{3} \overset{?}{=} -\dfrac{4}{3}$ true

So, the solution is $-\dfrac{10}{3}$.

39. $-\dfrac{3x}{8} = -\dfrac{9}{4}$

$\left(-\dfrac{8}{3}\right)\left(-\dfrac{3x}{8}\right) = \left(-\dfrac{8}{3}\right)\left(-\dfrac{9}{4}\right)$

$x = \dfrac{72}{12}$

$x = 6$

Check: $-\dfrac{3x}{8} = -\dfrac{9}{4}$

$-\dfrac{3(6)}{8} \overset{?}{=} -\dfrac{9}{4}$

$-\dfrac{18}{8} \overset{?}{=} -\dfrac{9}{4}$

$-\dfrac{9}{4} \overset{?}{=} -\dfrac{9}{4}$ true

So, the solution is 6.

41. $-x = 3$

$-1(-x) = -1(3)$

$x = -3$

Check: $-x = 3$

$-(-3) \overset{?}{=} 3$

$3 \overset{?}{=} 3$ true

So, the solution is -3.

43. $-\dfrac{1}{2} = -x$

$(-1)\left(-\dfrac{1}{2}\right) = (-1)(-x)$

$\dfrac{1}{2} = x$

Check: $-\dfrac{1}{2} = -x$

$-\dfrac{1}{2} \overset{?}{=} -\left(\dfrac{1}{2}\right)$

$-\dfrac{1}{2} \overset{?}{=} -\dfrac{1}{2}$ true

So, the solution is $\dfrac{1}{2}$.

45. $x + 4.3 = -6.8$

$x + 4.3 - 4.3 = -6.8 - 4.3$

$x = -11.1$

Check: $x + 4.3 = -6.8$

$-11.1 + 4.3 \overset{?}{=} -6.8$

$-6.8 \overset{?}{=} -6.8$ true

So, the solution is -11.1.

47. $25.17 = x - 16.59$

$25.17 + 16.59 = x - 16.59 + 16.59$

$41.76 = x$

Check: $25.17 = x - 16.59$

$25.17 \overset{?}{=} 41.76 - 16.59$

$25.17 \overset{?}{=} 25.17$ true

So, the solution is 41.76.

49. $-3.7r = -8.51$

$\dfrac{-3.7r}{-3.7} = \dfrac{-8.51}{-3.7}$

$r = 2.3$

Check: $-3.7r = -8.51$

$-3.7(2.3) \overset{?}{=} -8.51$

$-8.51 \overset{?}{=} -8.51$ true

So, the solution is 2.3.

51. The line $y = -\dfrac{1}{2}x + 1$ intersects the line $y = 3$ only at the point $(-4, 3)$. The intersection point, $(-4, 3)$, has x-coordinate -4. So, -4 is the solution of $-\dfrac{1}{2}x + 1 = 3$.

53. The line $y = -\dfrac{1}{2}x + 1$ intersects the line $y = -1$ only at the point $(4, -1)$. The intersection point, $(4, -1)$, has x-coordinate 4. So, 4 is the solution of $-\dfrac{1}{2}x + 1 = -1$.

55. $x + 2 = 7$

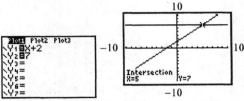

The intersection point, $(5, 7)$, has x-coordinate 5. So, 5 is the solution of $x + 2 = 7$.

57. $2x - 3 = 5$

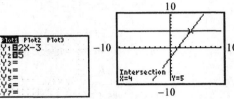

The intersection point, $(4, 5)$, has x-coordinate 4. So, 4 is the solution of $2x - 3 = 5$.

59. $-4(x - 1) = -8$

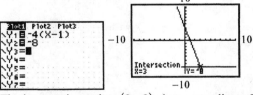

The intersection point, $(3, -8)$, has x-coordinate 3. So, 3 is the solution of $-4(x - 1) = -8$.

61. $\dfrac{2}{3}t - \dfrac{3}{2} = -\dfrac{7}{2}$

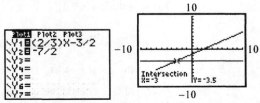

The intersection point $(-3, -3.5)$, has x-coordinate -3. So, -3 is the solution of $\dfrac{2}{3}t - \dfrac{3}{2} = -\dfrac{7}{2}$.

63. If we substitute 12 for y in the equation $y = 5x - 3$, the result is the equation $5x - 3 = 12$, which is what we are trying to solve. From the table, we see that the output $y = 12$ originates from the input $x = 3$. This means that 3 is a solution of the equation $5x - 3 = 12$.

65. If we substitute -13 for y in the equation $y = 5x - 3$, the result is the equation $5x - 3 = -13$, which is what we are trying to solve. From the table, we see that the output $y = -13$ originates from the input $x = -2$. This means that -2 is a solution of the equation $5x - 3 = -13$.

67. $2x + 5x = 14$

$(2 + 5)x = 14$

$7x = 14$

$\dfrac{7x}{7} = \dfrac{14}{7}$

$x = 2$

The solution is 2.

69. $4x - 5x = -2$

$(4 - 5)x = -2$

$-1x = -2$

$\dfrac{-1x}{-1} = \dfrac{-2}{-1}$

$x = 2$

The solution is 2.

71. The student subtracted two from both sides. Instead of dividing both sides by 2, and the subtraction itself was done incorrectly.

$2x = 10$

$\dfrac{2x}{2} = \dfrac{10}{2}$

$x = 5$

The solution is 5.

73. $4x = 12$ $\qquad$ $\dfrac{4x}{4} = \dfrac{12}{4}$ $\qquad$ $x = 3$

$4(3) \overset{?}{=} 12$ $\qquad$ $\dfrac{4(3)}{4} \overset{?}{=} \dfrac{12}{4}$ $\qquad$ $3 \overset{?}{=} 3$ true

$12 \overset{?}{=} 12$ true $\qquad$ $\dfrac{12}{4} \overset{?}{=} \dfrac{12}{4}$ true

75. Yes, the work is correct but the equation can be solved in fewer steps.

$$x + 2 = 7$$
$$x + 2 - 2 = 7 - 2$$
$$x = 5$$

The solution is 5.

77. Answers may vary. One possibility:

$$x - 7 = 2$$
$$x - 7 + 7 = 2 + 7$$
$$x = 9$$

79. Answers may vary. One possibility:

$$\frac{x}{7} = 4$$
$$7 \cdot \frac{x}{7} = 7 \cdot 4$$
$$x = 28$$

81.
$$\frac{x}{3} = 2 \qquad x - 1 = 4$$
$$\qquad\qquad x - 1 + 1 = 4 + 1$$
$$3 \cdot \frac{x}{3} = 3 \cdot 2 \qquad x = 5$$
$$x = 6$$

The equations do not have the same solution set so they are ***not*** equivalent.

83. Answers may vary.

85. Answers may vary.

87. Yes, the equations are equivalent. The second equation is obtained from the first by adding 6 to both sides. From the Addition Property of Equality, the two equations are equivalent.

89. a.
$$x + 2 = 7$$
$$x + 2 - 2 = 7 - 2$$
$$x = 5$$
The solution is 5.

b.
$$x + 5 = 9$$
$$x + 5 - 5 = 9 - 5$$
$$x = 4$$
The solution is 4.

c.
$$x + b = k$$
$$x + b - b = k - b$$
$$x = k - b$$
The solution is $k - b$.

91. Answers may vary.

Homework 4.4

1.
$$3x - 2 = 13$$
$$3x - 2 + 2 = 13 + 2$$
$$3x = 15$$
$$\frac{3x}{3} = \frac{15}{3}$$
$$x = 5$$
The solution is 5.

3.
$$-4x + 6 = 26$$
$$-4x + 6 - 6 = 26 - 6$$
$$-4x = 20$$
$$\frac{-4x}{-4} = \frac{20}{-4}$$
$$x = -5$$
The solution is -5.

5.
$$-5 = 6x + 3$$
$$-5 - 3 = 6x + 3 - 3$$
$$-8 = 6x$$
$$\frac{-8}{6} = \frac{6x}{6}$$
$$-\frac{4}{3} = x$$
The solution is $-\frac{4}{3}$.

7.
$$8 - x = -4$$
$$8 - x - 8 = -4 - 8$$
$$-x = -12$$
$$\frac{-x}{-1} = \frac{-12}{-1}$$
$$x = 12$$
The solution is 12.

9. $2x + 6 - 7x = -4$

$-5x + 6 = -4$

$-5x + 6 - 6 = -4 - 6$

$-5x = -10$

$\dfrac{-5x}{-5} = \dfrac{-10}{-5}$

$x = 2$

The solution is 2.

11. $5x + 4 = 3x + 16$

$5x + 4 - 3x = 3x + 16 - 3x$

$2x + 4 = 16$

$2x + 4 - 4 = 16 - 4$

$2x = 12$

$\dfrac{2x}{2} = \dfrac{12}{2}$

$x = 6$

The solution is 6.

13. $-3r - 1 = 2r + 24$

$-3r - 1 - 2r = 2r + 24 - 2r$

$-5r - 1 = 24$

$-5r - 1 + 1 = 24 + 1$

$-5r = 25$

$\dfrac{-5r}{-5} = \dfrac{25}{-5}$

$r = -5$

The solution is -5.

15. $9 - x - 5 = 2x - x$

$4 - x = x$

$4 - x + x = x + x$

$4 = 2x$

$\dfrac{4}{2} = \dfrac{2x}{2}$

$2 = x$

The solution is 2.

17. $2(x + 3) = 5x - 3$

$2x + 6 = 5x - 3$

$2x + 6 - 2x = 5x - 3 - 2x$

$6 = 3x - 3$

$6 + 3 = 3x - 3 + 3$

$9 = 3x$

$\dfrac{9}{3} = \dfrac{3x}{3}$

$3 = x$

The solution is 3.

19. $1 - 3(5b - 2) = 4 - (7b + 3)$

$1 - 15b + 6 = 4 - 7b - 3$

$-15b + 7 = 1 - 7b$

$-15b + 7 + 15b = 1 - 7b + 15b$

$7 = 1 + 8b$

$7 - 1 = 1 + 8b - 1$

$6 = 8b$

$\dfrac{6}{8} = \dfrac{8b}{8}$

$\dfrac{3}{4} = b$

The solution is $\dfrac{3}{4}$.

21. $4x = 3(2x - 1) + 5$

$4x = 6x - 3 + 5$

$4x = 6x + 2$

$4x - 6x = 6x + 2 - 6x$

$-2x = 2$

$\dfrac{-2x}{-2} = \dfrac{2}{-2}$

$x = -1$

The solution is -1.

23. $3(4x-5)-(2x+3)=2(x-4)$

$12x-15-2x-3=2x-8$

$10x-18=2x-8$

$10x-18-2x=2x-8-2x$

$8x-18=-8$

$8x-18+18=-8+18$

$8x=10$

$\dfrac{8x}{8}=\dfrac{10}{8}$

$x=\dfrac{5}{4}$

The solution is $\dfrac{5}{4}$.

25. $\dfrac{x}{2}-\dfrac{3}{4}=\dfrac{1}{2}$

The LCD is 4, so multiply both sides of the equation by 4 to clear the fractions.

$4\cdot\left(\dfrac{x}{2}-\dfrac{3}{4}\right)=4\cdot\dfrac{1}{2}$

$2x-3=2$

$2x-3+3=2+3$

$2x=5$

$\dfrac{2x}{2}=\dfrac{5}{2}$

$x=\dfrac{5}{2}$

The solution is $\dfrac{5}{2}$.

27. $\dfrac{5x}{6}+\dfrac{2}{3}=2$

The LCD is 6, so multiply both sides of the equation by 6 to clear the fractions.

$6\cdot\left(\dfrac{5x}{6}+\dfrac{2}{3}\right)=6\cdot 2$

$5x+4=12$

$5x+4-4=12-4$

$5x=8$

$\dfrac{5x}{5}=\dfrac{8}{5}$

$x=\dfrac{8}{5}$

The solution is $\dfrac{8}{5}$.

29. $\dfrac{5}{6}k=\dfrac{3}{4}k+\dfrac{1}{2}$

The LCD is 12, so multiply both sides of the equation by 12 to clear the fractions.

$12\cdot\dfrac{5}{6}k=12\cdot\left(\dfrac{3}{4}k+\dfrac{1}{2}\right)$

$10k=9k+6$

$10k-9k=9k+6-9k$

$k=6$

The solution is 6.

31. $\dfrac{7}{12}x-\dfrac{5}{3}=\dfrac{7}{4}+\dfrac{5}{6}x$

The LCD is 12, so multiply both sides of the equation by 12 to clear the fractions.

$12\cdot\left(\dfrac{7}{12}x-\dfrac{5}{3}\right)=12\cdot\left(\dfrac{7}{4}+\dfrac{5}{6}x\right)$

$7x-20=21+10x$

$7x-20-10x=21+10x-10x$

$-3x-20=21$

$-3x-20+20=21+20$

$-3x=41$

$\dfrac{-3x}{-3}=\dfrac{41}{-3}$

$x=-\dfrac{41}{3}$

The solution is $-\dfrac{41}{3}$.

33. $\dfrac{4}{3}x-2=3x+\dfrac{5}{2}$

The LCD is 6, so multiply both sides of the equation by 6 to clear the fractions.

$6\cdot\left(\dfrac{4}{3}x-2\right)=6\cdot\left(3x+\dfrac{5}{2}\right)$

$8x-12=18x+15$

$8x-12-18x=18x+15-18x$

$-10x-12=15$

$-10x-12+12=15+12$

$-10x=27$

$\dfrac{-10x}{-10}=\dfrac{27}{-10}$

$x=-\dfrac{27}{10}$

The solution is $-\dfrac{27}{10}$.

35. $\dfrac{3(x-4)}{5} = -2x$

The LCD is 5, so multiply both sides of the equation by 5 to clear the fractions.

$$5 \cdot \dfrac{3(x-4)}{5} = 5(-2x)$$

$$3(x-4) = 5(-2x)$$

$$3x - 12 = -10x$$

$$3x - 12 - 3x = -10x - 3x$$

$$-12 = -13x$$

$$\dfrac{-12}{-13} = \dfrac{-13x}{-13}$$

$$\dfrac{12}{13} = x$$

The solution is $\dfrac{12}{13}$.

37. $\dfrac{4x+3}{5} = \dfrac{2x-1}{3}$

The LCD is 15, so multiply both sides of the equation by 15 to clear the fractions.

$$15 \cdot \dfrac{4x+3}{5} = 15 \cdot \dfrac{2x-1}{3}$$

$$3(4x+3) = 5(2x-1)$$

$$12x + 9 = 10x - 5$$

$$12x + 9 - 10x = 10x - 5 - 10x$$

$$2x + 9 = -5$$

$$2x + 9 - 9 = -5 - 9$$

$$2x = -14$$

$$\dfrac{2x}{2} = \dfrac{-14}{2}$$

$$x = -7$$

The solution is -7.

39. $\dfrac{4m-5}{2} - \dfrac{3m+1}{3} = \dfrac{5}{6}$

The LCD is 6, so multiply both sides of the equation by 6 to clear the fractions.

$$6 \cdot \left(\dfrac{4m-5}{2} - \dfrac{3m+1}{3} \right) = 6 \cdot \dfrac{5}{6}$$

$$3(4m-5) - 2(3m+1) = 5$$

$$12m - 15 - 6m - 2 = 5$$

$$6m - 17 = 5$$

$$6m - 17 + 17 = 5 + 17$$

$$6m = 22$$

$$\dfrac{6m}{6} = \dfrac{22}{6}$$

$$m = \dfrac{11}{3}$$

The solution is $\dfrac{11}{3}$.

41.
$$0.3x + 0.2 = 0.7$$

$$0.3x + 0.2 - 0.2 = 0.7 - 0.2$$

$$0.3x = 0.5$$

$$\dfrac{0.3x}{0.3} = \dfrac{0.5}{0.3}$$

$$x = \dfrac{5}{3}$$

The solution is $\dfrac{5}{3}$.

43.
$$5.27x - 6.35 = 2.71x + 9.89$$

$$5.27x - 6.35 - 2.71x = 2.71x + 9.89 - 2.71x$$

$$2.56x - 6.35 = 9.89$$

$$2.56x - 6.35 + 6.35 = 9.89 + 6.35$$

$$2.56x = 16.24$$

$$\dfrac{2.56x}{2.56} = \dfrac{16.24}{2.56}$$

$$x \approx 6.34$$

The solution is approximately 6.34.

45. $0.4x - 1.6(2.5 - x) = 3.1(x - 5.4) - 11.3$

$0.4x - 4 + 1.6x = 3.1x - 16.74 - 11.3$

$2x - 4 = 3.1x - 28.04$

$2x - 4 - 3.1x = 3.1x - 28.04 - 3.1x$

$-1.1x - 4 = -28.04$

$-1.1x - 4 + 4 = -28.04 + 4$

$-1.1x = -24.04$

$$\dfrac{-1.1x}{-1.1} = \dfrac{-24.04}{-1.1}$$

$x \approx 21.85$

The solution is approximately 21.85.

47. a. An equation of the linear model can be written in slope-intercept form, $y = mx + b$.

Using t and a, we have $a = mt + b$. Since the slope (rate of change) is -5.6 and the a-intercept is $(0, 483)$, we have

$a = -5.6t + 483$.

b. Substitute $2011 - 2000 = 11$ for t in the equation from part (a).

$a = -5.6(11) + 483$

$= 421.4$

So, in 2011 there will be an average of 421.4 students per counselor.

c. Substitute 250 for a in the equation from part (a).

$250 = -5.6t + 483$

$250 - 483 = -5.6t + 483 - 483$

$-233 = -5.6t$

$$\dfrac{-233}{-5.6} = \dfrac{-5.6t}{-5.6}$$

$41.6 \approx t$

So, the ratio of students per counselor will be 250 in the year $2000 + 42 = 2042$.

49. a. An equation of the linear model can be written in slope-intercept form, $y = mx + b$.

Using t and v, we have $v = mt + b$. Since the slope (rate of change) is 3.36 and the v-intercept is $(0, 8.16)$, we have

$v = 3.36t + 8.16$.

b. Substitute 3 for t in the equation from part (a).

$v = 3.36(3) + 8.16$

$= 18.24$

So, the stock value was \$18.24 per share 3 months after her sentencing.

c. Substitute 28.32 for v in the equation from part (a).

$28.32 = 3.36t + 8.16$

$28.32 - 8.16 = 3.36t + 8.16 - 8.16$

$20.16 = 3.36t$

$$\dfrac{20.16}{3.36} = \dfrac{3.36t}{3.36}$$

$6 = t$

So, 6 months after her sentencing, the value of a share of the stock was \$28.32.

51. a. $p = 1.30t + 29.49$

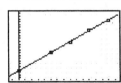

The line comes close to the data points.

b. The slope is 1.30; this means that the percent of college freshmen whose average grade in high school was an 'A' is increasing 1.30 percentage points each year.

c. Substitute 50 for p in the equation and solve for t.

$50 = 1.30t + 29.49$

$50 - 29.49 = 1.30t + 29.49 - 29.49$

$20.51 = 1.30t$

$$\dfrac{20.51}{1.30} = \dfrac{1.30t}{1.30}$$

$15.78 \approx t$

So, half of all college freshmen in $1990 + 16 = 2006$ earned an average grade of 'A' in high school.

d. Substitute $2010 - 1990 = 20$ for t in the equation.

$$p = 1.30(20) + 29.49$$
$$= 26 + 29.49$$
$$= 55.49$$

So, in 2010 the percentage of college freshmen with an average grade of 'A' in high school will be approximately 55.5%.

e. Answers may vary.

53. Let x be the number.

$$3 + 5x = 18$$
$$3 + 5x - 3 = 18 - 3$$
$$5x = 15$$
$$\frac{5x}{5} = \frac{15}{5}$$
$$x = 3$$

The number is 3.

55. Let x be the number.

$$2(x - 2) = -18$$
$$2x - 4 = -18$$
$$2x - 4 + 4 = -18 + 4$$
$$2x = -14$$
$$\frac{2x}{2} = \frac{-14}{2}$$
$$x = -7$$

The number is -7.

57. Let x be the number.

$$x - 3 = 2x + 1$$
$$x - 3 - 2x = 2x + 1 - 2x$$
$$-x - 3 = 1$$
$$-x - 3 + 3 = 1 + 3$$
$$-x = 4$$
$$-1(-x) = -1(4)$$
$$x = -4$$

The number is -4.

59. Let x be the number.

$$1 - 4(x + 5) = 9$$
$$1 - 4x - 20 = 9$$
$$-4x - 19 = 9$$
$$-4x - 19 + 19 = 9 + 19$$
$$-4x = 28$$
$$\frac{-4x}{-4} = \frac{28}{-4}$$
$$x = -7$$

The number is -7.

61. Three less than two times a number is 7 (answers may vary).

$$2x - 3 = 7$$
$$2x - 3 + 3 = 7 + 3$$
$$2x = 10$$
$$\frac{2x}{2} = \frac{10}{2}$$
$$x = 5$$

The number is 5.

63. Three less than six times a number is equal to four less than eight times the number (answers may vary).

$$6x - 3 = 8x - 4$$
$$6x - 3 - 8x = 8x - 4 - 8x$$
$$-2x - 3 = -4$$
$$-2x - 3 + 3 = -4 + 3$$
$$-2x = -1$$
$$\frac{-2x}{-2} = \frac{-1}{-2}$$
$$x = \frac{1}{2}$$

The number is $\frac{1}{2}$.

65. Twice the difference of a number and 4 is 10 (answers may vary).

$$2(x - 4) = 10$$
$$2x - 8 = 10$$
$$2x - 8 + 8 = 10 + 8$$
$$2x = 18$$
$$\frac{2x}{2} = \frac{18}{2}$$
$$x = 9$$

The number is 9.

67. Four, minus 7 times the sum of a number and 1, is 2 (answers may vary).

$$4 - 7(x+1) = 2$$
$$4 - 7x - 7 = 2$$
$$-7x - 3 = 2$$
$$-7x - 3 + 3 = 2 + 3$$
$$-7x = 5$$
$$\frac{-7x}{-7} = \frac{5}{-7}$$
$$x = -\frac{5}{7}$$

The number is $-\dfrac{5}{7}$.

69. $-3x + 7 = 5x + 15$

From tables 8 and 9, we see that for both of the equations $y = -3x + 7$ and $y = 5x + 15$, the input -1 leads to the output 10. Therefore, -1 is a solution of the equation $-3x + 7 = 5x + 15$.

71. $5x + 15 = 5$

Looking at table 9, we see that for the equation $y = 5x + 15$ the input -2 leads to the output 5. Therefore, -2 is a solution to the equation $5x + 15 = 5$.

73. $-4x + 8 = 2x - 9$

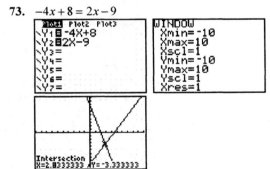

The approximate intersection point $(2.83, -3.33)$ has x-coordinate 2.83. So, the approximate solution of the original equation is 2.83.

75. $2.5x - 6.4 = -1.7x + 8.1$

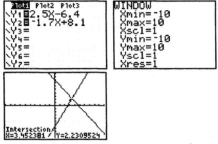

The approximate intersection point $(3.45, 2.23)$ has x-coordinate 3.45. So, the approximate solution of the original equation is 3.45.

77. $\dfrac{1}{3}x - \dfrac{7}{3} = -\dfrac{3}{5}x - \dfrac{15}{2}$

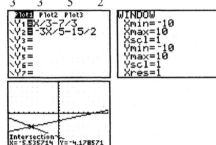

The approximate intersection point $(-5.54, -4.18)$ has x-coordinate -5.54. So, the approximate solution of the original equation is -5.54.

79. The two lines $y = -\dfrac{3}{2}x + 2$ and $y = \dfrac{1}{2}x - 2$ intersect only at $(2, -1)$, whose x-coordinate is 2. So, 2 is the solution of the equation $-\dfrac{3}{2}x + 2 = \dfrac{1}{2}x - 2$.

81. The two lines $y = -\dfrac{3}{2}x + 2$ and $y = 5$ intersect only at $(-2, 5)$, whose x-coordinate is -2. So, -2 is the solution of the equation $-\dfrac{3}{2}x + 2 = 5$.

83. The two lines $y = \dfrac{1}{2}x - 2$ and $y = 0$ intersect only at $(4, 0)$, whose x-coordinate is 4. So, 4 is the solution of the equation $\dfrac{1}{2}x - 2 = 0$.

85. The two lines $y = -\dfrac{1}{3}x + \dfrac{5}{3}$ and $y = \dfrac{3}{2}x + \dfrac{7}{2}$

intersect only at $(-1, 2)$, whose x-coordinate is

-1. So, -1 is the solution of the equation

$-\dfrac{1}{3}x + \dfrac{5}{3} = \dfrac{3}{2}x + \dfrac{7}{2}$.

87. The two lines $y = -\dfrac{1}{3}x + \dfrac{5}{3}$ and $y = 1$ intersect

only at $(2, 1)$, whose x-coordinate is 2. So, 2 is

the solution of the equation $-\dfrac{1}{3}x + \dfrac{5}{3} = 1$.

89. The two lines $y = \dfrac{3}{2}x + \dfrac{7}{2}$ and $y = -1$ intersect

only at $(-3, -1)$, whose x-coordinate is -3. So,

-3 is the solution of the equation $\dfrac{3}{2}x + \dfrac{7}{2} = -1$.

91.
$$3x + 4x = 7x$$
$$7x = 7x$$
$$7x - 7x = 7x - 7x$$
$$0 = 0 \text{ true}$$
Since $0 = 0$ is a true statement (of the form $a = a$), we conclude that the original equation is an identity and its solution set is the set of all real numbers.

93.
$$4x - 5 - 2x = 2x - 1$$
$$2x - 5 = 2x - 1$$
$$2x - 5 - 2x = 2x - 1 - 2x$$
$$-5 = -1 \text{ false}$$
Since $-5 = -1$ is a false statement, we conclude that the original equation is inconsistent and its solution set is the empty set.

95.
$$3k + 10 - 5k = 4k - 2$$
$$-2k + 10 = 4k - 2$$
$$-2k + 10 - 4k = 4k - 2 - 4k$$
$$-6k + 10 = -2$$
$$-6k + 10 - 10 = -2 - 10$$
$$-6k = -12$$
$$\dfrac{-6k}{-6} = \dfrac{-12}{-6}$$
$$k = 2$$
The number 2 is the only solution, so the equation is conditional.

97.
$$2(x + 3) - 2 = 2x + 4$$
$$2x + 6 - 2 = 2x + 4$$
$$2x + 4 = 2x + 4$$
$$2x + 4 - 2x = 2x + 4 - 2x$$
$$4 = 4 \text{ true}$$
Since $4 = 4$ is a true statement (of the form $a = a$), we conclude that the original equation is an identity and its solution set is the set of all real numbers.

99.
$$5(2x - 3) - 4x = 3(2x - 1) + 6$$
$$10x - 15 - 4x = 6x - 3 + 6$$
$$6x - 15 = 6x + 3$$
$$6x - 15 - 6x = 6x + 3 - 6x$$
$$-15 = 3 \text{ false}$$
Since $-15 = 3$ is a false statement, we conclude that the original equation is inconsistent and its solution set is the empty set.

101.
$$3(1 - 3) \stackrel{?}{=} 12 \text{ false}$$
$$3(1) - 9 \stackrel{?}{=} 12 \text{ false}$$
$$3(1) - 9 - 9 \stackrel{?}{=} 12 - 9 \text{ false}$$
$$3(1) \stackrel{?}{=} 3 \text{ true}$$
$$\dfrac{3(1)}{3} \stackrel{?}{=} \dfrac{3}{3} \text{ true}$$
$$1 \stackrel{?}{=} 1 \text{ true}$$
In going from the third line to the fourth, we went from a false statement to a true statement. This indicates that the mistake was made in the fourth line. In this step, the student subtracted 9 from both sides (and got the wrong result) when they should have added.
$$3(x - 3) = 12$$
$$3x - 9 = 12$$
$$3x - 9 + 9 = 12 + 9$$
$$3x = 21$$
$$\dfrac{3x}{3} = \dfrac{21}{3}$$
$$x = 7$$
The solution is 7.

103. The student has not isolated the variable on one side of the equation.

$$2(x-5) = x-3$$
$$2x-10 = x-3$$
$$2x-10-x = x-3-x$$
$$x-10 = -3$$
$$x-10+10 = -3+10$$
$$x = 7$$

The solution is 7.

105. Answers may vary.

107.
$$5(x-2) = 2x-1$$
$$5x-10 = 2x-1$$
$$5x-10-2x = 2x-1-2x$$
$$3x-10 = -1$$
$$3x-10+10 = -1+10$$
$$3x = 9$$
$$\frac{3x}{3} = \frac{9}{3}$$
$$x = 3$$

$$5(3-2) \overset{?}{=} 2(3)-1 \quad \text{true}$$
$$5(3)-10 \overset{?}{=} 2(3)-1 \quad \text{true}$$
$$5(3)-10-2(3) \overset{?}{=} 2(3)-1-2(3) \quad \text{true}$$
$$3(3)-10 \overset{?}{=} -1 \quad \text{true}$$
$$3(3)-10+10 \overset{?}{=} -1+10 \quad \text{true}$$
$$3(3) \overset{?}{=} 9 \quad \text{true}$$
$$\frac{3(3)}{3} \overset{?}{=} \frac{9}{3} \quad \text{true}$$
$$3 \overset{?}{=} 3 \quad \text{true}$$

109.
$$3(x+2)+1 = 16$$
$$3x+6+1 = 16$$
$$3x+7 = 16$$
$$3x+7-7 = 16-7$$
$$3x = 9$$
$$\frac{3x}{3} = \frac{9}{3}$$
$$x = 3$$

The solution is 3.

$$3(x+2) = 15$$
$$3x+6 = 15$$
$$3x+6-6 = 15-6$$
$$3x = 9$$
$$\frac{3x}{3} = \frac{9}{3}$$
$$x = 3$$

The solution is 3.
Answers may vary.

Homework 4.5

1. $3x+7x = 8$ is a linear equation because the statement contains an equal sign.

3. $3x+7x$ is a linear expression because the statement does not contain an equal sign.

5. $4-2(x-9)$ is a linear expression because the statement does not contain an equal sign.

7. $4-2(x-9) = 5$ is a linear equation because the statement contains an equal sign.

9.
$$3x+4x = 14$$
$$7x = 14$$
$$\frac{7x}{7} = \frac{14}{7}$$
$$x = 2$$

The solution is 2.

11. $3x+4x = (3+4)x = 7x$
The simplified expression is $7x$.

13. $b-5(b-1) = b-5b+5$
$$= -4b+5$$
The simplified expression is $-4b+5$.

15. $b - 5(b - 1) = 0$

$b - 5b + 5 = 0$

$-4b + 5 = 0$

$-4b + 5 - 5 = 0 - 5$

$-4b = -5$

$\dfrac{-4b}{-4} = \dfrac{-5}{-4}$

$b = \dfrac{5}{4}$

The solution is $\dfrac{5}{4}$.

17. $3(3x - 5) + 2(5x + 4) = 0$

$9x - 15 + 10x + 8 = 0$

$19x - 7 = 0$

$19x - 7 + 7 = 0 + 7$

$19x = 7$

$\dfrac{19x}{19} = \dfrac{7}{19}$

$x = \dfrac{7}{19}$

The solution is $\dfrac{7}{19}$.

19. $3(3x - 5) + 2(5x + 4) = 9x - 15 + 10x + 8$

$= 19x - 7$

The simplified expression is $19x - 7$.

21. $3(x - 2) - (7x + 2) = 4(3x + 1)$

$3x - 6 - 7x - 2 = 12x + 4$

$-4x - 8 = 12x + 4$

$-4x - 8 - 12x = 12x + 4 - 12x$

$-16x - 8 = 4$

$-16x - 8 + 8 = 4 + 8$

$-16x = 12$

$\dfrac{-16x}{-16} = \dfrac{12}{-16}$

$x = -\dfrac{3}{4}$

The solution is $-\dfrac{3}{4}$.

23. $3(x - 2) - (7x + 2) - 4(3x + 1)$

$= 3x - 6 - 7x - 2 - 12x - 4$

$= 3x - 7x - 12x - 6 - 2 - 4$

$= -16x - 12$

The simplified expression is $-16x - 12$.

25. $7.2p - 4.5 - 1.3p = 7.2p - 1.3p - 4.5$

$= 5.9p - 4.5$

The simplified expression is $5.9p - 4.5$.

27. $7.2k - 4.5 - 1.3k = 20.5 - 6.6k$

$5.9k - 4.5 = 20.5 - 6.6k$

$5.9k - 4.5 + 6.6k = 20.5 - 6.6k + 6.6k$

$12.5k - 4.5 = 20.5$

$12.5k - 4.5 + 4.5 = 20.5 + 4.5$

$12.5k = 25$

$\dfrac{12.5k}{12.5} = \dfrac{25}{12.5}$

$k = 2$

The solution is 2.

29. $-3.5(x - 8) - 2.6(x - 2.8) = 13.93$

$-3.5x + 28 - 2.6x + 7.28 = 13.93$

$-6.1x + 35.28 = 13.93$

$-6.1x + 35.28 - 35.28 = 13.93 - 35.28$

$-6.1x = -21.35$

$\dfrac{-6.1x}{-6.1} = \dfrac{-21.35}{-6.1}$

$x = 3.5$

The solution is 3.5.

31. $-3.5(x - 8) - 2.6(x - 2.8)$

$= -3.5x + 28 - 2.6x + 7.28$

$= -6.1x + 35.28$

The simplified expression is $-6.1x + 35.28$.

33. $-\dfrac{6w}{8} = -\dfrac{2 \cdot 3w}{2 \cdot 4} = -\dfrac{3w}{4}$

The simplified expression is $-\dfrac{3w}{4}$.

35.
$$-\frac{6w}{8} = \frac{3}{2}$$
$$-\frac{8}{6} \cdot \left(-\frac{6w}{8}\right) = -\frac{8}{6} \cdot \frac{3}{2}$$
$$w = -\frac{24}{12}$$
$$w = -2$$
The solution is -2.

37.
$$\frac{5x}{6} + \frac{1}{2} - \frac{3x}{4} = \frac{5x}{6} \cdot \frac{2}{2} - \frac{3x}{4} \cdot \frac{3}{3} + \frac{1}{2}$$
$$= \frac{10x}{12} - \frac{9x}{12} + \frac{1}{2}$$
$$= \frac{x}{12} + \frac{1}{2}$$
The simplified expression is $\frac{x}{12} + \frac{1}{2}$.

39.
$$\frac{5x}{6} + \frac{1}{2} - \frac{3x}{4} = 0$$
$$12\left(\frac{5x}{6} + \frac{1}{2} - \frac{3x}{4}\right) = 12 \cdot 0$$
$$10x + 6 - 9x = 0$$
$$x + 6 = 0$$
$$x + 6 - 6 = 0 - 6$$
$$x = -6$$
The solution is -6.

41.
$$\frac{7}{2}x - \frac{5}{6} = \frac{1}{3} + \frac{3}{4}x$$
$$12 \cdot \left(\frac{7}{2}x - \frac{5}{6}\right) = 12 \cdot \left(\frac{1}{3} + \frac{3}{4}x\right)$$
$$42x - 10 = 4 + 9x$$
$$42x - 10 - 9x = 4 + 9x - 9x$$
$$33x - 10 = 4$$
$$33x - 10 + 10 = 4 + 10$$
$$33x = 14$$
$$\frac{33x}{33} = \frac{14}{33}$$
$$x = \frac{14}{33}$$
The solution is $\frac{14}{33}$.

43.
$$\frac{7}{2}x - \frac{5}{6} - \frac{1}{3} + \frac{3}{4}x = \frac{7}{2}x + \frac{3}{4}x - \frac{5}{6} - \frac{1}{3}$$
$$= \frac{2}{2} \cdot \frac{7}{2}x + \frac{3}{4}x - \frac{5}{6} - \frac{1}{3} \cdot \frac{2}{2}$$
$$= \frac{14}{4}x + \frac{3}{4}x - \frac{5}{6} - \frac{2}{6}$$
$$= \frac{17}{4}x - \frac{7}{6}$$
The simplified expression is $\frac{17}{4}x - \frac{7}{6}$.

45. Answers may vary.

47. Answers may vary. Clearing fractions first is often less messy and requires fewer steps, but either method will work.

49. No, the student is not correct. A solution is a numeric value that satisfies the equation. While it is possible to represent a solution set using variables, a solution itself will not contain any variables.

51. When simplifying an expression, we cannot multiply through by the LCD. Instead, we rewrite the fractions so they contain the LCD but are equivalent to their original forms. The student should have multiplied the first term by $\frac{3}{3}$ and the second term by $\frac{4}{4}$ in order to have both terms with the LCD of 12.
$$\frac{1}{4}x + \frac{1}{3}x = \frac{3}{3} \cdot \frac{1}{4}x + \frac{4}{4} \cdot \frac{1}{3}x$$
$$= \frac{3}{12}x + \frac{4}{12}x$$
$$= \frac{7}{12}x$$

53. This means that the linear expressions $7 + 2(x+3)$ and $2x + 13$ are equivalent. It also means that every real number is a solution of the equation $7 + 2(x+3) = 2x + 13$.

55. When checking a simplification, it is not sufficient to check a single value, unless that value yields a contradiction (which indicates the simplification is incorrect). Since the simplification should be true for all real numbers, the student should try several values to be more confident that the simplification is correct.

57. Answers may vary.

59.
$$3 + 2x = -10$$
$$3 + 2x - 3 = -10 - 3$$
$$2x = -13$$
$$\frac{2x}{2} = \frac{-13}{2}$$
$$x = -\frac{13}{2}$$

The solution is $-\dfrac{13}{2}$.

61.
$$4 - 6(x - 2) = 4 - 6x + 12$$
$$= -6x + 16$$

The simplified expression is $-6x + 16$.

63.
$$-9x = x - 5$$
$$-9x - x = x - 5 - x$$
$$-10x = -5$$
$$\frac{-10x}{-10} = \frac{-5}{-10}$$
$$x = \frac{1}{2}$$

The solution is $\dfrac{1}{2}$.

65.
$$\frac{x}{2} = 3(x - 5)$$
$$2 \cdot \frac{x}{2} = 2 \cdot 3(x - 5)$$
$$x = 6(x - 5)$$
$$x = 6x - 30$$
$$x - 6x = 6x - 30 - 6x$$
$$-5x = -30$$
$$\frac{-5x}{-5} = \frac{-30}{-5}$$
$$x = 6$$

The solution is 6.

67.
$$x + x \cdot 6 = x + 6x$$
$$= 7x$$

The simplified expression is $7x$.

69.
$$x + \frac{x}{2} = \frac{x}{1} + \frac{x}{2}$$
$$= \frac{x}{1} \cdot \frac{2}{2} + \frac{x}{2}$$
$$= \frac{2x}{2} + \frac{x}{2}$$
$$= \frac{2x + x}{2}$$
$$= \frac{3x}{2}$$

The simplified expression is $\dfrac{3x}{2}$.

71. Answers may vary.

Homework 4.6

1. The polygon is a square. That is, a rectangle with equal length and width.
$$P = 2L + 2W$$
$$= 2(S) + 2(S)$$
$$= 2S + 2S$$
$$= 4S$$
The formula for the perimeter is $P = 4S$.

3. The perimeter is the sum of the lengths of all the sides of the polygon.
$$P = H + S + S + H + B$$
$$= 2H + 2S + B$$
The formula for the perimeter is
$P = 2H + 2S + B$.

5. The perimeter is the sum of the lengths of all the sides of the polygon. Note that the base of the polygon has length $A + C$ and the left side has length $B + D$ (see diagram).

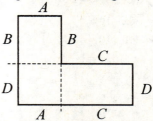

Therefore, the perimeter is given by
$$P = A + B + C + D + (A + C) + (B + D)$$
$$= A + B + C + D + A + C + B + D$$
$$= 2A + 2B + 2C + 2D$$
The formula for the perimeter is
$P = 2A + 2B + 2C + 2D$.

7. $P = VI$

$20 = V(4)$

$\dfrac{20}{4} = \dfrac{4V}{4}$

$5 = V$ or $V = 5$

9. $A = \dfrac{1}{2}BH$

$6 = \dfrac{1}{2}B(3)$

$6 = \dfrac{3}{2}B$

$\dfrac{2}{3} \cdot 6 = \dfrac{2}{3} \cdot \dfrac{3}{2}B$

$4 = B$ or $B = 4$

11. $v = gt + v_0$

$80 = (32.2)t + 20$

$80 - 20 = 32.2t + 20 - 20$

$60 = 32.2t$

$\dfrac{60}{32.2} = \dfrac{32.2t}{32.2}$

$1.86 \approx t$ or $t \approx 1.86$

13. $S = 2WL + 2WH + 2LH$

$52 = 2(2)L + 2(2)(4) + 2L(4)$

$52 = 4L + 16 + 8L$

$52 = 12L + 16$

$52 - 16 = 12L + 16 - 16$

$36 = 12L$

$\dfrac{36}{12} = \dfrac{12L}{12}$

$3 = L$ or $L = 3$

15. $A = \dfrac{a+b+c}{3}$

$5 = \dfrac{2+b+6}{3}$

$5 = \dfrac{8+b}{3}$

$3 \cdot 5 = 3 \cdot \dfrac{8+b}{3}$

$15 = 8 + b$

$15 - 8 = 8 + b - 8$

$7 = b$ or $b = 7$

17. $A = LW$

$116 = L(8)$

$\dfrac{116}{8} = \dfrac{8L}{8}$

$14.5 = L$

The length of the carpet is 14.5 feet.

19. $P = 2L + 2W$

$52 = 2L + 2(10)$

$52 = 2L + 20$

$52 - 20 = 2L + 20 - 20$

$32 = 2L$

$\dfrac{32}{2} = \dfrac{2L}{2}$

$16 = L$

The rectangle has a length of 16 inches.

21. $P = 2L + 2W$

$177 = 2L + 2(29.5)$

$177 = 2L + 59$

$177 - 59 = 2L + 59 - 59$

$118 = 2L$

$\dfrac{118}{2} = \dfrac{2L}{2}$

$59 = L$

The official length of the court is 59 feet.

23. a. $A = LW$

$A = x(3)$

$A = 3x$

b. $P = 2L + 2W$

$P = 2x + 2(3)$

$P = 2x + 6$

c. $\underbrace{P}_{\text{inches}} = \underbrace{2x}_{\text{inches}} + \underbrace{6}_{\text{inches}}$

The units for both of the expressions P and $2x + 6$ are inches.

25. a. Answers may vary. Some possibilities
follows:

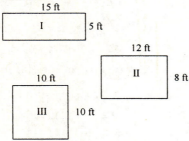

I: Length 15 feet; width 5 feet
II: Length 12 feet; width 8 feet
III: Length 10 feet; width 10 feet

b. I: $A = LW$
$$= (15)(5)$$
$$= 75$$
The area is 75 square feet.

II: $A = LW$
$$= (12)(8)$$
$$= 96$$
The area is 96 square feet.

III: $A = LW$
$$= (10)(10)$$
$$= 100$$
The area is 100 square feet.

c. Garden III would hold more flowers because
it has a larger area.

27. a. $3 \text{ dimes} \cdot \dfrac{10 \text{ cents}}{1 \text{ dime}} = 30 \text{ cents}$

b. $4 \text{ dimes} \cdot \dfrac{10 \text{ cents}}{1 \text{ dime}} = 40 \text{ cents}$

c. $T = d(10)$
$$T = 10d$$
The formula for the total value of d dimes is
$T = 10d$.

d. $\underset{\text{cents}}{T} = \underset{\substack{\text{cents} \\ \overline{\text{dime}}}}{10} \cdot \underset{\text{dimes}}{d}$
The units for both of the expressions T and
$10d$ are cents as required.

29. Total sales = (price per CD)·(number of CDs)
$$T = 12.95 \cdot n$$
$$T = 12.95n$$
The formula for the total price, in dollars, of n
CDs is $T = 12.95n$.

31. a. Total sales = (price per ticket)·(num. of people)
$$T = 32.50 \cdot x$$
$$T = 32.50x$$
The formula for the total sales, in dollars, if
x people bought tickets is $T = 32.50x$.

b. $T = 32.50x$
$$601,965 = 32.50x$$
$$\frac{601,965}{32.50} = \frac{32.50x}{32.50}$$
$$18,522 = x$$
If the total sales are \$601,965, then 18,522
people bought tickets.

33. a. Total cost = (ticket price)·(num. of tickets)
$$C = 10 \cdot k$$
$$C = 10k$$
The formula for total cost, in dollars, if k
tickets are sold for \$10 per ticket is
$C = 10k$.

b. Total cost = (ticket price)·(num. of tickets)
$$E = 95 \cdot n$$
$$E = 95n$$
The formula for total cost, in dollars, if n
tickets are sold for \$95 per ticket is
$E = 95n$.

c. $\underset{\text{cost}}{\text{total}} = \underset{\text{\$10 tickets}}{\text{cost of}} + \underset{\text{\$95 tickets}}{\text{cost of}}$
$$T = C + E$$
$$T = 10k + 95n$$
The total cost of k tickets at \$10 per ticket
and n tickets at \$95 per ticket is
$T = 10k + 95n$.

d.
$$T = 10k + 95n$$
$$270,000 = 10(8000) + 95n$$
$$270,000 = 80,000 + 95n$$
$$270,000 - 80,000 = 80,000 + 95n - 80,000$$
$$190,000 = 95n$$
$$\frac{190,000}{95} = \frac{95n}{95}$$
$$2000 = n$$
There were 2000 tickets sold at $95 per ticket.

35. a. $V = LWH$

b. $V = LWH$
$$48 = (3)(2)H$$
$$48 = 6H$$
$$\frac{48}{6} = \frac{6H}{6}$$
$$8 = H$$
The height of the box is 8 feet.

37. a. $I = P \cdot r \cdot t$
$$I = Prt$$

b. $I = (5000)(0.04)(3)$
$$= 600$$
$600 in interest was earned.

c. $B = P + I$
$$= P + Prt$$
The formula for the balance is $B = P + Prt$.

d. $B = 2000 + (2000)(0.05)(4)$
$$= 2000 + 400$$
$$= 2400$$
The balance after 4 years is $2400.

39.
$$A = \frac{t_1 + t_2 + t_3 + t_4 + t_5}{5}$$
$$80 = \frac{74 + 81 + 79 + 84 + t_5}{5}$$
$$80 = \frac{318 + t_5}{5}$$
$$5 \cdot 80 = 5 \cdot \frac{318 + t_5}{5}$$
$$400 = 318 + t_5$$
$$400 - 318 = 318 + t_5 - 318$$
$$82 = t_5$$
The student needs to score an 82 on the fifth exam to have her five-test average be an 80.

41. $A = LW$
$$\frac{A}{L} = \frac{LW}{L}$$
$$\frac{A}{L} = W \quad \text{or} \quad W = \frac{A}{L}$$

43. $PV = nRT$
$$\frac{PV}{nR} = \frac{nRT}{nR}$$
$$\frac{PV}{nR} = T \quad \text{or} \quad T = \frac{PV}{nR}$$

45.
$$U = -\frac{GmM}{r}$$
$$-\frac{r}{Gm} \cdot U = -\frac{r}{Gm}\left(-\frac{GmM}{r}\right)$$
$$-\frac{rU}{Gm} = M \quad \text{or} \quad M = -\frac{rU}{Gm}$$

47. $A = \frac{1}{2}BH$
$$2 \cdot A = 2 \cdot \frac{1}{2}BH$$
$$2A = BH$$
$$\frac{2A}{H} = \frac{BH}{H}$$
$$\frac{2A}{H} = B \quad \text{or} \quad B = \frac{2A}{H}$$

49.
$$v = gt + v_0$$
$$v - v_0 = gt + v_0 - v_0$$
$$v - v_0 = gt$$
$$\frac{v - v_0}{g} = \frac{gt}{g}$$
$$\frac{v - v_0}{g} = t \text{ or } t = \frac{v - v_0}{g}$$

51.
$$A = P + Prt$$
$$A - P = P + Prt - P$$
$$A - P = Prt$$
$$\frac{A - P}{Pt} = \frac{Prt}{Pt}$$
$$\frac{A - P}{Pt} = r \text{ or } r = \frac{A - P}{Pt}$$

53.
$$A = \frac{a + b + c}{3}$$
$$3 \cdot A = 3 \cdot \frac{a + b + c}{3}$$
$$3A = a + b + c$$
$$3A - a - c = a + b + c - a - c$$
$$3A - a - c = b \text{ or } b = 3A - a - c$$

55.
$$y - k = m(x - h)$$
$$y - k = mx - mh$$
$$y - k + mh = mx - mh + mh$$
$$y - k + mh = mx$$
$$\frac{y - k + mh}{m} = \frac{mx}{m}$$
$$\frac{y - k + mh}{m} = x \text{ or } x = \frac{y - k + mh}{m}$$

57.
$$\frac{x}{a} + \frac{y}{a} = 1$$
$$a \cdot \left(\frac{x}{a} + \frac{y}{a}\right) = a \cdot 1$$
$$x + y = a$$
$$x + y - x = a - x$$
$$y = a - x$$

59.
$$3x + 4y = 16$$
$$3x + 4y - 3x = 16 - 3x$$
$$4y = -3x + 16$$
$$\frac{4y}{4} = \frac{-3x + 16}{4}$$
$$y = \frac{-3x}{4} + \frac{16}{4}$$
$$y = -\frac{3}{4}x + 4$$

61.
$$2x + 4y - 8 = 0$$
$$2x + 4y - 8 - 2x + 8 = 0 - 2x + 8$$
$$4y = -2x + 8$$
$$\frac{4y}{4} = \frac{-2x + 8}{4}$$
$$y = \frac{-2x}{4} + \frac{8}{4}$$
$$y = -\frac{1}{2}x + 2$$

63.
$$5x - 2y = 6$$
$$5x - 2y - 5x = 6 - 5x$$
$$-2y = -5x + 6$$
$$\frac{-2y}{-2} = \frac{-5x + 6}{-2}$$
$$y = \frac{-5x}{-2} + \frac{6}{-2}$$
$$y = \frac{5}{2}x - 3$$

65.
$$-3x - 7y = 5$$
$$-3x - 7y + 3x = 5 + 3x$$
$$-7y = 3x + 5$$
$$\frac{-7y}{-7} = \frac{3x + 5}{-7}$$
$$y = \frac{3x}{-7} + \frac{5}{-7}$$
$$y = -\frac{3}{7}x - \frac{5}{7}$$

67. a. $c = 44.2t + 135.2$

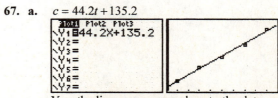

Yes, the line comes very close to the data.

b.
$$c = 44.2t + 135.2$$
$$c - 135.2 = 44.2t + 135.2 - 135.2$$
$$c - 135.2 = 44.2t$$
$$\frac{c - 135.2}{44.2} = \frac{44.2t}{44.2}$$
$$\frac{c - 135.2}{44.2} = t \quad \text{or} \quad t = \frac{c - 135.2}{44.2}$$

c. Substitute 800 for c in the equation from part (b).
$$t = \frac{800 - 135.2}{44.2} = \frac{664.8}{44.2} \approx 15.04$$
Federal flood insurance coverage was $800 billion in $1990 + 15 = 2005$.

d. Use $t = \dfrac{c - 135.2}{44.2}$.

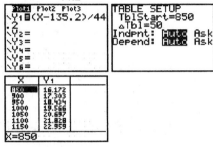

Federal flood insurance coverage is estimated to be
$850 billion in $1990 + 16 = 2006$,
$900 billion in $1990 + 17 = 2007$,
$950 billion in $1990 + 18 = 2008$,
$1000 billion in $1990 + 20 = 2010$, and
$1050 billion in $1990 + 21 = 2011$.

69. a. $s = 28.25t + 155$

b.
$$s = 28.25t + 155$$
$$s - 155 = 28.25t + 155 - 155$$
$$s - 155 = 28.25t$$
$$\frac{s - 155}{28.25} = \frac{28.25t}{28.25}$$
$$\frac{s - 155}{28.25} = t \quad \text{or} \quad t = \frac{s - 155}{28.25}$$

c. Substitute 500 for s in the equation from part (b).
$$t = \frac{500 - 155}{28.25} = \frac{345}{28.25} \approx 12.21$$
Kia will not reach their goal in 2010. They will reach sales of 500 thousand in $2000 + 12 = 2012$.

d. Use $t = \dfrac{s - 155}{28.25}$.

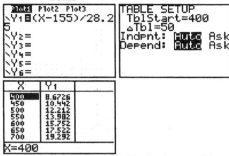

From the table, Kia's sales (in thousands) will be 400, 450, 500, 550, and 600 in the years $2000 + 9 = 2009$, $2000 + 10 = 2010$, $2000 + 12 = 2012$, $2000 + 14 = 2014$, and $2000 + 16 = 2016$, respectively.

71. a. $n = 1.3t + 17.7$

b.
$$32 = 1.3t + 17.7$$
$$32 - 17.7 = 1.3t + 17.7 - 17.7$$
$$14.3 = 1.3t$$
$$\frac{14.3}{1.3} = \frac{1.3t}{1.3}$$
$$11 = t$$
In $2000 + 11 = 2011$, there will be 32 thousand immigrant visas issued to orphans entering the U.S.

c.
$$n = 1.3t + 17.7$$
$$n - 17.7 = 1.3t + 17.7 - 17.7$$
$$n - 17.7 = 1.3t$$
$$\frac{n - 17.7}{1.3} = \frac{1.3t}{1.3}$$
$$\frac{n - 17.7}{1.3} = t \quad \text{or} \quad t = \frac{n - 17.7}{1.3}$$

d. Substitute 32 for n in the equation from part (c).
$$t = \frac{32 - 17.7}{1.3} = \frac{14.3}{1.3} = 11$$
In $2000 + 11 = 2011$, there will be 32 thousand immigrant visas issued to orphans entering the U.S.

e. The results are the same; answers may vary

f. The formula $n = 1.3t + 17.7$ is easier to use in this case because it is already solved for n.

$$n = 1.3(10) + 17.7$$
$$= 13 + 17.7$$
$$= 30.7$$

There will be 30.7 thousand visas issued to orphans entering the United States in 2010.

73. a.

Speed, s (miles per hour)	Time, t (hours)	Distance, d (miles)
50	4	$50 \cdot 4$
70	3	$70 \cdot 3$
65	2	$65 \cdot 2$
55	5	$55 \cdot 5$
s	t	$s \cdot t$

From the last row of the table, we see that the formula relating s, t, and d is $d = st$.

b. $\underset{\text{miles}}{d} = \underset{\substack{\text{miles} \\ \hline \text{hour}}}{s} \cdot \underset{\text{hours}}{t}$

The units for both of the expressions d and st are hours as required.

c. $d = st$

$$\frac{d}{s} = \frac{st}{s}$$

$$\frac{d}{s} = t \text{ or } t = \frac{d}{s}$$

d. $t = \dfrac{315}{70} = 4.5$

It will take 4.5 hours to travel 315 miles at a speed of 70 miles per hour.

e. The New Mexico portion will take

$t = \dfrac{229.9}{75} \approx 3.07$ hours.

The Colorado portion will take

$t = \dfrac{219.2}{65} \approx 3.37$ hours.

So, the total trip will take about $3.07 + 3.37 = 6.44$ hours (that is, 6 hours 26 minutes).

75. $\underset{\substack{\text{miles} \\ \hline \text{hour}}}{s} = \underset{\text{miles}}{d} \cdot \underset{\text{hours}}{t}$

The unit analysis show the units on the left to be $\dfrac{\text{miles}}{\text{hour}}$ while the units on the right are $(\text{miles}) \cdot (\text{hours})$. Since the units are not the same, the formula is incorrect.

77. a. $P = 2L + 2W$

Doubling the length and width, we get:
$2(2L) + 2(2W) = 2(2L + 2W) = 2P$

Therefore, doubling the length and width will double the perimeter.

b. $A = LW$

Doubling the length and width, we get:
$(2L)(2W) = 4LW = 4A$

Therefore, doubling the length and width will multiply the area by 4.

79. Answers may vary.

Chapter 4 Review Exercises

1. $-5(4x) = (-5 \cdot 4)x = -20x$

2. $-3(8x + 4) = -3(8x) + (-3)(4)$
$$= -24x + (-12)$$
$$= -24x - 12$$

3. $\dfrac{4}{5}(15y - 35) = \dfrac{4}{5} \cdot 15y - \dfrac{4}{5} \cdot 35$
$$= 12y - 28$$

4. $-(3x - 6y - 8) = -1(3x - 6y - 8)$
$$= -1 \cdot 3x - (-1)(6y) - (-1)(8)$$
$$= -3x - (-6y) - (-8)$$
$$= -3x + 6y + 8$$

5. $\dfrac{2}{9}x + \dfrac{5}{9}x = \left(\dfrac{2}{9} + \dfrac{5}{9}\right)x = \left(\dfrac{2+5}{9}\right)x = \dfrac{7}{9}x$

6. $5a + 2 - 13b - a + 4b - 9$
$$= 5a - a - 13b + 4b + 2 - 9$$
$$= 4a - 9b - 7$$

7. $-5y - 3(4x + y) - 6x$

$= -5y - 12x - 3y - 6x$

$= -12x - 6x - 5y - 3y$

$= -18x - 8y$

8. $-2.6(3.1x + 4.5) - 8.5 = -8.06x - 11.7 - 8.5$

$= -8.06x - 20.2$

9. $-(2m - 4) - (3m + 8) = -2m + 4 - 3m - 8$

$= -2m - 3m + 4 - 8$

$= -5m - 4$

10. $4(3a - 7b) - 3(5a + 4b) = 12a - 28b - 15a - 12b$

$= 12a - 15a - 28b - 12b$

$= -3a - 40b$

11. $-4(x - 7) = -4x - (-4)(7)$

$= -4x - (-28)$

$= -4x + 28$

12. $-7 + 2(x + 8) = -7 + 2x + 16$

$= 2x + 9$

13. Answers may vary. Some possibilities:
Let $a = 4$, $b = 3$, and $c = 2$.

$a(b + c) \overset{?}{=} ab + c$

$4(3 + 2) \overset{?}{=} 4(3) + 2$

$4(5) \overset{?}{=} 12 + 2$

$20 \overset{?}{=} 14$ false

Let $a = -2$, $b = 3$, and $c = -5$.

$a(b + c) \overset{?}{=} ab + c$

$-2(3 + (-5)) \overset{?}{=} (-2)(3) + (-5)$

$-2(-2) \overset{?}{=} -6 + (-5)$

$4 \overset{?}{=} -11$ false

$a(b + c) = a \cdot b + a \cdot c$

$= ab + ac$

14. Answers may vary. Some examples:
$3(x - 3)$, $2(x - 2) + (x - 5)$, $4x - (x + 9)$

15. $-5x + 20 = -5(x - 4)$

$= 5(4 - x)$

$= -2(x - 10) - 3x$

16. $y = 2x + 3 - 4x$

$= -2x + 3$

Since this is in the form $y = mx + b$, the graph is a line. The slope of the line is $m = -2$ and the y-intercept is $(0, 3)$. We plot the y-intercept and use the slope to obtain a second point, $(1, 1)$, and (to check) a third point, $(-1, 5)$.

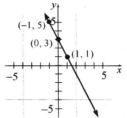

17. $y = -3(x - 2)$

$= -3x + 6$

Since this is in the form $y = mx + b$, the graph is a line. The slope of the line is $m = -3$ and the y-intercept is $(0, 6)$. We plot the y-intercept and use the slope to obtain a second point, $(1, 3)$, and (to check) a third point, $(2, 0)$.

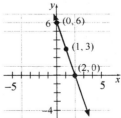

18. $2 - 5x = -3(4x - 7)$

$2 - 5(3) \overset{?}{=} -(4(3) - 7)$

$2 - 5(3) \overset{?}{=} -(12 - 7)$

$2 - 15 \overset{?}{=} -(5)$

$-13 \overset{?}{=} -5$ false

Since the result is a false statement, 3 is not a solution of the equation.

19.

$$a + 5 = 12$$
$$a + 5 - 5 = 12 - 5$$
$$a = 7$$

The solution is 7.

20. $-4x = 20$

$$\frac{-4x}{-4} = \frac{20}{-4}$$
$$x = -5$$

The solution is -5.

21. $-p = -3$

$$\frac{-p}{-1} = \frac{-3}{-1}$$
$$p = 3$$

The solution is 3.

22.

$$-\frac{7}{3}a = 14$$
$$-\frac{3}{7}\left(-\frac{7}{3}a\right) = -\frac{3}{7}\cdot 14$$
$$a = -6$$

The solution is -6.

23.

$$4.5x - 17.2 = -5.05$$
$$4.5x - 17.2 + 17.2 = -5.05 + 17.2$$
$$4.5x = 12.15$$
$$\frac{4.5x}{4.5} = \frac{12.15}{4.5}$$
$$x = 2.7$$

The solution is 2.7.

24. $5x - 9x + 3 = 17$

$$-4x + 3 = 17$$
$$-4x + 3 - 3 = 17 - 3$$
$$-4x = 14$$
$$\frac{-4x}{-4} = \frac{14}{-4}$$
$$x = -\frac{7}{2}$$

The solution is $-\frac{7}{2}$.

25.

$$8m - 3 - m = 2 - 4m$$
$$7m - 3 = 2 - 4m$$
$$7m - 3 + 4m = 2 - 4m + 4m$$
$$11m - 3 = 2$$
$$11m - 3 + 3 = 2 + 3$$
$$11m = 5$$
$$\frac{11m}{11} = \frac{5}{11}$$
$$m = \frac{5}{11}$$

The solution is $\frac{5}{11}$.

26.

$$8x = -7(2x - 3) + x$$
$$8x = -14x + 21 + x$$
$$8x = -13x + 21$$
$$8x + 13x = -13x + 21 + 13x$$
$$21x = 21$$
$$\frac{21x}{21} = \frac{21}{21}$$
$$x = 1$$

The solution is 1.

27. $6(4x - 1) - 3(2x + 5) = 2(5x - 3)$

$$24x - 6 - 6x - 15 = 10x - 6$$
$$18x - 21 = 10x - 6$$
$$18x - 21 - 10x = 10x - 6 - 10x$$
$$8x - 21 = -6$$
$$8x - 21 + 21 = -6 + 21$$
$$8x = 15$$
$$\frac{8x}{8} = \frac{15}{8}$$
$$x = \frac{15}{8}$$

The solution is $\frac{15}{8}$.

28.
$$\frac{w}{8} - \frac{3}{4} = \frac{5}{6}$$
$$24\left(\frac{w}{8} - \frac{3}{4}\right) = 24\left(\frac{5}{6}\right)$$
$$3w - 18 = 20$$
$$3w - 18 + 18 = 20 + 18$$
$$3w = 38$$
$$\frac{3w}{3} = \frac{38}{3}$$
$$w = \frac{38}{3}$$
The solution is $\frac{38}{3}$.

29.
$$\frac{3p-4}{2} = \frac{5p+2}{4} + \frac{7}{6}$$
$$12\left(\frac{3p-4}{2}\right) = 12\left(\frac{5p+2}{4} + \frac{7}{6}\right)$$
$$6(3p-4) = 3(5p+2) + 2(7)$$
$$18p - 24 = 15p + 6 + 14$$
$$18p - 24 = 15p + 20$$
$$18p - 24 - 15p = 15p + 20 - 15p$$
$$3p - 24 = 20$$
$$3p - 24 + 24 = 20 + 24$$
$$3p = 44$$
$$\frac{3p}{3} = \frac{44}{3}$$
$$p = \frac{44}{3}$$
The solution is $\frac{44}{3}$.

30. In the second line, the student only added 5 to one side of the equation instead of both as required.
$$x - 5 = 2$$
$$x - 5 + 5 = 2 + 5$$
$$x = 7$$
The solution is 7.

31. Answers will vary. One example:
$$3x - 5 = -23$$

32.
$$-2.5(3.8x - 1.9) = 83.7$$
$$-9.5x + 4.75 = 83.7$$
$$-9.5x + 4.75 - 4.75 = 83.7 - 4.75$$
$$-9.5x = 78.95$$
$$\frac{-9.5x}{-9.5} = \frac{78.95}{-9.5}$$
$$x = -\frac{78.95}{9.5}$$
$$x \approx -8.31$$

33. **a.** The rate of increase is constant, so we have a linear model. The linear model has the form $n = mt + b$, where n is the number of prisoners (in millions) and t is the number of years since 2000. The rate of change is the slope, so we have $m = 0.06$. The y-intercept is $(0, 1.9)$, so we have $b = 1.9$. Therefore, the model is $n = 0.06t + 1.9$.

b. For 2010, we let $t = 10$.
$$n = 0.06(10) + 1.9$$
$$= 0.6 + 1.9$$
$$= 2.5$$
In 2010, there will be about 2.5 million prisoners in the United States.

c. Substitute 2.6 for n and solve for t
$$2.6 = 0.06t + 1.9$$
$$2.6 - 1.9 = 0.06t + 1.9 - 1.9$$
$$0.7 = 0.06t$$
$$\frac{0.7}{0.06} = \frac{0.06t}{0.06}$$
$$11.67 \approx t$$
There will be about 2.6 million prisoners in $2000 + 12 = 2012$.

34. Let x be the number.
$$4(x - 6) = 15$$
$$4x - 24 = 15$$
$$4x - 24 + 24 = 15 + 24$$
$$4x = 39$$
$$\frac{4x}{4} = \frac{39}{4}$$
$$x = \frac{39}{4}$$
The number is $\frac{39}{4}$.

35. Let x be the number.

$$2 - 3(x + 8) = 95$$
$$2 - 3x - 24 = 95$$
$$-3x - 22 = 95$$
$$-3x - 22 + 22 = 95 + 22$$
$$-3x = 117$$
$$\frac{-3x}{-3} = \frac{117}{-3}$$
$$x = -39$$

The number is -39.

36. $\dfrac{1}{2}x + \dfrac{5}{3} = -\dfrac{2}{3}x - \dfrac{1}{4}$

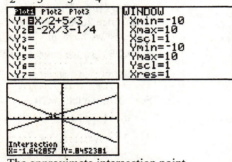

The approximate intersection point, $(-1.64, 0.85)$, has x-coordinate -1.64, so the approximate solution to the equation is -1.64.

37. $-2x + 17 = 5x - 4$

From tables 13 and 14, we see that for both of the equations $y = -2x + 17$ and $y = 5x - 4$, the input 3 leads to the output 11. Therefore, 3 is a solution of the equation $-2x + 17 = 5x - 4$.

38. $-2x + 17 = 15$

Looking at table 13, we see that for the equation $y = -2x + 17$ the input 1 leads to the output 15. Therefore, 1 is a solution to the equation $-2x + 17 = 15$.

39. $5x - 4 = 6$

Looking at table 14, we see that for the equation $y = 5x - 4$ the input 2 leads to the output 6. Therefore, 2 is a solution to the equation $5x - 4 = 6$.

40. $5x - 4 = -4$

Looking at table 14, we see that for the equation $y = 5x - 4$ the input 0 leads to the output -4. Therefore, 0 is a solution to the equation $5x - 4 = -4$.

41.
$$7x - 4 + 3x = 2 + 10x - 6$$
$$10x - 4 = 10x - 4$$
$$10x - 4 - 10x = 10x - 4 - 10x$$
$$-4 = -4 \quad \text{true}$$

Since $-4 = -4$ is a true statement (of the form $a = a$), we conclude that the original equation is an identity and its solution set is the set of all real numbers.

42.
$$6(2x - 3) - (5x + 2) = -2(4x - 1)$$
$$12x - 18 - 5x - 2 = -8x + 2$$
$$7x - 20 = -8x + 2$$
$$7x - 20 + 8x = -8x + 2 + 8x$$
$$15x - 20 = 2$$
$$15x - 20 + 20 = 2 + 20$$
$$15x = 22$$
$$\frac{15x}{15} = \frac{22}{15}$$
$$x = \frac{22}{15}$$

The number $\dfrac{22}{15}$ is the only solution, so the equation is conditional.

43.
$$2(x - 5) + 3 = 2x - 4$$
$$2x - 10 + 3 = 2x - 4$$
$$2x - 7 = 2x - 4$$
$$2x - 7 - 2x = 2x - 4 - 2x$$
$$-7 = -4 \quad \text{false}$$

Since $-7 = -4$ is a false statement, we conclude that the original equation is inconsistent and its solution set is the empty set.

44.
$$4(x - 5) = 28$$
$$4(2 - 5) \overset{?}{=} 28 \quad \text{false}$$
$$4(2) - 20 \overset{?}{=} 28 \quad \text{false}$$
$$4(2) - 20 - 20 \overset{?}{=} 28 - 20 \quad \text{false}$$
$$4(2) \overset{?}{=} 8 \quad \text{true}$$
$$\frac{4(2)}{4} \overset{?}{=} \frac{8}{4} \quad \text{true}$$
$$2 \overset{?}{=} 2 \quad \text{true}$$

In going from the fourth line to the fifth, we went from a false statement to a true statement.

This indicates that the mistake was made in going from the fourth line to the fifth. In the fourth step, the subtracted 20 from both sides when they should have added (and incorrectly simplified $-20-20$).

$$4(x-5)=28$$
$$4x-20=28$$
$$4x-20+20=28+20$$
$$4x=48$$
$$\frac{4x}{4}=\frac{48}{4}$$
$$x=12$$

The solution is 12.

45. $8-3(x+5)$ is a linear expression because the statement does not contain an equal sign.

46. $8-3(x+5)=4x$ is a linear equation because the statement contains an equal sign.

47. $6t-8t=(6-8)t$
$$=-2t$$
The simplified expression is $-2t$.

48. $0.1+0.5a-0.3a=0.7$
$$0.1+0.2a=0.7$$
$$0.1+0.2a-0.1=0.7-0.1$$
$$0.2a=0.6$$
$$\frac{0.2a}{0.2}=\frac{0.6}{0.2}$$
$$a=3$$
The solution is 3.

49. $6t-8t=10$
$$-2t=10$$
$$\frac{-2t}{-2}=\frac{10}{-2}$$
$$t=-5$$
The solution is -5 .

50. $0.1+0.5a-0.3a=0.1+(0.5-0.3)a$
$$=0.1+0.2a$$
$$=0.2a+0.1$$
The simplified expression is $0.2a+0.1$.

51. $9(2p-5)-3(7p+3)=18p-45-21p-9$
$$=18p-21p-45-9$$
$$=-3p-54$$
The simplified expression is $-3p-54$.

52. $$\frac{5}{6}r-\frac{3}{4}=\frac{1}{6}+\frac{7}{2}r$$
$$12\left(\frac{5}{6}r-\frac{3}{4}\right)=12\left(\frac{1}{6}+\frac{7}{2}r\right)$$
$$10r-9=2+42r$$
$$10r-9-42r=2+42r-42r$$
$$-32r-9=2$$
$$-32r-9+9=2+9$$
$$-32r=11$$
$$\frac{-32r}{-32}=\frac{11}{-32}$$
$$r=-\frac{11}{32}$$
The solution is $-\frac{11}{32}$.

53. $9(2p-5)-3(7p+3)=0$
$$18p-45-21p-9=0$$
$$-3p-54=0$$
$$-3p-54+54=0+54$$
$$-3p=54$$
$$\frac{-3p}{-3}=\frac{54}{-3}$$
$$p=-18$$
The solution is -18 .

54. $\frac{5}{6}r-\frac{3}{4}-\frac{1}{6}+\frac{7}{2}r=\frac{5}{6}r+\frac{7}{2}r-\frac{3}{4}-\frac{1}{6}$
$$=\frac{5}{6}r+\frac{3}{3}\cdot\frac{7}{2}r-\frac{3}{3}\cdot\frac{3}{4}-\frac{2}{2}\cdot\frac{1}{6}$$
$$=\frac{5}{6}r+\frac{21}{6}r-\frac{9}{12}-\frac{2}{12}$$
$$=\frac{26}{6}r-\frac{11}{12}$$
$$=\frac{13}{3}r-\frac{11}{12}$$
The simplified expression is $\frac{13}{3}r-\frac{11}{12}$.

55. No, the student is incorrect. Expressions do not have solutions. We solve equations and simplify expressions.

56. When simplifying an expression, we cannot multiply through by the LCD. Instead, we rewrite the fractions so they contain the LCD but are equivalent to their original forms.

The expression $\dfrac{2}{3}x + \dfrac{7}{5}$ is already simplified as much as possible.

57.
$$4(6-x) = 17$$
$$24 - 4x = 17$$
$$24 - 4x - 24 = 17 - 24$$
$$-4x = -7$$
$$\dfrac{-4x}{-4} = \dfrac{-7}{-4}$$
$$x = \dfrac{7}{4}$$

The solution is $\dfrac{7}{4}$.

58.
$$x - \dfrac{x}{2} = \dfrac{x}{1} - \dfrac{x}{2}$$
$$= \dfrac{2}{2} \cdot \dfrac{x}{1} - \dfrac{x}{2}$$
$$= \dfrac{2x}{2} - \dfrac{x}{2}$$
$$= \dfrac{x}{2}$$

The simplified expression is $\dfrac{x}{2}$, or $\dfrac{1}{2}x$.

59. We can start by labeling the remaining sides, but remember that the final answer must be in terms of A, B, C, D, and E.

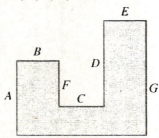

Now, note the following:
$$H = B + C + E$$
$$G = A + D - F$$

The perimeter is the sum of all the sides, so we get:
$$P = A + B + C + D + E + F + G + H$$
$$= A + B + C + D + E + F$$
$$\quad + (A + D - F) + (B + C + E)$$
$$= 2A + 2B + 2C + 2D + 2E$$

60. a. $T = 15n + 25w$

b.
$$T = 15n + 25w$$
$$11{,}050 = 15(370) + 25w$$
$$11{,}050 = 5550 + 25w$$
$$11{,}050 - 5550 = 5550 + 25w - 5550$$
$$5500 = 25w$$
$$\dfrac{5500}{25} = \dfrac{25w}{25}$$
$$220 = w$$

There were 220 tickets sold at $25 per ticket.

61.
$$C = 2\pi r$$
$$\dfrac{C}{2\pi} = \dfrac{2\pi r}{2\pi}$$
$$\dfrac{C}{2\pi} = r \text{ to } r = \dfrac{C}{2\pi}$$

62.
$$P = a + b + c$$
$$P - a - b = a + b + c - a - b$$
$$P - a - b = c \text{ or } c = P - a - b$$

63.
$$3x - 6y = 18$$
$$3x - 6y - 3x = 18 - 3x$$
$$-6y = -3x + 18$$
$$\dfrac{-6y}{-6} = \dfrac{-3x + 18}{-6}$$
$$y = \dfrac{-3x}{-6} + \dfrac{18}{-6}$$
$$y = \dfrac{1}{2}x - 3$$

64.
$$A = \frac{1}{2}H(B+T)$$
$$2 \cdot A = 2 \cdot \frac{1}{2}H(B+T)$$
$$2A = H(B+T)$$
$$2A = HB + HT$$
$$2A - HB = HB + HT - HB$$
$$2A - HB = HT$$
$$\frac{2A - HB}{H} = \frac{HT}{H}$$
$$\frac{2A - HB}{H} = T \quad \text{or} \quad T = \frac{2A - HB}{H}$$

65. a. A linear model has the form $y = mx + b$ where m is the slope, or average rate of change, and $(0,b)$ is the y-intercept. Using v and t, our model has the form $v = mt + b$ where v is the number of visits (in billions) and t is the number of years since 2000. The slope is $m = 0.04$ and the y-intercept is $(0, 1.17)$. Therefore, our model is $v = 0.04t + 1.17$.

b.
$$v = 0.04t + 1.17$$
$$v - 1.17 = 0.04t + 1.17 - 1.17$$
$$v - 1.17 = 0.04t$$
$$\frac{v - 1.17}{0.04} = \frac{0.04t}{0.04}$$
$$\frac{v - 1.17}{0.04} = t \quad \text{or} \quad t = \frac{v - 1.17}{0.04}$$

c. Substitute 1.6 for v and solve for t.
$$1.6 = 0.04t + 1.17$$
$$1.6 - 1.17 = 0.04t + 1.17 - 1.17$$
$$0.43 = 0.04t$$
$$\frac{0.43}{0.04} = \frac{0.04t}{0.04}$$
$$10.75 = t$$
The number of visits will be 1.6 billion in the year $2000 + 11 = 2011$.

d. Substitute 1.6 for v and solve for t.
$$t = \frac{v - 1.17}{0.04}$$
$$t = \frac{1.6 - 1.17}{0.04} = \frac{0.43}{0.04} = 10.75$$
The number of visits will be 1.6 billion in the year $2000 + 11 = 2011$.

e. The results are the same. The equation $t = \frac{v - 1.17}{0.04}$ was easier to use because it was already solved for t.

f. The equation $v = 0.04t + 1.17$ is easier to use when predicting the number of visits because it is already solved for v. Substitute 10 for t in the equation:
$$v = 0.04(10) + 1.17$$
$$= 0.4 + 1.17$$
$$= 1.57$$
In 2010, there will be 1.57 billion visits to U.S. libraries.

Chapter 4 Test

1.
$$-\frac{2}{3}(6x - 9) = -\frac{2}{3} \cdot 6x - \left(-\frac{2}{3}\right)(9)$$
$$= -4x - (-6)$$
$$= -4x + 6$$

2.
$$9.36 - 2.4(1.7x + 3.5) = 9.36 - 4.08x - 8.4$$
$$= -4.08x + 9.36 - 8.4$$
$$= -4.08x + 0.96$$

3.
$$-5(2w - 7) - 3(4w - 6) = -10w + 35 - 12w + 18$$
$$= -10w - 12w + 35 + 18$$
$$= -22w + 53$$

4.
$$-(3a + 7b) - (8a - 4b + 2) = -3a - 7b - 8a + 4b - 2$$
$$= -3a - 8a - 7b + 4b - 2$$
$$= -11a - 3b - 2$$

5.
$$3(3 - 2) - (5(3) + 4) = 3(1) - (15 + 4) = 3 - 19 = -16$$
$$3(3) - 6 - 5(3) + 4 = 9 - 6 - 15 + 4 = -8$$
$$3(3) - 5(3) - 6 + 4 = 9 - 15 - 6 + 4 = -8$$
$$-2(3) - 2 = -6 - 2 = -8$$
The mistake occurred in going from the first line to the second. The negative was incorrectly distributed.
$$3(x - 2) - (5x + 4) = 3x - 6 - 5x - 4$$
$$= 3x - 5x - 6 - 4$$
$$= -2x - 10$$

6. $y = -2(x+1)$

$\quad = -2x - 2$

Since this is in the form $y = mx + b$, the graph is a line. The slope of the line is $m = -2$ and the y-intercept is $(0, -2)$. We plot the y-intercept and use the slope to obtain a second point, $(1, -4)$, and (to check) a third point, $(-1, 0)$.

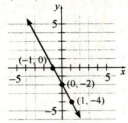

7. $\quad 6x - 3 = 19$

$\quad 6x - 3 + 3 = 19 + 3$

$\quad\quad 6x = 22$

$\quad\quad \dfrac{6x}{6} = \dfrac{22}{6}$

$\quad\quad\quad x = \dfrac{11}{3}$

The solution is $\dfrac{11}{3}$.

8. $\quad \dfrac{3}{5}x = 6$

$\quad \dfrac{5}{3} \cdot \dfrac{3}{5}x = \dfrac{5}{3} \cdot 6$

$\quad\quad x = \dfrac{30}{3}$

$\quad\quad x = 10$

The solution is 10.

9. $\quad 9a - 5 = 8a + 2$

$\quad 9a - 5 - 8a = 8a + 2 - 8a$

$\quad\quad a - 5 = 2$

$\quad a - 5 + 5 = 2 + 5$

$\quad\quad\quad a = 7$

The solution is 7.

10. $\quad 8 - 2(3t - 1) = 7t$

$\quad 8 - 6t + 2 = 7t$

$\quad 10 - 6t = 7t$

$\quad 10 - 6t + 6t = 7t + 6t$

$\quad\quad 10 = 13t$

$\quad\quad \dfrac{10}{13} = \dfrac{13t}{13}$

$\quad\quad \dfrac{10}{13} = t$

The solution is $\dfrac{10}{13}$.

11. $\quad 3(2x - 5) - 2(7x + 9) = 49$

$\quad 6x - 15 - 14x - 18 = 49$

$\quad\quad -8x - 33 = 49$

$\quad -8x - 33 + 33 = 49 + 33$

$\quad\quad -8x = 82$

$\quad\quad \dfrac{-8x}{-8} = \dfrac{82}{-8}$

$\quad\quad x = -\dfrac{41}{4}$

The solution is $-\dfrac{41}{4}$.

12. $\quad \dfrac{7}{8}x + \dfrac{3}{10} = \dfrac{1}{4}x - \dfrac{1}{2}$

$\quad 40\left(\dfrac{7}{8}x + \dfrac{3}{10}\right) = 40\left(\dfrac{1}{4}x - \dfrac{1}{2}\right)$

$\quad\quad 35x + 12 = 10x - 20$

$\quad 35x + 12 - 10x = 10x - 20 - 10x$

$\quad\quad 25x + 12 = -20$

$\quad 25x + 12 - 12 = -20 - 12$

$\quad\quad 25x = -32$

$\quad\quad \dfrac{25x}{25} = \dfrac{-32}{25}$

$\quad\quad x = -\dfrac{32}{25}$

The solution is $-\dfrac{32}{25}$.

13.
$$8.21x = 3.9(4.4x - 2.7)$$
$$8.21x = 17.16x - 10.53$$
$$8.21x - 17.16x = 17.16x - 10.53 - 17.16x$$
$$-8.95x = -10.53$$
$$\frac{-8.95x}{-8.95} = \frac{-10.53}{-8.95}$$
$$x \approx 1.18$$
The solution is approximately 1.18.

14. Let x be the number.
$$4 - 2(x + 7) = 54$$
$$4 - 2x - 14 = 54$$
$$-2x - 10 = 54$$
$$-2x - 10 + 10 = 54 + 10$$
$$-2x = 64$$
$$\frac{-2x}{-2} = \frac{64}{-2}$$
$$x = -32$$
The number is -32.

15. $9(3x + 2) - (4x - 6) = 27x + 18 - 4x + 6$
$$= 27x - 4x + 18 + 6$$
$$= 23x + 24$$
The simplified expression is $23x + 24$.

16.
$$9(3x + 2) - (4x - 6) = x$$
$$27x + 18 - 4x + 6 = x$$
$$23x + 24 = x$$
$$23x + 24 - 23x = x - 23x$$
$$24 = -22x$$
$$\frac{24}{-22} = \frac{-22x}{-22}$$
$$-\frac{12}{11} = x$$
The solution is $-\frac{12}{11}$.

17. No, the student is not correct. A solution is a numeric value that satisfies the equation. While it is possible to represent a solution set using variables, a solution itself will not contain any variables.

18. Answers may vary. Some examples:
$4x - 2(2x - 2)$, $10 - 15 \div 5 + (-3)$,
$3(x + 3) - (5 + 3x)$

19. Let x be the number.
$$5(x - 2) = 29$$
$$5x - 10 = 29$$
$$5x - 10 + 10 = 29 + 10$$
$$5x = 39$$
$$\frac{5x}{5} = \frac{39}{5}$$
$$x = \frac{39}{5}$$
The solution is $\frac{39}{5}$.

20. $2 + 4(3 + x) = 2 + 12 + 4x$
$$= 14 + 4x$$
$$= 4x + 14$$
The simplified expression is $4x + 14$.

21. The two lines $y = \frac{3}{2}x - 4$ and $y = \frac{1}{2}x - 2$ intersect only at $(2, -1)$, whose x-coordinate is 2. So, 2 is the solution of the equation $\frac{3}{2}x - 4 = \frac{1}{2}x - 2$.

22. The two lines $y = \frac{3}{2}x - 4$ and $y = 2$ intersect only at $(4, 2)$, whose x-coordinate is 4. So, 4 is the solution of the equation $\frac{3}{2}x - 4 = 2$.

23. The two lines $y = \frac{1}{2}x - 2$ and $y = -3$ intersect only at $(-2, -3)$, whose x-coordinate is -2. So, -2 is the solution of the equation $\frac{1}{2}x - 2 = -3$.

24. The two lines $y = \frac{1}{2}x - 2$ and $y = 0$ intersect only at $(4, 0)$, whose x-coordinate is 4. So, 4 is the solution of the equation $\frac{1}{2}x - 2 = 0$.

25. a. The rate of increase is constant, so we have a linear model. The linear model has the form $n = mt + b$, where n is the number of patent applications (in thousands) and t is the number of years since 2000. The rate of change is the slope, so we have $m = 30$. The y-intercept is $(0, 303)$, so we have $b = 303$. Therefore, the model is $n = 30t + 303$.

b. For 2010, we let $t = 10$.
$$n = 30(10) + 303$$
$$= 300 + 303$$
$$= 603$$
In 2010, there will be 603 thousand U.S. patent applications.

c. Substitute 650 for n and solve for t
$$650 = 30t + 303$$
$$650 - 303 = 30t + 303 - 303$$
$$347 = 30t$$
$$\frac{347}{30} = \frac{30t}{30}$$
$$11.57 \approx t$$
There will be about 650 thousand U.S. patent applications in $2000 + 12 = 2012$.

26. Start with the formula for the perimeter of a rectangle, substitute in the known values, and solve for the length.
$$P = 2L + 2W$$
$$52 = 2L + 2(8)$$
$$52 = 2L + 16$$
$$52 - 16 = 2L + 16 - 16$$
$$36 = 2L$$
$$\frac{36}{2} = \frac{2L}{2}$$
$$18 = L$$
The length of the garden is 18 feet.

27.
$$A = \frac{a+b}{2}$$
$$2 \cdot A = 2 \cdot \frac{a+b}{2}$$
$$2A = a + b$$
$$2A - b = a + b - b$$
$$2A - b = a \quad \text{or} \quad a = 2A - b$$

Cumulative Review of Chapters 1 – 4

1. Let n = the number of pages in a book. Examples of possible values: 275, 300 Examples of values not possible: 0, -150 (answers may vary)

2.

3. In 2006, there were 72.1 million unique visitors to the website YouTube.

4. $4 + 3(-2) = 4 + (-6) = -2$

5.
$$-8 \div 4 - 2(7 - 10) = -8 \div 4 - 2(-3)$$
$$= -2 - (-6)$$
$$= -2 + 6$$
$$= 4$$

6.
$$\frac{15}{8} \cdot \left(\frac{-4}{25}\right) = -\frac{15}{8} \cdot \frac{4}{25}$$
$$= -\frac{3 \cdot 5 \cdot 2 \cdot 2}{2 \cdot 2 \cdot 2 \cdot 5 \cdot 5}$$
$$= -\frac{2 \cdot 2 \cdot 5}{2 \cdot 2 \cdot 5} \cdot \frac{3}{2 \cdot 5}$$
$$= -\frac{3}{2 \cdot 5}$$
$$= -\frac{3}{10}$$

7.
$$\left(-\frac{3}{10}\right) + \left(-\frac{7}{8}\right) = \left(-\frac{3}{10} \cdot \frac{4}{4}\right) + \left(-\frac{7}{8} \cdot \frac{5}{5}\right)$$
$$= \left(-\frac{12}{40}\right) + \left(-\frac{35}{40}\right)$$
$$= \frac{-12 + (-35)}{40}$$
$$= -\frac{47}{40}$$

8. $\dfrac{27}{-45} = -\dfrac{27}{45} = -\dfrac{3 \cdot 3 \cdot 3}{3 \cdot 3 \cdot 5} = -\dfrac{3 \cdot 3}{3 \cdot 3} \cdot \dfrac{3}{5} = -\dfrac{3}{5}$

9. $P = 2L + 2W$

$$= 2\left(\frac{3}{4}\right) + 2\left(\frac{5}{6}\right)$$

$$= \frac{3}{2} + \frac{5}{3}$$

$$= \frac{3}{2} \cdot \frac{3}{3} + \frac{5}{3} \cdot \frac{2}{2}$$

$$= \frac{9}{6} + \frac{10}{6}$$

$$= \frac{19}{6}$$

The rectangle has a perimeter of $\frac{19}{6}$ feet.

10. $85 - 92 = -7$

The change in the scores was -7 points.

11. $a(b - c) = (-3)(5 - (-4))$

$$= -3(5 + 4)$$

$$= -3(9)$$

$$= -27$$

12. $m = \dfrac{-4 - (-2)}{-1 - (-5)} = \dfrac{-4 + 2}{-1 + 5} = \dfrac{-2}{4} = -\dfrac{1}{2}$

The line is decreasing because the slope is negative.

13. $m = \dfrac{3 - (-5)}{-4 - (-4)} = \dfrac{8}{0}$ undefined

The slope is undefined so the line is vertical.

14. Find the average rate of change for both roads.

Road A: $\dfrac{150 \text{ ft}}{5000 \text{ ft}} = \dfrac{3}{100} = \dfrac{0.03}{1}$

Road B: $\dfrac{95 \text{ ft}}{3500 \text{ ft}} = \dfrac{19}{700} \approx \dfrac{0.027}{1}$

The rate of change for road A is larger than for road B, so road A is steeper.

15. $y = -\dfrac{2}{3}x + 4$

Since this is in the form $y = mx + b$, the graph is a non-vertical line. The slope of the line is $m = -\dfrac{2}{3}$ and the y-intercept is $(0, 4)$. We plot the y-intercept and use the slope to obtain a second point, $(3, 2)$, and (to check) a third point, $(-3, 6)$.

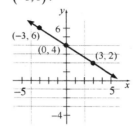

16. $y = 2x - 3$

Since this is in the form $y = mx + b$, the graph is a non-vertical line. The slope of the line is $m = 2$ and the y-intercept is $(0, -3)$. We plot the y-intercept and use the slope to obtain a second point, $(1, -1)$, and (to check) a third point, $(-1, -5)$.

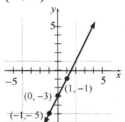

17. $x = -5$

Since this is in the form $x = a$, the graph is a vertical line (undefined slope).

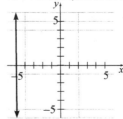

18. $y = 3$

Since this is in the form $y = b$, the graph is a horizontal line (0 slope).

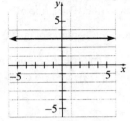

19. From the graph we see that the y-intercept is $(0, -2)$. Also, the point $(1, 1)$ is clearly on the graph. Using these points, we can compute the slope.

$$\frac{1 - (-2)}{1 - 0} = \frac{1 + 2}{1} = 3$$

Therefore, the slope is $m = \dfrac{3}{1} = 3$ and the equation of the line is $y = 3x - 2$.

20. $\dfrac{2840 - 7500}{5} = \dfrac{-4660}{5} = -932$

The average rate of change of the balance is -932 dollars per month.

21. $\dfrac{422 - 264}{2004 - 2000} = \dfrac{158}{4} = 39.5$

The rate of change is 39.5 thousand reports per year.

22. a. The slope is the average rate of change. In this situation, the slope is $m = 16.7$. This means that sales of convertibles increase by 16.7 thousand each year.

b. A linear model has the form $y = mx + b$. Using s and t, our equation will be $s = mt + b$. We found the slope to be $m = 16.7$ and are given the y-intercept as $(0, 242)$, so $b = 242$. Thus, our model is $s = 16.7t + 242$.

c. For 2011 we have $t = 11$. Substitute 11 for t in the equation from part (b).

$s = 16.7(11) + 242$

$\quad = 183.7 + 242$

$\quad = 425.7$

In 2011, there will be 425.7 thousand convertibles sold in the U.S.

d. Substitute 450 for s in the equation from part (b).

$450 = 16.7t + 242$

$450 - 242 = 16.7t + 242 - 242$

$208 = 16.7t$

$\dfrac{208}{16.7} = \dfrac{16.7t}{16.7}$

$12.46 \approx t$

The number of convertibles sold in the U.S. will be 450 thousand in $2000 + 12 = 2012$.

23. Consider the average rates of change:

Equation 1: $\dfrac{13 - 11}{1 - 0} = \dfrac{2}{1} = 2$

$\dfrac{16 - 13}{2 - 1} = \dfrac{3}{1} = 3$

$\dfrac{20 - 16}{3 - 2} = \dfrac{4}{1} = 4$

$\dfrac{25 - 20}{4 - 3} = \dfrac{5}{1} = 5$

Equation 2: $\dfrac{53 - 56}{1 - 0} = \dfrac{-3}{1} = -3$

$\dfrac{50 - 53}{2 - 1} = \dfrac{-3}{1} = -3$

$\dfrac{47 - 50}{3 - 2} = \dfrac{-3}{1} = -3$

$\dfrac{44 - 47}{4 - 3} = \dfrac{-3}{1} = -3$

Equation 3: $\dfrac{44 - 35}{4 - 3} = \dfrac{9}{1} = 9$

$\dfrac{53 - 44}{5 - 4} = \dfrac{9}{1} = 9$

$\dfrac{62 - 53}{6 - 5} = \dfrac{9}{1} = 9$

$\dfrac{71 - 62}{7 - 6} = \dfrac{9}{1} = 9$

Equation 4: $\dfrac{1 - 1}{2 - 1} = \dfrac{0}{1} = 0$

$\dfrac{1 - 1}{3 - 2} = \dfrac{0}{1} = 0$

$\dfrac{1 - 1}{4 - 3} = \dfrac{0}{1} = 0$

$\dfrac{1 - 1}{5 - 4} = \dfrac{0}{1} = 0$

To be a linear equation, the average rate of change must be constant. From the data, it appears that equations 2, 3, and 4 could be linear.

24. a. $n = 4.01t + 16.69$

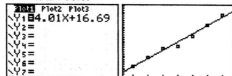

The line comes close to the data points.

b. The slope is 4.01. This means the number of live births from fertility treatments increases by 4.01 thousand each year.

c. The n-intercept is $(0, 16.69)$. This means that in 1995 there were 16.69 thousand live births from fertility treatment.

d. For 2010 we have $t = 15$. Substitute 15 for t in the equation $n = 4.01t + 16.69$.

$$n = 4.01(15) + 16.69$$
$$= 60.15 + 16.69$$
$$= 76.84$$

In 2010, there will be 76.84 thousand live births from fertility treatment.

e. Substitute 85 for n in the equation $n = 4.01t + 16.69$.

$$85 = 4.01t + 16.69$$
$$85 - 16.69 = 4.01t + 16.69 - 16.69$$
$$68.31 = 4.01t$$
$$\frac{68.31}{4.01} = \frac{4.01t}{4.01}$$
$$17.03 \approx t$$

The number of live births from fertility treatment will be 85 thousand in $1995 + 17 = 2012$.

25.
$$3r + 4 = 7r - 8$$
$$3r + 4 - 7r = 7r - 8 - 7r$$
$$-4r + 4 = -8$$
$$-4r + 4 - 4 = -8 - 4$$
$$-4r = -12$$
$$\frac{-4r}{-4} = \frac{-12}{-4}$$
$$r = 3$$

The solution is 3.

26.
$$2(3x - 2) = 4(3x + 5) - 3x$$
$$6x - 4 = 12x + 20 - 3x$$
$$6x - 4 = 9x + 20$$
$$6x - 4 - 9x = 9x + 20 - 9x$$
$$-3x - 4 = 20$$
$$-3x - 4 + 4 = 20 + 4$$
$$-3x = 24$$
$$\frac{-3x}{-3} = \frac{24}{-3}$$
$$x = -8$$

The solution is -8.

27. $4a - 5b + 6 - 2b - 7a = 4a - 7a - 5b - 2b + 6$
$$= -3a - 7b + 6$$

The simplified expression is $-3a - 7b + 6$.

28. $7 - 2(3p - 5) + 5(4p - 2)$
$$= 7 - 6p + 10 + 20p - 10$$
$$= -6p + 20p + 7 + 10 - 10$$
$$= 14p + 7$$

The simplified expression is $14p + 7$.

29.
$$\frac{2}{3}r - \frac{5}{6} = \frac{1}{2}$$
$$6\left(\frac{2}{3}r - \frac{5}{6}\right) = 6\left(\frac{1}{2}\right)$$
$$4r - 5 = 3$$
$$4r - 5 + 5 = 3 + 5$$
$$4r = 8$$
$$\frac{4r}{4} = \frac{8}{4}$$
$$r = 2$$

The solution is 2.

30. $-(2a + 5) - (4a - 1) = -2a - 5 - 4a + 1$
$$= -2a - 4a - 5 + 1$$
$$= -6a - 4$$

The simplified expression is $-6a - 4$.

31. $25.93 - 7.6(2.1x + 8.7) = 53.26$

$25.93 - 15.96x - 66.12 = 53.26$

$-15.96x - 40.19 = 53.26$

$-15.96x - 40.19 + 40.19 = 53.26 + 40.19$

$-15.96x = 93.45$

$$\frac{-15.96x}{-15.96} = \frac{93.45}{-15.96}$$

$x \approx -5.86$

The solution is approximately -5.86 .

32. $x + 9 \cdot \dfrac{x}{3} = x + 3x = 4x$

The simplified expression is $4x$.

33. $2(7 - 2x) = 87$

$14 - 4x = 87$

$14 - 4x - 14 = 87 - 14$

$-4x = 73$

$$\frac{-4x}{-4} = \frac{73}{-4}$$

$x = -\dfrac{73}{4}$

The solution is $-\dfrac{73}{4}$.

34. $A = 2\pi r h$

$$\frac{A}{2\pi r} = \frac{2\pi r h}{2\pi r}$$

$\dfrac{A}{2\pi r} = h$ or $h = \dfrac{A}{2\pi r}$

35. $4x - 6y = 12$

$4x - 6y - 4x = 12 - 4x$

$-6y = 12 - 4x$

$$\frac{-6y}{-6} = \frac{12 - 4x}{-6}$$

$y = \dfrac{12}{-6} + \dfrac{-4x}{-6}$

$y = -2 + \dfrac{2}{3}x$

$y = \dfrac{2}{3}x - 2$

Chapter 5
Linear Equations in Two Variables

Homework 5.1

1. The equation $y = 2x - 3$ is in slope-intercept form, so the slope is $m = 2 = \dfrac{2}{1}$, and the y-intercept is $(0, -3)$. We first plot $(0, -3)$. From this point we move 1 unit to the right and 2 units up, where we plot the point $(1, -1)$. We then sketch the line that contains these two points.

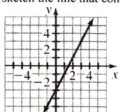

3. The equation $y = -3x + 5$ is in slope-intercept form, so the slope is $m = -3 = \dfrac{-3}{1}$, and the y-intercept is $(0, 5)$. We first plot $(0, 5)$. From this point we move 1 unit to the right and 3 units down, where we plot the point $(1, 2)$. We then sketch the line that contains these two points.

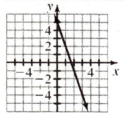

5. The equation $y = -\dfrac{3}{5}x - 2$ is in slope-intercept form, so the slope is $m = -\dfrac{3}{5}$, and the y-intercept is $(0, -2)$. We first plot $(0, -2)$. From this point we move 5 units to the right and 3 units down, where we plot the point $(5, -5)$. We then sketch the line that contains these two points.

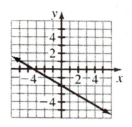

7. The equation $y = \dfrac{1}{2}x + 2$ is in slope-intercept form, so the slope is $m = \dfrac{1}{2}$, and the y-intercept is $(0, 2)$. We first plot $(0, 2)$. From this point we move 2 units to the right and 1 unit up, where we plot the point $(2, 3)$. We then sketch the line that contains these two points.

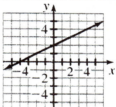

9. The equation $y = -3x$ is in slope-intercept form, so the slope is $m = -3 = \dfrac{-3}{1}$, and the y-intercept is $(0, 0)$. We first plot $(0, 0)$. From this point we move 1 unit to the right and 3 units down, where we plot the point $(1, -3)$. We then sketch the line that contains these two points.

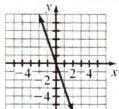

11. The equation $y = -4$ is in slope-intercept form, so the slope is $m = 0$ and the y-intercept is $(0, -4)$. The graph is a horizontal line that passes through the point $(0, -4)$.

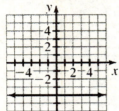

13. $y + x = 3$

$$y = -x + 3$$

The slope is $m = -1 = \dfrac{-1}{1}$, and the y-intercept is $(0, 3)$. We first plot $(0, 3)$. From this point we move 1 unit to the right and 1 unit down, where we plot the point $(1, 2)$. We then sketch the line that contains these two points.

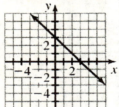

15. $y + 2x = 4$

$$y = -2x + 4$$

The slope is $m = -2 = \dfrac{-2}{1}$, and the y-intercept is $(0, 4)$. We first plot $(0, 4)$. From this point we move 1 unit to the right and 2 units down, where we plot the point $(2, 1)$. We then sketch the line that contains these two points.

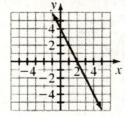

17. $y - 2x = -1$

$$y = 2x - 1$$

The slope is $m = 2 = \dfrac{2}{1}$, and the y-intercept is $(0, -1)$. We first plot $(0, -1)$. From this point we move 1 unit to the right and 2 units up, where we plot the point $(1, 1)$. We then sketch the line that contains these two points.

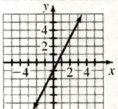

19. $3y = 2x$

$$y = \frac{2}{3}x$$

The slope is $m = \dfrac{2}{3}$, and the y-intercept is $(0, 0)$. We first plot $(0, 0)$. From this point we move 3 units to the right and 2 units up, where we plot the point $(3, 2)$. We then sketch the line that contains these two points.

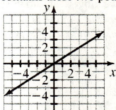

21. $2y = 5x - 6$

$$\frac{2y}{2} = \frac{5x}{2} - \frac{6}{2}$$

$$y = \frac{5}{2}x - 3$$

The slope is $m = \dfrac{5}{2}$, and the y-intercept is $(0, -3)$. We first plot $(0, -3)$. From this point we move 2 units to the right and 5 units up, where we plot the point $(2, 2)$. We then sketch the line that contains these two points.

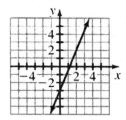

23. $5y = 4x - 15$

$$\frac{5y}{5} = \frac{4x}{5} - \frac{15}{5}$$

$$y = \frac{4}{5}x - 3$$

The slope is $m = \frac{4}{5}$, and the y-intercept is $(0, -3)$. We first plot $(0, -3)$. From this point we move 5 units to the right and 4 units up, where we plot the point $(5, 1)$. We then sketch the line that contains these two points.

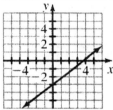

25. $3x - 4y = 8$

$$-4y = -3x + 8$$

$$\frac{-4y}{-4} = \frac{-3x}{-4} + \frac{8}{-4}$$

$$y = \frac{3}{4}x - 2$$

The slope is $m = \frac{3}{4}$, and the y-intercept is $(0, -2)$. We first plot $(0, -2)$. From this point we move 4 units to the right and 3 units up, where we plot the point $(4, 1)$. We then sketch the line that contains these two points.

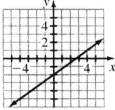

27. $6x - 15y = 30$

$$-15y = -6x + 30$$

$$\frac{-15y}{-15} = \frac{-6x}{-15} + \frac{30}{-15}$$

$$y = \frac{2}{5}x - 2$$

The slope is $m = \frac{2}{5}$, and the y-intercept is $(0, -2)$. We first plot $(0, -2)$. From this point we move 5 units to the right and 2 units up, where we plot the point $(5, 0)$. We then sketch the line that contains these two points.

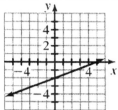

29. $x + 4y = 4$

$$4y = -x + 4$$

$$\frac{4y}{4} = \frac{-x}{4} + \frac{4}{4}$$

$$y = -\frac{1}{4}x + 1$$

The slope is $m = -\frac{1}{4}$, and the y-intercept is $(0, 1)$. We first plot $(0, 1)$. From this point we move 4 units to the right and 1 unit down, where we plot the point $(4, 0)$. We then sketch the line that contains these two points.

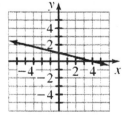

31. $-4 = x + 2y$

$-2y = x + 4$

$\dfrac{-2y}{-2} = \dfrac{x}{-2} + \dfrac{4}{-2}$

$y = -\dfrac{1}{2}x - 2$

The slope is $m = -\dfrac{1}{2}$, and the *y*-intercept is

$(0, -2)$. We first plot $(0, -2)$. From this point we move 2 units to the right and 1 unit down, where we plot the point $(2, -3)$. We then sketch the line that contains these two points.

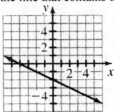

33. $4x + y + 2 = 0$

$y = -4x - 2$

The slope is $m = -4 = \dfrac{-4}{1}$, and the *y*-intercept is

$(0, -2)$. We first plot $(0, -2)$. From this point we move 1 unit to the right and 4 units down, where we plot the point $(1, -6)$. We then sketch the line that contains these two points.

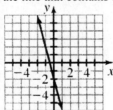

35. $6x - 4y + 8 = 0$

$-4y = -6x - 8$

$\dfrac{-4y}{-4} = \dfrac{-6x}{-4} - \dfrac{8}{-4}$

$y = \dfrac{3}{2}x + 2$

The slope is $m = \dfrac{3}{2}$, and the *y*-intercept is $(0, 2)$.

We first plot $(0, 2)$. From this point we move 2 units to the right and 3 units up, where we plot the point $(2, 5)$. We then sketch the line that contains these two points.

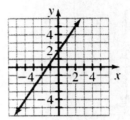

37. $0 = 5x + 3y$

$-3y = 5x$

$y = -\dfrac{5}{3}x$

The slope is $m = -\dfrac{5}{3}$, and the *y*-intercept is

$(0, 0)$. We first plot $(0, 0)$. From this point we move 3 units to the right and 5 units down, where we plot the point $(3, -5)$. We then sketch the line that contains these two points.

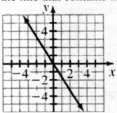

39. $y - 3 = 0$

$y = 3$

The slope is $m = 0$, and the *y*-intercept is $(0, 3)$. The graph is a horizontal line that passes through the point $(0, 3)$.

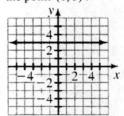

41. $x - 3y = 6$

To find the *x*-intercept, we substitute 0 for *y* and solve for *x*. To find the *y*-intercept, we substitute 0 for *x* and solve for *y*.

x-intercept	*y*-intercept
$x - 3(0) = 6$	$0 - 3y = 6$
$x = 6$	$-3y = 6$
	$y = -2$

The *x*-intercept is $(6, 0)$; the *y*-intercept is $(0, -2)$.

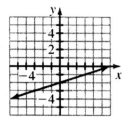

43. $15 = 3x + 5y$

To find the x-intercept, we substitute 0 for y and solve for x. To find the y-intercept, we substitute 0 for x and solve for y.

x-intercept	y-intercept
$15 = 3x + 5(0)$	$15 = 3(0) + 5y$
$15 = 3x$	$15 = 5y$
$5 = x$	$3 = y$

The x-intercept is $(5, 0)$; the y-intercept is $(0, 3)$.

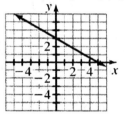

45. $2x - 3y + 12 = 0$

To find the x-intercept, we substitute 0 for y and solve for x. To find the y-intercept, we substitute 0 for x and solve for y.

x-intercept	y-intercept
$2x - 3(0) + 12 = 0$	$2(0) - 3y + 12 = 0$
$2x + 12 = 0$	$-3y + 12 = 0$
$2x = -12$	$-3y = -12$
$x = -6$	$y = 4$

The x-intercept is $(-6, 0)$; the y-intercept is $(0, 4)$.

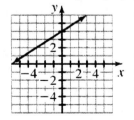

47. $y = -3x + 6$

To find the x-intercept, we substitute 0 for y and solve for x. To find the y-intercept, we substitute 0 for x and solve for y.

x-intercept	y-intercept
$0 = -3x + 6$	$y = -3(0) + 6$
$3x = 6$	$y = 6$
$x = 2$	

The x-intercept is $(2, 0)$; the y-intercept is $(0, 6)$.

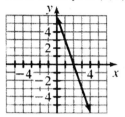

49. $\frac{1}{2}x + \frac{1}{3}y = 2$

To find the x-intercept, we substitute 0 for y and solve for x. To find the y-intercept, we substitute 0 for x and solve for y.

x-intercept	y-intercept
$\frac{1}{2}x + \frac{1}{3}(0) = 2$	$\frac{1}{2}(0) + \frac{1}{3}y = 2$
$\frac{1}{2}x = 2$	$\frac{1}{3}y = 2$
$x = 4$	$y = 6$

The x-intercept is $(4, 0)$; the y-intercept is $(0, 6)$.

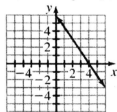

51. $\frac{x}{3} + \frac{y}{5} = 1$

To find the x-intercept, we substitute 0 for y and solve for x. To find the y-intercept, we substitute 0 for x and solve for y.

x-intercept	y-intercept
$\frac{x}{3} + \frac{0}{5} = 1$	$\frac{0}{3} + \frac{y}{5} = 1$
$\frac{x}{3} = 1$	$\frac{y}{5} = 1$
$x = 3$	$y = 5$

The *x*-intercept is $(3, 0)$; the *y*-intercept is $(0, 5)$.

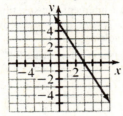

53. $6.2x + 2.8y = 7.5$

To find the *x*-intercept, we substitute 0 for *y* and solve for *x*. To find the *y*-intercept, we substitute 0 for *x* and solve for *y*.

x-intercept	*y*-intercept
$6.2x + 2.8(0) = 7.5$	$6.2(0) + 2.8y = 7.5$
$6.2x = 7.5$	$2.8y = 7.5$
$x \approx 1.21$	$x \approx 2.68$

The *x*-intercept is approximately $(1.21, 0)$;

the *y*-intercept is approximately $(0, 2.68)$.

55. $6.62x - 3.91y = -13.55$

To find the *x*-intercept, we substitute 0 for *y* and solve for *x*.

$6.62x - 3.91(0) = -13.55$

$6.62x = -13.55$

$x \approx -2.05$

To find the *y*-intercept, we substitute 0 for *x* and solve for *y*.

$6.62(0) - 3.91y = -13.55$

$-3.91y = -13.55$

$y \approx 3.47$

The *x*-intercept is approximately $(-2.05, 0)$;

the *y*-intercept is approximately $(0, 3.47)$.

57. $y = -4.5x + 9.32$

To find the *x*-intercept, we substitute 0 for *y* and solve for *x*. To find the *y*-intercept, we substitute 0 for *x* and solve for *y*.

x-intercept	*y*-intercept
$0 = -4.5x + 9.32$	$y = -4.5(0) + 9.32$
$4.5x = 9.32$	$y = 9.32$
$x \approx 2.07$	

The *x*-intercept is approximately $(2.07, 0)$;

the *y*-intercept is approximately $(0, 9.32)$.

59. $y = -2.49x - 37.21$

To find the *x*-intercept, we substitute 0 for *y* and solve for *x*. To find the *y*-intercept, we substitute 0 for *x* and solve for *y*.

x-intercept	*y*-intercept
$0 = -2.49x - 37.21$	$y = -2.49(0) - 37.21$
$2.49x = -37.21$	$y = -37.21$
$x \approx -14.94$	

The *x*-intercept is approximately $(-14.94, 0)$;

the *y*-intercept is approximately $(0, -37.21)$.

61. $2x - y = 5$

$-y = -2x + 5$

$\dfrac{-y}{-1} = \dfrac{-2x}{-1} + \dfrac{5}{-1}$

$y = 2x - 5$

The slope is $m = 2$, and the *y*-intercept is $(0, -5)$. We first plot $(0, -5)$. From this point we move 1 unit to the right and 2 units up, where we plot the point $(1, -3)$. We then sketch the line that contains these two points.

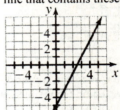

63. $3y = 4x - 3$

$\dfrac{3y}{3} = \dfrac{4x}{3} - \dfrac{3}{3}$

$y = \dfrac{4}{3}x - 1$

The slope is $m = \dfrac{4}{3}$, and the *y*-intercept is $(0, -1)$. We first plot $(0, -1)$. From this point we move 3 units to the right and 4 units up, where we plot the point $(3, 3)$. We then sketch the line that contains these two points.

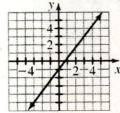

65. $4y - 3x = 0$

$$4y = 3x$$

$$\frac{4y}{4} = \frac{3x}{4}$$

$$y = \frac{3}{4}x$$

The slope is $m = \frac{3}{4}$, and the y-intercept is $(0, 0)$.
We first plot $(0, 0)$. From this point we move 4 units to the right and 3 units up, where we plot the point $(4, 3)$. We then sketch the line that contains these two points.

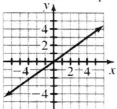

67. $2x - 3y - 12 = 0$

$$-3y = -2x + 12$$

$$\frac{-3y}{-3} = \frac{-2x}{-3} + \frac{12}{-3}$$

$$y = \frac{2}{3}x - 4$$

The slope is $m = \frac{2}{3}$, and the y-intercept is $(0, -4)$. We first plot $(0, -4)$. From this point we move 3 units to the right and 2 units up, where we plot the point $(3, -2)$. We then sketch the line that contains these two points.

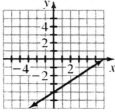

69. $6x + 5y = -13$

a. Substitute -3 for x and solve for y:

$$6(-3) + 5y = -13$$

$$-18 + 5y = -13$$

$$5y = 5$$

$$y = 1$$

b. Substitute -5 for y and solve for x:

$$6x + 5(-5) = -13$$

$$6x - 25 = -13$$

$$6x = 12$$

$$x = 2$$

c. We plot the points $(-3, 1)$ and $(2, -5)$ and sketch the line that contains the two points.

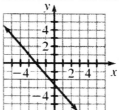

71. No, the slope is not 2 because the equation is not in slope-intercept form.

$$3y + 2x = 6$$

$$3y = -2x + 6$$

$$\frac{3y}{3} = \frac{-2x}{3} + \frac{6}{3}$$

$$y = -\frac{2}{3}x + 2$$

The slope is $m = -\frac{2}{3}$.

73. $3x - 5y = 10$

a. $3x - 5y = 10$

$$-5y = -3x + 10$$

$$\frac{-5y}{-5} = \frac{-3x}{-5} + \frac{10}{-5}$$

$$y = \frac{3}{5}x - 2$$

The slope is $m = \frac{3}{5}$ and the y-intercept is $(0, -2)$. We first plot $(0, -2)$. From this point we move 5 units to the right and 3 units up, where we plot the point $(5, 1)$. We then sketch the line that contains these two points.

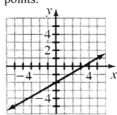

b. Answers may vary. One possibility follows:

x	y
-5	-5
0	-2
5	1

c. Answers may vary. One possibility follows: For each solution, the difference of three times the x-coordinate and five times the y-coordinate is equal to 10.

75. a. $\dfrac{x}{5}+\dfrac{y}{7}=1$

<u>x-intercept</u>
Let $y=0$:

$\dfrac{x}{5}+\dfrac{0}{7}=1$

$\dfrac{x}{5}=1$

$x=5$

<u>y-intercept</u>
Let $x=0$:

$\dfrac{0}{5}+\dfrac{y}{7}=1$

$\dfrac{y}{7}=1$

$y=7$

The x-intercept is $(5,0)$; the y-intercept is $(0,7)$.

b. $\dfrac{x}{4}+\dfrac{y}{6}=1$

<u>x-intercept</u>
Let $y=0$:

$\dfrac{x}{4}+\dfrac{0}{6}=1$

$\dfrac{x}{4}=1$

$x=4$

<u>y-intercept</u>
Let $x=0$:

$\dfrac{0}{4}+\dfrac{y}{6}=1$

$\dfrac{y}{6}=1$

$y=6$

The x-intercept is $(4,0)$; the y-intercept is $(0,6)$.

c. $\dfrac{x}{a}+\dfrac{y}{b}=1$

<u>x-intercept</u>
Let $y=0$:

$\dfrac{x}{a}+\dfrac{0}{b}=1$

$\dfrac{x}{a}=1$

$x=a$

<u>y-intercept</u>
Let $x=0$:

$\dfrac{0}{a}+\dfrac{y}{b}=1$

$\dfrac{y}{b}=1$

$y=b$

The x-intercept is $(a,0)$; the y-intercept is $(0,b)$.

d. The x-intercept is $(2,0)$, so $a=2$ in the equation from part (c). Likewise, the y-intercept is $(0,5)$, so $b=5$ in the equation from part (c). The equation is $\dfrac{x}{2}+\dfrac{y}{5}=1$.

77. Answers may vary. One possibility follows: The y-coordinate of the x-intercept is zero because the x-intercept is a point on the x-axis, which must have a y-coordinate of zero. The x-coordinate of the y-intercept is zero because the y-intercept is a point on the y-axis, which must have an x-coordinate of zero.

79. Answers will vary.

81. The graph contains the point $(6,1)$, so $y=1$ when $x=6$.

83. The graph contains the point $(2,-1)$, so $x=2$ when $y=-1$.

85. The graph and the x-axis intersect at the point $(4,0)$, so the x-intercept is $(4,0)$.

87. The line contains the points $(4,0)$ and $(6,1)$, so the slope is $m=\dfrac{1-0}{6-4}=\dfrac{1}{2}$.

89. $5-3(2x-7)=5-6x+21$
$=-6x+26$
This is a linear expression in one variable.

91. $5-3(2x-7)=8$
$5-6x+21=8$
$-6x+26=8$
$-6x=-18$
$x=3$
This is a linear equation in one variable.

Homework 5.2

1. We substitute $m=2$ and $(x_1,y_1)=(3,5)$ into the point slope form and solve for y:
$y-y_1=m(x-x_1)$
$y-5=2(x-3)$
$y-5=2x-6$
$y=2x-1$

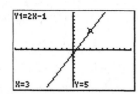

3. We substitute $m = -3$ and $(x_1, y_1) = (1, -2)$ into the point slope form and solve for y:

$$y - y_1 = m(x - x_1)$$
$$y - (-2) = -3(x - 1)$$
$$y + 2 = -3x + 3$$
$$y = -3x + 1$$

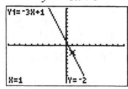

5. We substitute $m = 2$ and $(x_1, y_1) = (-4, -6)$ into the point slope form and solve for y:

$$y - y_1 = m(x - x_1)$$
$$y - (-6) = 2(x - (-4))$$
$$y + 6 = 2(x + 4)$$
$$y + 6 = 2x + 8$$
$$y = 2x + 2$$

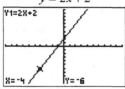

7. We substitute $m = -6$ and $(x_1, y_1) = (-2, -3)$ into the point slope form and solve for y:

$$y - y_1 = m(x - x_1)$$
$$y - (-3) = -6(x - (-2))$$
$$y + 3 = -6(x + 2)$$
$$y + 3 = -6x - 12$$
$$y = -6x - 15$$

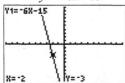

9. We substitute $m = \dfrac{2}{5}$ and $(x_1, y_1) = (3, 1)$ into the point slope form and solve for y:

$$y - y_1 = m(x - x_1)$$
$$y - 1 = \frac{2}{5}(x - 3)$$
$$y - 1 = \frac{2}{5}x - \frac{6}{5}$$
$$y = \frac{2}{5}x - \frac{1}{5}$$

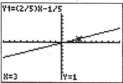

11. We substitute $m = -\dfrac{3}{4}$ and $(x_1, y_1) = (-2, -5)$ into the point slope form and solve for y:

$$y - y_1 = m(x - x_1)$$
$$y - (-5) = -\frac{3}{4}(x - (-2))$$
$$y + 5 = -\frac{3}{4}(x + 2)$$
$$y + 5 = -\frac{3}{4}x - \frac{3}{2}$$
$$y = -\frac{3}{4}x - \frac{13}{2}$$

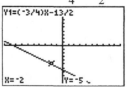

13. We substitute $m = 0$ and $(x_1, y_1) = (5, 3)$ into the point slope form and solve for y:

$$y - y_1 = m(x - x_1)$$
$$y - 3 = 0(x - 5)$$
$$y - 3 = 0$$
$$y = 3$$

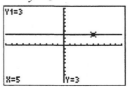

15. Since the slope of the line is undefined, the line must be vertical. Since the x-coordinate of the point on the line is -2, an equation of the line is $x = -2$.

17. We substitute $m = 2.1$ and $(x_1, y_1) = (3.7, -5.9)$ into the point slope form and solve for y:

$$y - y_1 = m(x - x_1)$$
$$y - (-5.9) = 2.1(x - 3.7)$$
$$y + 5.9 = 2.1x - 7.77$$
$$y = 2.1x - 13.67$$

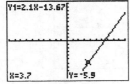

19. We substitute $m = -5.6$ and $(x_1, y_1) = (-4.5, 2.8)$ into the point slope form and solve for y:

$$y - y_1 = m(x - x_1)$$
$$y - 2.8 = -5.6(x - (-4.5))$$
$$y - 2.8 = -5.6(x + 4.5)$$
$$y - 2.8 = -5.6x - 25.2$$
$$y = -5.6x - 22.4$$

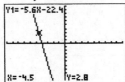

21. We substitute $m = -6.59$ and $(x_1, y_1) = (-2.48, -1.61)$ into the point slope form and solve for y:

$$y - y_1 = m(x - x_1)$$
$$y - (-1.61) = -6.59(x - (-2.48))$$
$$y + 1.61 = -6.59(x + 2.48)$$
$$y + 1.61 \approx -6.59x - 16.34$$
$$y \approx -6.59x - 17.95$$

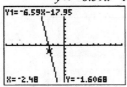

23. We begin by finding the slope of the line:

$$m = \frac{6 - 2}{5 - 3} = \frac{4}{2} = 2$$

Then we substitute $m = 2$ and $(x_1, y_1) = (3, 2)$ into the point slope form and solve for y:

$$y - y_1 = m(x - x_1)$$
$$y - 2 = 2(x - 3)$$
$$y - 2 = 2x - 6$$
$$y = 2x - 4$$

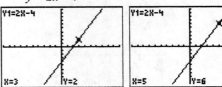

25. We begin by finding the slope of the line:

$$m = \frac{2 - 6}{3 - 1} = \frac{-4}{2} = -2$$

Then we substitute $m = -2$ and $(x_1, y_1) = (1, 6)$ into the point slope form and solve for y:

$$y - y_1 = m(x - x_1)$$
$$y - 6 = -2(x - 1)$$
$$y - 6 = -2x + 2$$
$$y = -2x + 8$$

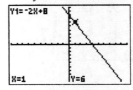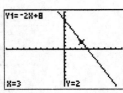

27. We begin by finding the slope of the line:

$$m = \frac{8 - (-7)}{2 - (-1)} = \frac{8 + 7}{2 + 1} = \frac{15}{3} = 5$$

Then we substitute $m = 5$ and $(x_1, y_1) = (-1, -7)$ into the point slope form and solve for y:

$$y - y_1 = m(x - x_1)$$
$$y - (-7) = 5(x - (-1))$$
$$y + 7 = 5(x + 1)$$
$$y + 7 = 5x + 5$$
$$y = 5x - 2$$

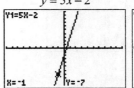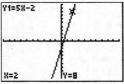

29. We begin by finding the slope of the line:
$$m = \frac{-4-(-7)}{-2-(-5)} = \frac{-4+7}{-2+5} = \frac{3}{3} = 1$$

Then we substitute $m = 1$ and $(x_1, y_1) = (-5, -7)$ into the point slope form and solve for y:
$$y - y_1 = m(x - x_1)$$
$$y - (-7) = 1(x - (-5))$$
$$y + 7 = (x + 5)$$
$$y = x - 2$$

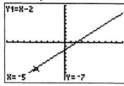 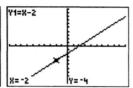

31. We begin by finding the slope of the line:
$$m = \frac{1-0}{1-0} = \frac{1}{1} = 1$$

Then we substitute $m = 1$ and $(x_1, y_1) = (0, 0)$ into the point slope form and solve for y:
$$y - y_1 = m(x - x_1)$$
$$y - 0 = 1(x - 0)$$
$$y = x$$

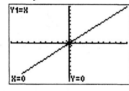 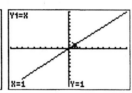

33. We begin by finding the slope of the line:
$$m = \frac{1-9}{2-0} = \frac{-8}{2} = -4$$

Then we substitute $m = -4$ and $(x_1, y_1) = (0, 9)$ into the point slope form and solve for y:
$$y - y_1 = m(x - x_1)$$
$$y - 9 = -4(x - 0)$$
$$y - 9 = -4x$$
$$y = -4x + 9$$

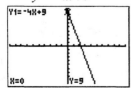 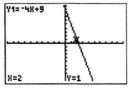

35. We begin by finding the slope of the line:
$$m = \frac{2-2}{5-3} = \frac{0}{2} = 0$$

Then we substitute $m = 0$ and $(x_1, y_1) = (3, 2)$ into the point slope form and solve for y:
$$y - y_1 = m(x - x_1)$$
$$y - 2 = 0(x - 3)$$
$$y - 2 = 0$$
$$y = 2$$

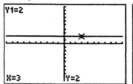 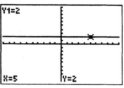

37. Since the x-coordinates of the given points are equal (both -4), the line that contains the points is vertical. An equation of the line is $x = -4$.

39. We begin by finding the slope of the line:
$$m = \frac{5-3}{8-4} = \frac{2}{4} = \frac{1}{2}$$

Then we substitute $m = \frac{1}{2}$ and $(x_1, y_1) = (4, 3)$ into the point slope form and solve for y:
$$y - y_1 = m(x - x_1)$$
$$y - 3 = \frac{1}{2}(x - 4)$$
$$y - 3 = \frac{1}{2}x - 2$$
$$y = \frac{1}{2}x + 1$$

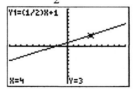 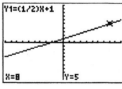

41. We begin by finding the slope of the line:

$$m = \frac{3-(-3)}{5-(-4)} = \frac{3+3}{5+4} = \frac{6}{9} = \frac{2}{3}$$

Then we substitute $m = \frac{2}{3}$ and $(x_1, y_1) = (-4, -3)$

into the point slope form and solve for y:

$$y - y_1 = m(x - x_1)$$

$$y - (-3) = \frac{2}{3}(x - (-4))$$

$$y + 3 = \frac{2}{3}(x + 4)$$

$$y + 3 = \frac{2}{3}x + \frac{8}{3}$$

$$y = \frac{2}{3}x - \frac{1}{3}$$

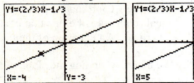

43. We begin by finding the slope of the line:

$$m = \frac{1-2}{3-(-3)} = \frac{1-2}{3+3} = \frac{-1}{6} = -\frac{1}{6}$$

Then we substitute $m = -\frac{1}{6}$ and $(x_1, y_1) = (-3, 2)$

into the point slope form and solve for y:

$$y - y_1 = m(x - x_1)$$

$$y - 2 = -\frac{1}{6}(x - (-3))$$

$$y - 2 = -\frac{1}{6}(x + 3)$$

$$y - 2 = -\frac{1}{6}x - \frac{1}{2}$$

$$y = -\frac{1}{6}x + \frac{3}{2}$$

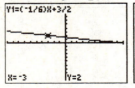 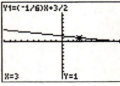

45. We begin by finding the slope of the line:

$$m = \frac{-1-1}{5-(-2)} = \frac{-1-1}{5+2} = \frac{-2}{7} = -\frac{2}{7}$$

Then we substitute $m = -\frac{2}{7}$ and $(x_1, y_1) = (-2, 1)$

into the point slope form and solve for y:

$$y - y_1 = m(x - x_1)$$

$$y - 1 = -\frac{2}{7}(x - (-2))$$

$$y - 1 = -\frac{2}{7}(x + 2)$$

$$y - 1 = -\frac{2}{7}x - \frac{4}{7}$$

$$y = -\frac{2}{7}x + \frac{3}{7}$$

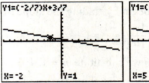 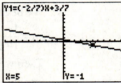

47. We begin by finding the slope of the line:

$$m = \frac{4-(-2)}{6-(-4)} = \frac{4+2}{6+4} = \frac{6}{10} = \frac{3}{5}$$

Then we substitute $m = \frac{3}{5}$ and $(x_1, y_1) = (-4, -2)$

into the point slope form and solve for y:

$$y - y_1 = m(x - x_1)$$

$$y - (-2) = \frac{3}{5}(x - (-4))$$

$$y + 2 = \frac{3}{5}(x + 4)$$

$$y + 2 = \frac{3}{5}x + \frac{12}{5}$$

$$y = \frac{3}{5}x + \frac{2}{5}$$

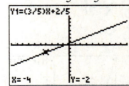 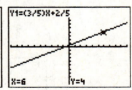

49. We begin by finding the slope of the line:
$$m = \frac{-5-(-8)}{-2-(-4)} = \frac{-5+8}{-2+4} = \frac{3}{2}$$

Then we substitute $m = \frac{3}{2}$ and $(x_1, y_1) = (-4, -8)$
into the point slope form and solve for y:
$$y - y_1 = m(x - x_1)$$
$$y - (-8) = \frac{3}{2}(x - (-4))$$
$$y + 8 = \frac{3}{2}(x + 4)$$
$$y + 8 = \frac{3}{2}x + 6$$
$$y = \frac{3}{2}x - 2$$

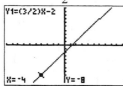 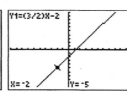

51. We begin by finding the slope of the line:
$$m = \frac{3.1 - 5.8}{4.5 - 2.3} = \frac{-2.7}{2.2} \approx -1.227 \,.$$

So, the equation of the line is of the form
$y = -1.227x + b$. To find b, we substitute the
coordinates of the point $(2.3, 5.8)$ into the
equation and solve for b:
$$y = -1.227x + b$$
$$5.8 = -1.227(2.3) + b$$
$$5.8 = -2.8221 + b$$
$$8.6221 = b$$
Rounding both m and b to two decimal places,
the approximate equation is $y = -1.23x + 8.62$.

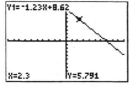 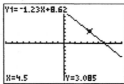

53. We begin by finding the slope of the line:
$$m = \frac{-7.5 - 2.2}{1.2 - (-4.5)} = \frac{-7.5 - 2.2}{1.2 + 4.5} = \frac{-9.7}{5.7} \approx -1.702 \,.$$

So, the equation of the line is of the form
$y = -1.702x + b$. To find b, we substitute the
coordinates of the point $(-4.5, 2.2)$ into the
equation and solve for b:

$$y = -1.702x + b$$
$$2.2 = -1.702(-4.5) + b$$
$$2.2 = 7.659 + b$$
$$-5.459 = b$$
Rounding both m and b to two decimal places,
the approximate equation is $y = -1.70x - 5.46$.

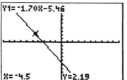 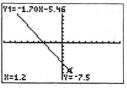

55. We begin by finding the slope of the line:
$$m = \frac{-5.29 - (-1.84)}{5.87 - 2.46} = \frac{-3.45}{3.41} \approx -1.012 \,.$$
So, the equation of the line is of the form
$y = -1.012x + b$. To find b, we substitute the
coordinates of the point $(2.46, -1.84)$ into the
equation and solve for b:
$$y = -1.012x + b$$
$$-1.84 = -1.012(2.46) + b$$
$$-1.84 = -2.48952 + b$$
$$0.64952 = b$$
Rounding both m and b to two decimal places,
the approximate equation is $y = -1.01x + 0.65$.

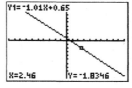 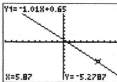

57. We begin by finding the slope of the line:
$$m = \frac{-2.69 - (-8.29)}{7.17 - (-4.57)} = \frac{5.60}{11.74} \approx 0.477 \,.$$

So, the equation of the line is of the form
$y = 0.477x + b$. To find b, we substitute the
coordinates of the point $(-4.57, -8.29)$ into the
equation and solve for b:
$$y = 0.477x + b$$
$$-8.29 = 0.477(-4.57) + b$$
$$-8.29 = -2.17989 + b$$
$$-6.11011 = b$$
Rounding both m and b to two decimal places,
the approximate equation is $y = 0.48x - 6.11$.

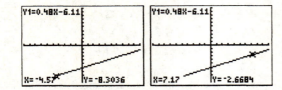

59. The slope of the line $y = 2x - 7$ is 2. The line we need to find is parallel to $y = 2x - 7$, so it also has slope $m = 2$. So, the equation of the line is of the form $y = 2x + b$. To find b, we substitute the coordinates of the point $(2, 3)$ into the equation and solve for b:

$$y = 2x + b$$
$$3 = 2(2) + b$$
$$3 = 4 + b$$
$$-1 = b$$

So, the equation is $y = 2x - 1$.

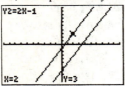

61. $4x + y = 1$

$$y = -4x + 1$$

So, the slope of the line $4x + y = 1$ is -4. The line we need to find is parallel to $4x + y = 1$, so it also has slope $m = -4$. So, the equation of the line is of the form $y = -4x + b$. To find b, we substitute the coordinates of the point $(-3, 5)$ into the equation and solve for b:

$$y = -4x + b$$
$$5 = -4(-3) + b$$
$$5 = 12 + b$$
$$-7 = b$$

So, the equation is $y = -4x - 7$.

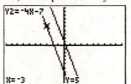

63. The slope of the line $y = \frac{1}{2}x + 5$ is $\frac{1}{2}$. The line we need to find is perpendicular to $y = \frac{1}{2}x + 5$, so its slope is the opposite reciprocal, which is $m = -2$. So, the equation of the line is of the form $y = -2x + b$. To find b, we substitute the coordinates of the point $(3, 4)$ into the equation and solve for b:

$$y = -2x + b$$
$$4 = -2(3) + b$$
$$4 = -6 + b$$
$$10 = b$$

So, the equation is $y = -2x + 10$.

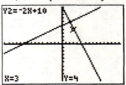

65. $x + 4y = 12$

$$4y = -x + 12$$
$$\frac{4y}{4} = \frac{-x}{4} + \frac{12}{4}$$
$$y = -\frac{1}{4}x + 3$$

So, the slope of the line $x + 4y = 12$ is $-\frac{1}{4}$. The line we need to find is perpendicular to $x + 4y = 12$, so its slope is the opposite reciprocal, which is $m = 4$. So, the equation of the line is of the form $y = 4x + b$. To find b, we substitute the coordinates of the point $(-2, -5)$ into the equation and solve for b:

$$y = 4x + b$$
$$-5 = 4(-2) + b$$
$$-5 = -8 + b$$
$$3 = b$$

So, the equation is $y = 4x + 3$.

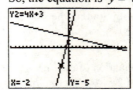

67. The line contains the points (2, 5) and (5, 1), so the slope of the line is $m = \dfrac{1-5}{5-2} = \dfrac{-4}{3} = -\dfrac{4}{3}$.

So, the equation of the line is of the form $y = -\dfrac{4}{3}x + b$. To find b, we substitute the coordinates of the point $(2, 5)$ into the equation and solve for b:

$$y = -\dfrac{4}{3}x + b$$

$$5 = -\dfrac{4}{3}(2) + b$$

$$5 = -\dfrac{8}{3} + b$$

$$\dfrac{23}{3} = b$$

So, the equation is $y = -\dfrac{4}{3}x + \dfrac{23}{3}$.

69. a. We begin by finding the slope of the line:

$$m = \dfrac{3-(-6)}{4-(-2)} = \dfrac{3+6}{4+2} = \dfrac{9}{6} = \dfrac{3}{2}$$

Then we substitute $m = \dfrac{3}{2}$ and $(x_1, y_1) = (4, 3)$ into the point slope form and solve for y:

$$y - y_1 = m(x - x_1)$$

$$y - 3 = \dfrac{3}{2}(x - 4)$$

$$y - 3 = \dfrac{3}{2}x - 6$$

$$y = \dfrac{3}{2}x - 3$$

b. The slope is $\dfrac{3}{2}$ and the y-intercept is $(0, -3)$.

We first plot $(0, -3)$. From this point we move 2 units to the right and 3 units up, where we plot the point $(2, 0)$. We then sketch the line that contains these two points.

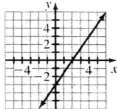

c. Answers may vary. Five possible points are included in the following table.

x	y
-4	$\dfrac{3}{2}(-4) - 3 = -9$
-2	$\dfrac{3}{2}(-2) - 3 = -6$
0	$\dfrac{3}{2}(0) - 3 = -3$
2	$\dfrac{3}{2}(2) - 3 = 0$
4	$\dfrac{3}{2}(4) - 3 = 3$

71. a. Yes, it is possible for a line to have no x-intercepts. Examples may vary. One possible equation is $y = 2$.

b. Yes, it is possible for a line to have exactly one x-intercept. Examples may vary. One possible equation is $y = x + 1$.

c. No, it is not possible for a line to have two x-intercepts. Explanations may vary. One possibility follows: Since lines are straight, they cannot intersect the x-axis in more than one place.

d. Yes, it is possible for a line to have an infinite number of x-intercepts. An equation of such a line is $y = 0$.

73. We use the points (2, 9) and (4, 13) to find the equation of the line that relates t and E. We begin by finding the slope of the line:

$$m = \dfrac{13-9}{4-2} = \dfrac{4}{2} = 2$$

Then we substitute $m = 2$ and $(t_1, E_1) = (2, 9)$ into the point slope form and solve for y:

$$E - E_1 = m(t - t_1)$$

$$E - 9 = 2(t - 2)$$

$$E - 9 = 2t - 4$$

$$E = 2t + 5$$

Substituting the remaining points into the equation will show that all satisfy the equation.

75. a. We begin by finding the slope of the line:

$$m = \frac{7-1}{4-2} = \frac{6}{2} = 3$$

(i) The equation of the line is of the form $y = 3x + b$. To find b, we substitute the coordinates of the point $(2, 1)$ into the equation and solve for b:

$$y = 3x + b$$
$$1 = 3(2) + b$$
$$1 = 6 + b$$
$$-5 = b$$

So, the equation is $y = 3x - 5$.

(ii) We substitute $m = 3$ and $(x_1, y_1) = (2, 1)$ into the point slope form and solve for y:

$$y - y_1 = m(x - x_1)$$
$$y - 1 = 3(x - 2)$$
$$y - 1 = 3x - 6$$
$$y = 3x - 5$$

b. The results from parts (ai) and (aii) are the same.

77. a. Answers may vary. One possible equation with slope -2 is $y = -2x + 1$.

b. Answers may vary. One possible equation with y-intercept $(0, 4)$ is $y = x + 4$.

c. Answers may vary. One possible equation that contains the point $(3, 5)$ is $y = x + 2$.

d. Such a line is not possible. Explanations may vary. One explanation follows: The equation of a line with slope -2 and y-intercept $(0, 4)$ is $y = -2x + 4$, but this line does not contain the point $(3, 5)$.

79. Set 1: We use the points $(0, 25)$ and $(1, 23)$. We begin by finding the slope of the line:

$$m = \frac{23-25}{1-0} = \frac{-2}{1} = -2$$

The y-intercept is $(0, 25)$, so $b = 25$. Thus, the equation is $y = -2x + 25$. Substituting the remaining points into the equation will show that all satisfy the equation.

Set 2: We use the points $(0, 12)$ and $(1, 16)$. We begin by finding the slope of the line:

$$m = \frac{16-12}{1-0} = \frac{4}{1} = 4$$

The y-intercept is $(0, 12)$, so $b = 12$. Thus, the

equation is $y = 4x + 12$. Substituting the remaining points into the equation will show that all satisfy the equation.

Set 3: We use the points $(0, 77)$ and $(1, 72)$. We begin by finding the slope of the line:

$$m = \frac{72-77}{1-0} = \frac{-5}{1} = -5$$

The y-intercept is $(0, 77)$, so $b = 77$. Thus, the equation is $y = -5x + 77$. Substituting the remaining points into the equation will show that all satisfy the equation.

Set 4: We use the points $(0, 3)$ and $(1, 3)$. We begin by finding the slope of the line:

$$m = \frac{3-3}{1-0} = \frac{0}{1} = 0$$

The y-intercept is $(0, 3)$, so $b = 3$. Thus, the equation is $y = 0x + 3$, or simply $y = 3$. Since the y-coordinate of the remaining points are all 3, all satisfy the equation.

81. Answers will vary.

83. a. $y = 2x - 3$

We first plot $(0, -3)$. From this point we move 1 unit to the right and 2 units up, where we plot the point $(1, -1)$. We then sketch the line that contains these two points.

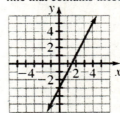

b. Points selected may vary. One possibility is to choose points $(-1, -5)$ and $(2, 1)$. The slope is $m = \frac{1-(-5)}{2-(-1)} = \frac{1+5}{2+1} = \frac{6}{3} = 2$. So, the equation is of the form $y = 2x + b$. We substitute $(2, 1)$ into this equation and solve for b.

$$y = 2x + b$$
$$1 = 2(2) + b$$
$$1 = 4 + b$$
$$-3 = b$$

So, the equation is $y = 2x - 3$, which is the same equation.

85. $3x + 2y = 6$

$$2y = -3x + 6$$

$$\frac{2y}{2} = \frac{-3x}{2} + \frac{6}{2}$$

$$y = -\frac{3}{2}x + 3$$

The slope is $m = -\dfrac{3}{2}$ and the y-intercept is $(0, 3)$.
We first plot $(0, 3)$. From this point we move 2
units to the right and 3 units down, where we plot
the point $(2, 0)$. We then sketch the line that
contains these two points.

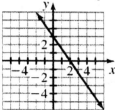

This is a linear equation in two variables.

87. Substitute 4 for x and -5 for y.
$$3x + 2y = 3(4) + 2(-5)$$
$$= 12 + (-10)$$
$$= 2$$
This is an expression in two variables.

Homework 5.3

1. Since the variables t and n are approximately
linearly related and n depends on t, we will find
an equation of the form $n = mt + b$. The year
1970 is represented by $t = 0$ and the year 2001 is
represented by $t = 31$. Thus, we have the data
points $(0, 179)$ and $(31, 633)$, which can use to
find m and b.
$$m = \frac{633 - 179}{31 - 0} = \frac{454}{31} \approx 14.65$$
So, the equation of the line is of the form
$n = 14.65t + b$. To find b, we substitute the
coordinates of the point $(0, 179)$ into the
equation and solve for b:
$$n = 14.65t + b$$
$$179 = 14.65(0) + b$$
$$179 = b$$
So, the equation is $n = 14.65t + 179$.

3. Since the variables t and n are approximately
linearly related and n depends on t, we will find
an equation of the form $n = mt + b$. The year
2001 is represented by $t = 1$ and the year 2005 is
represented by $t = 5$. Thus, we have the data
points $(1, 3147)$ and $(5, 2266)$, which can use to
find m and b.
$$m = \frac{2266 - 3147}{5 - 1} = \frac{-881}{4} = -220.25$$
So, the equation of the line is of the form
$n = -220.25t + b$. To find b, we substitute the
coordinates of the point $(1, 3147)$ into the
equation and solve for b:
$$n = 14.65t + b$$
$$3147 = -220.25(1) + b$$
$$3147 = -220.25 + b$$
$$3367.25 = b$$
So, the equation is $n = -220.25t + 3367.25$.

5. a. Since the variables t and p are approximately
linearly related and p depends on t, we will
find an equation of the form $p = mt + b$.
The year 1992 is represented by $t = 2$ and
the year 2003 is represented by $t = 13$.
Thus, we have the data points $(2, 9.1)$ and
$(13, 14.7)$, which can use to find m and b.
$$m = \frac{14.7 - 9.1}{13 - 2} = \frac{5.6}{11} \approx 0.51$$
So, the equation of the line is of the form
$p = 0.51t + b$. To find b, we substitute the
coordinates of the point $(2, 9.1)$ into the
equation and solve for b:
$$p = 0.51t + b$$
$$9.1 = 0.51(2) + b$$
$$9.1 = 1.02 + b$$
$$8.08 = b$$
So, the equation is $p = 0.51t + 8.08$.

b. The slope is 0.51. It means that the
percentage of sexual harassment charges
filed by men is increasing by 0.51
percentage point per year.

c. The p-intercept is $(0, 8.08)$. It means that
8.08%of sexual harassment charges in 1990
were filed by men.

7. a. Since the variables t and n are approximately linearly related and n depends on t, we will find an equation of the form $n = mt + b$. The year 2001 is represented by $t = 1$ and the year 2004 is represented by $t = 4$. Thus, we have the data points $(1, 14)$ and $(4, 3)$, which can use to find m and b.

$$m = \frac{3 - 14}{4 - 1} = \frac{-11}{3} \approx -3.67$$

So, the equation of the line is of the form $n = -3.67t + b$. To find b, we substitute the coordinates of the point $(1, 14)$ into the equation and solve for b:

$$n = -3.67t + b$$
$$14 = -3.67(1) + b$$
$$14 = -3.67 + b$$
$$17.67 = b$$

So, the equation is $n = -3.67t + 17.67$.

b. The slope is -3.67. It means that the number of U.S. flag desecrations is decreasing by 3.67 desecrations per year.

c. The n-intercept is $(0, 17.67)$. It means that there were about 18 U.S. flag desecrations in 2000.

9.

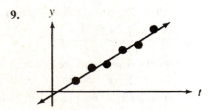

The value of m must increase; the value of b must decrease.

11. Answers may vary. One possibility follows: We begin by creating a scattergram of the data.

From the scattergram, we see that a line through the points $(4, 8)$ and $(8, 18)$ will come close to the rest of the data points. We use these points to find an equation of the form $y = mx + b$. We begin by finding the slope:

$$m = \frac{18 - 8}{8 - 4} = \frac{10}{4} = 2.5$$

So, the equation of the line is of the form $y = 2.5x + b$. To find b, we substitute the coordinates of the point $(4, 8)$ into the equation and solve for b:

$$y = 2.5x + b$$
$$8 = 2.5(4) + b$$
$$8 = 10 + b$$
$$-2 = b$$

So, the equation is $y = 2.5x - 2$. It comes close to the data points, as can be seen in the graph.

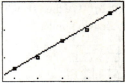

13. Answers may vary. One possibility follows: We begin by creating a scattergram of the data.

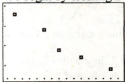

From the scattergram, we see that a line through the points $(3, 18)$ and $(16, 4)$ will come close to the rest of the data points. We use these points to find an equation of the form $y = mx + b$. We begin by finding the slope:

$$m = \frac{4 - 18}{16 - 3} = \frac{-14}{13} \approx -1.08$$

So, the equation of the line is of the form $y = -1.08x + b$. To find b, we substitute the coordinates of the point $(3, 18)$ into the equation and solve for b:

$$y = -1.08x + b$$
$$18 = -1.08(3) + b$$
$$18 = -3.24 + b$$
$$21.34 = b$$

So, the equation is $y = -1.08x + 21.34$. It comes close to the data points, as can be seen in the graph.

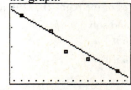

15. Student C has made the best choice. The line through the points (6, 10.5) and (9, 4.5) comes closer to the data points than the lines chosen by the other students.

17. a.

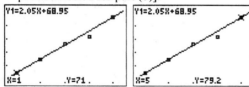

b. Answers may vary depending on the data points selected. We use the points (1, 71.0) and (5, 79.5) to find an equation of the form $p = mt + b$. We begin by finding the slope:
$$m = \frac{79.2 - 71.0}{5 - 1} = \frac{8.2}{4} = 2.05$$
So, the equation of the line is of the form $p = 2.05t + b$. To find b, we substitute the coordinates of the point (1, 71.0) into the equation and solve for b:
$$p = 2.05t + b$$
$$71.0 = 2.05(1) + b$$
$$71.0 = 2.05 + b$$
$$68.95 = b$$
So, the equation is $p = 2.05t + 68.95$.

c. Answers may vary [depending on the data points selected for part (b)].

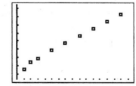

The figures show that the line passes through our selected points, (1, 71.0) and (5, 79.5), and comes close to all of the data points.

19. a.

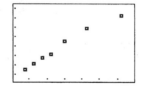

b. Answers may vary depending on the data points selected. We use the points (3, 29) and (13, 74) to find an equation of the form $H = md + b$. We begin by finding the slope:

$$m = \frac{74 - 29}{13 - 3} = \frac{45}{10} = 4.5$$
So, the equation of the line is of the form $H = 4.5d + b$. To find b, we substitute the coordinates of the point (3, 29) into the equation and solve for b:
$$H = 4.5d + b$$
$$29 = 4.5(3) + b$$
$$29 = 13.5 + b$$
$$15.5 = b$$
So, the equation is $H = 4.5d + 15.5$.

c. Answers may vary [depending on the data points selected for part (b)].

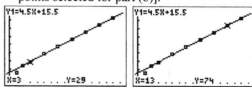

The figures show that the line passes through our selected points, (3, 29) and (13, 74), and comes close to all of the data points.

21. a.

b. Answers may vary depending on the data points selected. We use the points (12.5, 22) and (62.5, 73) to find an equation of the form $p = md + b$. We begin by finding the slope:
$$m = \frac{73 - 22}{62.5 - 12.5} = \frac{51}{50} = 1.02$$
So, the equation of the line is of the form $p = 1.02d + b$. To find b, we substitute the coordinates of the point (12.5, 22) into the equation and solve for b:
$$p = 1.02d + b$$
$$22 = 1.02(12.5) + b$$
$$22 = 12.75 + b$$
$$9.25 = b$$
So, the equation is $p = 1.02d + 9.25$.

c. Answers may vary [depending on the data points selected for part (b)].

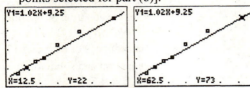

The figures show that the line passes through our selected points, $(12.5, 22)$ and $(62.5, 73)$, and comes close to all of the data points.

23. a.

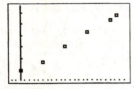

b. Answers may vary depending on the data points selected. We use the points $(20, 16.2)$ and $(40, 25.6)$ to find an equation of the form $p = mt + b$. We begin by finding the slope:

$$m = \frac{25.6 - 16.2}{40 - 20} = \frac{9.4}{20} = 0.47$$

So, the equation of the line is of the form $p = 0.47t + b$. To find b, we substitute the coordinates of the point $(20, 16.2)$ into the equation and solve for b:

$$p = 0.47t + b$$
$$16.2 = 0.47(20) + b$$
$$16.2 = 9.4 + b$$
$$6.8 = b$$

So, the equation is $p = 0.47t + 6.8$.

c. Answers may vary [depending on the data points selected for part (b)].

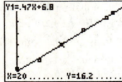 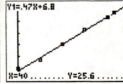

The figures show that the line passes through our selected points, $(20, 16.2)$ and $(40, 25.6)$, and comes close to all of the data points.

d. Answers may vary [depending on the data points selected for part (b)]. The slope is 0.47. It means that the percentage of Americans who have a college degree increases by approximately 0.47 percentage points per year.

e. Answers may vary [depending on the data points selected for part (b)]. The p-intercept is $(0, 6.8)$. It means that approximately 6.8% of Americans earned a college degree in 1960.

25. a.

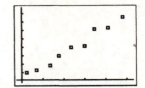

b. Answers may vary depending on the data points selected. We use the points $(14, 124)$ and $(97, 763)$ to find an equation of the form $r = mt + b$. We begin by finding the slope:

$$m = \frac{763 - 124}{97 - 14} = \frac{639}{83} \approx 7.70$$

So, the equation of the line is of the form $r = 7.70t + b$. To find b, we substitute the coordinates of the point $(14, 124)$ into the equation and solve for b:

$$r = 7.70t + b$$
$$124 = 7.70(14) + b$$
$$124 = 107.8 + b$$
$$16.2 = b$$

So, the equation is $r = 7.70t + 16.22$.

c. Answers may vary [depending on the data points selected for part (b)].

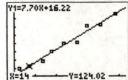 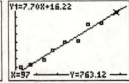

The figures show that the line passes through our selected points, $(14, 124)$ and $(97, 763)$, and comes close to all of the data points.

d. Answers may vary [depending on the data points selected for part (b).] The slope is 7.70. It means that the land speed record is increasing by approximately 7.70 mile per hour each year.

e. The p-intercept is $(0, 16.22)$. It means that the land speed record was approximately 16.22 miles per hour in 1900.

27. a.

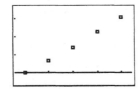

b. Answers may vary depending on the data points selected. We use the points $(5, 7)$ and $(7, 23)$ to find an equation of the form $p = mr + b$. We begin by finding the slope:

$$m = \frac{23 - 7}{7 - 5} = \frac{16}{2} = 8$$

So, the equation of the line is of the form $p = 8r + b$. To find b, we substitute the coordinates of the point $(5, 7)$ into the equation and solve for b:

$$p = 8r + b$$
$$7 = 8(5) + b$$
$$7 = 40 + b$$
$$-33 = b$$

So, the equation is $p = 8r - 33$.

c. Answers may vary [depending on the data points selected for part (b)].

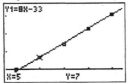 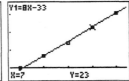

The figures show that the line passes through our selected points, $(5, 7)$ and $(7, 23)$, and comes close to all of the data points.

29. Answers may vary. One possibility follows: To find an equation of a linear model, begin by creating a scattergram of the data. Then, determine whether there is a line that comes close to the data points. If so, choose two points (not necessarily data points) that you can use to find the equation of a linear model. Next, find an equation of the line. Finally, use a graphing calculator to verify that the graph of the equation comes close to the points of the scattergram.

31. a. The value of the stock is increasing by $2 per year, so the slope of the equation is $m = 2$. Since t represents the number of years since 2005, the V-intercept is $(0, 10)$, the value of the stock in 2005. Thus, $b = 10$. An equation that describes the situation is $V = 2t + 10$.

$$\underset{\text{dollars}}{V} = \underset{\substack{\text{dollars} \\ \hline \text{year}}}{2} \cdot \underset{\text{years}}{t} + \underset{\text{dollars}}{10}$$

We use the fact that $\dfrac{\text{years}}{\text{years}} = 1$ to simplify the right-hand side of the equation:

$$\frac{\text{dollars}}{\text{year}} \cdot \text{years} + \text{dollars} = \text{dollars} + \text{dollars}.$$

So, the units on both sides of the equation are dollars, suggesting the equation is correct.

b. To graph the line, we first plot the V-intercept $(0, 10)$. The slope is $2 = \dfrac{2}{1}$, so the run is 1 and the rise is 2. From $(0, 10)$, we count 1 unit to the right and 2 units up, where we plot the point $(1, 12)$. We then sketch the line that contains these two points.

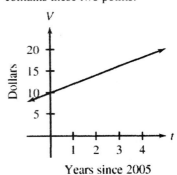

Years since 2005

c. Answers may vary. One possible table follows:

t	V
0	$2 \cdot 0 + 10 = 10$
1	$2 \cdot 1 + 10 = 12$
2	$2 \cdot 2 + 10 = 14$
3	$2 \cdot 3 + 10 = 16$
4	$2 \cdot 4 + 10 = 18$

33. a. The number of spectators is decreasing by 298 spectators per year, so the slope of the equation is $m = -298$. Since t represents the number of years since 2003, the n-intercept is $(0, 50575)$, the number of spectators in 2003. Thus, $b = 50,575$. An equation that describes the situation is $n = -298t + 50,575$

b. The slope is -298. It means that the average attendance is decreasing by 298 spectators per year.

c.
$$\underbrace{n}_{\text{spectators}} = \underbrace{-298}_{\frac{\text{spectators}}{\text{year}}} \cdot \underbrace{t}_{\text{years}} + \underbrace{50{,}575}_{\text{spectators}}$$

We use the fact that $\dfrac{\text{years}}{\text{years}} = 1$ to simplify the

right-hand side of the equation:

$$\frac{\text{spectators}}{\text{year}} \cdot \text{years} + \text{spectators}$$

$$= \text{spectators} + \text{spectators}$$

So, the units on both sides of the equation are spectators, suggesting the equation is correct.

35.
$$3 = -\frac{2}{5}x + 4$$

$$-1 = -\frac{2}{5}x$$

$$-\frac{5}{2}(-1) = -\frac{5}{2}\left(-\frac{2}{5}x\right)$$

$$\frac{5}{2} = x$$

This is a linear equation in one variable.

37. $y = -\dfrac{2}{5}x + 4$

We first plot the y-intercept $(0, 4)$. The slope is

$-\dfrac{2}{5} = \dfrac{-2}{5}$, so the run is 5 and the rise is -2.

From $(0, 4)$, we count 5 units to the right and 2 units down, where we plot the point $(5, 2)$. We then sketch the line that contains these two points.

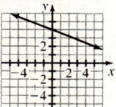

This is a linear equation is two variables.

Homework 5.4

1. a. The year 2011 is represented by $t = 11$, so we substitute 11 for t in the equation $p = 1.98t + 68.98$:

$$p = 1.98(11) + 68.98 = 90.76$$

In May of 2011, about 90.8% of seats will be filled for the six largest airlines.

b. We substitute 100 for p and solve for t:
$$p = 1.98t + 68.98$$
$$100 = 1.98t + 68.98$$
$$31.02 = 1.98t$$
$$t = \frac{31.02}{1.98} \approx 16$$

Now, $t = 16$ represents the year 2016. Thus, the model predicts that 100% will be filled in May of 2016. Model breakdown has occurred.

c. Since $p = 1.98t + 68.98$ is in slope-intercept form, the p-intercept is $(0, 68.98)$. It means that about 69% of all seats were filled in May of 2000 for the six largest airlines.

d. The slope is 1.98. It means that, each year, the percentage of all seats that are filled in May increases by 1.98 percentage points for the six larges airlines.

3. a. We substitute 50 for p in the equation $p = 1.08d + 8.81$ and solve for d:
$$50 = 1.08d + 8.81$$
$$41.19 = 1.08d$$
$$d = \frac{41.19}{1.08} \approx 38.1$$

For an income of $38.1 thousand, half of households own a computer.

b. Since $p = 1.08d + 8.81$ is in slope-intercept form, the slope is 1.08. It means that the percentage of households that own a computer increases by 1.08 percentage points for each $1 thousand increase in income.

c. The income group $0 to $5 thousand is represented by $d = 2.5$, so we substitute 2.5 for d in the equation $p = 1.08d + 8.81$:

$$p = 1.08(2.5) + 8.81 = 11.51 \approx 12$$

The model predicts that about 12% of households with incomes between $0 and $5 thousand own a computer.
The error is $12\% - 22\% = -10\%$.
Explanations may vary.

5. a. We substitute 78 for H in the equation
$H = 4.5d + 15.3$ and solve for d:

$78 = 4.5d + 15.3$

$62.7 = 4.5d$

$d = \dfrac{62.7}{4.5} \approx 13.93$

The model predicts that the life expectancy for dogs is about 14 years.

b. We substitute 29 for d in the equation:

$H = 4.5d + 15.3$

$H = 4.5(29) + 15.3 = 145.8$

Bluey lived to be about 146 in human years.

c. We substitute 119 for H in the equation
$H = 4.5d + 15.3$ and solve for d:

$119 = 4.5d + 15.3$

$103.7 = 4.5d$

$d = \dfrac{103.7}{4.5} \approx 23.04$

Sarah Knauss lived to be about 23 in dog years.

d. Since $H = 4.5d + 15.3$ is in slope-intercept form, the slope is 4.5. It means that a dog aging 1 year is equivalent to a human aging 4.5 years.

e. If each dog year is equivalent to 7 human years, the model will be $H = 7d$.

f. The slope is 7. It means that a dog aging 1 year is equivalent to a human aging 7 years. The slope of the equation in part (e) is greater than the slope found in part (d).

7. a.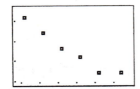

b. Answers may vary depending on the data points selected. We use the points $(11, 47)$ and $(29, 37)$ to find an equation of the form $p = mt + b$. [Note: $t = 29$ represents the year 1999 which we obtained by averaging the years 1996 and 2002 since both years had the same percentage. We did this to find a line that lies close to the all of the data points.] We begin by finding the slope:

$m = \dfrac{37 - 47}{29 - 11} = \dfrac{-10}{18} \approx -0.56$

So, the equation of the line is of the form $p = -0.56t + b$. To find b, we substitute the coordinates of the point $(11, 47)$ into the equation and solve for b:

$p = -0.56t + b$

$47 = -0.56(11) + b$

$47 = -6.16 + b$

$53.16 = b$

So, the equation is $p = -0.56t + 53.16$.

c. Answers may vary [depending on the equation found in part (b)]. Substitute 31 for p in the equation and solve for t:

$p = -0.56t + 53.16$

$31 = -0.56t + 53.16$

$-22.16 = -0.56t$

$t = \dfrac{-22.16}{-0.56} \approx 39.57$

The year will be about 40 years after 1970 which is 2010.

d. Answers may vary [depending on the equation found in part (b)]. Since $p = -0.56t + 53.16$ is in slope-intercept form, the p-intercept is $(0, 53.16)$. It means that about 53.2% of children lived with parents in "very happy" marriages in 1970.

e. Answers may vary [depending on the equation found in part (b)]. We substitute 0 for p and solve for t:

$0 = -0.56t + 53.16$

$0.56t = 53.16$

$t = \dfrac{53.16}{0.56} \approx 94.93$

The t-intercept is approximately $(94.93, 0)$. It means that no children will live with parents in "very happy" marriages in the year $1970 + 94.93 \approx 2065$. Model breakdown has likely occurred.

9. a.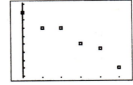

b. Answers may vary depending on the data points selected. We use the points $(0, 159.0)$ and $(3, 155.0)$ to find an equation of the form $w = mt + b$.

We begin by finding the slope:
$$m = \frac{155.0 - 159.0}{3 - 0} = \frac{-4.0}{3} \approx -1.33$$
So, the equation of the line is of the form
$w = -1.33t + b$. To find b, we substitute the
coordinates of the point $(0, 159.0)$ into the
equation and solve for b:
$$w = -1.33t + b$$
$$159.0 = -1.33(0) + b$$
$$159.0 = b$$
So, the equation is $w = -1.33t + 159.0$.

c. Answers may vary [depending on the
equation found in part (b)]. Since
$w = -1.33t + 159.0$ is in slope-intercept
form, the slope is -1.33. It means that the
math professor lost approximately 1.33
pounds per week.

d. Answers may vary [depending on the equation
found in part (b)]. Substitute 145 for w in the
equation and solve for t:
$$w = -1.33t + 159.0$$
$$145 = -1.33t + 159.0$$
$$-14 = -1.33t$$
$$t = \frac{-14}{-1.33} \approx 10.53$$
It took the math professor approximately 11
week to reach goal weight of 145 pounds.

e. Answers may vary [depending on the
equation found in part (b)]. We substitute 0
for w and solve for t:
$$0 = -1.33t + 159.0$$
$$1.33t = 159.0$$
$$t = \frac{159.0}{1.33} \approx 119.55$$
The t-intercept is approximately $(119.55, 0)$.
It means that the math professor will be
weightless in about 120 weeks. Model
breakdown has occurred.

11. a.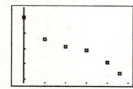

b. Answers may vary depending on the data
points selected. We use the points $(0, 12.0)$
and $(23, 8.4)$ to find an equation of the form
$p = mt + b$.

We begin by finding the slope:
$$m = \frac{8.4 - 12.0}{23 - 0} = \frac{-3.6}{23} \approx -0.16$$
So, the equation of the line is of the form
$p = -0.16t + b$. To find b, we substitute the
coordinates of the point $(0, 12.0)$ into the
equation and solve for b:
$$p = -0.16t + b$$
$$12.0 = -0.16(0) + b$$
$$12.0 = b$$
So, the equation is $p = -0.16t + 12.0$.

c. Answers may vary [depending on the equation
found in part (b)]. Substitute 7.5 for p in the
equation and solve for t:
$$p = -0.16t + 12.0$$
$$7.5 = -0.16t + 12.0$$
$$-4.5 = -0.16t$$
$$t = \frac{-4.5}{-0.16} \approx 28.13$$
The year will be about 28 years after 1980
which is 2008.

d. Answers may vary [depending on the
equation found in part (b)]. Since
$p = -0.16t + 12.0$ is in slope-intercept form,
the p-intercept is $(0, 12.0)$. It means that
12.0% of high school students dropped out
of school in 1980.

e. Answers may vary [depending on the
equation found in part (b)]. We substitute 0
for p and solve for t:
$$0 = -0.16t + 12.0$$
$$0.16t = 12.0$$
$$t = \frac{12}{0.16} = 75$$
The t-intercept is $(75, 0)$. It means no high
school students will drop out of school in
the year $1980 + 75 = 2055$. Model
breakdown has likely occurred.

13 a. Since $p = 7.8r - 31.8$ is in slope-intercept
form, the slope is 7.8. It means that the
percent increase in patients who would die
increases by 7.8 percentage points if the
patient-to-nurse ratio is increased by 1.

b. We substitute 10 for r in the equation:
$$p = 7.8r - 31.8$$
$$p = 7.8(10) - 31.8 = 46.2$$
The percent increase in patient mortality
would be 46.2%.

c. We substitute 5 for r in the equation:
$$p = 7.8r - 31.8$$
$$p = 7.8(5) - 31.8 = 7.2$$
The percent increase in patient mortality would be 7.2%. This prediction is a slight overestimate since our prediction is slightly above the actual percent increase of 7%.

d. Answers will vary.

15. Let n represent the number of cable TV networks at t years after 1990. Then 1994 is represented by $t = 4$ and 2003 is represented by $t = 13$. We find a linear model by using the data points (4, 106) and (13, 339). We begin by finding the slope:
$$m = \frac{339 - 106}{13 - 4} = \frac{233}{9} \approx 25.89$$
So, the equation of the line is of the form $n = 25.89t + b$. To find b, we substitute the coordinates of the point (4, 106) into the equation and solve for b:
$$n = 25.89t + b$$
$$106 = 25.89(4) + b$$
$$106 = 103.56 + b$$
$$2.44 = b$$
So, the equation is $n = 25.89t + 2.44$. To find when there will be 520 cable networks, we substitute 520 for n in the equation and solve for t:
$$520 = 25.89t + 2.44$$
$$517.56 = 25.89t$$
$$t = \frac{517.56}{25.89} \approx 19.99$$
We predict that there will be 520 cable networks in the year $1990 + 19.99 \approx 2010$.

17. Let p represent the percentage of mothers who smoke cigarettes during pregnancy at t years after 1995. Then 1995 is represented by $t = 0$ and 2001 is represented by $t = 6$. We find a linear model by using the data points (0, 13.9) and (6, 12.0). We begin by finding the slope:
$$m = \frac{12.0 - 13.9}{6 - 0} = \frac{-1.9}{6} \approx -0.32$$
So, the equation of the line is of the form $p = -0.32t + b$. To find b, we substitute the coordinates of the point (0, 13.9) into the equation and solve for b:
$$p = -0.32t + b$$
$$13.9 = -0.32(0) + b$$
$$13.9 = b$$

So, the equation is $p = -0.32t + 13.9$. To find the percentage for the year 2010, we substitute 15 for t in the equation:
$$p = -0.32(15) + 13.9 = 9.1$$
We predict that 9.1% of mothers will smoke cigarettes during pregnancy in the year 2010.

19. Let s represent the salary of a U.S. senator at t years after 2000. Then 2000 is represented by $t = 0$ and 2005 is represented by $t = 5$. We find a linear model by using the data points (0, 141300) and (5, 162100). We begin by finding the slope:
$$m = \frac{162,100 - 141,300}{5 - 0} = \frac{20,800}{5} = 4160$$
So, the equation of the line is of the form $s = 4160t + b$. To find b, we substitute the coordinates of the point (0, 141300) into the equation and solve for b:
$$s = 4160t + b$$
$$141,300 = 4160(0) + b$$
$$141,300 = b$$
So, the equation is $s = 4160t + 141,300$. To find when the salary will be \$200,000, we substitute 200,000 for s in the equation and solve for t:
$$200,000 = 4160t + 141,300$$
$$58,700 = 4160t$$
$$t = \frac{58,700}{4160} \approx 14.11$$
We predict that the salary will be \$200,000 in the year $2000 + 14.11 \approx 2014$.

21. Let p represent the price for a 30-second ad during the Academy Awards at t years after 1990. Then 1994 is represented by $t = 4$ and 2005 is represented by $t = 15$. We find a linear model by using the data points (4, 0.6) and (15, 1.6). We begin by finding the slope:
$$m = \frac{1.6 - 0.6}{15 - 4} = \frac{1}{11} \approx 0.09$$
So, the equation of the line is of the form $p = 0.09t + b$. To find b, we substitute the coordinates of the point (4, 0.6) into the equation and solve for b:
$$p = 0.09t + b$$
$$0.6 = 0.09(4) + b$$
$$0.6 = 0.36 + b$$
$$0.24 = b$$

So, the equation is $p = 0.09t + 0.24$. To find when the price will be \$2.2 million, we substitute 2.2 for p in the equation and solve for t:

$$2.2 = 0.09t + 0.24$$

$$1.96 = 0.09t$$

$$t = \frac{1.96}{0.09} \approx 21.78$$

We predict that the price will be \$2.2 million in the year $1990 + 21.78 \approx 2012$.

23. a. The number of Americans who live alone is increasing by 0.55 million people per year, so the slope of the equation is $m = 0.55$. Since t represents the number of years since 2000, the n-intercept is (0, 27.2), the number living alone in 2000. Thus, $b = 27.7$. An equation that describes the situation is $n = 0.55t + 27.2$.

b. The n-intercept is (0, 27.2). This means that there were 27.2 million Americans who lived alone in the year 2000.

c.
$$\underset{\substack{\smile \\ \text{million people}}}{n} = \underset{\substack{\underbrace{} \\ \frac{\text{million people}}{\text{year}}}}{0.55} \cdot \underset{\text{years}}{t} + \underset{\text{million people}}{27.2}$$

We use the fact that $\dfrac{\text{years}}{\text{years}} = 1$ to simplify the right-hand side of the equation:

$$\frac{\text{million people}}{\text{year}} \cdot \text{years} + \text{million people} =$$

million people + million people

So, the units on both sides of the equation are millions of people, suggesting the equation is correct.

d. We substitute 41.4 for n in the equation and solve for t:

$$41.4 = 0.55t + 27.2$$

$$14.2 = 0.55t$$

$$t = \frac{14.2}{0.55} \approx 25.82$$

We predict that 41.4 million people will live alone in the year $2000 + 25.81 \approx 2026$.

25. a. The sales of Echinacea are decreasing by \$20.67 million per year, so the slope of the equation is $m = -20.67$. Since t represents the number of years since 2004, the s-intercept is (0, 152). Thus, $b = 152$. An equation that describes the situation is $s = -20.67t + 152$.

b. The s-intercept is (0, 152). This means that annual sales of Echinacea were \$152 million in the year 2004.

c.
$$\underset{\substack{\smile \\ \text{million dollars}}}{s} = \underset{\substack{\underbrace{} \\ \frac{\text{million dollars}}{\text{year}}}}{-20.67} \cdot \underset{\text{years}}{t} + \underset{\text{million dollars}}{152}$$

We use the fact that $\dfrac{\text{years}}{\text{years}} = 1$ to simplify the right-hand side of the equation:

$$\frac{\text{million dollars}}{\text{year}} \cdot \text{years} + \text{million dollars} =$$

million dollars + million dollars

So, the units on both sides of the equation are millions of dollars, suggesting the equation is correct.

d. We substitute 111 for s in the equation and solve for t:

$$111 = -20.67t + 152$$

$$-41 = -20.67t$$

$$t = \frac{-41}{-20.67} \approx 1.98$$

We predict that annual Echinacea sales were \$111 million in the year $2004 + 1.98 \approx 2006$.

27.
$$-4(3x - 5) = 3(2x + 1)$$

$$-12x + 20 = 6x + 3$$

$$-18x + 20 = 3$$

$$-18x = -17$$

$$x = \frac{-17}{-18} = \frac{17}{18}$$

This is a linear equation in one variable.

29.
$$-4(3x - 5) - 3(2x + 1) = -12x + 20 - 6x - 3$$
$$= -18x + 17$$

This is a linear expression in one variable.

Homework 5.5

1. Since -3 is to the right of -5 on the number line, the statement $-3 > -5$ is true.

3. Since $4 = 4$, the statement $4 \geq 4$ is true.

5. $x < 4$

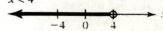

---

7. $x \geq -1$

9. $x \leq -2$

11. $x > 6$

13. **In words**: numbers greater than or equal to 4

Inequality: $x \geq 4$

Graph:

Interval Notation: $[4, \infty)$

In words: numbers less than or equal to -2

Inequality: $x \leq -2$

Graph:

Interval Notation: $(-\infty, 2]$

In words: numbers less than 1

Inequality: $x < 1$

Graph:

Interval Notation: $(-\infty, 1)$

In words: numbers greater than -5

Inequality: $x > -5$

Graph:

Interval Notation: $(-5, \infty)$

15. We substitute 2 for x in the inequality $3x + 5 \geq 14$:

$$3(2) + 5 \overset{?}{\geq} 14$$

$$6 + 5 \overset{?}{\geq} 14$$

$$11 \overset{?}{\geq} 14 \leftarrow \text{False}$$

So, 2 does not satisfy the inequality $3x + 5 \geq 14$.

We substitute 3 for x in the inequality $3x + 5 \geq 14$:

$$3(3) + 5 \overset{?}{\geq} 14$$

$$9 + 5 \overset{?}{\geq} 14$$

$$14 \overset{?}{\geq} 14 \leftarrow \text{True}$$

So, 3 satisfies the inequality $3x + 5 \geq 14$.

We substitute 6 for x in the inequality $3x + 5 \geq 14$:

$$3(6) + 5 \overset{?}{\geq} 14$$

$$18 + 5 \overset{?}{\geq} 14$$

$$23 \overset{?}{\geq} 14 \leftarrow \text{True}$$

So, 6 satisfies the inequality $3x + 5 \geq 14$.

17. We substitute -4 for x in the inequality $2x < x + 2$:

$$2(-4) \overset{?}{<} (-4) + 2$$

$$-8 \overset{?}{<} -2 \leftarrow \text{True}$$

So, -4 satisfies the inequality $2x < x + 2$.

We substitute 2 for x in the inequality $2x < x + 2$:

$$2(2) \overset{?}{<} (2) + 2$$

$$4 \overset{?}{<} 4 \leftarrow \text{False}$$

So, 2 does not satisfy the inequality $2x < x + 2$.

We substitute 3 for x in the inequality $2x < x + 2$:

$$2(3) \overset{?}{<} (3) + 2$$

$$6 \overset{?}{<} 5 \leftarrow \text{False}$$

So, 3 does not satisfy the inequality $2x < x + 2$.

19. $$x + 2 > 3$$
$$x + 2 - 2 > 3 - 2$$
$$x > 1$$

We graph the solution set, $(1, \infty)$.

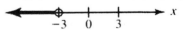

21. $$x - 1 < -4$$
$$x - 1 + 1 < -4 + 1$$
$$x < -3$$

We graph the solution set, $(-\infty, -3)$.

23. $$2x \leq 6$$
$$\frac{2x}{2} \leq \frac{6}{2}$$
$$x \leq 3$$

We graph the solution set, $(-\infty, 3]$.

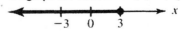

25. $4x \geq -8$

$$\frac{4x}{4} \geq \frac{-8}{4}$$

$$x \geq -2$$

We graph the solution set, $[-2, \infty)$.

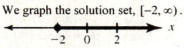

27. $-3t \geq 6$

$$\frac{-3t}{-3} \leq \frac{6}{-3}$$

$$t \leq -2$$

We graph the solution set, $(-\infty, -2]$.

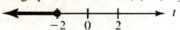

29. $-5x < 20$

$$\frac{-5x}{-5} > \frac{20}{-5}$$

$$x > -4$$

We graph the solution set, $(-4, \infty)$.

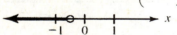

31. $-2x > 1$

$$\frac{-2x}{-2} < \frac{1}{-2}$$

$$x < -\frac{1}{2}$$

We graph the solution set, $\left(-\infty, -\frac{1}{2}\right)$.

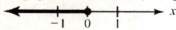

33. $5x \leq 0$

$$\frac{5x}{5} \leq \frac{0}{5}$$

$$x \leq 0$$

We graph the solution set, $(-\infty, 0]$.

35. $-x < 2$

$$\frac{-x}{-1} > \frac{2}{-1}$$

$$x > -2$$

We graph the solution set, $(-2, \infty)$.

37. $\frac{1}{2}a < 3$

$$2\left(\frac{1}{2}a\right) < 2(3)$$

$$a < 6$$

We graph the solution set, $(-\infty, 6)$.

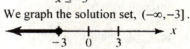

39. $-\frac{2}{3}x \geq 2$

$$-\frac{3}{2}\left(-\frac{2}{3}x\right) \leq -\frac{3}{2}(2)$$

$$x \leq -3$$

We graph the solution set, $(-\infty, -3]$.

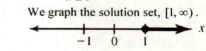

41. $3x - 1 \geq 2$

$$3x - 1 + 1 \geq 2 + 1$$

$$3x \geq 3$$

$$\frac{3x}{3} \geq \frac{3}{3}$$

$$x \geq 1$$

We graph the solution set, $[1, \infty)$.

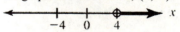

43. $5 - 3x < -7$

$$5 - 3x - 5 < -7 - 5$$

$$-3x < -12$$

$$\frac{-3x}{-3} > \frac{-12}{-3}$$

$$x > -4$$

We graph the solution set, $(4, \infty)$.

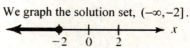

45. $-4x - 3 \geq 5$

$$-4x - 3 + 3 \geq 5 + 3$$

$$-4x \geq 8$$

$$\frac{-4x}{-4} \leq \frac{8}{-4}$$

$$x \leq -2$$

We graph the solution set, $(-\infty, -2]$.

47.
$$3c - 6 \le 5c$$
$$3c - 6 - 5c \le 5c - 5c$$
$$-2c - 6 \le 0$$
$$-2c - 6 + 6 \le 0 + 6$$
$$-2c \le 6$$
$$\frac{-2c}{-2} \ge \frac{6}{-2}$$
$$c \ge -3$$
We graph the solution set, $[-3, \infty)$.

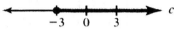

49.
$$5x \ge x - 12$$
$$5x - x \ge x - 12 - x$$
$$4x \ge -12$$
$$\frac{4x}{4} \ge \frac{-12}{4}$$
$$x \ge -3$$
We graph the solution set, $[-3, \infty)$.

51.
$$-3.8x + 1.9 > -7.6$$
$$-3.8x + 1.9 - 1.9 > -7.6 - 1.9$$
$$-3.8x > -9.5$$
$$\frac{-3.8x}{-3.8} < \frac{-9.5}{-3.8}$$
$$x < 2.5$$
We graph the solution set, $(-\infty, 2.5)$.

53.
$$3b + 2 > 7b - 6$$
$$3b + 2 - 7b > 7b - 6 - 7b$$
$$-4b + 2 > -6$$
$$-4b + 2 - 2 > -6 - 2$$
$$-4b > -8$$
$$\frac{-4b}{-4} < \frac{-8}{-4}$$
$$b < 2$$
We graph the solution set, $(-\infty, 2)$.

55.
$$4 - 3x < 9 - 2x$$
$$4 - 3x + 2x < 9 - 2x + 2x$$
$$4 - x < 9$$
$$4 - x - 4 < 9 - 4$$
$$-x < 5$$
$$\frac{-x}{-1} > \frac{5}{-1}$$
$$x > -5$$
We graph the solution set, $(-5, \infty)$.

57.
$$2(x + 3) \le 8$$
$$2x + 6 \le 8$$
$$2x + 6 - 6 \le 8 - 6$$
$$2x \le 2$$
$$\frac{2x}{2} \le \frac{2}{2}$$
$$x \le 1$$
We graph the solution set, $(-\infty, 1]$.

59.
$$-(a - 3) > 4$$
$$-a + 3 > 4$$
$$-a + 3 - 3 > 4 - 3$$
$$-a > 1$$
$$\frac{-a}{-1} < \frac{1}{-1}$$
$$a < -1$$
We graph the solution set, $(-\infty, -1)$.

61.
$$3(2x - 1) \le 2(2x + 1)$$
$$6x - 3 \le 4x + 2$$
$$6x - 3 - 4x \le 4x + 2 - 4x$$
$$2x - 3 \le 2$$
$$2x - 3 + 3 \le 2 + 3$$
$$2x \le 5$$
$$\frac{2x}{2} \le \frac{5}{2}$$
$$x \le \frac{5}{2}$$
We graph the solution set, $\left(-\infty, \frac{5}{2}\right]$.

63. $4(2x-3)+1 \geq 3(4x-5)-x$

$8x-12+1 \geq 12x-15-x$

$8x-11 \geq 11x-15$

$8x-11-11x \geq 11x-15-11x$

$-3x-11 \geq -15$

$-3x-11+11 \geq -15+11$

$-3x \geq -4$

$\dfrac{-3x}{-3} \leq \dfrac{-4}{-3}$

$x \leq \dfrac{4}{3}$

We graph the solution set, $\left(-\infty, \dfrac{4}{3}\right]$.

65. $4.3(1.5-x) \geq 13.76$

$6.45-4.3x \geq 13.76$

$6.45-4.3x-6.45 \geq 13.76-6.45$

$-4.3x \geq 7.31$

$\dfrac{-4.3x}{-4.3} \leq \dfrac{7.31}{-4.3}$

$x \leq -1.7$

We graph the solution set, $(-\infty, -1.7]$.

67. $\dfrac{1}{2}y + \dfrac{2}{3} \geq \dfrac{3}{2}$

$6\left(\dfrac{1}{2}y + \dfrac{2}{3}\right) \geq 6\left(\dfrac{3}{2}\right)$

$3y+4 \geq 9$

$3y+4-4 \geq 9-4$

$3y \geq 5$

$\dfrac{3y}{3} \geq \dfrac{5}{3}$

$y \geq \dfrac{5}{3}$

We graph the solution set, $\left[\dfrac{5}{3}, \infty\right)$.

69. $\dfrac{5}{3} - \dfrac{1}{6}x < \dfrac{1}{2}$

$6\left(\dfrac{5}{3} - \dfrac{1}{6}x\right) < 6\left(\dfrac{1}{2}\right)$

$10-x < 3$

$10-x-10 < 3-10$

$-x < -7$

$\dfrac{-x}{-1} > \dfrac{-7}{-1}$

$x > 7$

We graph the solution set, $(7, \infty)$.

71. $\dfrac{2(3-x)}{3} > -4x$

$3\left(\dfrac{2(3-x)}{3}\right) > 3(-4x)$

$2(3-x) > -12x$

$6-2x > -12x$

$6-2x+12x > -12x+12x$

$6+10x > 0$

$6+10x-6 > 0-6$

$10x > -6$

$\dfrac{10x}{10} > \dfrac{-6}{10}$

$x > -\dfrac{3}{5}$

We graph the solution set, $\left(-\dfrac{3}{5}, \infty\right)$.

73. $\dfrac{3r-5}{4} \leq \dfrac{2r-3}{3}$

$12\left(\dfrac{3r-5}{4}\right) \leq 12\left(\dfrac{2r-3}{3}\right)$

$3(3r-5) \leq 4(2r-3)$

$9r-15 \leq 8r-12$

$9r-15-8r \leq 8r-12-8r$

$r-15 \leq -12$

$r-15+15 \leq -12+15$

$r \leq 3$

We graph the solution set, $(-\infty, 3]$.

75. a. The slope is 0.83. It means that the percentage of teachers who are "very satisfied" with teaching is increasing by 0.83 percentage points per year.

b. We find where $p > 60$:
$$0.83t + 37.43 > 60$$
$$0.83t + 37.43 - 37.43 > 60 - 37.43$$
$$0.83t > 22.57$$
$$\frac{0.83t}{0.83} > \frac{22.57}{0.83}$$
$$t > 27.19 \quad \text{(rounded)}$$
Since $1980 + 27.19 \approx 2007$, more than 60% of teachers will be "very satisfied" after the year 2007.

77. a. Answers may vary depending on the data points selected to create the linear model. One possibility follows: We begin by creating a scattergram of the data.

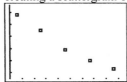

From the scattergram, we see that a line through the points $(3, 74.0)$ and $(33, 55.1)$ will come close to the rest of the data points. We use these points to find an equation of the form $p = mt + b$. We begin by finding the slope:
$$m = \frac{55.1 - 74.0}{33 - 3} = \frac{-18.9}{30} = -0.63$$
So, the equation of the line is of the form $p = -0.63t + b$. To find b, we substitute the coordinates of the point $(3, 74.0)$ into the equation and solve for b:
$$p = -0.63t + b$$
$$74 = -0.63(3) + b$$
$$74 = -1.89 + b$$
$$75.89 = b$$
So, the equation is $p = -0.63t + 75.89$. It comes close to the data points, as can be seen in the graph.

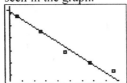

b. Answers may vary [depending on the equation found in part (a)].
We find where $p > 53$:
$$-0.63t + 75.89 > 53$$
$$-0.63t + 75.89 - 75.89 > 53 - 75.89$$
$$-0.63t > -22.89$$
$$\frac{-0.63t}{-0.63} < \frac{-22.89}{-0.63}$$
$$t < 36.33 \quad \text{(rounded)}$$
Since $1960 + 36.33 \approx 1996$, more than 53% of households were married-couple households prior to the year 1996.

c. Answers may vary. One possibility follows: This is possible because the number of nonmarried households is growing at a greater rate than the number of married households.

79. a. Answers may vary depending on the data points selected to create the linear model. One possibility follows: We begin by creating a scattergram of the data.

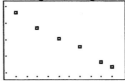

From the scattergram, we see that a line through the points $(4, 58.2)$ and $(13, 41.7)$ will come close to the rest of the data points. We use these points to find an equation of the form $r = mt + b$. We begin by finding the slope:
$$m = \frac{41.7 - 58.2}{13 - 4} = \frac{-16.5}{9} \approx -1.83$$
So, the equation of the line is of the form $r = -1.83t + b$. To find b, we substitute the coordinates of the point $(4, 58.2)$ into the equation and solve for b:
$$r = -1.83t + b$$
$$58.2 = -1.83(4) + b$$
$$58.2 = -7.32 + b$$
$$65.52 = b$$
So, the equation is $r = -1.83t + 65.52$. It comes close to the data points, as can be seen in the graph.

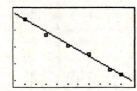

b. Answers may vary [depending on the equation found in part (a)]. Since the year 2010 is represented by $t = 20$, we substitute 20 for t in the linear equation:
$r = -1.83(20) + 65.52 \approx 28.9$ births per 1000 women.

The number of births to women ages 15-19 in 2010 will be:
$\dfrac{28.9}{1000} \cdot 10,398,000 \approx 300,502$ births.

c. Answers may vary [depending on the equation found in part (a)].
We find where $r < 10$:
$$-1.83t + 65.52 < 10$$
$$-1.83t + 65.52 - 65.52 < 10 - 65.52$$
$$-1.83t < -55.52$$
$$\dfrac{-1.83t}{-1.83} > \dfrac{-55.52}{-1.83}$$
$$t > 30.34 \quad \text{(rounded)}$$

Since $1990 + 30.34 \approx 2020$, we predict that the American birthrate will be less than 10 births pre 1000 women ages 15-19 after the year 2020.

81. The student failed to reverse the inequality symbol (from $<$ to $>$) when he or she divided both sides of the inequality by a negative.
$$-3x < 15$$
$$\dfrac{-3x}{-3} > \dfrac{15}{-3}$$
$$x > -5$$

83. a. Answers may vary. The solution to the inequality is:
$$3x - 7 < 5$$
$$3x - 7 + 7 < 5 + 7$$
$$3x < 12$$
$$\dfrac{3x}{3} < \dfrac{12}{3}$$
$$x < 4$$
Any number less than 4 will satisfy the inequality.

b. Answers may vary. Any number greater than or equal to 4 will not satisfy the inequality.

85. Answers will vary.

87.
$$-2x + 6 = 3x - 14$$
$$-2x + 6 - 3x = 3x - 14 - 3x$$
$$-5x + 6 = -14$$
$$-5x + 6 - 6 = -14 - 6$$
$$-5x = -20$$
$$\dfrac{-5x}{-5} = \dfrac{-20}{-5}$$
$$x = 4$$

89.
$$-2x + 6 > 3x - 14$$
$$-2x + 6 - 3x > 3x - 14 - 3x$$
$$-5x + 6 > -14$$
$$-5x + 6 - 6 > -14 - 6$$
$$-5x > -20$$
$$\dfrac{-5x}{-5} < \dfrac{-20}{-5}$$
$$x < 4$$
We graph the solution set, $(-\infty, 4)$.

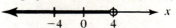

91.
$$x + 5 > 2$$
$$x + 5 - 5 > 2 - 5$$
$$x > -3$$
We graph the solution set, $(-3, \infty)$.

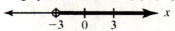

93.
$$2x \leq 5x - 6$$
$$2x - 5x \leq 5x - 6 - 5x$$
$$-3x \leq -6$$
$$\dfrac{-3x}{-3} \geq \dfrac{-6}{-3}$$
$$x \geq 2$$
We graph the solution set, $[2, \infty)$.

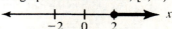

95. Answers will vary.

97. Answers will vary.

Chapter 5 Review Exercises

1. The equation $y = 4x - 5$ is in slope-intercept form, so the slope is $m = 4 = \frac{4}{1}$, and the y-intercept is $(0, -5)$. We first plot $(0, -5)$. From this point we move 1 unit to the right and 4 units up, where we plot the point $(1, -1)$. We then sketch the line that contains these two points.

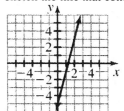

2. The equation $y = -3x + 4$ is in slope-intercept form, so the slope is $m = -3 = \frac{-3}{1}$, and the y-intercept is $(0, 4)$. We first plot $(0, 4)$. From this point we move 1 unit to the right and 3 units down, where we plot the point $(1, 1)$. We then sketch the line that contains these two points.

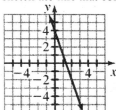

3. The equation $y = \frac{1}{2}x + 1$ is in slope-intercept form, so the slope is $m = \frac{1}{2}$, and the y-intercept is $(0, 1)$. We first plot $(0, 1)$. From this point we move 2 units to the right and 1 unit up, where we plot the point $(2, 2)$. We then sketch the line that contains these two points.

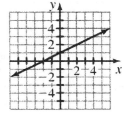

4. The equation $y = -\frac{2}{3}x - 2$ is in slope-intercept form, so the slope is $m = -\frac{2}{3}$, and the y-intercept is $(0, -2)$. We first plot $(0, -2)$. From this point we move 3 units to the right and 2 units down, where we plot the point $(3, -4)$. We then sketch the line that contains these two points.

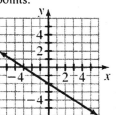

5. The equation $y = \frac{5}{3}x$ is in slope-intercept form, so the slope is $m = \frac{5}{3}$, and the y-intercept is $(0, 0)$. We first plot $(0, 0)$. From this point we move 3 units to the right and 5 units up, where we plot the point $(3, 5)$. We then sketch the line that contains these two points.

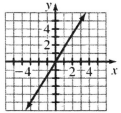

6. The equation $y = -5$ is in slope-intercept form, so the slope is $m = 0$ and the y-intercept is $(0, -5)$. The graph is a horizontal line that passes through the point $(0, -5)$.

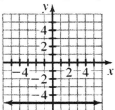

7. $2x - y = 5$

$$-y = -2x + 5$$

$$\frac{-y}{-1} = \frac{-2x}{-1} + \frac{5}{-1}$$

$$y = 2x - 5$$

The slope is $m = 2 = \dfrac{2}{1}$, and the y-intercept is

$(0, -5)$. We first plot $(0, -5)$. From this point we move 1 unit to the right and 2 units up, where we plot the point $(1, -3)$. We then sketch the line that contains these two points.

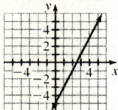

8. $3x - 2y = -6$

$$-2y = -3x - 6$$

$$\frac{-2y}{-2} = \frac{-3x}{-2} - \frac{6}{-2}$$

$$y = \frac{3}{2}x + 3$$

The slope is $m = \dfrac{3}{2}$, and the y-intercept is $(0, 3)$.

We first plot $(0, 3)$. From this point we move 2 units to the right and 3 units up, where we plot the point $(2, 6)$. We then sketch the line that contains these two points.

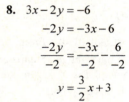

9. $4x + 5y = 10$

$$5y = -4x + 10$$

$$\frac{5y}{5} = \frac{-4x}{5} + \frac{10}{5}$$

$$y = -\frac{4}{5}x + 2$$

The slope is $m = -\dfrac{4}{5}$, and the y-intercept is

$(0, 2)$. We first plot $(0, 2)$. From this point we move 5 units to the right and 4 units down, where we plot the point $(5, -2)$. We then sketch the line that contains these two points.

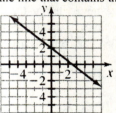

10. $x + 3y = 6$

$$3y = -x + 6$$

$$\frac{3y}{3} = \frac{-x}{3} + \frac{6}{3}$$

$$y = -\frac{1}{3}x + 2$$

The slope is $m = -\dfrac{1}{3}$, and the y-intercept is $(0, 2)$.

We first plot $(0, 2)$. From this point we move 3 units to the right and 1 unit down, where we plot the point $(3, 1)$. We then sketch the line that contains these two points.

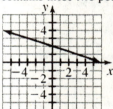

11. $2x + 5y - 20 = 0$

$$5y = -2x + 20$$

$$\frac{5y}{5} = \frac{-2x}{5} + \frac{20}{5}$$

$$y = -\frac{2}{5}x + 4$$

The slope is $m = -\dfrac{2}{5}$, and the y-intercept is

$(0, 4)$. We first plot $(0, 4)$. From this point we move 5 units to the right and 2 units down, where we plot the point $(5, 2)$. We then sketch the line that contains these two points.

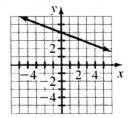

12. $y - 4 = 0$

$y = 4$

The slope is $m = 0$, and the y-intercept is $(0, 4)$. The graph is a horizontal line that passes through the point $(0, 4)$.

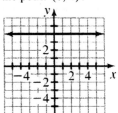

13. $4x - 5y = 20$

To find the x-intercept, we substitute 0 for y and solve for x. To find the y-intercept, we substitute 0 for x and solve for y.

x-intercept	y-intercept
$4x - 5(0) = 20$	$4(0) - 5y = 20$
$4x = 20$	$-5y = 20$
$x = 5$	$y = -4$

The x-intercept is $(5, 0)$; the y-intercept is $(0, -4)$.

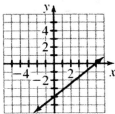

14. $3x + 2y = 6$

To find the x-intercept, we substitute 0 for y and solve for x. To find the y-intercept, we substitute 0 for x and solve for y.

x-intercept	y-intercept
$3x + 2(0) = 6$	$3(0) + 2y = 6$
$3x = 6$	$2y = 6$
$x = 2$	$y = 3$

The x-intercept is $(2, 0)$; the y-intercept is $(0, 3)$.

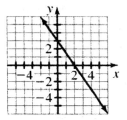

15. $3x + 4y + 12 = 0$

To find the x-intercept, we substitute 0 for y and solve for x. To find the y-intercept, we substitute 0 for x and solve for y.

x-intercept	y-intercept
$3x + 4(0) + 12 = 0$	$3(0) + 4y + 12 = 0$
$3x + 12 = 0$	$4y + 12 = 0$
$3x = -12$	$4y = -12$
$x = -4$	$y = -3$

The x-intercept is $(-4, 0)$; the y-intercept is $(0, -3)$.

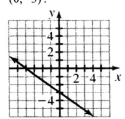

16. $y = 2x - 4$

To find the x-intercept, we substitute 0 for y and solve for x. To find the y-intercept, we substitute 0 for x and solve for y.

x-intercept	y-intercept
$0 = 2x - 4$	$y = 2(0) - 4$
$-2x = -4$	$y = -4$
$x = 2$	

The x-intercept is $(2, 0)$; the y-intercept is $(0, -4)$.

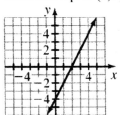

17. $y = -x + 3$

To find the x-intercept, we substitute 0 for y and solve for x. To find the y-intercept, we substitute 0 for x and solve for y.

x-intercept	y-intercept
$0 = -x + 3$	$y = -(0) + 3$
$x = 3$	$y = 3$

The x-intercept is $(3, 0)$; the y-intercept is $(0, 3)$.

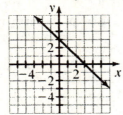

18. $\frac{1}{3}x - \frac{1}{2}y = 1$

To find the x-intercept, we substitute 0 for y and solve for x. To find the y-intercept, we substitute 0 for x and solve for y.

x-intercept	y-intercept
$\frac{1}{3}x - \frac{1}{2}(0) = 1$	$\frac{1}{3}(0) - \frac{1}{2}y = 1$
$\frac{1}{3}x = 1$	$-\frac{1}{2}y = 1$
$x = 3$	$y = -2$

The x-intercept is $(3, 0)$; the y-intercept is $(0, -2)$.

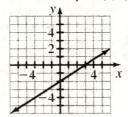

19. $9.2x - 3.8y = 87.2$

To find the x-intercept, we substitute 0 for y and solve for x. To find the y-intercept, we substitute 0 for x and solve for y.

x-intercept	y-intercept
$9.2x - 3.8(0) = 87.2$	$9.2(0) - 3.8y = 87.2$
$9.2x = 87.2$	$-3.8y = 87.2$
$x \approx 9.48$	$x \approx -22.95$

The x-intercept is approximately $(9.48, 0)$;

the y-intercept is approximately $(0, -22.95)$.

20. $y = 2.56x + 97.25$

To find the x-intercept, we substitute 0 for y and solve for x. To find the y-intercept, we substitute 0 for x and solve for y.

x-intercept	y-intercept
$0 = 2.56x + 97.25$	$y = 2.56(0) + 97.25$
$-2.56x = 97.25$	$y = 97.25$
$x \approx -37.99$	

The x-intercept is approximately $(-37.99, 0)$;

the y-intercept is approximately $(0, 97.25)$.

21. For each case, to find the x-intercept, we substitute 0 for y and solve for x. To find the y-intercept, we substitute 0 for x and solve for y.

a. $y = 3x + 7$

x-intercept	y-intercept
$0 = 3x + 7$	$y = 3(0) + 7$
$-3x = 7$	$y = 7$
$x = -\frac{7}{3}$	

The x-intercept is $\left(-\frac{7}{3}, 0\right)$; the y-intercept is $(0, 7)$.

b. $y = 2x + 9$

x-intercept	y-intercept
$0 = 2x + 9$	$y = 2(0) + 9$
$-2x = 9$	$y = 9$
$x = -\frac{9}{2}$	

The x-intercept is $\left(-\frac{9}{2}, 0\right)$; the y-intercept is $(0, 9)$.

c. $y = mx + b$

x-intercept	y-intercept
$0 = mx + b$	$y = m(0) + b$
$-mx = b$	$y = b$
$x = -\frac{b}{m}$	

The x-intercept is $\left(-\frac{b}{m}, 0\right)$; the y-intercept is $(0, b)$.

22. $2x - 4y = 8$

    a. $-4y = -2x + 8$

$$\frac{-4y}{-4} = \frac{-2x}{-4} + \frac{8}{-4}$$

$$y = \frac{1}{2}x - 2$$

The slope is $m = \frac{1}{2}$, and the y-intercept is $(0, -2)$. We first plot $(0, -2)$. From this point we move 2 units to the right and 1 unit up, where we plot the point $(2, -1)$. We then sketch the line that contains these two points.

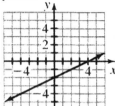

    b. To find the x-intercept, we substitute 0 for y and solve for x. To find the y-intercept, we substitute 0 for x and solve for y.

x-intercept	y-intercept
$2x - 4(0) = 8$	$2(0) - 4y = 8$
$2x = 8$	$-4y = 8$
$x = 4$	$y = -2$

The x-intercept is $(4, 0)$; the y-intercept is $(0, -2)$.

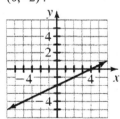

    c. The graphs are the same. Preferences for graphing the equation $2x - 4y = 8$ may vary.

23. We substitute $m = -4$ and $(x_1, y_1) = (2, -1)$ into the point slope form and solve for y:

$$y - y_1 = m(x - x_1)$$
$$y - (-1) = -4(x - 2)$$
$$y + 1 = -4x + 8$$
$$y = -4x + 7$$

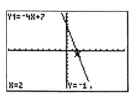

24. We substitute $m = 2$ and $(x_1, y_1) = (-9, -3)$ into the point slope form and solve for y:

$$y - y_1 = m(x - x_1)$$
$$y - (-3) = 2(x - (-9))$$
$$y + 3 = 2(x + 9)$$
$$y + 3 = 2x + 18$$
$$y = 2x + 15$$

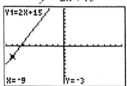

25. We substitute $m = -3$ and $(x_1, y_1) = (-4, 5)$ into the point slope form and solve for y:

$$y - y_1 = m(x - x_1)$$
$$y - 5 = -3(x - (-4))$$
$$y - 5 = -3(x + 4)$$
$$y - 5 = -3x - 12$$
$$y = -3x - 7$$

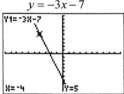

26. We substitute $m = -\frac{2}{5}$ and $(x_1, y_1) = (-3, 6)$ into the point slope form and solve for y:

$$y - y_1 = m(x - x_1)$$
$$y - 6 = -\frac{2}{5}(x - (-3))$$
$$y - 6 = -\frac{2}{5}(x + 3)$$
$$y - 6 = -\frac{2}{5}x - \frac{6}{5}$$
$$y = -\frac{2}{5}x + \frac{24}{5}$$

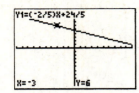

27. We substitute $m = \dfrac{3}{7}$ and $(x_1, y_1) = (2, 9)$ into

the point slope form and solve for *y*:

$$y - y_1 = m(x - x_1)$$

$$y - 9 = \frac{3}{7}(x - 2)$$

$$y - 9 = \frac{3}{7}x - \frac{6}{7}$$

$$y = \frac{3}{7}x + \frac{57}{7}$$

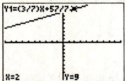

28. We substitute $m = -\dfrac{2}{3}$ and $(x_1, y_1) = (-6, -4)$

into the point slope form and solve for *y*:

$$y - y_1 = m(x - x_1)$$

$$y - (-4) = -\frac{2}{3}(x - (-6))$$

$$y + 4 = -\frac{2}{3}(x + 6)$$

$$y + 4 = -\frac{2}{3}x - 4$$

$$y = -\frac{2}{3}x - 8$$

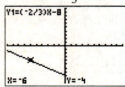

29. Since the slope of the line is undefined, the line must be vertical. Since the *x*-coordinate of the point on the line is 2, an equation of the line is $x = 2$.

30. We substitute $m = 0$ and $(x_1, y_1) = (-1, -4)$ into the point slope form and solve for *y*:

$$y - y_1 = m(x - x_1)$$

$$y - (-4) = 0(x - (-1))$$

$$y + 4 = 0(x + 1)$$

$$y + 4 = 0$$

$$y = -4$$

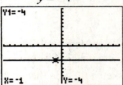

31. We substitute $m = -5.29$ and $(x_1, y_1) = (-4.93, 8.82)$ into the point slope form and solve for *y*:

$$y - y_1 = m(x - x_1)$$

$$y - 8.82 = -5.29(x - (-4.93))$$

$$y - 8.82 = -5.29(x + 4.93)$$

$$y - 8.82 \approx -5.29x - 26.08$$

$$y \approx -5.29x - 17.26$$

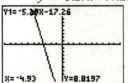

32. We substitute $m = 1.45$ and $(x_1, y_1) = (-2.79, -7.13)$ into the point slope form and solve for *y*:

$$y - y_1 = m(x - x_1)$$

$$y - (-7.13) = 1.45(x - (-2.79))$$

$$y + 7.13 = 1.45(x + 2.79)$$

$$y + 7.13 \approx 1.45x + 4.05$$

$$y \approx 1.45x - 3.08$$

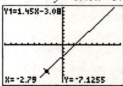

33. We begin by finding the slope of the line:
$$m = \frac{7-1}{5-2} = \frac{6}{3} = 2$$

Then we substitute $m = 2$ and $(x_1, y_1) = (2, 1)$ into the point slope form and solve for y:
$$y - y_1 = m(x - x_1)$$
$$y - 1 = 2(x - 2)$$
$$y - 1 = 2x - 4$$
$$y = 2x - 3$$

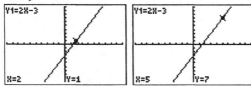

34. We begin by finding the slope of the line:
$$m = \frac{3-9}{1-(-2)} = \frac{3-9}{1+2} = \frac{-6}{3} = -2$$

Then we substitute $m = -2$ and $(x_1, y_1) = (-2, 9)$ into the point slope form and solve for y:
$$y - y_1 = m(x - x_1)$$
$$y - 9 = -2(x - (-2))$$
$$y - 9 = -2(x + 2)$$
$$y - 9 = -2x - 4$$
$$y = -2x + 5$$

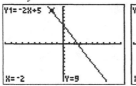 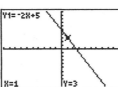

35. We begin by finding the slope of the line:
$$m = \frac{2-(-7)}{1-(-2)} = \frac{2+7}{1+2} = \frac{9}{3} = 3$$

Then we substitute $m = 3$ and $(x_1, y_1) = (-2, -7)$ into the point slope form and solve for y:
$$y - y_1 = m(x - x_1)$$
$$y - (-7) = 3(x - (-2))$$
$$y + 7 = 3(x + 2)$$
$$y + 7 = 3x + 6$$
$$y = 3x - 1$$

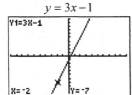 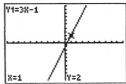

36. We begin by finding the slope of the line:
$$m = \frac{5-(-5)}{4-2} = \frac{5+5}{4-2} = \frac{10}{2} = 5$$

Then we substitute $m = 1$ and $(x_1, y_1) = (2, -5)$ into the point slope form and solve for y:
$$y - y_1 = m(x - x_1)$$
$$y - (-5) = 5(x - 2)$$
$$y + 5 = 5x - 10$$
$$y = 5x - 15$$

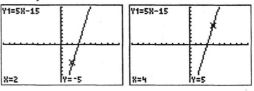

37. We begin by finding the slope of the line:
$$m = \frac{2-8}{-1-(-5)} = \frac{2-8}{-1+5} = \frac{-6}{4} = -\frac{3}{2}$$

Then we substitute $m = -\frac{3}{2}$ and $(x_1, y_1) = (-5, 8)$ into the point slope form and solve for y:
$$y - y_1 = m(x - x_1)$$
$$y - 8 = -\frac{3}{2}(x - (-5))$$
$$y - 8 = -\frac{3}{2}(x + 5)$$
$$y - 8 = -\frac{3}{2}x - \frac{15}{2}$$
$$y = -\frac{3}{2}x + \frac{1}{2}$$

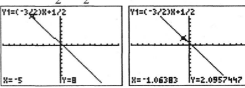

38. We begin by finding the slope of the line:
$$m = \frac{-2-(-5)}{4-(-8)} = \frac{-2+5}{4+8} = \frac{3}{12} = \frac{1}{4}$$

Then we substitute $m = \frac{1}{4}$ and $(x_1, y_1) = (-8, -5)$ into the point slope form and solve for y:
$$y - y_1 = m(x - x_1)$$
$$y - (-5) = \frac{1}{4}(x - (-8))$$
$$y + 5 = \frac{1}{4}(x + 8)$$

$$y + 5 = \frac{1}{4}x + 2$$

$$y = \frac{1}{4}x - 3$$

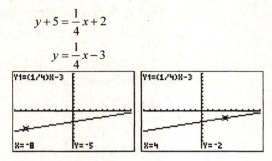

39. We begin by finding the slope of the line:
$$m = \frac{-6 - 9}{6 - (-3)} = \frac{-6 - 9}{6 + 3} = \frac{-15}{9} = -\frac{5}{3}$$

Then we substitute $m = -\frac{5}{3}$ and $(x_1, y_1) = (-3, 9)$

into the point slope form and solve for y:
$$y - y_1 = m(x - x_1)$$

$$y - 9 = -\frac{5}{3}(x - (-3))$$

$$y - 9 = -\frac{5}{3}(x + 3)$$

$$y - 9 = -\frac{5}{3}x - 5$$

$$y = -\frac{5}{3}x + 4$$

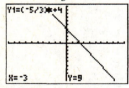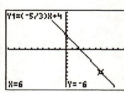

40. We begin by finding the slope of the line:
$$m = \frac{-7 - (-10)}{-2 - (-4)} = \frac{-7 + 10}{-2 + 4} = \frac{3}{2}$$

Then we substitute $m = \frac{3}{2}$ and $(x_1, y_1) = (-3, 9)$

into the point slope form and solve for y:
$$y - y_1 = m(x - x_1)$$

$$y - (-10) = \frac{3}{2}(x - (-4))$$

$$y + 10 = \frac{3}{2}(x + 4)$$

$$y + 10 = \frac{3}{2}x + 6$$

$$y = \frac{3}{2}x - 4$$

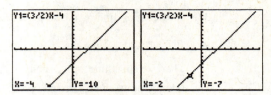

41. Since the x-coordinates of the given points are equal (both 5), the line that contains the points is vertical. An equation of the line is $x = 5$.

42. We begin by finding the slope of the line:
$$m = \frac{-3 - (-3)}{-1 - (-4)} = \frac{-3 + 3}{-1 + 4} = \frac{0}{3} = 0$$

Then we substitute $m = 0$ and $(x_1, y_1) = (-4, -3)$

into the point slope form and solve for y:
$$y - y_1 = m(x - x_1)$$

$$y - (-3) = 0(x - (-4))$$

$$y + 3 = 0(x + 4)$$

$$y + 3 = 0$$

$$y = -3$$

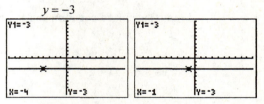

43. We begin by finding the slope of the line:
$$m = \frac{4.8 - 9.2}{8.7 - 3.5} = \frac{-4.4}{5.2} \approx -0.846.$$

So, the equation of the line is of the form $y = -0.846x + b$. To find b, we substitute the coordinates of the point $(3.5, 9.2)$ into the equation and solve for b:
$$y = -0.846x + b$$

$$9.2 = -0.846(3.5) + b$$

$$9.2 = -2.961 + b$$

$$12.161 = b$$

Rounding both m and b to two decimal places, the approximate equation is $y = -0.85x + 12.16$.

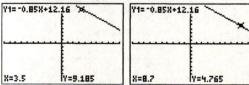

44. We begin by finding the slope of the line:
$$m = \frac{-3.99 - 2.49}{1.83 - (-5.22)} = \frac{-6.48}{7.05} \approx -0.919.$$
So, the equation of the line is of the form
$y = -0.919x + b$. To find b, we substitute the
coordinates of the point $(-5.22, 2.49)$ into the
equation and solve for b:
$$y = -0.919x + b$$
$$2.49 = -0.919(-5.22) + b$$
$$2.49 = 4.79718 + b$$
$$-2.30718 = b$$
Rounding both m and b to two decimal places,
the approximate equation is $y = -0.92x - 2.31$.

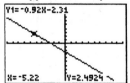 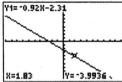

45. The line contains the points $(-2, -3)$ and $(3, -4)$,
so the slope of the line is
$$m = \frac{-4 - (-3)}{3 - (-2)} = \frac{-4 + 3}{3 + 2} = \frac{-1}{5} = -\frac{1}{5}.$$
So, the equation of the line is of the form
$y = -\frac{1}{5}x + b$. To find b, we substitute the
coordinates of the point $(-2, -3)$ into the equation
and solve for b:
$$y = -\frac{1}{5}x + b$$
$$-3 = -\frac{1}{5}(-2) + b$$
$$-3 = \frac{2}{5} + b$$
$$-\frac{17}{5} = b$$
So, the equation is $y = -\frac{1}{5}x - \frac{17}{5}$.

46. Answers may vary. One possibility follows:
We begin by creating a scattergram of the data.

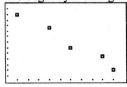

From the scattergram, we see that a line through
the points $(1, 28)$ and $(10, 8)$ will come close to
the rest of the data points. We use these points
to find an equation of the form $y = mx + b$. We
begin by finding the slope:
$$m = \frac{8 - 28}{10 - 1} = \frac{-20}{9} \approx -2.22$$
So, the equation of the line is of the form
$y = -2.22x + b$. To find b, we substitute the
coordinates of the point $(1, 28)$ into the equation
and solve for b:
$$y = -2.22x + b$$
$$28 = -2.22(1) + b$$
$$28 = -2.21 + b$$
$$30.21 = b$$
So, the equation is $y = -2.22x + 30.21$. It comes
close to the data points, as can be seen in the
graph.

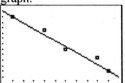

47. a.

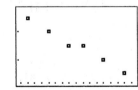

b. Answers may vary depending on the data
points selected. We use the points $(12, 14)$
and $(20, 12)$ to find an equation of the form
$p = mt + b$. We begin by finding the slope:
$$m = \frac{12 - 14}{20 - 12} = \frac{-2}{8} = -0.25$$
So, the equation of the line is of the form
$p = -0.25t + b$. To find b, we substitute the
coordinates of the point $(12, 14)$ into the
equation and solve for b:
$$p = -0.25t + b$$
$$14 = -0.25(12) + b$$
$$14 = -3 + b$$
$$17 = b$$
So, the equation is $p = -0.25t + 17$.

c. Answers may vary [depending on the equation found in part (b)]. We substitute 9 for p and solve for t:

$$p = -0.25t + 17$$
$$9 = -0.25t + 17$$
$$-8 = -0.25t$$
$$t = \frac{-8}{-0.25} = 32$$

Now, $t = 32$ represents the year 2012. Thus, the model predicts that 9% of restaurant patrons will order desert in 2012.

d. Answers may vary [depending on the equation found in part (b)]. To find the t-intercept, we substitute 0 for p and solve for t.

$$p = -0.25t + 17$$
$$0 = -0.25t + 17$$
$$0.25t = 17$$
$$t = 68$$

Thus, the t-intercept is $(68, 0)$. It means that no restaurant patrons will order desert in the year $1980 + 68 = 2048$. Model breakdown has likely occurred.

e. Answers may vary [depending on the equation found in part (b)]. The p-intercept is $(0, 17)$. It means that 17% of restaurant patrons ordered desert in the year 1980.

48. a.

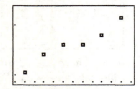

b. Answers may vary depending on the data points selected. We use the points $(2, 2.1)$ and $(12, 4.3)$ to find an equation of the form $s = mt + b$. We begin by finding the slope:

$$m = \frac{4.3 - 2.1}{12 - 2} = \frac{2.2}{10} = 0.22$$

So, the equation of the line is of the form $s = 0.22t + b$. To find b, we substitute the coordinates of the point $(2, 2.1)$ into the equation and solve for b:

$$s = 0.22t + b$$
$$2.1 = 0.22(2) + b$$
$$2.1 = 0.44 + b$$
$$1.66 = b$$

So, the equation is $s = 0.22t + 1.66$.

c. Answers may vary [depending on the equation found in part (b).] The slope is 0.22. It means that the sales are increasing by 220 million units per year.

d. Answers may vary [depending on the equation found in part (b).] The year 2009 is represented by $t = 19$. We substitute 19 for t in the equation:

$$s = 0.22t + 1.66$$
$$s = 0.22(19) + 1.66 = 5.84$$

We predict that approximately 5.84 billion units will be sold in 2009.

e. Answers may vary [depending on the equation found in part (b)]. We substitute 6 for s and solve for t:

$$s = 0.22t + 1.66$$
$$6 = 0.22t + 1.66$$
$$4.34 = 0.22t$$
$$t = \frac{4.34}{0.22} \approx 19.73$$

We predict that sales will be 6 billion units in the year $1990 + 19.73 \approx 2010$.

49. a. The number of living Americans who have be diagnosed with cancer is increasing by 0.3 million Americans per year, so the slope of the equation is $m = 0.3$. Since t represents the number of years since 2000, the n-intercept is $(0, 9.6)$. Thus, $b = 9.6$. An equation that describes the situation is $n = 0.3t + 9.6$.

b. The slope is 0.3. This means that the number of living Americans who have been diagnosed with cancer increases by 0.3 million Americans per year.

c. The year 2010 is represented by $t = 10$. We substitute 10 for t in the equation:

$$n = 0.3t + 9.6$$
$$n = 0.3(10) + 9.6 = 12.6$$

We predict that, in 2010, 12.6 million living Americans will have been diagnosed with cancer.

d. We substitute 14 for n in the equation and solve for t:

$$14 = 0.3t + 9.6$$
$$4.4 = 0.3t$$
$$t = \frac{4.4}{0.3} \approx 14.67$$

We predict that 14 million living Americans will have been diagnosed with cancer in the year $2000 + 14.67 \approx 2015$.

50. Let p represent the percentage of counselors who consider admission tests as being "considerably important" at t years after 1990. Then 1992 is represented by $t = 2$ and 2002 is represented by $t = 12$. We find a linear model by using the data points (2, 39) and (12, 57). We begin by finding the slope:

$$m = \frac{57-39}{12-2} = \frac{18}{10} = 1.8$$

So, the equation of the line is of the form $p = 1.8t + b$. To find b, we substitute the coordinates of the point (2, 39) into the equation and solve for b:

$$p = 1.8t + b$$
$$39 = 1.8(2) + b$$
$$39 = 3.6 + b$$
$$35.4 = b$$

So, the equation is $s = 1.8t + 35.4$. To find when the percentage will be 75, we substitute 75 for p in the equation and solve for t:

$$75 = 1.8t + 35.4$$
$$39.6 = 1.8t$$
$$t = \frac{39.6}{1.8} = 22$$

We predict that 75% of counselors will in the consider admission tests "considerable important" in the year $1990 + 22 = 2012$.

51. $x + 7 > 10$
$$x + 7 - 7 > 10 - 7$$
$$x > 3$$
We graph the solution set, $(3, \infty)$.

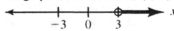

52. $x - 3 \geq -4$
$$x - 3 + 3 \geq -4 + 3$$
$$x \geq -1$$
We graph the solution set, $[-1, \infty)$.

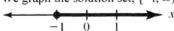

53. $3w \leq -15$
$$\frac{3w}{3} \leq \frac{-15}{3}$$
$$w \leq -5$$
We graph the solution set, $(-\infty, -5]$.

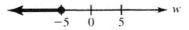

54. $-\frac{4}{3}p < -8$
$$-\frac{3}{4}\left(-\frac{4}{3}p\right) > -\frac{3}{4}(-8)$$
$$p > 6$$
We graph the solution set, $(6, \infty)$.

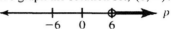

55. $-4x < 8$
$$\frac{-4x}{-4} > \frac{8}{-4}$$
$$x > -2$$
We graph the solution set, $(-2, \infty)$.

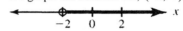

56. $5x - 3 > 3x - 9$
$$5x - 3 - 3x > 3x - 9 - 3x$$
$$2x - 3 > -9$$
$$2x - 3 + 3 > -9 + 3$$
$$2x > -6$$
$$\frac{2x}{2} > \frac{-6}{2}$$
$$x > -3$$
We graph the solution set, $(-3, \infty)$.

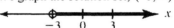

57. $-3(2a + 5) + 5a \geq 2(a - 3)$
$$-6a - 15 + 5a \geq 2a - 6$$
$$-a - 15 \geq 2a - 6$$
$$-a - 15 - 2a \geq 2a - 6 - 2a$$
$$-3a - 15 \geq -6$$
$$-3a - 15 + 15 \geq -6 + 15$$
$$-3a \geq 9$$
$$\frac{-3a}{-3} \leq \frac{9}{-3}$$
$$a \leq -3$$
We graph the solution set, $(-\infty, -3]$.

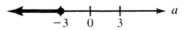

58. $\dfrac{2b-4}{3} \le \dfrac{3b-4}{4}$

$12\left(\dfrac{2b-4}{3}\right) \le 12\left(\dfrac{3b-4}{4}\right)$

$4(2b-4) \le 3(3b-4)$

$8b-16 \le 9b-12$

$8b-16-9b \le 9b-12-9b$

$-b-16 \le -12$

$-b-16+16 \le -12+16$

$-b \le 4$

$\dfrac{-b}{-1} \ge \dfrac{4}{-1}$

$b \ge -4$

We graph the solution set, $[-4, \infty)$.

59. a. Answers may vary depending on the data points selected to create the linear model. One possibility follows: We begin by creating a scattergram of the data.

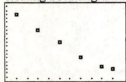

From the scattergram, we see that a line through the points $(6, 42)$ and $(12, 23)$ will come close to the rest of the data points. We use these points to find an equation of the form $r = mt + b$. We begin by finding the slope:

$m = \dfrac{23-42}{12-6} = \dfrac{-19}{6} \approx -3.17$

So, the equation of the line is of the form $r = -3.17t + b$. To find b, we substitute the coordinates of the point $(6, 42)$ into the equation and solve for b:

$r = -3.17t + b$

$42 = -3.17(6) + b$

$42 = -19.02 + b$

$61.02 = b$

So, the equation is $r = -3.17t + 61.02$. It comes close to the data points, as can be seen in the graph.

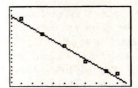

b. Answers may vary [depending on the equation found in part (a)]. The slope is -3.17. It means that the violent crime rate is decreasing by 3.17 violent crimes per 1000 people age 12 or older per year.

c. Answers may vary [depending on the equation found in part (a)]. We find where $r < 11$:

$-3.17t + 61.02 < 11$

$-3.17t + 61.02 - 61.02 < 11 - 61.02$

$-3.17t < -50.02$

$\dfrac{-3.17t}{-3.17} > \dfrac{-50.02}{-3.17}$

$t > 15.78$ (rounded)

Since $1990 + 15.78 \approx 2006$, we predict that the crime rate will be less than 11 violent crimes per 1000 people age 12 or older after the year 2006.

Chapter 5 Test

1. The equation $y = -3x - 1$ is in slope-intercept form, so the slope is $m = -3 = \dfrac{-3}{1}$, and the y-intercept is $(0, -1)$. We first plot $(0, -1)$. From this point we move 1 unit to the right and 3 units down, where we plot the point $(1, -4)$. We then sketch the line that contains these two points.

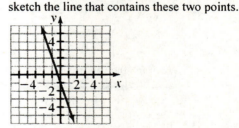

2. $2x - 5y = 10$

$-5y = -2x + 10$

$\dfrac{-5y}{-5} = \dfrac{-2x}{-5} + \dfrac{10}{-5}$

$y = \dfrac{2}{5}x - 2$

The slope is $m = \dfrac{2}{5}$, and the y-intercept is

$(0, -2)$. We first plot $(0, -2)$. From this point we move 5 units to the right and 2 units up, where we plot the point $(5, 0)$. We then sketch the line that contains these two points.

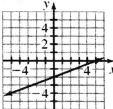

3. $y - 5 = 0$

$\quad\quad y = 5$

The slope is $m = 0$ and the y-intercept is $(0, 5)$. The graph is a horizontal line that passes through the point $(0, 5)$.

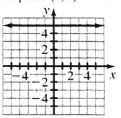

4. $6x - 3y = 18$

To find the x-intercept, we substitute 0 for y and solve for x. To find the y-intercept, we substitute 0 for x and solve for y.

x-intercept	y-intercept
$6x - 3(0) = 18$	$6(0) - 3y = 18$
$6x = 18$	$-3y = 18$
$x = 3$	$y = -6$

The x-intercept is $(3, 0)$; the y-intercept is $(0, -6)$.

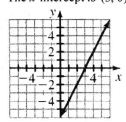

5. $5.93x - 4.81y = 43.79$

To find the x-intercept, we substitute 0 for y and solve for x. To find the y-intercept, we substitute 0 for x and solve for y.

x-intercept	y-intercept
$5.93x - 4.81(0) = 43.79$	$5.93(0) - 4.81y = 43.79$
$5.93x = 43.79$	$-4.81y = 43.79$
$x \approx 7.38$	$y \approx -9.10$

The x-intercept is approximately $(7.38, 0)$; the y-intercept is approximately $(0, -9.10)$.

6. $\dfrac{x}{2} + \dfrac{y}{7} = 1$

To find the x-intercept, we substitute 0 for y and solve for x. To find the y-intercept, we substitute 0 for x and solve for y.

x-intercept	y-intercept
$\dfrac{x}{2} + \dfrac{0}{7} = 1$	$\dfrac{0}{2} + \dfrac{y}{7} = 1$
$\dfrac{x}{2} = 1$	$\dfrac{y}{7} = 1$
$x = 2$	$y = 7$

The x-intercept is $(2, 0)$; the y-intercept is $(0, 7)$.

7. No, the student is not correct because the equation is not in slope-intercept form.

$$5x - 4y = 20$$
$$-4y = -5x + 20$$
$$\dfrac{-4y}{-4} = \dfrac{-5x}{-4} + \dfrac{20}{-4}$$
$$y = \dfrac{5}{4}x - 5$$

The slope is $m = \dfrac{5}{4}$.

8. We substitute $m = 7$ and $(x_1, y_1) = (-2, -4)$ into the point slope form and solve for y:

$$y - y_1 = m(x - x_1)$$
$$y - (-4) = 7(x - (-2))$$
$$y + 4 = 7(x + 2)$$
$$y + 4 = 7x + 14$$
$$y = 7x + 10$$

9. We substitute $m = -\dfrac{2}{3}$ and $(x_1, y_1) = (6, -1)$ into

   the point slope form and solve for y:

   $$y - y_1 = m(x - x_1)$$

   $$y - (-1) = -\frac{2}{3}(x - 6)$$

   $$y + 1 = -\frac{2}{3}x + 4$$

   $$y = -\frac{2}{3}x + 3$$

10. Since the slope of the line is undefined, the line must be vertical. Since the x-coordinate of the point on the line is 2, an equation of the line is $x = 2$.

11. We begin by finding the slope of the line:

    $$m = \frac{3 - 6}{2 - (-4)} = \frac{3 - 6}{2 + 4} = \frac{-3}{6} = -\frac{1}{2}$$

    Then we substitute $m = -\dfrac{1}{2}$ and $(x_1, y_1) = (-4, 6)$

    into the point slope form and solve for y:

    $$y - y_1 = m(x - x_1)$$

    $$y - 6 = -\frac{1}{2}(x - (-4))$$

    $$y - 6 = -\frac{1}{2}(x + 4)$$

    $$y - 6 = -\frac{1}{2}x - 2$$

    $$y = -\frac{1}{2}x + 4$$

12. We begin by finding the slope of the line:

    $$m = \frac{-7.1 - 2.9}{1.8 - (-3.4)} = \frac{-7.1 - 2.9}{1.8 + 3.4} = \frac{-10.0}{5.2} \approx -1.923 .$$

    So, the equation of the line is of the form $y = -1.923x + b$. To find b, we substitute the coordinates of the point $(-3.4, 2.9)$ into the equation and solve for b:

    $$y = -1.923x + b$$

    $$2.9 = -1.923(-3.4) + b$$

    $$2.9 = 6.5382 + b$$

    $$-3.6382 = b$$

    Rounding both m and b to two decimal places, the approximate equation is $y = -1.92x - 3.64$.

13. The line contains the points $(2, -3)$ and $(5, -4)$, so the slope of the line is

    $$m = \frac{-4 - (-3)}{5 - 2} = \frac{-4 + 3}{5 - 2} = \frac{-1}{3} = -\frac{1}{3} .$$

    So, the equation of the line is of the form $y = -\dfrac{1}{3}x + b$. To find b, we substitute the coordinates of the point $(2, -3)$ into the equation and solve for b:

    $$y = -\frac{1}{3}x + b$$

    $$-3 = -\frac{1}{3}(2) + b$$

    $$-3 = -\frac{2}{3} + b$$

    $$-\frac{7}{3} = b$$

    So, the equation is $y = -\dfrac{1}{3}x - \dfrac{7}{3}$.

14.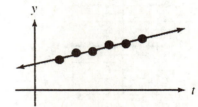

    The value of m must decrease; the value of b must increase.

15. Answers may vary. One possibility follows:
    We begin by creating a scattergram of the data.

    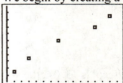

    From the scattergram, we see that a line through the points $(3, 11)$ and $(14, 23)$ will come close to the rest of the data points. We use these points to find an equation of the form $y = mx + b$. We begin by finding the slope:

    $$m = \frac{23 - 11}{14 - 3} = \frac{12}{11} \approx 1.09$$

    So, the equation of the line is of the form $y = 1.09x + b$. To find b, we substitute the coordinates of the point $(3, 11)$ into the equation and solve for b:

$$y = 1.09x + b$$
$$11 = 1.09(3) + b$$
$$11 = 3.27 + b$$
$$7.73 \approx b$$

So, the equation is $y = 1.09x + 7.73$. It comes close to the data points, as can be seen in the graph.

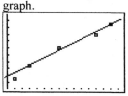

16. a.

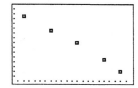

b. Answers may vary depending on the data points selected. We use the points $(10, 85)$ and $(20, 73)$ to find an equation of the form $p = mt + b$. We begin by finding the slope:
$$m = \frac{73 - 85}{20 - 10} = \frac{-12}{10} = -1.2$$

So, the equation of the line is of the form $p = -1.2t + b$. To find b, we substitute the coordinates of the point $(10, 85)$ into the equation and solve for b:
$$p = -1.2t + b$$
$$85 = -1.2(10) + b$$
$$85 = -12 + b$$
$$97 = b$$

So, the equation is $p = -1.2t + 97$.

c. Answers may vary [depending on the equation found in part (b).] For our linear model, the slope is -1.2. It means that the percentage of employers who offer pensions is decreasing by 1.2 percentage points per year.

d. Answers may vary [depending on the equation found in part (b)]. For our linear model, the p-intercept is $(0, 97)$. It means that 97% of employers offered pensions in the year 1980.

e. Answers may vary [depending on the equation found in part (b)]. To find the t-intercept, we substitute 0 for p and solve for t.

$$p = -1.2t + 97$$
$$0 = -1.2t + 97$$
$$1.2t = 97$$
$$t \approx 80.8$$

Thus, for our linear model, the t-intercept is $(80.8, 0)$. It means that no company will offer a pension in the year $1980 + 80.8 \approx 2061$. Model breakdown has likely occurred.

f. Answers may vary [depending on the equation found in part (b)]. We substitute 100 for p and solve for t:
$$p = -1.2t + 97$$
$$100 = -1.2t + 97$$
$$3 = -1.2t$$
$$t = -2.5$$

Thus, the model predicts that 100% of companies offered a pension in the year $1980 + (-2.5) \approx 1977$. A little research would show that this is false. Model breakdown has occurred.

g. Answers may vary [depending on the equation found in part (b)]. We find where $p < 50$:
$$-1.2t + 97 < 50$$
$$-1.2t + 97 - 97 < 50 - 97$$
$$-1.2t < -47$$
$$\frac{-1.2t}{-1.2} > \frac{-47}{-1.2}$$
$$t > 39.17$$

Since $1980 + 39.17 \approx 2020$, we predict that less than half of all employers will offer pensions after the year 2020.

17.
$$3(2x + 1) \leq 4(x + 2) - 1$$
$$6x + 3 \leq 4x + 8 - 1$$
$$6x + 3 \leq 4x + 7$$
$$6x + 3 - 4x \leq 4x + 7 - 4x$$
$$2x + 3 \leq 7$$
$$2x + 3 - 3 \leq 7 - 3$$
$$2x \leq 4$$
$$\frac{2x}{2} \leq \frac{4}{2}$$
$$x \leq 2$$

We graph the solution set, $(-\infty, 2]$.

18.

$$\frac{5}{6} - \frac{2}{3}x > -\frac{1}{2}$$

$$6\left(\frac{5}{6} - \frac{2}{3}x\right) > 6\left(-\frac{1}{2}\right)$$

$$5 - 4x > -3$$

$$5 - 4x - 5 > -3 - 5$$

$$-4x > -8$$

$$\frac{-4x}{-4} < \frac{-8}{-4}$$

$$x < 2$$

We graph the solution set, $(-\infty, 2)$.

Chapter 6
Systems of Linear Equations

Homework 6.1

1. To be a solution to the system, an ordered pair must satisfy both equations in the system.

Check $(2,3)$:

$$y = 4x - 5 \qquad\qquad y = -2x + 1$$

$$3 \overset{?}{=} 4(2) - 5 \qquad\qquad 3 \overset{?}{=} -2(2) + 1$$

$$3 \overset{?}{=} 3 \;\; \text{true} \qquad\qquad 3 \overset{?}{=} -3 \;\; \text{false}$$

Since the ordered pair $(2,3)$ did not satisfy both equations, it is not a solution to the system.

Check $(1,-1)$:

$$y = 4x - 5 \qquad\qquad y = -2x + 1$$

$$-1 \overset{?}{=} 4(1) - 5 \qquad\qquad -1 \overset{?}{=} -2(1) + 1$$

$$-1 \overset{?}{=} -1 \;\; \text{true} \qquad\qquad -1 \overset{?}{=} -1 \;\; \text{true}$$

Since the ordered pair $(1,-1)$ satisfies both equations, it is a solution to the system.

Check $(-4,6)$:

$$y = 4x - 5 \qquad\qquad y = -2x + 1$$

$$6 \overset{?}{=} 4(-4) - 5 \qquad\qquad 6 \overset{?}{=} -2(-4) + 1$$

$$6 \overset{?}{=} -21 \;\; \text{false} \qquad\qquad 6 \overset{?}{=} 9 \;\; \text{false}$$

Since the ordered pair $(-4,6)$ did not satisfy both equations, it is not a solution to the system.

3. To be a solution to the system, an ordered pair must satisfy both equations in the system.

Check $(-1,8)$:

$$5x + 2y = 11 \qquad\qquad 3x - 4y = 17$$

$$5(-1) + 2(8) \overset{?}{=} 11 \qquad 3(-1) - 4(8) \overset{?}{=} 17$$

$$11 \overset{?}{=} 11 \;\; \text{true} \qquad\qquad -35 \overset{?}{=} 17 \;\; \text{false}$$

Since the ordered pair $(-1,8)$ did not satisfy both equations, it is not a solution to the system.

Check $(3,-2)$:

$$5x + 2y = 11 \qquad\qquad 3x - 4y = 17$$

$$5(3) + 2(-2) \overset{?}{=} 11 \qquad 3(3) - 4(-2) \overset{?}{=} 17$$

$$11 \overset{?}{=} 11 \;\; \text{true} \qquad\qquad 17 \overset{?}{=} 17 \;\; \text{true}$$

Since the ordered pair $(3,-2)$ satisfies both equations, it is a solution to the system.

Check $(7,1)$:

$$5x + 2y = 11 \qquad\qquad 3x - 4y = 17$$

$$5(7) + 2(1) \overset{?}{=} 11 \qquad 3(7) - 4(1) \overset{?}{=} 17$$

$$37 \overset{?}{=} 11 \;\; \text{false} \qquad\qquad 17 \overset{?}{=} 17 \;\; \text{true}$$

Since the ordered pair $(7,1)$ did not satisfy both equations, it is not a solution to the system.

5. $y = 3x - 5$

$y = -2x + 5$

To begin, we graph both equations in the same coordinate system.

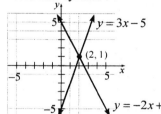

The intersection point is $(2,1)$. So, the solution is the ordered pair $(2,1)$.

Check:

 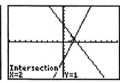

7. $y = \dfrac{1}{2}x + 2$

$y = -\dfrac{3}{2}x - 2$

To begin, we graph both equations in the same coordinate system.

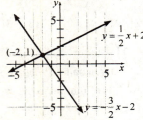

The intersection point is $(-2, 1)$. So, the solution is the ordered pair $(-2, 1)$.

Check:

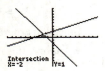

9. $y = -3x$

$y = 4x$

To begin, we graph both equations in the same coordinate system.

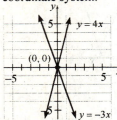

The intersection point is $(0, 0)$. So, the solution is the ordered pair $(0, 0)$.

Check:

$y = -3x$ $y = 4x$

$0 \overset{?}{=} -3(0)$ $0 \overset{?}{=} 4(0)$

$0 = 0$ true $0 = 0$ true

11. $3x + y = 2$

$2x - y = 8$

To begin, we rewrite each equation in slope-intercept form.

$3x + y = 2$ $2x - y = 8$

$\quad y = -3x + 2$ $-y = -2x + 8$

                        $y = 2x - 8$

Next, we graph both equations in the same coordinate system.

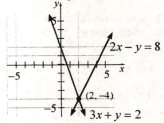

The intersection point is $(2, -4)$. So, the solution is the ordered pair $(2, -4)$.

Check:

$3x + y = 2$ $2x - y = 8$

$3(2) + (-4) \overset{?}{=} 2$ $2(2) - (-4) \overset{?}{=} 8$

$\quad\quad 2 = 2$ true $8 = 8$ true

13. $4y = -3x$

$x - 2y = 10$

To begin, we rewrite each equation in slope-intercept form.

$4y = -3x$ $x - 2y = 10$

$\quad y = -\dfrac{3}{4}x$ $-2y = -x + 10$

                      $y = \dfrac{1}{2}x - 5$

Next, we graph both equations in the same coordinate system.

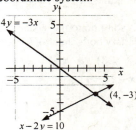

The intersection point is $(4, -3)$. So, the solution is the ordered pair $(4, -3)$.

Check:

$4y = -3x$ $x - 2y = 10$

$4(-3) \overset{?}{=} -3(4)$ $(4) - 2(-3) \overset{?}{=} 10$

$-12 \overset{?}{=} -12$ true $10 \overset{?}{=} 10$ true

15. $4x + 3y = 6$

$2x - 3y = 12$

To begin, we rewrite each equation in slope-intercept form.

$4x + 3y = 6$ $\qquad$ $2x - 3y = 12$

$\quad 3y = -4x + 6$ $\qquad$ $-3y = -2x + 12$

$\quad\quad y = -\dfrac{4}{3}x + 2$ $\qquad$ $y = \dfrac{2}{3}x - 4$

Next, we graph both equations in the same coordinate system.

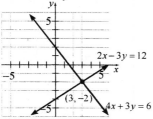

The intersection point is $(3, -2)$. So, the solution is the ordered pair $(3, -2)$.

Check:

$4x + 3y = 6$ $\qquad$ $2x - 3y = 12$

$4(3) + 3(-2) \overset{?}{=} 6$ $\qquad$ $2(3) - 3(-2) \overset{?}{=} 12$

$6 = 6$ true $\qquad\qquad$ $12 = 12$ true

17. $6y + 15x = 12$

$2y - x = -8$

To begin, we rewrite each equation in slope-intercept form.

$6y + 15x = 12$ $\qquad$ $2y - x = -8$

$\quad 6y = -15x + 12$ $\qquad$ $2y = x - 8$

$\quad\quad y = -\dfrac{5}{2}x + 2$ $\qquad$ $y = \dfrac{1}{2}x - 4$

Next, we graph both equations in the same coordinate system.

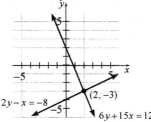

The intersection point is $(2, -3)$. So, the solution is the ordered pair $(2, -3)$.

Check:

$6y + 15x = 12$ $\qquad\qquad$ $2y - x = -8$

$6(-3) + 15(2) \overset{?}{=} 12$ $\qquad$ $2(-3) - (2) \overset{?}{=} -8$

$12 = 12$ true $\qquad\qquad$ $-8 = -8$ true

19. $x = 3$

$y = -2$

To begin, we graph both equations in the same coordinate system.

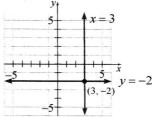

The intersection point is $(3, -2)$. So, the solution is the ordered pair $(3, -2)$.

Check:

$x = 3$ $\qquad\qquad$ $y = -2$

$3 \overset{?}{=} 3$ true $\qquad$ $-2 \overset{?}{=} -2$ true

21. $x = \dfrac{1}{3}y$

$y = -2x + 5$

To begin, we rewrite each equation in slope-intercept form.

$x = \dfrac{1}{3}y$ $\qquad$ $y = -2x + 5$

$y = 3x$

Next, we graph both equations in the same coordinate system.

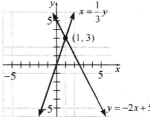

The intersection point is $(1, 3)$. So, the solution is the ordered pair $(1, 3)$.

Check:

$$x = \frac{1}{3}y \qquad\qquad y = -2x + 5$$

$$1 \overset{?}{=} \frac{1}{3}(3) \qquad\qquad 3 \overset{?}{=} -2(1) + 5$$

$$\qquad\qquad\qquad\qquad 3 = 3 \text{ true}$$

$$1 = 1 \text{ true}$$

23. We use "intersect" to find the approximate
solution $(1.16, -2.81)$.

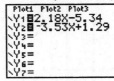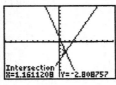

Check:

$$y = 2.18x - 5.34$$

$$-2.81 = 2.18(1.16) - 5.34$$

$$-2.81 \approx -2.8112$$

$$y = -3.53x + 1.29$$

$$-2.81 = -3.53(1.16) + 1.29$$

$$-2.81 \approx -2.8048$$

25. We use "intersect" to find the approximate
solution $(-4.67, -3.83)$.

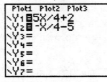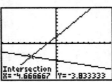

Check:

$$y = \frac{5}{4}x + 2$$

$$-3.83 = \frac{5}{4}(-4.67) + 2$$

$$-3.83 \approx -3.8375$$

$$y = -\frac{1}{4}x - 5$$

$$-3.83 = -\frac{1}{4}(-4.67) - 5$$

$$-3.83 \approx -3.8325$$

27. We first write both equations in slope-intercept
form.

$$y = \frac{3}{4}x - 8 \qquad\qquad 5x + 3y = 6$$

$$\qquad\qquad\qquad\qquad 3y = -5x + 6$$

$$\qquad\qquad\qquad\qquad y = -\frac{5}{3}x + 2$$

We use "intersect" to find the approximate
solution $(4.14, -4.90)$.

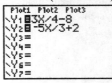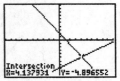

Check:

$$y = \frac{3}{4}x - 8$$

$$-4.90 = \frac{3}{4}(4.14) - 8$$

$$-4.90 \approx -4.895$$

$$5x + 3y = 6$$

$$5(4.14) + 3(-4.90) = 6$$

$$6 = 6$$

29. We first write both equations in slope-intercept
form.

$$-2x + 5y = 15 \qquad\qquad 6x + 14y = -14$$

$$5y = 2x + 15 \qquad\qquad 14y = -6x - 14$$

$$y = \frac{2}{5}x + 3 \qquad\qquad y = -\frac{3}{7}x - 1$$

We use "intersect" to find the approximate
solution $(-4.83, 1.07)$.

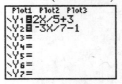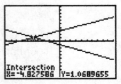

Check:

$$-2x + 5y = 15$$

$$-2(-4.83) + 5(1.07) = 15$$

$$15.01 \approx 15$$

$$6x + 14y = -14$$

$$6(-4.83) + 14(1.07) = -14$$

$$-14 = -14$$

31. $y = -2x + 4$

$6x + 3y = 12$

We first write both equations in slope-intercept form.

$y = -2x + 4 \qquad 6x + 3y = 12$

$\qquad\qquad\qquad 3y = -6x + 12$

$\qquad\qquad\qquad y = -2x + 4$

The two equations are equivalent, so the graph of $y = -2x + 4$ and the graph of $6x + 3y = 12$ are the same line. The solution set of the system is the infinite set of ordered pairs represented by points on the line $y = -2x + 4$. The system is dependent.

33. $y = -2x - 1$

$3x - y = 6$

We first write both equations in slope-intercept form.

$y = -2x - 1 \qquad 3x - y = 6$

$\qquad\qquad\qquad -y = -3x + 6$

$\qquad\qquad\qquad y = 3x - 6$

Next, we graph both equations in the same coordinate system.

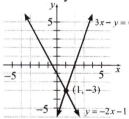

The intersection point is $(1, -3)$. So, the solution is the ordered pair $(1, -3)$.

35. $y = 3x - 4$

$6x - 2y = 6$

We first write both equations in slope-intercept form.

$y = 3x - 4 \qquad 6x - 2y = 6$

$\qquad\qquad\qquad -2y = -6x + 6$

$\qquad\qquad\qquad y = 3x - 3$

Next, we graph both equations in the same coordinate system.

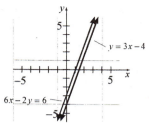

Since the distinct lines have equal slopes, the lines are parallel. Parallel lines do not intersect, so there are no ordered pairs that satisfy both equations. The solution set is the empty set. The system is inconsistent.

37. $n = 13.28t + 440.09$

$n = 3.42t + 468.14$

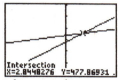

The models predict that there was the same number of women as men earning a bachelor's degree (478 thousand of each) in the year $1980 + 3 = 1983$.

39. a. Using the regression feature of a graphing calculator, we obtain the model

$r = -0.064t + 27.00$.

b. Using the regression feature of a graphinc calculator, we obtain the model

$r = -0.029t + 22.10$.

c. $r = -0.064t + 27.00$

$r = -0.029t + 22.10$

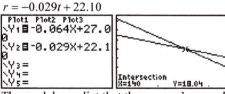

The models predict that the women's record time will equal the men's record time (18.04 seconds) in the year $1900 + 140 = 2040$.

41. a. The ordered pairs B and D are on the graph of the line $y = ax + b$, so they are solutions to the equation.

b. The ordered pairs B and C are on the graph of the line $y = cx + d$, so they are solutions to the equation.

c. The ordered pair B is on the graph of both lines, so it satisfies both equations.

d. The points A, E, and F, do not lie on either line, so they are not solutions to either equation.

43. The solution of the system is the ordered pair that corresponds to the intersection point of the graphs. The graphs appear to intersect at approximately $(2.8, -2.4)$, so $(2.8, -2.4)$ is the approximate solution to the system.

45. First estimate the slope of the lines.

Blue:
The line appears to pass through the points $(-2, 5)$ and $(2, 4)$. The slope is

$$m = \frac{4-5}{2-(-2)} = -\frac{1}{4}.$$

Red:
The line appears to pass through the points $(-2, -3)$ and $(2, -2)$. The slope is

$$m = \frac{-2-(-3)}{2-(-2)} = \frac{1}{4}.$$

Next, use the slopes to find additional points on each line.

Blue	Red
$(6, 3)$	$(6, -1)$
$(10, 2)$	$(10, 0)$
$(14, 1)$	$(14, 1)$
$(18, 0)$	$(18, 2)$

Since the point $(14, 1)$ lies on both lines, the ordered pair $(14, 1)$ is the solution to th system.

47. The input 3 leads to the output 9 for both equations. Therefore, the ordered pair $(3, 9)$ is the solution to the system.

49. Answers may vary. One possibility:
$$y = 2x + 1$$
$$3x - 5y = -12$$

51. The student is not correct. The mistake was made because the student did not check the ordered pair in both equations. Checking in the second equation yields
$$y = 2x - 4$$
$$1 \overset{?}{=} 2(4) - 4$$
$$1 \overset{?}{=} 4 \quad \text{false}$$

53. a. The lines have the same slope, but different y-intercepts. Therefore, the lines are parallel.

b. Parallel lines do not cross so they do not have an intersection point.

c. Since the lines have the same slope, but different y-intercepts, the lines do not have an intersection point. Therefore, the system has no solution.

55. Answers may vary. One possibility:
$$2x + y = -6$$
$$-x - 3y = 8$$

57. $y = 2x + 5$
$y = -3x - 5$

To begin, we graph both equations in the same coordinate system.

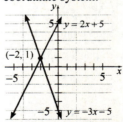

The intersection point is $(-2, 1)$. So, the solution is the ordered pair $(-2, 1)$.

Description: *system of linear equations in two variables*

59. $2x + 5 = -3x - 5$
$$2x = -3x - 10$$
$$5x = -10$$
$$x = -2$$
The solution is -2.
Description: *linear equation in one variable*

Homework 6.2

1. $y = 2x$

$3x + y = 10$

Substitute $2x$ for y in the second equation and solve for x.

$3x + (2x) = 10$

$5x = 10$

$x = 2$

Substitute 2 for x in the equation $y = 2x$ and solve for y.

$y = 2(2) = 4$

The solution is $(2, 4)$.

3. $x - 4y = -3$

$x = 2y - 1$

Substitute $2y - 1$ for x in the first equation and solve for y.

$(2y - 1) - 4y = -3$

$2y - 1 - 4y = -3$

$-2y - 1 = -3$

$-2y = -2$

$y = 1$

Substitute 1 for y in the equation $x = 2y - 1$ and solve for x.

$x = 2(1) - 1 = 2 - 1 = 1$

The solution is $(1, 1)$.

5. $2x + 3y = 5$

$y = x + 5$

Substitute $x + 5$ for y in the first equation and solve for x.

$2x + 3(x + 5) = 5$

$2x + 3x + 15 = 5$

$5x + 15 = 5$

$5x = -10$

$x = -2$

Substitute -2 for x in the equation $y = x + 5$ and solve for y.

$y = (-2) + 5 = 3$

The solution is $(-2, 3)$.

7. $-5x - 2y = 17$

$x = 4y + 1$

Substitute $4y + 1$ for x in the first equation and solve for y.

$-5(4y + 1) - 2y = 17$

$-20y - 5 - 2y = 17$

$-22y - 5 = 17$

$-22y = 22$

$y = -1$

Substitute -1 for y in the equation $x = 4y + 1$ and solve for x.

$x = 4(-1) + 1 = -4 + 1 = -3$

The solution is $(-3, -1)$.

9. $2x - 5y - 3 = 0$

$y = 2x - 7$

Substitute $2x - 7$ for y in the first equation and solve for x.

$2x - 5(2x - 7) - 3 = 0$

$2x - 10x + 35 - 3 = 0$

$-8x + 32 = 0$

$-8x = -32$

$x = 4$

Substitute 4 for x in the equation $y = 2x - 7$ and solve for y.

$y = 2(4) - 7 = 8 - 7 = 1$

The solution is $(4, 1)$.

11. $x = 2y + 6$

$-4x + 5y + 12 = 0$

Substitute $2y + 6$ for x in the second equation and solve for y.

$-4(2y + 6) + 5y + 12 = 0$

$-8y - 24 + 5y + 12 = 0$

$-3y - 12 = 0$

$-3y = 12$

$y = -4$

Substitute -4 for y in the equation $x = 2y + 6$ and solve for x.

$x = 2(-4) + 6 = -8 + 6 = -2$

The solution is $(-2, -4)$.

13. $y = -2x - 1$

$y = 3x + 9$

Substitute $3x + 9$ for y in the first equation and solve for x.

$3x + 9 = -2x - 1$

$5x + 9 = -1$

$5x = -10$

$x = -2$

Substitute -2 for x in the equation $y = 3x + 9$ and solve for y.

$y = 3(-2) + 9 = -6 + 9 = 3$

The solution is $(-2, 3)$.

15. $y = 2x$

$y = 3x$

Substitute $3x$ for y in the first equation and solve for x.

$3x = 2x$

$3x - 2x = 2x - 2x$

$x = 0$

Substitute 0 for x in the equation $y = 3x$ and solve for y.

$y = 3(0) = 0$

The solution is $(0, 0)$.

17. $y = 2(x - 4)$

$y = -3(x + 1)$

Substitute $-3(x + 1)$ for y in the first equation and solve for x.

$-3(x + 1) = 2(x - 4)$

$-3x - 3 = 2x - 8$

$-5x - 3 = -8$

$-5x = -5$

$x = 1$

Substitute 1 for x in the equation $y = -3(x + 1)$ and solve for y.

$y = -3(1 + 1) = -3(2) = -6$

The solution $(1, -6)$.

19. $y = 2.57x + 7.09$

$y = -3.61x - 5.72$

Substitute $-3.61x - 5.72$ for y in the first equation and solve for x.

$-3.61x - 5.72 = 2.57x + 7.09$

$-6.18x - 5.72 = 7.09$

$-6.18x = 12.81$

$x = \dfrac{12.81}{-6.18} \approx -2.073$

Substitute -2.073 for x in the equation $y = -3.61x - 5.72$ and solve for y.

$y = -3.61(-2.073) - 5.72 \approx 1.764$

The approximate solution is $(-2.07, 1.76)$.

21. $y = -3.17x + 8.92$

$y = 1.65x - 7.24$

Substitute $1.65x - 7.24$ for y in the first equation and solve for x.

$1.65x - 7.24 = -3.17x + 8.92$

$4.82x - 7.24 = 8.92$

$4.82x = 16.16$

$x = \dfrac{16.16}{4.82} \approx 3.353$

Substitute 3.353 for x in the equation $y = 1.65x - 7.24$ and solve for y.

$y = 1.65(3.353) - 7.24 \approx -1.708$

The approximate solution is $(3.35, -1.71)$.

23. $y = -1.82x + 3.95$

$y = 1.57x + 4.68$

Substitute $1.57x + 4.68$ for y in the first equation and solve for x.

$1.57x + 4.68 = -1.82x + 3.95$

$3.39x + 4.68 = 3.95$

$3.39x = -0.73$

$x = \dfrac{-0.73}{3.39} \approx -0.215$

Substitute -0.215 for x in the equation $y = 1.57x + 4.68$ and solve for y.

$y = 1.57(-0.215) + 4.68 \approx 4.342$

The approximate solution is $(-0.22, 4.34)$.

25. $2x + y = -9$
$5x - 3y = 5$
Solve the first equation for y.
$2x + y = -9$
$\quad y = -2x - 9$
Substitute $-2x - 9$ for y in the second equation and solve for x.
$5x - 3(-2x - 9) = 5$
$\quad 5x + 6x + 27 = 5$
$\quad\quad 11x + 27 = 5$
$\quad\quad\quad 11x = -22$
$\quad\quad\quad\quad x = -2$
Substitute -2 for x in the equation $y = -2x - 9$ and solve for y.
$y = -2(-2) - 9 = 4 - 9 = -5$
The solution is $(-2, -5)$.

27. $4x - 7y = 15$
$x - 3y = 5$
Solve the second equation for x.
$x - 3y = 5$
$\quad x = 3y + 5$
Substitute $3y + 5$ for x in the first equation and solve for y.
$4(3y + 5) - 7y = 15$
$\quad 12y + 20 - 7y = 15$
$\quad\quad 5y + 20 = 15$
$\quad\quad\quad 5y = -5$
$\quad\quad\quad y = -1$
Substitute -1 for y in the equation $x = 3y + 5$ and solve for x.
$x = 3(-1) + 5 = -3 + 5 = 2$
The solution is $(2, -1)$.

29. $4x + 3y = 5$
$x - 2y = -7$
Solve the second equation for x.
$x - 2y = -7$
$\quad x = 2y - 7$
Substitute $2y - 7$ for x in the first equation and solve for y.

$4(2y - 7) + 3y = 5$
$\quad 8y - 28 + 3y = 5$
$\quad\quad 11y - 28 = 5$
$\quad\quad\quad 11y = 33$
$\quad\quad\quad\quad y = 3$
Substitute 3 for y in the equation $x = 2y - 7$ and solve for x.
$x = 2(3) - 7 = 6 - 7 = -1$
The solution is $(-1, 3)$.

31. $2x - y = 1$
$5x - 3y = 5$
Solve the first equation for y.
$2x - y = 1$
$\quad -y = -2x + 1$
$\quad\quad y = 2x - 1$
Substitute $2x - 1$ for y in the second equation and solve for x.
$5x - 3(2x - 1) = 5$
$\quad 5x - 6x + 3 = 5$
$\quad\quad -x + 3 = 5$
$\quad\quad\quad -x = 2$
$\quad\quad\quad\quad x = -2$
Substitute -2 for x in the equation $y = 2x - 1$ and solve for y.
$y = 2(-2) - 1 = -4 - 1 = -5$
The solution is $(-2, -5)$.

33. $3x + 2y = -3$
$\quad 2x = y + 5$
Solve the second equation for y.
$\quad 2x = y + 5$
$2x - 5 = y$
Substitute $2x - 5$ for y in the first equation and solve for x.
$3x + 2(2x - 5) = -3$
$\quad 3x + 4x - 10 = -3$
$\quad\quad 7x - 10 = -3$
$\quad\quad\quad 7x = 7$
$\quad\quad\quad\quad x = 1$
Substitute 1 for x in the equation $y = 2x - 5$ and solve for y.
$y = 2(1) - 5 = 2 - 5 = -3$
The solution is $(1, -3)$.

35.
$$x = 4 - 3y$$
$$2x + 6y = 8$$
Substitute $4 - 3y$ for x in the second equation and solve for y.
$$2(4 - 3y) + 6y = 8$$
$$8 - 6y + 6y = 8$$
$$8 = 8 \text{ true}$$
The result is a true statement, so the system is dependent. The solution is the infinite set of ordered pairs that satisfy the equation $x = 4 - 3y$.

37. $5x - 2y = 18$
$$y = -3x + 2$$
Substitute $-3x + 2$ for y in the first equation and solve for x.
$$5x - 2(-3x + 2) = 18$$
$$5x + 6x - 4 = 18$$
$$11x - 4 = 18$$
$$11x = 22$$
$$x = 2$$
Substitute 2 for x in the equation $y = -3x + 2$ and solve for y.
$$y = -3(2) + 2 = -6 + 2 = -4$$
The solution is $(2, -4)$.

39.
$$y = -5x + 3$$
$$15x + 3y = 6$$
Substitute $-5x + 3$ for y in the second equation and solve for x.
$$15x + 3(-5x + 3) = 6$$
$$15x - 15x + 9 = 6$$
$$9 = 6 \text{ false}$$
The result is a false statement, so the system is inconsistent. The solution is the empty set, $\varnothing$.

41. $-4x + 12y = 4$
$$x = 3y - 1$$
Substitute $3y - 1$ for x in the first equation and solve for y.
$$-4(3y - 1) + 12y = 4$$
$$-12y + 4 + 12y = 4$$
$$4 = 4 \text{ true}$$
The result is a true statement, so the system is dependent. The solution is the infinite set of ordered pairs that satisfy the equation $x = 3y - 1$.

43.
$$y = 3x + 2$$
$$12x - 4y = 9$$
Substitue $3x + 2$ for y in the second equation and solve for x.
$$12x - 4(3x + 2) = 9$$
$$12x - 12x - 8 = 9$$
$$-8 = 9 \text{ false}$$
The result is a false statement, so the system is inconsistent. The solution is the empty set, $\varnothing$.

45. We begin by finding the equations for each table.
Table 13:
The input $x = 0$ yields the output $y = 5$, so the y-intercept is $(0, 5)$.
$$m = \frac{8 - 5}{1 - 0} = \frac{3}{1} = 3$$, so the slope is 3.
The first equation is $y = 3x + 5$.

Table 14:
The input $x = 0$ yields the output $y = 90$, so the y-intercept is $(0, 90)$.
$$m = \frac{88 - 90}{1 - 0} = \frac{-2}{1} = -2$$, so the slope is -2.
The second equation is $y = -2x + 90$.

Now we solve the system
$$y = 3x + 5$$
$$y = -2x + 90$$
Substitute $-2x + 90$ for y in the first equation and solve for x.
$$-2x + 90 = 3x + 5$$
$$-5x + 90 = 5$$
$$-5x = -85$$
$$x = 17$$
Substitute 17 for x in the equation $y = -2x + 90$ and solve for y.
$$y = -2(17) + 90 = -34 + 90 = 56$$
The solution is $(17, 56)$.

47. A: This point is the origin so it has coordinates $(0, 0)$.

B: This point is the y-intercept for line l_1. Let $x = 0$ in the equation and solve for y.
$$y = -(0) + 8 = 0 + 8 = 8$$
The coordinates of the point are $(0, 8)$.

C : This point is the intersection of the two lines. Therefore, it is the solution of the system
$$y = -x + 8$$
$$y = 2x - 7$$
Substitute $2x - 7$ for y in the first equation and solve x.
$$2x - 7 = -x + 8$$
$$3x - 7 = 8$$
$$3x = 15$$
$$x = 5$$
Substitute 5 for x in the equation $y = 2x - 7$ and solve for y.
$$y = 2(5) - 7 = 10 - 7 = 3$$
The coordinates of the point are $(5,3)$.

D : This point is the x-intercept of line l_2. Let $y = 0$ in the equation and solve for x.
$$0 = 2x - 7$$
$$7 = 2x$$
$$\frac{7}{2} = x$$
The coordinates of the point are $\left(\frac{7}{2}, 0 \right)$.

49. **a.** Solve the first equation for x.
$$x + y = 3$$
$$x = -y + 3$$
Substitute $-y + 3$ for x in the second equation and solve for y.
$$3(-y + 3) + 2y = 7$$
$$-3y + 9 + 2y = 7$$
$$-y + 9 = 7$$
$$-y = -2$$
$$y = 2$$
Substitute 2 for y in the equation $x = -y + 3$ and solve for x.
$$x = -(2) + 3 = -2 + 3 = 1$$
The solution is $(1,2)$.

b. Solve the first equation for y.
$$x + y = 3$$
$$y = -x + 3$$
Substitute $-x + 3$ for y in the second equation and solve for x.

$$3x + 2(-x + 3) = 7$$
$$3x - 2x + 6 = 7$$
$$x + 6 = 7$$
$$x = 1$$
Substitute 1 for x in the equation $y = -x + 3$ and solve for y.
$$y = -(1) + 3 = -1 + 3 = 2$$
The solution is $(1,2)$.

c. The results are the same. The solution to the system is $(1,2)$.

51. Answers may vary.

53. **a.** $y = 2x - 3$
$$y = -3x + 7$$

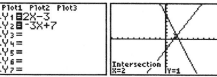

The intersection point is $(2,1)$, so the ordered pair $(2,1)$ is the solution to the system.

b. Substitute $-3x + 7$ for y in the first equation and solve for x.
$$-3x + 7 = 2x - 3$$
$$-5x + 7 = -3$$
$$-5x = -10$$
$$x = 2$$
Substitute 2 for x in the equation $y = -3x + 7$ and solve for y.
$$y = -3(2) + 7 = -6 + 7 = 1$$
The solution is $(2,1)$.

c. $2x - 3 = -3x + 7$
$$5x - 3 = 7$$
$$5x = 10$$
$$x = 2$$
The solution is 2.

d. The solution to part (c) is the same as the x-coordinate of the solution in part (b).

e. Set each side of the equation equal to y to create a system of equations.
$$y = x - 1$$
$$y = -2x + 8$$
Use "intersect" to find the intersection point of the two lines.

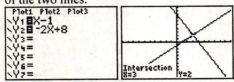

The intersection point is $(3, 2)$, so the x-coordinate is 3. The solution of the equation $x - 1 = -2x + 8$ is 3.
$$x - 1 = -2x + 8$$
$$3x - 1 = 8$$
$$3x = 9$$
$$x = 3$$
The solution is 3.

The answers are the same.

f. Answers may vary.

55. The equation $y = -3x + 5$ is in slope-intercept form, so the slope is $m = -3$ and the y-intercept is $(0, 5)$. We first plot $(0, 5)$. From this point we move 1 unit to the right and 3 units down, plotting the point $(1, 2)$. We then draw the line that connects the two points, extending in both directions.

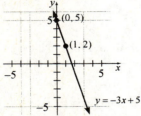

Description: *linear equation in two variables*

57. $9x + 2y = 1$
$$y = -3x + 5$$
Substitute $-3x + 5$ for y in the first equation and solve for x.
$$9x + 2(-3x + 5) = 1$$
$$9x - 6x + 10 = 1$$
$$3x + 10 = 1$$
$$3x = -9$$
$$x = -3$$
Substitute -3 for x in the equation $y = -3x + 5$ and solve for y.
$$y = -3(-3) + 5 = 9 + 5 = 14$$
The solution is $(-3, 14)$. Description: *system of two linear equations in two variables*

Homework 6.3

1. $2x + 3y = 7$ Equation (1)
   $-2x + 5y = 1$ Equation (2)
   The coefficients of the x terms are equal in absolute value and opposite in sign. Add the left sides and the right sides of the equations and solve for y.
$$2x + 3y = 7$$
$$\underline{-2x + 5y = 1}$$
$$8y = 8$$
$$y = 1$$
   Substitute 1 for y in equation (1) and solve for x.
$$2x + 3(1) = 7$$
$$2x + 3 = 7$$
$$2x = 4$$
$$x = 2$$
   The solution is $(2, 1)$.

3. $5x - 2y = 2$ Equation (1)
   $-3x + 2y = 2$ Equation (2)
   The coefficients of the y terms are equal in absolute value and opposite in sign. Add the left sides and the right sides of the equations and solve for x.
$$5x - 2y = 2$$
$$\underline{-3x + 2y = 2}$$
$$2x = 4$$
$$x = 2$$

Substitute 2 for x in equation (2) and solve for y.

$$-3(2) + 2y = 2$$
$$-6 + 2y = 2$$
$$2y = 8$$
$$y = 4$$

The solution is $(2, 4)$.

5. $\quad x + 2y = -4 \quad$ Equation (1)

$\quad 3x - 4y = 18 \quad$ Equation (2)

To eliminate the y terms, we multiply both sides of equation (1) by 2, yielding the system

$$2x + 4y = -8$$
$$3x - 4y = 18$$

The coefficients of the y terms are equal in absolute value and opposite in sign. Add the left sides and the right sides of the equations and solve for x.

$$2x + 4y = -8$$
$$\underline{3x - 4y = 18}$$
$$5x = 10$$
$$x = 2$$

Substitute 2 for x in equation (1) and solve for y.

$$(2) + 2y = -4$$
$$2y = -6$$
$$y = -3$$

The solution is $(2, -3)$.

7. $\quad 2x - 3y = 8 \quad$ Equation (1)

$\quad 5x + 6y = -7 \quad$ Equation (2)

To eliminate the y terms, we multiply both sides of equation (1) by 2, yielding the system

$$4x - 6y = 16$$
$$5x + 6y = -7$$

The coefficients of the y terms are equal in absolute value and opposite in sign. Add the left sides and the right sides of the equations and solve for x.

$$4x - 6y = 16$$
$$\underline{5x + 6y = -7}$$
$$9x = 9$$
$$x = 1$$

Substitute 1 for x in equation (2) and solve for y.

$$5(1) + 6y = -7$$
$$5 + 6y = -7$$
$$6y = -12$$
$$y = -2$$

The solution is $(1, -2)$.

9. $\quad 6x - 5y = 4 \quad$ Equation (1)

$\quad 2x + 3y = -8 \quad$ Equation (2)

To eliminate the x terms, we multiply both sides of equation (2) by -3, yielding the system

$$6x - 5y = 4$$
$$-6x - 9y = 24$$

The coefficients of the x terms are equal in absolute value and opposite in sign. Add the left sides and the right sides of the equations and solve for y.

$$6x - 5y = 4$$
$$\underline{-6x - 9y = 24}$$
$$-14y = 28$$
$$y = -2$$

Substitute -2 for y in equation (2) and solve for x.

$$2x + 3(-2) = -8$$
$$2x - 6 = -8$$
$$2x = -2$$
$$x = -1$$

The solution is $(-1, -2)$.

11. $\quad 5x + 7y = -16 \quad$ Equation (1)

$\quad 2x - 5y = 17 \quad$ Equation (2)

To eliminate the y terms, we multiply both sides of equation (1) by 5 and both sides of equation (2) by 7, yielding the system

$$25x + 35y = -80$$
$$14x - 35y = 119$$

The coefficients of the y terms are equal in absolute value and opposite in sign. Add the left sides and the right sides of the equations and solve for x.

$$25x + 35y = -80$$
$$\underline{14x - 35y = 119}$$
$$39x = 39$$
$$x = 1$$

Substitute 1 for x in equation (2) and solve for y.

$$2(1) - 5y = 17$$
$$2 - 5y = 17$$
$$-5y = 15$$
$$y = -3$$

The solution is $(1, -3)$.

13. $-8x + 3y = 1$ Equation (1)

$3x - 4y = 14$ Equation (2)

To eliminate the y terms, we multiply both sides of equation (1) by 4 and both sides of equation (2) by 3, yielding the system

$$-32x + 12y = 4$$
$$9x - 12y = 42$$

The coefficients of the y terms are equal in absolute value and opposite in sign. Add the left sides and the right sides of the equations and solve for x.

$$-32x + 12y = 4$$
$$\underline{9x - 12y = 42}$$
$$-23x = 46$$
$$x = -2$$

Substitute -2 for x in equation (1) and solve for y.

$$-8(-2) + 3y = 1$$
$$16 + 3y = 1$$
$$3y = -15$$
$$y = -5$$

The solution is $(-2, -5)$.

15. $y = 3x - 6$ → $3x - y = 6$ Equation (1)

$y = -4x + 1$ $4x + y = 1$ Equation (2)

The coefficients of the y terms are equal in absolute value and opposite in sign. Add the left sides and the right sides of the equations and solve for x.

$$3x - y = 6$$
$$\underline{4x + y = 1}$$
$$7x = 7$$
$$x = 1$$

Substitute 1 for x in equation (1) and solve for y.

$$y = 3(1) - 6$$
$$= 3 - 6$$
$$= -3$$

The solution is $(1, -3)$.

17. $3x + 6y - 18 = 0$ → $3x + 6y = 18$ Equation (1)

$17 = 7x + 9y$ $7x + 9y = 17$ Equation (2)

To eliminate the y terms, we multiply both sides of equation (1) by 3 and both sides of equation (2) by -2, yielding the system

$$9x + 18y = 54$$
$$-14x - 18y = -34$$

The coefficients of the y terms are equal in absolute value and opposite in sign. Add the left sides and the right sides of the equations and solve for x.

$$9x + 18y = 54$$
$$\underline{-14x - 18y = -34}$$
$$-5x = 20$$
$$x = -4$$

Substitute -4 for x in equation (1) and solve for y.

$$3(-4) + 6y - 18 = 0$$
$$-12 + 6y - 18 = 0$$
$$6y - 30 = 0$$
$$6y = 30$$
$$y = 5$$

The solution is $(-4, 5)$

19. $\dfrac{2}{9}x + \dfrac{1}{3}y = 4$ → $2x + 3y = 36$ Equation (1)

$\dfrac{1}{2}x - \dfrac{2}{5}y = -\dfrac{5}{2}$ → $5x - 4y = -25$ Equation (2)

To eliminate the y terms, we multiply both sides of equation (1) by 4 and both sides of equation (2) by 3, yielding the system

$$8x + 12y = 144$$
$$15x - 12y = -75$$

The coefficients of the y terms are equal in absolute value and opposite in sign. Add the left sides and the right sides of the equations and solve for x.

$$8x + 12y = 144$$
$$\underline{15x - 12y = -75}$$
$$23x = 69$$
$$x = 3$$

Substitute 3 for x in equation (1) and solve for y.

$$2(3) + 3y = 36$$
$$6 + 3y = 36$$
$$3y = 30$$
$$y = 10$$

The solution is $(3, 10)$.

21. $4x - 7y = 3$ Equation (1)

$8x - 14y = 6$ Equation (2)

To eliminate the x terms, we multiply both sides of equation (1) by -2, yielding the system

$-8x + 14y = -6$

$8x - 14y = 6$

The coefficients of the x terms are equal in absolute value and opposite in sign. Add the left sides and the right sides of the equations and solve for y.

$-8x + 14y = -6$

$\underline{8x - 14y = 6}$

$\qquad 0 = 0$ true

The result is a true statement of the form $a = a$. Therefore, the system is dependent. The solution set is the infinite set of ordered pairs that satisfy the equation $4x - 7y = 3$.

23. $8x - 6y = 4$ Equation (1)

$12x - 9y = 5$ Equation (2)

To eliminate the x terms, we multiply both sides of equation (1) by 3 and both sides of equation (2) by -2, yielding the system

$24x - 18y = 12$

$-24x + 18y = -10$

The coefficients of the x terms are equal in absolute value and opposite in sign. Add the left sides and the right sides of the equations and solve for y.

$24x - 18y = 12$

$\underline{-24x + 18y = -10}$

$\qquad 0 = 2$ false

The result is a false statement. Therefore, the system is inconsistent. The solution set is the empty set, $\varnothing$.

25. $3x - 2y = -14$ Equation (1)

$6x + 5y = -19$ Equation (2)

To eliminate the x terms, we multiply both sides of equation (1) by -2, yielding the system

$-6x + 4y = 28$

$6x + 5y = -19$

The coefficients of the x terms are equal in absolute value and opposite in sign. Add the left sides and the right sides of the equations and solve for y.

$-6x + 4y = 28$

$\underline{6x + 5y = -19}$

$\qquad 9y = 9$

$\qquad y = 1$

Substitute 1 for y in equation (1) and solve for x.

$3x - 2(1) = -14$

$3x - 2 = -14$

$3x = -12$

$x = -4$

The solution is $(-4, 1)$.

27. $6x - 15y = 7$

$-4x + 10y = -5$

To eliminate the x terms, we multiply both sides of equation (1) by 2 and both sides of equation (2) by 3, yielding the system

$12x - 30y = 14$

$-12x + 30y = -15$

The coefficients of the x terms are equal in absolute value and opposite in sign. Add the left sides and the right sides of the equations and solve for y.

$12x - 30y = 14$

$\underline{-12x + 30y = -15}$

$\qquad 0 = -1$ false

The result is a false statement. Therefore, the system is inconsistent. The solution set is the empty set, $\varnothing$.

29. $3x - 9y = 12$

$-4x + 12y = -16$

To eliminate the x terms, we multiply both sides of equation (1) by 4 and both sides of equation (2) by 3, yielding the system

$12x - 36y = 48$

$-12x + 36y = -48$

The coefficients of the x terms are equal in absolute value and opposite in sign. Add the left sides and the right sides of the equations and solve for y.

$12x - 36y = 48$

$\underline{-12x + 36y = -48}$

$\qquad 0 = 0$ true

The result is a true statement of the form $a = a$. Therefore, the system is dependent. The solution set is the infinite set of ordered pairs that satisfy the equation $3x - 9y = 12$.

31. $y = 4.29x - 8.91$ Equation (1)

$y = -1.26x + 9.75$ Equation (2)

$\rightarrow 4.29x - y = 8.91$

$1.26x + y = 9.75$

The coefficients of the y terms are equal in absolute value and opposite in sign. Add the left sides and the right sides of the equations and solve for x.

$4.29x - y = 8.91$

$\underline{1.26x + y = 9.75}$

$5.55x = 18.66$

$x = \dfrac{18.66}{5.55} \approx 3.362$

Substitute 3.362 for x in equation (1) and solve for y.

$y = 4.29(3.362) - 8.91 \approx 5.513$

The approximate solution is $(3.36, 5.51)$.

33. $y = -2.15x + 8.38$ Equation (1)

$y = 1.67x + 2.57$ Equation (2)

$\rightarrow 2.15x + y = 8.38$

$1.67x - y = -2.57$

The coefficients of the y terms are equal in absolute value and opposite in sign. Add the left sides and the right sides of the equations and solve for x.

$2.15x + y = 8.38$

$\underline{1.67x - y = -2.57}$

$3.82x = 5.81$

$x = \dfrac{5.81}{3.82} \approx 1.521$

Substitute 1.521 for x in equation (1) and solve for y.

$y = -2.15(1.521) + 8.38 \approx 5.110$

The approximate solution is $(1.52, 5.11)$.

35. $y = -2.62x + 7.24$ Equation (1)

$y = 1.89x - 6.44$ Equation (2)

$\rightarrow 2.62x + y = 7.24$

$1.89x - y = 6.44$

The coefficients of the y terms are equal in absolute value and opposite in sign. Add the left sides and the right sides of the equations and solve for x.

$2.62x + y = 7.24$

$\underline{1.89x - y = 6.44}$

$4.51x = 13.68$

$x = \dfrac{13.68}{4.51} \approx 3.033$

Substitute 3.033 for x in equation (1) and solve for y.

$y = -2.62(3.033) + 7.24 \approx -0.706$

The approximate solution is $(3.03, -0.71)$.

37. $4x - y = -12$ Equation (1)

$3x + 5y = 14$ Equation (2)

To eliminate the y terms, we multiply both sides of equation (1) by 5, yielding the system

$20x - 5y = -60$

$3x + 5y = 14$

The coefficients of the y terms are equal in absolute value and opposite in sign. Add the left sides and the right sides of the equations and solve for x.

$20x - 5y = -60$

$\underline{3x + 5y = 14}$

$23x = -46$

$x = -2$

Substitute -2 for x in equation (1) and solve for y.

$4(-2) - y = -12$

$-8 - y = -12$

$-y = -4$

$y = 4$

The solution is $(-2, 4)$.

39. $-2x + 7y = -3$

$x = 3y + 2$

Substitute $3y + 2$ for x in the first equation and solve for y.

$-2(3y + 2) + 7y = -3$

$-6y - 4 + 7y = -3$

$y - 4 = -3$

$y = 1$

Substitute 1 for y in the equation $x = 3y + 2$ and solve for x.

$x = 3(1) + 2 = 3 + 2 = 5$

The solution is $(5, 1)$.

41. $2x + 7y = 13$ Equation (1)

$3x - 4y = -24$ Equation (2)

To eliminate the x terms, we multiply both sides of equation (1) by 3 and both sides of equation (2) by -2, yielding the system

$6x + 21y = 39$

$-6x + 8y = 48$

The coefficients of the x terms are equal in absolute value and opposite in sign. Add the left sides and the right sides of the equations and solve for y.

$6x + 21y = 39$

$\underline{-6x + 8y = 48}$

$\qquad 29y = 87$

$\qquad\quad y = 3$

Substitute 3 for y in equation (1) and solve for x.

$2x + 7(3) = 13$

$2x + 21 = 13$

$2x = -8$

$x = -4$

The solution is $(-4, 3)$.

43. $2x - 7y = -1$ Equation (1)

$-x - 3y = 7$ Equation (2)

To eliminate the x terms, we multiply both sides of equation (2) by 2, yielding the system

$2x - 7y = -1$

$-2x - 6y = 14$

The coefficients of the x terms are equal in absolute value and opposite in sign. Add the left sides and the right sides of the equations and solve for y.

$2x - 7y = -1$

$\underline{-2x - 6y = 14}$

$-13y = 13$

$\qquad y = -1$

Substitute -1 for y in equation (1) and solve for x.

$2x - 7(-1) = -1$

$2x + 7 = -1$

$2x = -8$

$x = -4$

The solution is $(-4, -1)$.

45. $y = -2x - 3$

$y = 3x + 7$

Substitute $3x + 7$ for y in the first equation and solve for x.

$3x + 7 = -2x - 3$

$5x + 7 = -3$

$5x = -10$

$x = -2$

Substitute -2 for x in the equation $y = 3x + 7$ and solve for y.

$y = 3(-2) + 7 = -6 + 7 = 1$

The solution is $(-2, 1)$.

47. $8x + 5y = 7$ Equation (1)

$7y = -6x + 15$ Equation (2)

$\rightarrow\ 8x + 5y = 7$

$6x + 7y = 15$

To eliminate the x terms, we multiply both sides of equation (1) by 3 and both sides of equation (2) by -4, yielding the system

$24x + 15y = 21$

$-24x - 28y = -60$

The coefficients of the x terms are equal in absolute value and opposite in sign. Add the left sides and the right sides of the equations and solve for y.

$24x + 15y = 21$

$\underline{-24x - 28y = -60}$

$\qquad -13y = -39$

$\qquad\quad y = 3$

Substitute 3 for y in equation (1) and solve for x.

$8x + 5(3) = 7$

$8x + 15 = 7$

$8x = -8$

$x = -1$

The solution is $(-1, 3)$.

49. $3(2x - 5) + 4y = 11$

$5x - 2(3y + 1) = 1$

Begin by writing each equation in standard form.

$3(2x - 5) + 4y = 11 \qquad 5x - 2(3y + 1) = 1$

$6x - 15 + 4y = 11 \qquad\ 5x - 6y - 2 = 1$

$6x + 4y = 26 \qquad\qquad 5x - 6y = 3$

This yields the system

$6x + 4y = 26$ Equation (1)

$5x - 6y = 3$ Equation (2)

To eliminate the y terms, we multiply both sides of equation (1) by 3 and both sides of equation (2) by 2, yielding the system

$18x + 12y = 78$

$10x - 12y = 6$

The coefficients of the y terms are equal in absolute value and opposite in sign. Add the left sides and the right sides of the equations and solve for x.

$18x + 12y = 78$

$\underline{10x - 12y = 6}$

$\qquad 28x = 84$

$\qquad\quad x = 3$

Substitute 3 for x in equation (1) and solve for y.

$6(3) + 4y = 26$

$\quad 18 + 4y = 26$

$\qquad\quad 4y = 8$

$\qquad\quad\ y = 2$

The solution is $(3,2)$.

51. $3x + 2y = 8$

$\quad 2x - y = 3$

Graphing by hand:

Begin by writing each equation in slope-intercept form.

$3x + 2y = 8 \qquad\qquad 2x - y = 3$

$\quad 2y = -3x + 8 \qquad\quad -y = -2x + 3$

$\qquad y = -\dfrac{3}{2}x + 4 \qquad\quad y = 2x - 3$

Next we graph both equations in the same coordinate system.

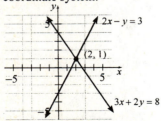

The intersection point is $(2,1)$. So, the ordered pair $(2,1)$ is the solution.

Substitution:

$3x + 2y = 8$

$\quad 2x - y = 3$

Solve the second equation for y.

$2x - y = 3$

$\qquad y = 2x - 3$

Substitute $2x - 3$ for y in the first equation and solve for x.

$3x + 2(2x - 3) = 8$

$\quad 3x + 4x - 6 = 8$

$\qquad\quad 7x - 6 = 8$

$\qquad\qquad\ 7x = 14$

$\qquad\qquad\ \ x = 2$

Substitute 2 for x in the equation $y = 2x - 3$ and solve for y.

$y = 2(2) - 3 = 4 - 3 = 1$

The solution is $(2,1)$.

Elimination:

$3x + 2y = 8$ Equation (1)

$\quad 2x - y = 3$ Equation (2)

To eliminate the y terms, we multiply both sides of equation (2) by 2, yielding the system

$3x + 2y = 8$

$4x - 2y = 6$

The coefficients of the y terms are equal in absolute value and opposite in sign. Add the left sides and the right sides of the equations and solve for x.

$3x + 2y = 8$

$\underline{4x - 2y = 6}$

$\qquad 7x = 14$

$\qquad\ x = 2$

Substitute 2 for x in equation (2) and solve for y.

$2(2) - y = 3$

$\quad 4 - y = 3$

$\qquad -y = -1$

$\qquad\ \ y = 1$

The solution is $(2,1)$.

Preference may vary.

53. $5x + 3y = 11$ Equation (1)

$2x - 4y = -6$ Equation (2)

a. To eliminate the x terms, we multiply both sides of equation (1) by 2 and both sides of equation (2) by -5, yielding the system
$$10x + 6y = 22$$
$$-10x + 20y = 30$$
The coefficients of the x terms are equal in absolute value and opposite in sign. Add the left sides and the right sides of the equations and solve for y.
$$10x + 6y = 22$$
$$\underline{-10x + 20y = 30}$$
$$26y = 52$$
$$y = 2$$
Substitute 2 for y in equation (2) and solve for x.
$$2x - 4(2) = -6$$
$$2x - 8 = -6$$
$$2x = 2$$
$$x = 1$$
The solution is $(1, 2)$.

b. To eliminate the y terms, we multiply both sides of equation (1) by 4 and both sides of equation (2) by 3, yielding the system
$$20x + 12y = 44$$
$$6x - 12y = -18$$
The coefficients of the y terms are equal in absolute value and opposite in sign. Add the left sides and the right sides of the equations and solve for x.
$$20x + 12y = 44$$
$$\underline{6x - 12y = -18}$$
$$26x = 26$$
$$x = 1$$
Substitute 1 for x in equation (2) and solve for y.
$$2(1) - 4y = -6$$
$$2 - 4y = -6$$
$$-4y = -8$$
$$y = 2$$
The solution is $(1, 2)$.

c. The results are the same.

55. We begin by finding the equations for each table.

Table 17:

The input $x = 0$ yields the output $y = 93$, so the y-intercept is $(0, 93)$.

$$m = \frac{96 - 99}{-1 - (-2)} = \frac{-3}{1} = -3, \text{ so the slope is } -3.$$

The first equation is $y = -3x + 93$.

Table 18:

The input $x = 0$ yields the output $y = -22$, so the y-intercept is $(0, -22)$.

$$m = \frac{-24 - (-26)}{-1 - (-2)} = \frac{2}{1} = 2, \text{ so the slope is 2.}$$

The second equation is $y = 2x - 22$.

Now we solve the system
$$y = -3x + 93$$
$$y = 2x - 22$$

Writing in standard form, we get

$3x + y = 93$ Equation (1)

$2x - y = 22$ Equation (2)

The coefficients of the y terms are equal in absolute value and opposite in sign. Add the left sides and the right sides of the equations and solve for x.
$$3x + y = 93$$
$$\underline{2x - y = 22}$$
$$5x = 115$$
$$x = 23$$
Substitute 23 for x in equation (1) and solve for y.
$$3(23) + y = 93$$
$$69 + y = 93$$
$$y = 24$$
The solution is $(23, 24)$.

57. A : This point is the origin so it has coordinates $(0,0)$.

B : This point is the y-intercept for line l_1. Let $x = 0$ in the equation and solve for y.

$$2(0) + 3y = 12$$
$$3y = 12$$
$$y = 4$$

The coordinates of the point are $(0, 4)$.

C : This point is the intersection of the two lines. Therefore, it is the solution of the system

$$2x + 3y = 12 \quad \text{Equation (1)}$$
$$5x - 3y = 9 \quad \text{Equation (2)}$$

The coefficients of the y terms are equal in absolute value and opposite in sign. Add the left sides and the right sides of the equations and solve for x.

$$2x + 3y = 12$$
$$\underline{5x - 3y = 9}$$
$$7x = 21$$
$$x = 3$$

Substitute 3 for x in equation (1) and solve for y.

$$2(3) + 3y = 12$$
$$6 + 3y = 12$$
$$3y = 6$$
$$y = 2$$

The coordinates of the point are $(3, 2)$.

D : This point is the x-intercept of line l_2. Let $y = 0$ in the equation and solve for x.

$$5x - 3(0) = 9$$
$$5x = 9$$
$$x = \frac{9}{5}$$

The coordinates of the point are $\left(\frac{9}{5}, 0\right)$.

59. Answers may vary

61. a.
$$y = mx + b$$
$$5 = m(2) + b$$
$$2m + b = 5$$

b.
$$y = mx + b$$
$$9 = m(4) + b$$
$$4m + b = 9$$

c. $2m + b = 5$ Equation (1)
$4m + b = 9$ Equation (2)

To eliminate the b terms, we multiply both sides of equation (1) by -1, yielding the system

$$-2m - b = -5$$
$$4m + b = 9$$

The coefficients of the b terms are equal in absolute value and opposite in sign. Add the left sides and the right sides of the equations and solve for m.

$$-2m - b = -5$$
$$\underline{4m + b = 9}$$
$$2m = 4$$
$$m = 2$$

Substitute 2 for m in equation (2) and solve for b.

$$4(2) + b = 9$$
$$8 + b = 9$$
$$b = 1$$

The solution of the system is $(m, b) = (2, 1)$.

d. The equation of the line is $y = 2x + 1$.

e. Graph the equation and check if the points $(2, 5)$ and $(4, 9)$ are on the graph.

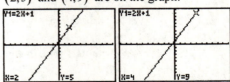

63. $2(5x+4) - 6(3x+2) = 10x + 8 - 18x - 12$

$$= 10x - 18x + 8 - 12$$

$$= -8x - 4$$

Description: *linear expression in one variable*

65. $2(5x+4) - 6(3x+2) = 0$

$$10x + 8 - 18x - 12 = 0$$

$$-8x - 4 = 0$$

$$-8x = 4$$

$$x = -\frac{1}{2}$$

The solution is $-\frac{1}{2}$.

Description: *linear equation in one variable*

Homework 6.4

1. $n = 13.28t + 440.09$

$n = 3.42t + 468.14$

Substitute $3.42t + 468.14$ for n in the first equation and solve for t.

$3.42t + 468.14 = 13.28t + 440.09$

$-9.86t + 468.14 = 440.09$

$-9.86t = -28.05$

$t = \dfrac{-28.05}{-9.86} \approx 2.845$

Substitute 2.845 for t in the equation $n = 3.42t + 468.14$ and solve for n.

$n = 3.42(2.845) + 468.14 \approx 477.870$

The approximate solution of the system is $(2.85, 477.87)$.

According to the models, there will be the same number of men who earned a bachelor's degree as women who earned a bachelor's degree (478 thousand of each) in 1983.

3. $r = -0.064t + 27.00$

$r = -0.029t + 22.10$

Substitute $-0.029t + 22.10$ for r in the first equation and solve for t.

$-0.029t + 22.10 = -0.064t + 27.00$

$0.035t + 22.10 = 27.00$

$0.035t = 4.90$

$t = \dfrac{4.90}{0.035} = 140$

Substitute 140 for t in the equation $r = -0.029t + 22.10$ and solve for r.

$r = -0.029(140) + 22.10 = 18.04$

The solution of the system is $(140, 18.04)$.

According to the models, the women's record time will equal the men's record time (18.04 seconds) in 2040.

5. a. Start by creating a scatter diagram of the data.

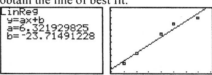

The data appear fairly linear so we use the regression feature of a graphing utility to obtain the line of best fit.

A reasonable model for the data is $p = 6.32t - 23.71$.

b. Start by creating a scatter diagram of the data.

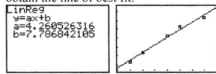

The data appear fairly linear so we use the regression feature of a graphing utility to obtain the line of best fit.

A reasonable model for the data is $p = 4.26t + 7.79$.

c. To estimate when the percentages will be equal, we solve the system

$p = 6.32t - 23.71$

$p = 4.26t + 7.79$

Substitute $4.26t + 7.79$ for p in the first equation and solve for t.

$4.26t + 7.79 = 6.32t - 23.71$

$-2.06t + 7.79 = -23.71$

$-2.06t = -31.5$

$t = \dfrac{-31.5}{-2.06} \approx 15.291$

Substitute 15.291 for t in the equation $p = 4.26t + 7.79$ and solve for p.

$p = 4.26(15.291) + 7.79 \approx 72.930$

The approximate solution of the system is $(15.29, 72.93)$.

According to the models, the percentage of households with a computer was equal to the percentage of households with an Internet connection (72.9%) in 2005.

7. a. Start by creating a scatter diagram of the data.

The data appear fairly linear so we use the regression feature of a graphing utility to obtain the line of best fit.

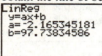 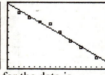

A reasonable model for the data is $s = -2.17t + 97.74$.

b. Start by creating a scatter diagram of the data.

The data appear fairly linear so we use the regression feature of a graphing utility to obtain the line of best fit.

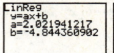 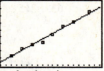

A reasonable model for the data is $s = 2.02t - 4.84$.

c. To estimate when the viewing shares will be equal, we solve the system

$s = -2.17t + 97.74$

$s = 2.02t - 4.84$

Substitute $2.02t - 4.84$ for s in the first equation and solve for t.

$2.02t - 4.84 = -2.17t + 97.74$

$4.19t - 4.84 = 97.74$

$4.19t = 102.58$

$t = \dfrac{102.58}{4.19} \approx 24.482$

Substitute 24.482 for t in the equation $s = 2.02t - 4.84$ and solve for s.

$s = 2.02(24.482) - 4.84 \approx 44.614$

The approximate solution of the system is $(24.48, 44.61)$.

According to the models, the primetime household viewing shares for all broadcast stations and all ad-supported cable stations were equal (44.6%) in 2004.

d. We use our models to predict the viewing shares of broadcast stations and ad-supported cable stations in 2011 ($t = 31$). Broadcast stations:

$s = -2.17(31) + 97.74 = 30.47$

Ad-supported stations:

$s = 2.02(31) - 4.84 = 57.78$

The combined viewing share of broadcast stations and ad-supported stations is predicted to be $30.47 + 57.78 = 88.25$ percent. Therefore, the viewing share for PBS, pay cable, and other types of cable stations is predicted to be $100 - 88.25 = 11.75\%$.

9. a. $n = 2.34t + 9.54$

$n = -2.29t + 424.40$

Substitute $-2.29t + 424.40$ for n in the first equation and solve for t.

$-2.29t + 424.40 = 2.34t + 9.54$

$-4.63t + 424.40 = 9.54$

$-4.63t = -414.86$

$$t = \frac{-414.86}{-4.63} \approx 89.603$$

Substitute 89.603 for t in the equation $n = -2.29t + 424.40$ and solve for n.

$n = -2.29(89.603) + 424.40 \approx 219.209$

The approximate solution of the system is $(89.60, 219.21)$.

According to the models, the number of women and men in the House of Representatives will be the same (219 each) in 2070. Technically this will happen in 2071 since each new Congress occurs in an odd year.

b. The greatest change in women from one Congress to the next was $47 - 28 = 19$ women. This occurred between the 102nd Congress and the 103rd Congress. No, this change is much larger.

c. $n = 1.73t + 22.66$

$n = -1.73t + 412.34$

Substitute $-1.73t + 412.34$ for n in the first equation and solve for t.

$-1.73t + 412.34 = 1.73t + 22.66$

$-3.46t + 412.34 = 22.66$

$-3.46t = -389.68$

$$t = \frac{-389.68}{-3.46} \approx 112.624$$

Substitute 112.624 for t in the equation $n = -1.73t + 412.34$ and solve for n.

$n = -1.73(112.624) + 412.34 \approx 217.500$

The approximate solution of the system is $(112.62, 217.50)$.

According to the models, the number of women and men in the House of Representatives will be the same (218 each) in 2093.

d. Since the slopes in the models for part (a) are larger in magnitude than those in part (c), the graphs in part (a) will rise and fall more quickly. Therefore, the intersection point in (a) will occur sooner than in (c).

11. a. Since the number of students who earned a bachelor's degree in communications increases by 1.6 thousand each year, we can model the situation by a linear equation. The slope is 1.6 thousand degrees per year. Since the number of degrees earned in 2000 was 56.9 thousand, the d-intercept is $(0, 56.9)$. So, a reasonable model is

$d = 1.6t + 56.9$.

b. By the same reasoning as in part (a), the model is $d = 2.4t + 36.2$.

c. $d = 1.6t + 56.9$

$d = 2.4t + 36.2$

Substitute $2.4t + 36.2$ for d in the first equation and solve for t.

$2.4t + 36.2 = 1.6t + 56.9$

$0.8t + 36.2 = 56.9$

$0.8t = 20.7$

$$t = \frac{20.7}{0.8} = 25.875$$

Substitute 25.875 for t in the equation $d = 2.4t + 36.2$ and solve for d.

$d = 2.4(25.875) + 36.2 = 98.3$

The approximate solution of the system is $(25.88, 98.3)$.

According to the models, the same number of bachelor's degrees (98.3 thousand each) will be earned in 2026.

13. a. Since the sales of digital cameras have increased by $2.5 million each year, we can model the situation by a linear equation. The slope is 2.5 million dollars per year. Since digital camera sales were $4.6 million in 2000, the s-intercept is $(0, 4.6)$. So, a reasonable model is

$s = 2.5t + 4.6$.

b. By the same reasoning as in part (a), the model is $s = -2.6t + 20$.

c. $s = 2.5t + 4.6$

$s = -2.6t + 20$

Substitute $-2.6t + 20$ for s in the first equation and solve for t.

$-2.6t + 20 = 2.5t + 4.6$

$-5.1t + 20 = 4.6$

$-5.1t = -15.4$

$t = \dfrac{-15.4}{-5.1} \approx 3.020$

Substitute 3.020 for t in the equation $s = -2.6t + 20$ and solve for s.

$x = -2.6(3.020) + 20 = 12.148$

The approximate solution to the system is $(3.0, 12.1)$.

According to the models, the sales of digital cameras were equal to sales of traditional cameras ($12.1 million) in 2003.

15. **a.** Since a one-year-old Acura CL coupe depreciates about $1903 per year, we can model the situation by a linear equation. The slope is -1903 dollars per year. Since a one-year-old Acura CL coupe was worth $18,249 in 2002, the V-intercept is $(0, 18249)$. So, a reasonable model is

$V = -1903t + 18,249$.

b. By the same reasoning as in part (a), the model is $V = -1225t + 14,564$.

c. $V = -1903t + 18,249$

$V = -1225t + 14,564$

Substitute $-1225t + 14,564$ for V in the first equation and solve for t.

$-1225t + 14,564 = -1903t + 18,249$

$678t + 14,564 = 18,249$

$678t = 3685$

$t = \dfrac{3685}{678} \approx 5.435$

Substitute 5.435 for t in the equation $V = -1225t + 14,564$ and solve for V.

$V = -1225(5.435) + 14,564 \approx 7906.13$.

The approximate solution to the system is $(5.44, 7906.13)$.

According to the models, the two cars will have the same value ($7906) in 2007.

17. **a.** Since sales for company A have declined by $1.2 million each year, we can model the situation by a linear equation. The slope is -1.2 million dollars per year. Since sales for company A were $29.5 million in 2005, the s-intercept is $(0, 29.5)$. So, a reasonable model is $s = -1.2t + 29.5$.

b. By the same reasoning as in part (a), the model is $s = 1.7t + 12.1$.

c. $s = -1.2t + 29.5$

$s = 1.7t + 12.1$

Substitute $1.7t + 12.1$ for s in the first equation and solve for t.

$1.7t + 12.1 = -1.2t + 29.5$

$2.9t + 12.1 = 29.5$

$2.9t = 17.4$

$t = \dfrac{17.4}{2.9} = 6$

Substitute 6 for t in the equation $s = 1.7t + 12.1$ and solve for s.

$s = 1.7(6) + 12.1 = 22.3$.

The solution to the system of equations is $(6, 22.3)$.

According to the models, sales for both companies will be the same ($22.3 million) in 2011.

19. **a.** Since tuition at college A increases by $850 per year, we can model the situation by a linear equation. The slope is 850 dollars per year. Since tuition at college A was $5425 in 2005, the c-intercept is $(0, 5425)$. So, a reasonable model is $c = 850t + 5425$.

b. By the same reasoning as in part (a), the model is $c = 570t + 6557$.

c. $c = 850t + 5425$

$c = 570t + 6557$

Substitute $570t + 6557$ for c in the first equation and solve for t.

$570t + 6557 = 850t + 5425$

$-280t + 6557 = 5425$

$-280t = -1132$

$t = \dfrac{-1132}{-280} \approx 4.043$

Substitute 4.043 for t in the equation $c = 570t + 6557$ and solve for c.

$c = 570(4.043) + 6557 = 8861.51$.

The approximate solution to the system of equations is $(4.04, 8862)$.

According to the models, both colleges will have the same tuition ($8862 per year) in 2009.

21. We are given that the average daily calorie consumption is approximately linear for both men and women. Let t = the number of years since 1980, and c = the average daily calorie consumption.

Men:
When $t = 0$, we have $c = 2450$, so the c-intercept is $(0, 2450)$.

$$m = \frac{2618 - 2450}{20 - 0} = \frac{168}{20} = 8.4$$

Therefore, the linear equation for men is $c = 8.4t + 2450$.

Women:
When $t = 0$, we have $c = 1521$, so the c-intercept is $(0, 1521)$.

$$m = \frac{1877 - 1521}{20 - 0} = \frac{356}{20} = 17.8$$

Therefore, the linear equation for women is $c = 17.8t + 1521$.

The two equations yield the system
$c = 8.4t + 2450$

$c = 17.8t + 1521$
Substitute $17.8t + 1521$ for c in the first equation and solve for t.
$17.8t + 1521 = 8.4t + 2450$

$9.4t + 1521 = 2450$

$9.4t = 929$

$$t = \frac{929}{9.4} \approx 98.830$$

Substitute 98.830 for t in the equation $c = 17.8t + 1521$ and solve for c.

$c = 17.8(98.830) + 1521 = 3280.174$

The approximate solution to the system is $(99, 3280)$. According to the models, the

average daily calorie consumption will be the same for men and women (3280 calories per day) in 2079.

23. $y = -2x + 6$

$y = 3x + 1$
Substitute $3x + 1$ for y in the first equation and solve for x.
$3x + 1 = -2x + 6$

$5x + 1 = 6$

$5x = 5$

$x = 1$
Substitute 1 for x in the equation $y = 3x + 1$ and solve for y.
$y = 3(1) + 1 = 3 + 1 = 4$

The solution is $(1, 4)$.

Description: *system of two linear equations in two variables*

25. $-2x + 6 = 3x + 1$

$-5x + 6 = 1$

$-5x = -5$

$x = 1$
The solution is 1.
Description: *linear equation in one variable*

Homework 6.5

1. Let L = the length and W = the width.
$L = 1.62W \qquad P = 2L + 2W$

$$600 = 2L + 2W$$
To determine the dimensions of the rectangle, we solve the system
$L = 1.62W$

$600 = 2L + 2W$
Substitute $1.62W$ for L in the second equation and solve for w.
$600 = 2(1.62W) + 2W$

$600 = 3.24W + 2W$

$600 = 5.24W$

$$W = \frac{600}{5.24} \approx 114.504$$
Substitute 114.504 for W in the equation $L = 1.62W$ and solve for L.

$L = 1.62(114.504) \approx 185.496$

The width of the rectangle is approximately 114.50 feet and the length is approximately 185.50 feet.

3. L = the length and W = the width.

$L = W + 5 \qquad P = 2L + 2W$
$\qquad\qquad\quad 42 = 2L + 2W$
$\qquad\qquad\quad 21 = L + W$

To determine the dimensions of the garden, we solve the system

$L = W + 5$
$21 = L + W$

Substitute $W + 5$ for L in the second equation and solve for W.

$21 = (W + 5) + W$
$21 = 2W + 5$
$16 = 2W$
$8 = W$

Substitute 8 for W in the equation $L = W + 5$ and solve for L.

$L = 8 + 5 = 13$

The width of the garden is 8 feet and the length is 13 feet.

5. L = the length and W = the width.

$L = 2W + 6 \qquad P = 2L + 2W$
$\qquad\qquad\quad 228 = 2L + 2W$
$\qquad\qquad\quad 114 = L + W$

To determine the dimensions of the court, we solve the system

$L = 2W + 6$
$114 = L + W$

Substitute $2W + 6$ for L in the second equation and solve for W.

$114 = (2W + 6) + W$
$114 = 3W + 6$
$108 = 3W$
$36 = W$

Substitute 36 for W in the equation $L = 2W + 6$ and solve for L.

$L = 2(36) + 6 = 72 + 6 = 78$

The width of the court is 36 feet and the length is 78 feet.

7. L = the length and W = the width.

$L = 2W - 3 \qquad P = 2L + 2W$
$\qquad\qquad\quad 108 = 2L + 2W$
$\qquad\qquad\quad 54 = L + W$

To determine the dimensions of the rectangle, we solve the system

$L = 2W - 3$
$54 = L + W$

Substitute $2W - 3$ for L in the second equation and solve for W.

$54 = (2W - 3) + W$
$54 = 3W - 3$
$57 = 3W$
$19 = W$

Substitute 19 for W in the equation $L = 2W - 3$ and solve for L.

$L = 2(19) - 3 = 38 - 3 = 35$

The width of the rectangle is 19 inches and the length is 35 inches.

9. L = the length and W = the width.

$L = 3W + 1 \qquad P = 2L + 2W$
$\qquad\qquad\quad 146 = 2L + 2W$
$\qquad\qquad\quad 73 = L + W$

To determine the dimensions of the rectangle, we solve the system

$L = 3W + 1$
$73 = L + W$

Substitute $3W + 1$ for L in the second equation and solve for W.

$73 = (3W + 1) + W$
$73 = 4W + 1$
$72 = 4W$
$18 = W$

Substitute 18 for W in the equation $L = 3W + 1$ and solve for L.

$L = 3(18) + 1 = 54 + 1 = 55$

The width of the rectangle is 18 yards and the length is 55 yards.

11. Let x = the number of \$15 tickets and y = the number of \$22 tickets.

There are a total of 2000 tickets, so our first equation is $x + y = 2000$.

The total revenue is \$33,500. The revenue for each ticket type is obtained by multiplying the ticket price by the number of tickets sold at that price. Therefore, our second equation is

$15x + 22y = 33,500$

The system is

$\quad x + y = 2000 \qquad$ Equation (1)
$15x + 22y = 33,500 \qquad$ Equation (2)

To eliminate the x terms, multiply both sides of equation (1) by -15, yielding the system

$-15x - 15y = -30,000$
$\quad 15x + 22y = 33,500$

Add the left sides and right sides and solve for y.

$-15x - 15y = -30,000$

$\underline{15x + 22y = 33,500}$

$7y = 3,500$

$y = 500$

Substitute 500 for y in equation (1) and solve for x.

$x + (500) = 2000$

$x = 1500$

The theater should sell 1500 of the $15 tickets and 500 of the $22 tickets.

13. Let x = the number sold of *Amnesiac* and y = the number sold of *Hail to the Thief*.
A total of 253 CDs were sold, so our first equation is $x + y = 253$.

The total revenue was $4762.94. The revenue for each CD is obtained by multiplying the CD price by the number of CDs sold at that price. Therefore, our second equation is
$17.98x + 18.98y = 4762.94$

The system is

$x + y = 253$

$17.98x + 18.98y = 4762.94$

Solve the first equation for y.

$x + y = 253$

$y = -x + 253$

Substitute $-x + 253$ for y in the second equation and solve for x.

$17.98x + 18.98(-x + 253) = 4762.94$

$17.98x - 18.98x + 4801.94 = 4762.94$

$-x + 4801.94 = 4762.94$

$-x = -39$

$x = 39$

Substitute 39 for x in the equation $y = -x + 253$ and solve for y.

$y = -39 + 253 = 214$

The store sold 39 *Amnesiac* CDs and 214 *Hail to the Thief* CDs.

15. Let x = the number of books and y = the number of audio CDs.
A total of 1500 books and CDs were sold, so our first equation is $x + y = 1500$.

The total revenue was $53,987. The revenue for each product is obtained by multiplying the product price by the number sold of the product. Therefore, our second equation is

$29.99x + 75.00y = 53,987$.

The system is

$x + y = 1500$ Equation (1)

$29.99x + 75.00y = 53,987$ Equation (2)

To eliminate the y terms, multiply both sides of equation (1) by -75.00, yielding the system

$-75.00x - 75.00y = -112,500$

$29.99x + 75.00y = 53,987$

Add the left sides and right sides and solve for x.

$-75.00x - 75.00y = -112,500$

$\underline{29.99x + 75.00y = 53,987}$

$-45.01x = -58,513$

$x = 1300$

Substitute 1300 for x in equation (1) and solve for y.

$1300 + y = 1500$

$y = 200$

The bookstore sold 1300 hardcover books and 200 audio CDs.

17. Let x = the price of balcony seats and y = the price of main-level seats.
Tickets for the balcony are $12 less than tickets for the main-level, so our first equation is $x = y - 12$.

The total revenue is $40,600. To obtain the revenue for each seat type, we multiply the price of each seat type by the number of seats at that price. Therefore, our second equation is
$300x + 1400y = 40,600$.

The system is

$x = y - 12$

$300x + 1400y = 40,600$

Substitute $y - 12$ for x in the second equation and solve for y.

$300(y - 12) + 1400y = 40,600$

$300y - 3600 + 1400y = 40,600$

$1700y - 3600 = 40,600$

$1700y = 44,200$

$y = 26$

Substitute 26 for y in the equation $x = y - 12$ and solve for x.

$x = 26 - 12 = 14$

Each balcony seat should cost $14 and each main-level seat should cost $26.

19. Let x = the price of general seats and y = the price of reserved seats.

Tickets for the general are $25 less than tickets for the reserved, so our first equation is $x = y - 25$.

The total revenue is $544,000. To obtain the revenue for each seat type, we multiply the price of each seat type by the number of seats at that price. Therefore, our second equation is $8000x + 4000y = 544,000$.

The system is
$$x = y - 25$$
$$8000x + 4000y = 544,000$$

Substitute $y - 25$ for x in the second equation and solve for y.
$$8000(y - 25) + 4000y = 544,000$$
$$8000y - 200,000 + 4000y = 544,000$$
$$12,000y - 200,000 = 544,000$$
$$12,000y = 744,000$$
$$y = 62$$

Substitute 62 for y in the equation $x = y - 25$ and solve for x.
$$x = 62 - 25 = 37$$
Each general seat should cost $37 and each reserved seat should cost $62.

21. a. We find 8% of 2500.
$$(0.08)(2500) = 200$$
The interest in one year would be $200.

b. We find 8% of 3500.
$$(0.08)(3500) = 280$$
The interest in one year would be $280.

c. We find 8% of d.
$$(0.08)(d) = 0.08d$$
The interest in one year would be $0.08d$ dollars.

23. a. $0.03(2000) + 0.06(7000) = 60 + 420$
$$= 480$$
The total interest would be $480.

b. $0.03(4000) + 0.06(5000) = 120 + 300$
$$= 420$$
The total interest would be $420.

c. $0.03(x) + 0.06(y) = 0.03x + 0.06y$
The total interest would be $0.03x + 0.06y$.

25. Let x = the amount invested in the First Funds TN Tax-Free I account and y = the amount invested in the W&R International Growth C account.

The total amount invested is $20,000 so our first equation is $x + y = 20,000$.

The total interest after one year is $1500. Therefore, our second equation is $0.06x + 0.11y = 1500$.

The system is
$$x + y = 20,000$$
$$0.06x + 0.11y = 1500$$

Solve the first equation for y.
$$x + y = 20,000$$
$$y = -x + 20,000$$

Substitute $-x + 20,000$ for y in the second equation and solve for x.
$$0.06x + 0.11(-x + 20,000) = 1500$$
$$0.06x - 0.11x + 2200 = 1500$$
$$-0.05x + 2200 = 1500$$
$$-0.05x = -700$$
$$x = 14,000$$

Substitute 14,000 for x in the equation $y = -x + 20,000$ and solve for y.
$$y = -(14,000) + 20,000 = 6,000$$

She should invest $14,000 in the First Funds TN Tax-Free I account and $6,000 in the W&R International Growth C account.

27. Let x = the amount invested in the Middlesex Savings Bank CD account and y = the amount invested in the First Funds Growth & Income I account.

The total amount invested is $8500, so our first equation is $x + y = 8500$.

The total interest after one year is $990, so our second equation is $0.036x + 0.15y = 990$.

The system is
$$x + y = 8500 \quad \text{Equation (1)}$$
$$0.036x + 0.15y = 990 \quad \text{Equation (2)}$$

To eliminate the y terms, multiply both sides of equation (1) by -0.15, yielding the system
$$-0.15x - 0.15y = -1275$$
$$0.036x + 0.15y = 990$$

Add the left sides and the right sides, and solve for x.

$$-0.15x - 0.15y = -1275$$
$$\underline{0.036x + 0.15y = 990}$$
$$-0.114x = -285$$
$$x = \frac{-285}{-0.114} = 2500$$

Substitute 2500 for x in equation (1) and solve for y.
$$2500 + y = 8500$$
$$y = 6000$$

He should invest \$2500 in the Middlesex Savings Bank CD account and \$6000 in the First Funds Growth & Income I account.

29. Let x = the amount invested in the Limited Term NY Municipal X account and y = the amount invested in the Calvert Income A account. The amount invested in the Limited Term NY Municipal X account will be three times as much as the amount in the Calvert Income A account, so our first equation is $x = 3y$.

The total interest for one year is \$625, so our second equation is $0.05x + 0.10y = 625$.

The system is
$$x = 3y$$
$$0.05x + 0.10y = 625$$

Substitute $3y$ for x in the second equation and solve for y.
$$0.05(3y) + 0.10y = 625$$
$$0.15y + 0.10y = 625$$
$$0.25y = 625$$
$$y = 2500$$

Substitute 2500 for y in the equation $x = 3y$ and solve for x.
$$x = 3(2500) = 7500$$

The person should invest \$7500 in the Limited Term NY Municipal X account and \$2500 in the Calvert Income A account.

31. Let x = the amount invested in the USAA Tax Exempt Short-Term account and y = the amount invested in the Putnam Global Equity B account. The amounts invested in each account are equal, so our first equation is $x = y$.

The total interest for one year is \$700, so our second equation is $0.05x + 0.09y = 700$.

The system is
$$x = y$$
$$0.05x + 0.09y = 700$$

Substitute x for y in the second equation and solve for x.
$$0.05x + 0.09x = 700$$
$$0.14x = 700$$
$$x = 5000$$

Since the amounts are equal, $y = 5000$.
The person should invest \$5000 in each account.

33. a. We compute 65% of 2.
$$0.65(2) = 1.3$$
There are 1.3 ounces of oil in the solution.

b. We compute 65% of 3.
$$0.65(3) = 1.95$$
There are 1.95 ounces of oil in the solution.

c. We compute 65% of x.
$$0.65(x) = 0.65x$$
There are $0.65x$ ounces of oil in the solution.

35. a. $0.35(4) + 0.10(3) = 1.4 + 0.3$
$$= 1.7$$
There are 1.7 ounces of pure alcohol in the mixture.

b. Let x = the number of ounces of the 35% solution and y = the number of ounces of the 10% solution.
The total number of ounces is 15, so our first equation is $x + y = 15$.

The total amount of acid between the two solutions must be the same as the total amount of acid in the mixture, so our second equation is $0.35x + 0.10y = 0.20(15)$.

The system is
$$x + y = 15$$
$$0.35x + 0.10y = 0.20(15)$$

Solve the first equation for y.
$$x + y = 15$$
$$y = -x + 15$$

Substitute $-x + 15$ for y in the second equation and solve for x.
$$0.35x + 0.10(-x + 15) = 0.20(15)$$
$$0.35x - 0.10x + 1.50 = 3$$
$$0.25x + 1.50 = 3$$
$$0.25x = 1.50$$
$$x = 6$$

Substitute 6 for x in the equation
$y = -x + 15$ and solve for y.

$$y = -(6) + 15 = 9$$

The mixture should contain 6 ounces of the 35% solution and 9 ounces of the 10% solution.

37. Let x = the number of quarts of the 20% solution and y = the number of quarts of the 30% solution.
The total number of quarts is 5, so our first equation is $x + y = 5$.

The total amount of acid between the two solutions must be the same as the total amount of acid in the mixture, so our second equation is
$0.20x + 0.30y = 0.22(5)$.

The system is
$$x + y = 5$$
$$0.20x + 0.30y = 0.22(5)$$

Solve the first equation for y.
$$x + y = 5$$
$$y = -x + 5$$

Substitute $-x + 5$ for y in the second equation and solve for x.
$$0.20x + 0.30(-x + 5) = 0.22(5)$$
$$0.20x - 0.30x + 1.50 = 1.1$$
$$-0.1x + 1.50 = 1.1$$
$$-0.1x = -0.4$$
$$x = 4$$

Substitute 4 for x in the equation $y = -x + 5$ and solve for y.
$$y = -(4) + 5 = 1$$

The mixture should contain 4 quarts of the 20% solution and 1 quart of the 30% solution.

39. Let x = the number of gallons of the 10% solution and y = the number of gallons of 25% solution.
The total number of gallons is 3, so our first equation is $x + y = 3$.

The total amount of pure antifreeze between the two solutions must be the same as the total amount of pure antifreeze in the mixture, so our second equation is $0.10x + 0.25y = 0.20(3)$.

The system is
$$x + y = 3$$
$$0.10x + 0.25y = 0.20(3)$$

Solve the first equation for y.

$$x + y = 3$$
$$y = -x + 3$$

Substitute $-x + 3$ for y in the second equation and solve for x.
$$0.10x + 0.25(-x + 3) = 0.20(3)$$
$$0.10x - 0.25x + 0.75 = 0.60$$
$$-0.15x + 0.75 = 0.60$$
$$-0.15x = -0.15$$
$$x = 1$$

Substitute 1 for x in the equation $y = -x + 3$ and solve for y.
$$y = -(1) + 3 = 2$$

The mixture should contain 1 gallon of the 10% antifreeze solution and 2 gallons of the 25% antifreeze solution.

41. Let x = the number of quarts of the 15% solution and y = the number of quarts of the 35% solution.
The total number of quarts is 4, so our first equation is $x + y = 4$.

The total amount of acid between the two solutions must be the same as the total amount of acid in the mixture, so our second equation is
$0.15x + 0.35y = 0.30(4)$.

The system is
$$x + y = 4$$
$$0.15x + 0.35y = 0.30(4)$$

Solve the first equation for y.
$$x + y = 4$$
$$y = -x + 4$$

Substitute $-x + 4$ for y in the second equation and solve for x.
$$0.15x + 0.35(-x + 4) = 0.30(4)$$
$$0.15x - 0.35x + 1.40 = 1.20$$
$$-0.2x + 1.40 = 1.20$$
$$-0.2x = -0.20$$
$$x = 1$$

Substitute 1 for x in the equation $y = -x + 4$ and solve for y.
$$y = -(1) + 4 = 3$$

The mixture should contain 1 quart of the 15% solution and 3 quarts of the 35% solution.

43. Let x = the number of gallons of the 12% acid solution and y = the number of gallons of the 24% acid solution.
The total number of gallons is 8, so our first equation is $x + y = 8$.
The total amount of acid between the two solutions must be the same as the total amount of acid in the mixture, so our second equation is $0.12x + 0.24y = 0.21(8)$.
The system is
$$x + y = 8$$
$$0.12x + 0.24y = 0.21(8)$$
Solve the first equation for y.
$$x + y = 8$$
$$y = -x + 8$$
Substitute $-x + 8$ for y in the second equation and solve for x.
$$0.12x + 0.24(-x + 8) = 0.21(8)$$
$$0.12x - 0.24x + 1.92 = 1.68$$
$$-0.12x + 1.92 = 1.68$$
$$-0.12x = -0.24$$
$$x = 2$$
Substitute 2 for x in the equation $y = -x + 8$ and solve for y.
$$y = -(2) + 8 = 6$$
The mixture should contain 2 gallons of the 12% acid solution and 6 gallons of the 24% acid solution.

45. Let x = the number of ounces of the 20% alcohol solution and y = the number of ounces of pure water.
The total number of ounces is 5, so our first equation is $x + y = 5$.
The total amount of acid in the 20% solution must be the same as the total amount of acid in the mixture since there is no acid in pure water. Therefore, our second equation is
$$0.20x = 0.12(5).$$
The system is
$$x + y = 5$$
$$0.20x = 0.12(5)$$
Solve the second equation for x.
$$0.20x = 0.12(5)$$
$$0.20x = 0.60$$
$$x = 3$$
Substitute 3 for x in the equation $x + y = 5$ and solve for y.

$$3 + y = 5$$
$$y = 2$$
The mixture should contain 3 ounces of the 20% solution and 2 ounces of pure water.

47. a. $10(300) = 3000$
The revenue is $3000.

b. $15(300) = 4500$
The revenue is $4500.

c. i. Since all tickets cost at least $10, the minimum revenue will be $3000. Therefore, it is not possible to have revenue of $2500.

ii. Since all tickets cost at most $15, the maximum revenue will be $4500. Therefore, it is not possible to have revenue of $4875.

iii. Since $3250 is between $3000 and $4500, it is possible to achieve this revenue.
Let x = the number of $10 tickets and y = the number of $15 tickets.
The total number of tickets is 300, so our first equation is $x + y = 300$.
The total revenue is $3250, so our second equation is $10x + 15y = 3250$.
The system is
$$x + y = 300 \quad \text{Equation (1)}$$
$$10x + 15y = 3250 \quad \text{Equation (2)}$$
To eliminate the x terms, multiply both sides of equation (1) by -10, yielding the system
$$-10x - 10y = -3000$$
$$10x + 15y = 3250$$
Add the left sides and right sides, and solve for y.
$$-10x - 10y = -3000$$
$$\underline{10x + 15y = 3250}$$
$$5y = 250$$
$$y = 50$$
Substitute 50 for y in equation (1) and solve for x.
$$x + 50 = 300$$
$$x = 250$$
The theater should sell 250 of the $10 tickets and 50 of the $15 tickets.

197

49. Descriptions may vary.

$$y = x + 3$$
$$2x + 2y = 50$$

Substitute $x + 3$ for y in the second equation and solve for x.

$$2x + 2(x + 3) = 50$$
$$2x + 2x + 6 = 50$$
$$4x + 6 = 50$$
$$4x = 44$$
$$x = 11$$

Substitute 11 for x in the equation $y = x + 3$ and solve for y.

$$y = 11 + 3 = 14$$

The solution is $(11, 14)$.

51. Descriptions may vary.

$$y = 2x$$
$$0.03x + 0.07y = 340$$

Substitute $2x$ for y in the second equation and solve for x.

$$0.03x + 0.07(2x) = 340$$
$$0.03x + 0.14x = 340$$
$$0.17x = 340$$
$$x = 2000$$

Substitute 2000 for x in the equation $y = 2x$ and solve for y.

$$y = 2(2000) = 4000.$$

The solution is $(2000, 4000)$.

53.
$$P = 2(L + W)$$
$$P = 2L + 2W$$
$$P - 2W = 2L$$
$$\frac{P - 2W}{2} = L \quad \text{or} \quad L = \frac{P - 2W}{2}$$

55.
$$A = P(1 + rt)$$
$$A = P + Prt$$
$$A - P = Prt$$
$$\frac{A - P}{Pt} = r \quad \text{or} \quad r = \frac{A - P}{Pt}$$

57. $-4(7x - 3) = -4(7x) - (-4)(3)$
$$= -28x - (-12)$$
$$= -28x + 12$$

Description: *linear expression in one variable*

59. $-4(7x - 3) = 5x + 2$
$$-28x + 12 = 5x + 2$$
$$-33x + 12 = 2$$
$$-33x = -10$$
$$x = \frac{10}{33}$$

The solution is $\frac{10}{33}$.

Description: *linear equation in one variable*

Homework 6.6

1. Check $(4, 1)$:

$$y < 2x - 3$$
$$1 \overset{?}{<} 2(4) - 3$$
$$1 \overset{?}{<} 8 - 3$$
$$1 < 5 \quad \text{true}$$

The ordered pair $(4, 1)$ satisfies the inequality.

Check $(-2, 3)$:

$$y < 2x - 3$$
$$3 \overset{?}{<} 2(-2) - 3$$
$$3 \overset{?}{<} -4 - 3$$
$$3 < -7 \quad \text{false}$$

The ordered pair $(-2, 3)$ does not satisfy the inequality.

Check $(-3, -1)$:

$$y < 2x - 3$$
$$-1 \overset{?}{<} 2(-3) - 3$$
$$-1 \overset{?}{<} -6 - 3$$
$$-1 < -9 \quad \text{false}$$

The ordered pair $(-3, -1)$ does not satisfy the inequality.

3. Check $(-3,-1)$:

$$2x - 5y \geq 10$$

$$2(-3) - 5(-1) \overset{?}{\geq} 10$$

$$-6 - (-5) \overset{?}{\geq} 10$$

$$-6 + 5 \overset{?}{\geq} 10$$

$$-1 \overset{?}{\geq} 10 \quad \text{false}$$

The ordered pair $(-3,-1)$ does not satisfy the inequality.

Check $(-1,-4)$:

$$2x - 5y \geq 10$$

$$2(-1) - 5(-4) \overset{?}{\geq} 10$$

$$-2 - (-20) \overset{?}{\geq} 10$$

$$-2 + 20 \overset{?}{\geq} 10$$

$$18 \overset{?}{\geq} 10 \quad \text{true}$$

The ordered pair $(-1,-4)$ satisfies the inequality.

Check $(5,0)$:

$$2x - 5y \geq 10$$

$$2(5) - 5(0) \overset{?}{\geq} 10$$

$$10 - 0 \overset{?}{\geq} 10$$

$$10 \overset{?}{\geq} 10 \quad \text{true}$$

The ordered pair $(5,0)$ satisfies the inequality.

5. The graph of $y > x - 3$ is the region above the line $y = x - 3$. We use a dashed line along the border to indicate that the points on the line $y = x - 3$ are not solutions of $y > x - 3$.

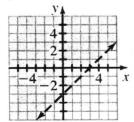

7. The graph of $y \leq -2x + 3$ is the line $y = -2x + 3$ and the region below that line. We use a solid line along the border to indicate that the points on the line $y = -2x + 3$ are solutions of $y \leq -2x + 3$.

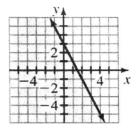

9. The graph of $y < \frac{1}{3}x + 1$ is the region below the line $y = \frac{1}{3}x + 1$. We use a dashed line along the border to indicate that the points on the line $y = \frac{1}{3}x + 1$ are not solutions of $y < \frac{1}{3}x + 1$.

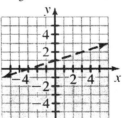

11. The graph of $y \geq -\frac{3}{5}x - 1$ is the line $y = -\frac{3}{5}x - 1$ and the region above that line. We use a solid line along the border to indicate that the points on the line $y = -\frac{3}{5}x - 1$ are solutions of $y \geq -\frac{3}{5}x - 1$.

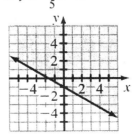

13. The graph of $y > x$ is the region above the line $y = x$. We use a dashed line along the border to indicate that the points on the line $y = x$ are not solutions of $y > x$.

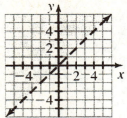

15. First, we get y alone on one side of the inequality.
$$y - 2x < 0$$
$$y < 2x$$

The graph of $y < 2x$ is the region below the line $y = 2x$. We use a dashed line along the border to indicate that the points on the line $y = 2x$ are not solutions of $y - 2x < 0$.

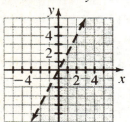

17. First, we get y alone on one side of the inequality.
$$3x + y \leq 2$$
$$y \leq -3x + 2$$

The graph of $y \leq -3x + 2$ is the line $y = -3x + 2$ and the region below that line. We use a solid line along the border to indicate that the points on the line are solutions to $3x + y \leq 2$.

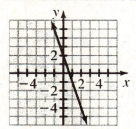

19. First, we get y alone on one side of the inequality.
$$4x - y > 1$$
$$-y > -4x + 1$$
$$y < 4x - 1$$

The graph of $y < 4x - 1$ is the region below the line $y = 4x - 1$. We use a dashed line along the border to indicate that the points on the line are not solutions to $4x - y > 1$.

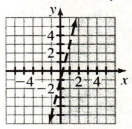

21. First, we get y alone on one side of the inequality.
$$4x - 3y \geq 0$$
$$-3y \geq -4x$$
$$y \leq \frac{4}{3}x$$

The graph of $y \leq \frac{4}{3}x$ is the line $y = \frac{4}{3}x$ and the region below that line. We use a solid line along the border to indicate that the points on the line are solutions to $4x - 3y \geq 0$.

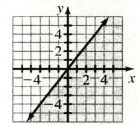

23. First, we get y alone on one side of the inequality.

$$4x + 5y \geq 10$$

$$5y \geq -4x + 10$$

$$y \geq -\frac{4}{5}x + 2$$

The graph of $y \geq -\frac{4}{5}x + 2$ is the line

$y = -\frac{4}{5}x + 2$ and the region above that line. We use a solid line along the border to indicate that the points on the line are solutions of $4x + 5y \geq 10$.

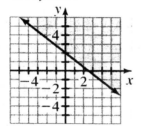

25. First, we get y alone on one side of the inequality.

$$2x - 5y < 5$$

$$-5y < -2x + 5$$

$$y > \frac{2}{5}x - 1$$

The graph of $y > \frac{2}{5}x - 1$ is the region above the

line $y = \frac{2}{5}x - 1$. We use a dashed line along the

border to indicate that the points on the line are not solutions of $2x - 5y < 5$.

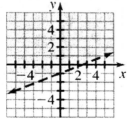

27. The graph of $y > 3$ is the region above the line $y = 3$. We use a dashed line along the border to indicate that the points on the line are not solutions to $y > 3$.

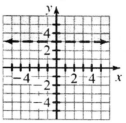

29. The graph of $x \geq -3$ is the vertical line $x = -3$ and the region to the right of that line. We use a solid line along the border to indicate that the points on the line are solutions to $x \geq -3$.

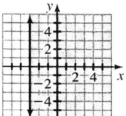

31. The graph of $x < -2$ is the region to the left of the vertical line $x = -2$. We use a dashed line along the border to indicate that the points on the line are not solutions to $x < -2$.

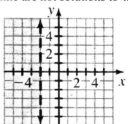

33. The graph of $y \leq 0$ is the line $y = 0$ and the region below that line. We use a solid line along the border to indicate that the points on the line are solutions to $y \leq 0$

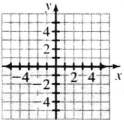

35. First we sketch the graph of $y \geq \frac{1}{3}x - 3$ and the graph of $y \leq -\frac{3}{2}x + 2$. The graph of $y \geq \frac{1}{3}x - 3$ is the line $y = \frac{1}{3}x - 3$ (graph the line with a solid line) and the region above the line. The graph of $y \leq -\frac{3}{2}x + 2$ is the line $y = -\frac{3}{2}x + 2$ (graph the line with a solid line) and the region below that line. The graph of the solution set of the system is the intersection of the graphs of the inequalities.

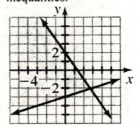

37. First we sketch the graph of $y > 2x - 3$ and the graph of $y > -\frac{3}{4}x + 1$. The graph of $y > 2x - 3$ is the region above the line $y = 2x - 3$ (graph the line with a dashed line). The graph of $y > -\frac{3}{4}x + 1$ is the region above the line $y = -\frac{3}{4}x + 1$ (graph the line with a dashed line). The graph of the solution set of the system is the intersection of the graphs of the inequalities.

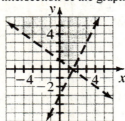

39. First we sketch the graph of $y \leq -\frac{2}{3}x - 3$ and the graph of $y > 2x + 1$. The graph of $y \leq -\frac{2}{3}x - 3$ is the line $y = -\frac{2}{3}x - 3$ (graph the line with a solid line) and the region below that line. The graph of $y > 2x + 1$ is the region above

the line $y = 2x + 1$ (graph the line with a dashed line). The graph of the solution set of the system is the intersection of the graphs of the inequalities.

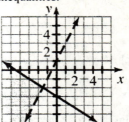

41. First we sketch the graph of $y \geq -2x - 1$ and the graph of $y > \frac{1}{3}x + 2$. The graph of $y \geq -2x - 1$ is the line $y = -2x - 1$ (graph the line with a solid line) and the region above that line. The graph of $y > \frac{1}{3}x + 2$ is the region above the line $y = \frac{1}{3}x + 2$ (graph the line with a dashed line). The graph of the solution set of the system is the intersection of the graphs of the inequalities.

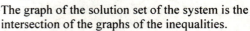
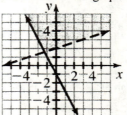

43. First we sketch the graph of $y > -3x + 4$ and the graph of $y \leq 2x + 3$. The graph of $y > -3x + 4$ is the region above the line $y = -3x + 4$ (graph the line with a dashed line). The graph of $y \leq 2x + 3$ is the line $y = 2x + 3$ (graph the line with a solid line) and the region below that line. The graph of the solution set of the system is the intersection of the graphs of the inequalities.

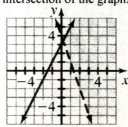

45. First we sketch the graph of $y \leq \frac{2}{3}x$ and the

graph of $y < -\frac{2}{5}x$. The graph of $y \leq \frac{2}{3}x$ is the

line $y = \frac{2}{3}x$ (graph the line with a solid line)

and the region below that line. The graph of

$y < -\frac{2}{5}x$ is the region below the line $y = -\frac{2}{5}x$

(graph the line with a dashed line). The graph of
the solution set of the system is the intersection
of the graphs of the inequalities.

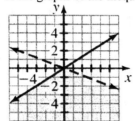

47. First we get y alone on one side of each inequality.

$5x - 3y \leq 12$ $-2y < x$

$-3y \leq -5x + 12$ $y > -\frac{1}{2}x$

$\dfrac{-3y}{-3} \geq \dfrac{-5x + 12}{-3}$

$y \geq \frac{5}{3}x - 4$

Next we sketch the graph of $y \geq \frac{5}{3}x - 4$ and the

graph of $y > -\frac{1}{2}x$. The graph of $y \geq \frac{5}{3}x - 4$ is the

line $y = \frac{5}{3}x - 4$ (graph the line with a solid line)

and the region above that line. The graph of

$y > -\frac{1}{2}x$ is the region above the line $y = -\frac{1}{2}x$

(graph the line with a dashed line). The graph of
the solution set of the system is the intersection of
the graphs of the inequalities.

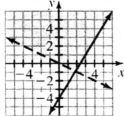

49. First we get y alone on one side of each
inequality.

$x - y \leq 2$ $2x + y < 1$

$-y \leq -x + 2$ $y < -2x + 1$

$y \geq x - 2$

Next we sketch the graph of $y \geq x - 2$ and the

graph of $y < -2x + 1$. The graph of $y \geq x - 2$ is

the line $y = x - 2$ (graph the line with a solid

line) and the region above that line. The graph

of $y < -2x + 1$ is the region below the line

$y = -2x + 1$ (graph the line with a dashed line).

The graph of the solution set of the system is the
intersection of the graphs of the inequalities.

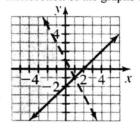

51. First we get y alone on one side of each
inequality.

$2x - 3y > 3$ $3x + 5y \geq 10$

$-3y > -2x + 3$ $5y \geq -3x + 10$

$y < \frac{2}{3}x - 1$ $y \geq -\frac{3}{5}x + 2$

Next we sketch the graph of $y < \frac{2}{3}x - 1$ and the

graph of $y \geq -\frac{3}{5}x + 2$. The graph of $y < \frac{2}{3}x - 1$

is the region below the line $y = \frac{2}{3}x - 1$ (graph

the line with a dashed line). The graph of

$y \geq -\frac{3}{5}x + 2$ is the line $y = -\frac{3}{5}x + 2$ (graph the

line with a solid line) and the region above that
line. The graph of the solution set of the system
is the intersection of the graphs of the
inequalities.

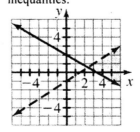

53. First we sketch the graph of $y < 3$ and the graph of $x \geq -2$. The graph of $y < 3$ is the region below the line $y = 3$ (graph the line with a dashed line). The graph of $x \geq -2$ is the vertical line $x = -2$ (graph the line with a solid line) and the rgion to the right of that line. The graph of the solution set of the system is the intersection of the graphs of the inequalities.

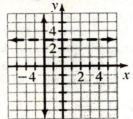

55. First we sketch the graph of $y \leq 1$, the graph of $y \geq -2$, the graph of $x \geq -3$, and the graph of $x \leq 4$. The graph of $y \leq 1$ is the horizontal line $y = 1$ (graph the line with a solid line) and the region below that line. The graph of $y \geq -2$ is the horizontal line $y = -2$ (graph the line with a solid line) and the region above that line. The graph of $x \geq -3$ is the vertical line $x = -3$ (graph the line with a solid line) and the region to the right of that line. The graph of $x \leq 4$ is the vertical line $x = 4$ (graph the line with a solid line) and the region to the left of that line. The graph of the solution set of the system is the intersection of the graphs of the inequalities.

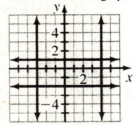

57. First we get y alone on one side of each of the first two inequalities.

$2x - 5y \geq -5 \qquad\qquad 2x - 5y \leq 15$

$\quad -5y \geq -2x - 5 \qquad\quad -5y \leq -2x + 15$

$\qquad y \leq \dfrac{2}{5}x + 1 \qquad\qquad y \geq \dfrac{2}{5}x - 3$

Next, we sketch the graph of $y \leq \dfrac{2}{5}x + 1$, the

graph of $y \geq \dfrac{2}{5}x - 3$, the graph of $x \geq -1$, and

the graph of $x \leq 3$. The graph of $y \leq \dfrac{2}{5}x + 1$ is

the line $y = \dfrac{2}{5}x + 1$ (graph the line with a solid

line) and the region below that line. The graph

of $y \geq \dfrac{2}{5}x - 3$ is the line $y = \dfrac{2}{5}x - 3$ (graph the

line with a solid line) and the region above that line. The graph of $x \geq -1$ is the vertical line $x = -1$ (graph the line with a solid line) and the region to the right of that line. The graph of $x \leq 3$ is the vertical line $x = 3$ (graph the line with a solid line) and the region to the left of that line. The graph of the solution set of the system is the intersection of the graphs of the inequalities.

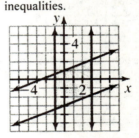

59. a. We are given equations for the upper and lower limits on w and we know that the height is between 63 and 78 inches, inclusive. The system is

$w \leq 3.50h - 80.97$

$w \geq 3.08h - 64.14$

$h \geq 63$

$h \leq 78$

b. First we sketch the graph of $w \leq 3.50h - 80.97$, the graph of $w \geq 3.08h - 64.14$, the graph of $h \geq 63$, and the graph of $h \leq 78$. The graph of $w \leq 3.50h - 80.97$ is the line $w = 3.50h - 80.97$ (graph the line with a solid line) and the region below that line. The graph of $w \geq 3.08h - 64.14$ is the line $w = 3.08h - 64.14$ (graph the line with a solid line) and the region above that line. The graph of $h \geq 63$ is the vertical line $h = 63$ (graph the line with a solid line) and the region to the right of that line. The graph of $h \leq 78$ is the vertical line $h = 78$

(graph the line with a solid line) and the region to the left of that line. The graph of the solution set of the system is the intersection of the graphs of the inequalities.

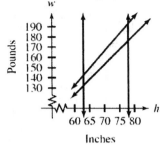

c. Lower limit:

$w = 3.08(68) - 64.14$

$= 209.44 - 64.14$

$= 145.3$

Upper limit:

$w = 3.50(68) - 80.97$

$= 238 - 80.97$

$= 157.03$

The ideal weights are approximately between 145 pounds and 157 pounds, inclusive.

61. First we find the equation of the border line. From the graph we see the y-intercept is $(0, 2)$, the x-intercept is $(4, 0)$, and so the slope is

$m = \dfrac{-2}{4} = -\dfrac{1}{2}$. Therefore, the equation of the

border line is $y = -\dfrac{1}{2}x + 2$. Since the border line

is solid, and the region below the line is shaded,

the inequality is $y \le -\dfrac{1}{2}x + 2$.

63. Start by isolating y on one side of the inequality.
$5x - 2y < 6$

$-2y < -5x + 6$

$y > \dfrac{5}{2}x - 3$

The student is incorrect. The graph of
$5x - 2y < 6$ is the region above the line
$5x - 2y = 6$.

65. a. The points A, B, C, and D satisfy the inequality $y > ax + b$ because the points lie above the line $y = ax + b$.

b. The points A, B, C, F, G, and H satisfy the inequality $y \le cx + d$ because the points either lie on the line $y = cx + d$ or in the region below the line.

c. Comparing the results from parts (a) and (b), we see that the points A, B, and C satisfy both $y > ax + b$ and $y \le cx + d$.

d. Combining the results from parts (a) and (b), we see that the remaining point, E, satisfies neither $y > ax + b$ nor $y \le cx + d$.

67. Answers may vary. Some examples are given below.

Solutions: $(1, 3)$, $(2, 5)$, $(0, 0)$

Not solutions: $(4, 2)$, $(4, 4)$, $(5, 3)$

69. Answers may vary.

71. The equation $y = 2x - 1$ is in slope-intercept form, so the slope is $m = 2$ (or $m = \dfrac{2}{1}$)and the y-intercept is $(0, -1)$. We first plot $(0, -1)$. From this point we move 1 unit to the right and 2 units up, where we plot the point $(1, 1)$. We then sketch the line that contains these two points.

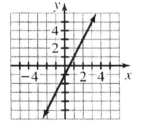

73. The graph of $y \le 2x - 1$ is the line $y = 2x - 1$ and the region below that line. We use a solid line along the border to indicate that the points on the line $y = 2x - 1$ are solutions of $y \le 2x - 1$.

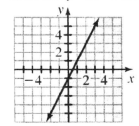

75.
$$3 \le 2x - 1$$
$$3 + 1 \le 2x - 1 + 1$$
$$4 \le 2x$$
$$\frac{4}{2} \le \frac{2x}{2}$$
$$2 \le x \quad \text{or} \quad x \ge 2$$

In interval notation, the solution set is $[2, \infty)$.

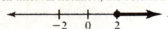

77. First we sketch the graph of $y \le 2x - 1$ and the graph of $y \ge -x + 5$. The graph of $y \le 2x - 1$ is the line $y = 2x - 1$ (graph the line with a solid line) and the region below that line. The graph of $y \ge -x + 5$ is the line $y = -x + 5$ (graph the line with a solid line) and the region above that line. The graph of the solution set of the system is the intersection of the graphs of the inequalities.

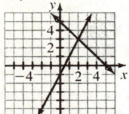

79. $y = 2x - 1$
$y = -x + 5$

Substitute $-x + 5$ for y in the first equation and solve for x.
$$-x + 5 = 2x - 1$$
$$-3x + 5 = -1$$
$$-3x = -6$$
$$x = 2$$

Substitute 2 for x in the equation $y = -x + 5$ and solve for y.
$$y = -(2) + 5 = 3$$

The solution is $(2, 3)$.

81. Answers may vary.

83. Answers may vary.

85. Answers may vary.

Chapter 6 Review Exercises

1. $y = 2x - 3$
$y = -3x + 7$

To begin, we graph both equations in the same coordinate system.

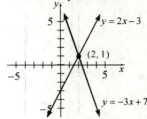

The intersection point is $(2, 1)$. So, the solution is the ordered pair $(2, 1)$.

2. $y = \frac{3}{2}x + 4$

$y = -\frac{1}{2}x - 4$

To begin, we graph both equations in the same coordinate system.

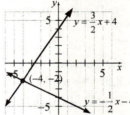

The intersection point is $(-4, -2)$. So, the solution is the ordered pair $(-4, -2)$.

3. $y = \frac{2}{5}x$

$y = -2x$

To begin, we graph both equations in the same coordinate system.

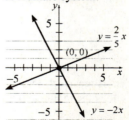

The intersection point is $(0, 0)$. So, the solution is the ordered pair $(0, 0)$.

4. $4x + y = 3$

$-3x + y = -4$

To begin, we rewrite each equation in slope-intercept form.

$4x + y = 3$ $-3x + y = -4$

$y = -4x + 3$ $y = 3x - 4$

Next, we graph both equations in the same coordinate system.

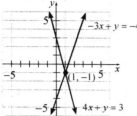

The intersection point is $(1, -1)$. So, the solution is the ordered pair $(1, -1)$.

5. $-x + y = 4$

$2x + y = -5$

To begin, we rewrite each equation in slope-intercept form.

$-x + y = 4$ $2x + y = -5$

$y = x + 4$ $y = -2x - 5$

Next, we graph both equations in the same coordinate system.

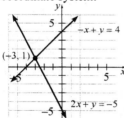

The intersection point is $(-3, 1)$. So, the solution is the ordered pair $(-3, 1)$.

6. $x - 3y = 3$

$2x + 3y = -12$

To begin, we rewrite each equation in slope-intercept form.

$x - 3y = 3$ $2x + 3y = -12$

$-3y = -x + 3$ $3y = -2x - 12$

$y = \dfrac{1}{3}x - 1$ $y = -\dfrac{2}{3}x - 4$

Next, we graph both equations in the same coordinate system.

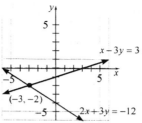

The intersection point is $(-3, -2)$. So, the solution is the ordered pair $(-3, -2)$.

7. $3x - 2y = 11$

$y = 5x - 16$

Substitute $5x - 16$ for y in the first equation and solve for x.

$3x - 2(5x - 16) = 11$

$3x - 10x + 32 = 11$

$-7x + 32 = 11$

$-7x = -21$

$x = 3$

Substitute 3 for x in the equation $y = 5x - 16$ and solve for y.

$y = 5(3) - 16 = -1$

The solution is $(3, -1)$.

8. $4x - 3y - 5 = 0$

$x = 4 - 2y$

Substitute $4 - 2y$ for x in the first equation and solve for y.

$4(4 - 2y) - 3y - 5 = 0$

$16 - 8y - 3y - 5 = 0$

$-11y + 11 = 0$

$-11y = -11$

$y = 1$

Substitute 1 for y in the equation $x = 4 - 2y$ and solve for x.

$x = 4 - 2(1) = 4 - 2 = 2$

The solution is $(2, 1)$.

9. $y = -5x$

$y = 2x$

Substitute $2x$ for y in the first equation and solve for x.

$2x = -5x$

$7x = 0$

$x = 0$

Substitute 0 for x in the equation $y = 2x$ and solve for y.

$y = 2(0) = 0$

The solution is $(0,0)$.

10. $y = -3(x+2)$

$y = 4(x-5)$

Substitute $4(x-5)$ for y in the first equation and solve for x.

$4(x-5) = -3(x+2)$

$4x - 20 = -3x - 6$

$7x - 20 = -6$

$7x = 14$

$x = 2$

Substitute 2 for x in the equation $y = 4(x-5)$ and solve for y.

$y = 4(2-5) = 4(-3) = -12$

The solution is $(2,-12)$.

11. $x + y = -1$

$2x - y = 4$

Solve the first equation for y.

$x + y = -1$

$y = -x - 1$

Substitute $-x-1$ for y in the second equation and solve for x.

$2x - (-x-1) = 4$

$2x + x + 1 = 4$

$3x + 1 = 4$

$3x = 3$

$x = 1$

Substitute 1 for x in the equation $y = -x - 1$ and solve for y.

$y = -(1) - 1 = -2$

The solution is $(1,-2)$.

12. $x + 2y = 5$

$4x + 2y = -4$

Solve the first equation for x.

$x + 2y = 5$

$x = -2y + 5$

Substitute $-2y+5$ for x in the second equation and solve for y.

$4(-2y+5) + 2y = -4$

$-8y + 20 + 2y = -4$

$-6y + 20 = -4$

$-6y = -24$

$y = 4$

Substitute 4 for y in the equation $x = -2y+5$ and solve for x.

$x = -2(4) + 5 = -8 + 5 = -3$

The solution is $(-3,4)$.

13. $y = -2.19x + 3.51$

$y = 1.54x - 6.22$

Substitute $1.54x - 6.22$ for y in the first equation and solve for x.

$1.54x - 6.22 = -2.19x + 3.51$

$3.73x - 6.22 = 3.51$

$3.73x = 9.73$

$x = \dfrac{9.73}{3.73} \approx 2.609$

Substitute 2.609 for x in the equation $y = 1.54x - 6.22$ and solve for y.

$y = 1.54(2.609) - 6.22 \approx -2.202$

The approximate solution is $(2.61, -2.20)$.

14. $y = -4.98x - 1.18$

$y = -0.57x + 4.08$

Substitute $-0.57x + 4.08$ for y in the first equation and solve for x.

$-0.57x + 4.08 = -4.98x - 1.18$

$4.41x + 4.08 = -1.18$

$4.41x = -5.26$

$x = \dfrac{-5.26}{4.41} \approx -1.193$

Substitute -1.193 for x in the equation $y = -0.57x + 4.08$ and solve for y.

$y = -0.57(-1.193) + 4.08 \approx 4.760$

The approximate solution is $(-1.19, 4.76)$.

15. $x - 2y = -1$ Equation (1)

$3x + 5y = 19$ Equation (2)

To eliminate the x terms, we multiply both sides of equation (1) by -3, yielding the system

$-3x + 6y = 3$

$3x + 5y = 19$

The coefficients of the x terms are equal in absolute value and opposite in sign. Add the left sides and the right sides of the equations and solve for y.

$-3x + 6y = 3$

$\underline{3x + 5y = 19}$

$11y = 22$

$y = 2$

Substitute 2 for y in equation (1) and solve for x.

$x - 2(2) = -1$

$x - 4 = -1$

$x = 3$

The solution is $(3, 2)$.

16. $2x - 5y = -3$ Equation (1)

$4x + 3y = -19$ Equation (2)

To eliminate the x terms, we multiply both sides of equation (1) by -2, yielding the system

$-4x + 10y = 6$

$4x + 3y = -19$

The coefficients of the x terms are equal in absolute value and opposite in sign. Add the left sides and the right sides of the equations and solve for y.

$-4x + 10y = 6$

$\underline{4x + 3y = -19}$

$13y = -13$

$y = -1$

Substitute -1 for y in equation (1) and solve for x.

$2x - 5(-1) = -3$

$2x + 5 = -3$

$2x = -8$

$x = -4$

The solution is $(-4, -1)$.

17. $3x + 8y = 2$ Equation (1)

$5x - 2y = -12$ Equation (2)

To eliminate the y terms, we multiply both sides of equation (2) by 4, yielding the system

$3x + 8y = 2$

$20x - 8y = -48$

The coefficients of the y terms are equal in absolute value and opposite in sign. Add the left sides and the right sides of the equations and solve for x.

$3x + 8y = 2$

$\underline{20x - 8y = -48}$

$23x = -46$

$x = -2$

Substitute -2 for x in equation (2) and solve for y.

$5(-2) - 2y = -12$

$-10 - 2y = -12$

$-2y = -2$

$y = 1$

The solution is $(-2, 1)$.

18. $4x - 3y = -6$ Equation (1)

$-7x + 5y = 11$ Equation (2)

To eliminate the y terms, we multiply both sides of equation (1) by 5 and both sides of equation (2) by 3, yielding the system

$20x - 15y = -30$

$-21x + 15y = 33$

The coefficients of the y terms are equal in absolute value and opposite in sign. Add the left sides and the right sides of the equations and solve for x.

$20x - 15y = -30$

$\underline{-21x + 15y = 33}$

$-x = 3$

$x = -3$

Substitute -3 for x in equation (1) and solve for y.

$4(-3) - 3y = -6$

$-12 - 3y = -6$

$-3y = 6$

$y = -2$

The solution is $(-3, -2)$.

19. $-2x - 5y = 2$ Equation (1)

     $3x + 6y = 0$ Equation (2)

To eliminate the x terms, we multiply both sides of equation (1) by 3 and both sides of equation (2) by 2, yielding the system

$-6x - 15y = 6$

   $6x + 12y = 0$

The coefficients of the x terms are equal in absolute value and opposite in sign. Add the left sides and the right sides of the equations and solve for y.

$-6x - 15y = 6$

$\underline{6x + 12y = 0}$

     $-3y = 6$

        $y = -2$

Substitute -2 for y in equation (2) and solve for x.

$3x + 6(-2) = 0$

   $3x - 12 = 0$

       $3x = 12$

         $x = 4$

The solution is $(4, -2)$.

20. $y = 3x - 5$ Equation (1)

   $y = -2x + 5$ Equation (2)

$\rightarrow$ $3x - y = 5$

     $2x + y = 5$

The coefficients of the y terms are equal in absolute value and opposite in sign. Add the left sides and the right sides of the equations and solve for x.

$3x - y = 5$

$\underline{2x + y = 5}$

$5x = 10$

   $x = 2$

Substitute 2 for x in equation (2) and solve for y.

$y = -2(2) + 5 = -4 + 5 = 1$

The solution is $(2, 1)$.

21. $2(x + 3) + y = 6$

   $x - 3(y - 2) = -1$

Begin by writing each equation in standard form.

$2(x + 3) + y = 6$ $x - 3(y - 2) = -1$

  $2x + 6 + y = 6$ $x - 3y + 6 = -1$

     $2x + y = 0$ $x - 3y = -7$

This yields the system

$2x + y = 0$ Equation (1)

$x - 3y = -7$ Equation (2)

To eliminate the y terms, we multiply both sides of equation (1) by 3, yielding the system

$6x + 3y = 0$

  $x - 3y = -7$

The coefficients of the y terms are equal in absolute value and opposite in sign. Add the left sides and the right sides of the equations and solve for x.

$6x + 3y = 0$

$\underline{x - 3y = -7}$

   $7x = -7$

    $x = -1$

Substitute -1 for x in equation (1) and solve for y.

$2(-1) + y = 0$

   $-2 + y = 0$

        $y = 2$

The solution is $(-1, 2)$.

22. $\dfrac{1}{2}x - \dfrac{2}{3}y = -\dfrac{5}{3}$ Equation (1)

   $\dfrac{1}{3}x - \dfrac{3}{2}y = -\dfrac{13}{6}$ Equation (2)

To eliminate the x terms (and clear fractions), we multiply both sides of equation (1) by -12 and both sides of equation (2) by 18, yielding the system

$-6x + 8y = 20$

$6x - 27y = -39$

The coefficients of the x terms are equal in absolute value and opposite in sign. Add the left sides and the right sides of the equations and solve for y.

$-6x + 8y = 20$

$\underline{6x - 27y = -39}$

   $-19y = -19$

       $y = 1$

Substitute 1 for y in equation (2) and solve for y.

$$\frac{1}{3}x - \frac{3}{2}(1) = -\frac{13}{6}$$

$$\frac{1}{3}x - \frac{3}{2} = -\frac{13}{6}$$

$$\frac{1}{3}x = -\frac{13}{6} + \frac{9}{6}$$

$$\frac{1}{3}x = -\frac{4}{6}$$

$$x = -\frac{4}{2} = -2$$

The solution is $(-2,1)$.

23. $y = 4.59x + 1.25$ Equation (1)

$\quad y = 0.52x + 4.39$ Equation (2)

$\rightarrow \quad 4.59x - y = -1.25$

$\quad\quad -0.52x + y = 4.39$

The coefficients of the y terms are equal in absolute value and opposite in sign. Add the left sides and the right sides of the equations and solve for x.

$$4.59x - y = -1.25$$
$$\underline{-0.52x + y = 4.39}$$
$$4.07x = 3.14$$

$$x = \frac{3.14}{4.07} \approx 0.771$$

Substitute 0.771 for x in equation (1) and solve for y.

$$y = 4.59(0.771) + 1.25 \approx 4.789$$

The approximate solution is $(0.77, 4.79)$.

24. $y = 0.91x - 3.57$ Equation (1)

$\quad y = -3.58x + 6.05$ Equation (2)

$\rightarrow \quad 0.91x - y = 3.57$

$\quad\quad 3.58x + y = 6.05$

The coefficients of the y terms are equal in absolute value and opposite in sign. Add the left sides and the right sides of the equations and solve for x.

$$0.91x - y = 3.57$$
$$\underline{3.58x + y = 6.05}$$
$$4.49x = 9.62$$

$$x = \frac{9.62}{4.49} \approx 2.143$$

Substitute 2.143 for x in equation (1) and solve for y.

$$y = 0.91(2.143) - 3.57 \approx -1.620$$

The approximate solution is $(2.14, -1.62)$.

25. $2x - 7y = -13$

$\quad 5x + 3y = -12$

To eliminate the x terms, we multiply both sides of equation (1) by 5 and both sides of equation (2) by -2, yielding the system

$$10x - 35y = -65$$
$$-10x - 6y = 24$$

The coefficients of the x terms are equal in absolute value and opposite in sign. Add the left sides and the right sides of the equations and solve for y.

$$10x - 35y = -65$$
$$\underline{-10x - 6y = 24}$$
$$-41y = -41$$
$$y = 1$$

Substitute 1 for y in equation (1) and solve for x.

$$2x - 7(1) = -13$$
$$2x - 7 = -13$$
$$2x = -6$$
$$x = -3$$

The solution is $(-3,1)$.

26. $4x + 7y = 8$

$\quad\quad x = 3 - 2y$

Substitute $3 - 2y$ for x in the first equation and solve for y.

$$4(3 - 2y) + 7y = 8$$
$$12 - 8y + 7y = 8$$
$$12 - y = 8$$
$$-y = -4$$
$$y = 4$$

Substitute 4 for y in the equation $x = 3 - 2y$ and solve for x.

$$x = 3 - 2(4) = 3 - 8 = -5$$

The solution is $(-5,4)$.

27. $-3x + 7y = 6$

$\quad\quad 6x + 2y = -12$

To eliminate the x terms, we multiply both sides of equation (1) by 2, yielding the system

$$-6x + 14y = 12$$
$$6x + 2y = -12$$

The coefficients of the x terms are equal in absolute value and opposite in sign. Add the left sides and the right sides of the equations and solve for y.

$$-6x+14y=12$$
$$\underline{6x+2y=-12}$$
$$16y=0$$
$$y=0$$

Substitute 0 for y in equation (1) and solve for x.
$$-3x+7(0)=6$$
$$-3x=6$$
$$x=-2$$

The solution is $(-2,0)$.

28. $y=-x+7$

$y=2x-5$

Substitute $2x-5$ for y in the first equation and solve for x.
$$2x-5=-x+7$$
$$3x-5=7$$
$$3x=12$$
$$x=4$$

Substitute 4 for x in the equation $y=2x-5$ and solve for y.
$$y=2(4)-5=8-5=3$$

The solution is $(4,3)$.

29. $4x+5y=-6$

$2y=-3x-8$

Solve the second equation for y.
$$2y=-3x-8$$
$$y=-\frac{3}{2}x-4$$

Substitute $-\frac{3}{2}x-4$ for y in the first equation and solve for x.
$$4x+5\left(-\frac{3}{2}x-4\right)=-6$$
$$4x-\frac{15}{2}x-20=-6$$
$$-\frac{7}{2}x-20=-6$$
$$-\frac{7}{2}x=14$$
$$x=-4$$

Substitute -4 for x in the equation $y=-\frac{3}{2}x-4$ and solve for y.
$$y=-\frac{3}{2}(-4)-4=6-4=2$$

The solution is $(-4,2)$.

30. $y=x-2$

$3x+5y-30=0$

Substitute $x-2$ for y in the second equation and solve for x.
$$3x+5(x-2)-30=0$$
$$3x+5x-10-30=0$$
$$8x-40=0$$
$$8x=40$$
$$x=5$$

Substitute 5 for x in the equation $y=x-2$ and solve for y.
$$y=5-2=3$$

The solution is $(5,3)$.

31. $2x-3y=0$

$5x-7y=-1$

To eliminate the x terms, we multiply both sides of equation (1) by 5 and both sides of equation (2) by -2, yielding the system
$$10x-15y=0$$
$$-10x+14y=2$$

The coefficients of the x terms are equal in absolute value and opposite in sign. Add the left sides and the right sides of the equations and solve for y.
$$10x-15y=0$$
$$\underline{-10x+14y=2}$$
$$-y=2$$
$$y=-2$$

Substitute -2 for y in equation (1) and solve for x.
$$2x-3(-2)=0$$
$$2x+6=0$$
$$2x=-6$$
$$x=-3$$

The solution is $(-3,-2)$.

32. $2(4x-3)-5y=12$

$5(3x-1)+2y=6$

Begin by writing each equation in standard form.

$$2(4x-3)-5y=12 \qquad 5(3x-1)+2y=6$$
$$8x-6-5y=12 \qquad 15x-5+2y=6$$
$$8x-5y=18 \qquad 15x+2y=11$$

This yields the system

$8x - 5y = 18$ Equation (1)

$15x + 2y = 11$ Equation (2)

To eliminate the y terms, we multiply both sides of equation (1) by 2 and both sides of equation (2) by 5, yielding the system

$16x - 10y = 36$

$75x + 10y = 55$

The coefficients of the y terms are equal in absolute value and opposite in sign. Add the left sides and the right sides of the equations and solve for x.

$16x - 10y = 36$

$\underline{75x + 10y = 55}$

$\qquad 91x = 91$

$\qquad x = 1$

Substitute 1 for x in equation (2) and solve for y.

$15(1) + 2y = 11$

$15 + 2y = 11$

$2y = -4$

$y = -2$

The solution is $(1, -2)$.

33. $2x - 6y = 4$ Equation (1)

$-3x + 9y = -3$ Equation (2)

To eliminate the x terms, we multiply both sides of equation (1) by 3 and both sides of equation (2) by 2, yielding the system

$6x - 18y = 12$

$-6x + 18y = -6$

The coefficients of the x terms are equal in absolute value and opposite in sign. Add the left sides and the right sides of the equations and solve for y.

$6x - 18y = 12$

$\underline{-6x + 18y = -6}$

$\qquad 0 = 6$ false

The result is a false statement. Therefore, the system is inconsistent. The solution set is the empty set, $\varnothing$.

34. $y = -4x + 3 \rightarrow$ $4x + y = 3$ Equation (1)

$8x + 2y = 6$ $8x + 2y = 6$ Equation (2)

To eliminate the x terms, we multiply both sides of equation (1) by -2, yielding the system

$-8x - 2y = -6$

$8x + 2y = 6$

The coefficients of the x terms are equal in absolute value and opposite in sign. Add the left sides and the right sides of the equations and solve for y.

$-8x - 2y = -6$

$\underline{8x + 2y = 6}$

$\qquad 0 = 0$ true

The result is a true statement of the form $a = a$. Therefore, the system is dependent. The solution set is the infinite set of ordered pairs that satisfy the equation $y = -4x + 3$.

35. a.   Answers may vary. One possibility: $(1, 5)$

    b.   Answers may vary. One possibility: $(1, 0)$

    c.   Answers may vary. One possibility: $(1, 1)$

    d.   A point that satisfies both equations would be a solution to the system.

$y = -2x + 7$

$y = 3x - 3$

Substitute $3x - 3$ for y in the first equation and solve for x.

$3x - 3 = -2x + 7$

$5x - 3 = 7$

$5x = 10$

$x = 2$

Substitute 2 for x in the equation $y = 3x - 3$ and solve for y.

$y = 3(2) - 3 = 6 - 3 = 3$

The solution of the system is $(2, 3)$.

Therefore, an ordered pair that satisfies both equations is $(2, 3)$.

36. We can start by writing equations for the two lines.

Red Line:

Examining the graph, it appears that the points $(-1,3)$ and $(4,5)$ are on the graph.

$$m = \frac{5-3}{4-(-1)} = \frac{2}{5}$$

$$y - y_1 = m(x - x_1)$$

$$y - 5 = \frac{2}{5}(x-4)$$

$$y - 5 = \frac{2}{5}x - \frac{8}{5}$$

$$y = \frac{2}{5}x + \frac{17}{5}$$

Blue Line:

Examining the graph, it appears that the points $(-2,3)$ and $(2,2)$ are on the graph.

$$m = \frac{2-3}{2-(-2)} = \frac{-1}{4} = -\frac{1}{4}$$

$$y - y_1 = m(x - x_1)$$

$$y - 2 = -\frac{1}{4}(x-2)$$

$$y - 2 = -\frac{1}{4}x + \frac{1}{2}$$

$$y = -\frac{1}{4}x + \frac{5}{2}$$

Using the two equations, we get the system

$$y = \frac{2}{5}x + \frac{17}{5}$$

$$y = -\frac{1}{4}x + \frac{5}{2}$$

Substitute $-\frac{1}{4}x + \frac{5}{2}$ for y in the first equation and solve for x.

$$-\frac{1}{4}x + \frac{5}{2} = \frac{2}{5}x + \frac{17}{5}$$

$$-\frac{13}{20}x + \frac{5}{2} = \frac{17}{5}$$

$$-\frac{13}{20}x = \frac{9}{10}$$

$$x = -\frac{18}{13} \approx -1.385$$

Substitute -1.385 for x in the equation

$$y = -\frac{1}{4}x + \frac{5}{2} \text{ and solve for } y.$$

$$y = -\frac{1}{4}(-1.385) + \frac{5}{2} \approx 2.846$$

The approximate solution is $(-1.4, 2.8)$.

37. We begin by finding the equations for each table.

Table 37:

The input $x = 0$ yields the output $y = 75$, so the y-intercept is $(0, 75)$.

$$m = \frac{73 - 75}{1 - 0} = \frac{-2}{1} = -2 \text{, so the slope is } -2.$$

The first equation is $y = -2x + 75$.

Table 38:

The input $x = 0$ yields the output $y = 5$, so the y-intercept is $(0, 5)$.

$$m = \frac{8-5}{1-0} = \frac{3}{1} = 3 \text{, so the slope is } 3.$$

The second equation is $y = 3x + 5$.

Now we solve the system

$$y = -2x + 75$$

$$y = 3x + 5$$

Substitute $3x + 5$ for y in the first equation and solve for x.

$$3x + 5 = -2x + 75$$

$$5x + 5 = 75$$

$$5x = 70$$

$$x = 14$$

Substitute 14 for x in the equation $y = 3x + 5$ and solve for y.

$$y = 3(14) + 5 = 42 + 5 = 47$$

The solution is $(14, 47)$.

38.
$$3x + 4y = 15$$
$$2y = -5x + 11$$

Graphing by hand:

Begin by writing each equation in slope-intercept form.

$$3x + 4y = 15 \qquad\qquad 2y = -5x + 11$$

$$4y = -3x + 15 \qquad\qquad y = -\frac{5}{2}x + \frac{11}{2}$$

$$y = -\frac{3}{4}x + \frac{15}{4}$$

Next we graph both equations in the same coordinate system.

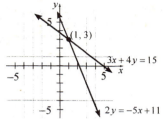

The intersection point is $(1,3)$. So, the ordered pair $(1,3)$ is the solution.

Substitution:
$$3x + 4y = 15$$
$$2y = -5x + 11$$

Solve the second equation for y.
$$2y = -5x + 11$$
$$y = -\frac{5}{2}x + \frac{11}{2}$$

Substitute $-\frac{5}{2}x + \frac{11}{2}$ for y in the first equation and solve for x.
$$3x + 4\left(-\frac{5}{2}x + \frac{11}{2}\right) = 15$$
$$3x - 10x + 22 = 15$$
$$-7x + 22 = 15$$
$$-7x = -7$$
$$x = 1$$

Substitute 1 for x in the equation $y = -\frac{5}{2}x + \frac{11}{2}$ and solve for y.
$$y = -\frac{5}{2}(1) + \frac{11}{2} = \frac{6}{2} = 3$$
The solution is $(1,3)$.

Elimination:
$$3x + 4y = 15$$
$$2y = -5x + 11$$

Start by writing both equations in standard form.
$$3x + 4y = 15 \qquad\qquad 2y = -5x + 11$$
$$5x + 2y = 11$$

This yields the system
$$3x + 4y = 15 \quad \text{Equation (1)}$$
$$5x + 2y = 11 \quad \text{Equation (2)}$$

To eliminate the y terms, we multiply both sides of equation (2) by -2, yielding the system
$$3x + 4y = 15$$
$$-10x - 4y = -22$$

The coefficients of the y terms are equal in absolute value and opposite in sign. Add the left sides and the right sides of the equations and solve for x.
$$3x + 4y = 15$$
$$\underline{-10x - 4y = -22}$$
$$-7x = -7$$
$$x = 1$$

Substitute 1 for x in equation (1) and solve for y.
$$3(1) + 4y = 15$$
$$3 + 4y = 15$$
$$4y = 12$$
$$y = 3$$

The solution is $(1,3)$.

Preference may vary.

39. Start by creating a scatter diagram of the data.

The data appear fairly linear so we use the regression feature of a graphing utility to obtain the line of best fit.

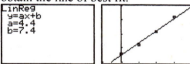

A reasonable model for the data is
$p = 4.4t + 7.4$.

b. Start by creating a scatter diagram of the data.

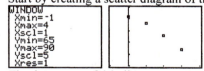

The data appear fairly linear so we use the regression feature of a graphing utility to obtain the line of best fit.

A reasonable model for the data is
$p = -4.7t + 86.8$.

c. To estimate when the percentages will be equal, we solve the system

$p = 4.4t + 7.4$

$p = -4.7t + 86.8$

Substitute $-4.7t + 86.8$ for p in the first equation and solve for t.

$-4.7t + 86.8 = 4.4t + 7.4$

$-9.1t + 86.8 = 7.4$

$-9.1t = -79.4$

$t = \dfrac{-79.4}{-9.1} \approx 8.725$

Substitute 8.725 for t in the equation $p = -4.7t + 86.8$ and solve for p.

$p = -4.7(8.725) + 86.8 \approx 45.793$

The approximate solution of the system is $(8.73, 45.79)$.

According to the models, the percentage of airplanes that are regional jets will be equal to the percentage that are turboprops or large jets (45.8%) in 2009.

40. a. Since the percentage of car and light-truck sales that were car sales decreases by 1.25 percentage points each year, we can model the situation by a linear equation. The slope is -1.25 percentage points per year. Since the percentage that were car sales in 1995 was 58 percentage points, the p-intercept is $(0, 58)$. So, a reasonable model is $p = -1.25t + 58$.

b. By the same reasoning as in part (a), the model is $p = 1.25t + 42$.

c. $p = -1.25t + 58$

$p = 1.25t + 42$

Substitute $1.25t + 42$ for p in the first equation and solve for t.

$1.25t + 42 = -1.25t + 58$

$2.50t + 42 = 58$

$2.50t = 16$

$t = \dfrac{16}{2.5} = 6.4$

Substitute 6.4 for t in the equation $p = 1.25t + 42$ and solve for d.

$p = 1.25(6.4) + 42 = 50$

The solution of the system is $(6.4, 50)$.

According to the models, the sales of cars and light trucks were equal (50% each) in 2001.

d. When the sales are equal, the percentage will be 50% each. Substitute 50 for p in the equation $p = -1.25t + 58$ and solve for t.

$50 = -1.25t + 58$

$-8 = -1.25t$

$\dfrac{-8}{-1.25} = t$

$6.4 = t$

The sales of cars and light trucks were equal (50% each) in 2001.

e. The slopes are equal in absolute value, but opposite in sign. Since there are only two types of vehicles in the group, any increase in the percentage of sales of one type must correspond to an equal decrease in the other type. Therefore, it makes sense that the slopes (average rate of change) are equal in absolute value, but opposite in sign.

41. Let L = the length of the rectangle in feet and W = the width of the rectangle in feet.

$L = 3W + 2 \qquad\qquad P = 2L + 2W$

$\qquad\qquad\qquad\qquad 44 = 2L + 2W$

To determine the dimensions of the rectangle, we solve the system

$L = 3W + 2$

$44 = 2L + 2W$

Substitute $3W + 2$ for L in the second equation and solve for w.

$44 = 2(3W + 2) + 2W$

$44 = 8W + 4$

$40 = 8W$

$5 = W$

Substitute 5 for W in the equation $L = 3W + 2$ and solve for L.

$L = 3(5) + 2 = 17$

The width of the rectangle is 5 feet and the length is 17 feet.

42. Let x = the number of $22 tickets and y = the number of $39 tickets.

There are a total of 8000 tickets, so our first equation is $x + y = 8000$.

The total revenue is $201,500. The revenue for each ticket type is obtained by multiplying the ticket price by the number of tickets sold at that price. Therefore, our second equation is

$22x + 39y = 201,500$

The system is

$x + y = 8000$ Equation (1)

$22x + 39y = 201,500$ Equation (2)

To eliminate the x terms, multiply both sides of equation (1) by -22, yielding the system

$-22x - 22y = -176,000$

$22x + 39y = 201,500$

Add the left sides and right sides and solve for y.

$-22x - 22y = -176,000$

$\underline{22x + 39y = 201,500}$

$17y = 25,500$

$y = 1500$

Substitute 1500 for y in equation (1) and solve for x.

$x + (1500) = 8000$

$x = 6500$

The theater should sell 6500 of the $22 tickets and $1500 of the $39 tickets.

43. The graph of $y \le 3x - 5$ is the line $y = 3x - 5$ and the region below that line. We use a solid line along the border to indicate that the points on the line $y = 3x - 5$ are solutions of $y \le 3x - 5$.

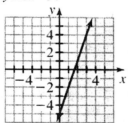

44. The graph of $y \ge -2x + 4$ is the line $y = -2x + 4$ and the region above that line. We use a solid line along the border to indicate that the points on the line $y = -2x + 4$ are solutions of $y \ge -2x + 4$.

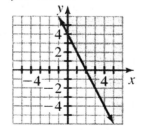

45. First, we get y alone on one side of the inequality.

$3x - 2y > 4$

$-2y > -3x + 4$

$y < \dfrac{3}{2}x - 2$

The graph of $y < \dfrac{3}{2}x - 2$ is the region below that line $y = \dfrac{3}{2}x - 2$. We use a dashed line along the border to indicate that the points on the line $y = \dfrac{3}{2}x - 2$ are not solutions of $3x - 2y > 4$.

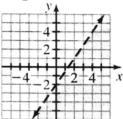

46. First, we get y alone on one side of the inequality.

$2y - 5x < 0$

$2y < 5x$

$y < \dfrac{5}{2}x$

The graph of $y < \dfrac{5}{2}x$ is the region below that line $y = \dfrac{5}{2}x$. We use a dashed line along the border to indicate that the points on the line $y = \dfrac{5}{2}x$ are not solutions of $2y - 5x < 0$.

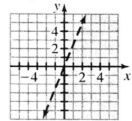

47. $x \geq 3$

 The graph of $x \geq 3$ is the vertical line $x = 3$ and the region to the right of that line. We use a solid line to indicate that the points on the line $x = 3$ are solutions of $x \geq 3$.

 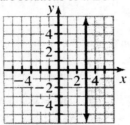

48. $y < -2$

 The graph of $y < -2$ is the region below the horizontal line $y = -2$. We use a dashed line to indicate that the points on the line $y = -2$ are not solutions of $y < -2$.

 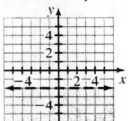

49. First we sketch the graph of $y > x + 1$ and the graph of $y \leq -2x + 5$. The graph of $y > x + 1$ is the region above the line $y = x + 1$ (graph the line with a dashed line). The graph of $y \leq -2x + 5$ is the line $y = -2x + 5$ (graph the line with a solid line) and the region below that line. The graph of the solution set of the system is the intersection of the graphs of the inequalities.

 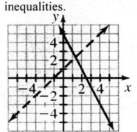

50. First we sketch the graph of $y \geq \frac{3}{5}x + 1$ and the graph of $x < -1$. The graph of $y \geq \frac{3}{5}x + 1$ is the line $y = \frac{3}{5}x + 1$ (graph the line with a solid line) and the region above the line. The graph of $x < -1$ is the region to the left of the vertical line $x = -1$ (graph the line with a dashed line). The graph of the solution set of the system is the intersection of the graphs of the inequalities.

 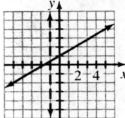

51. First we get y alone on one side of each inequality.

 $$3x - 4y \geq 12 \qquad\qquad 5y \leq -3x$$
 $$-4y \geq -3x + 12 \qquad\quad y \leq -\frac{3}{5}x$$
 $$y \leq \frac{3}{4}x - 3$$

 Next we sketch the graph of $y \leq \frac{3}{4}x - 3$ and the graph of $y \leq -\frac{3}{5}x$. The graph of $y \leq \frac{3}{4}x - 3$ is the line $y = \frac{3}{4}x - 3$ (graph the line with a solid line) and the region below that line. The graph of $y \leq -\frac{3}{5}x$ is the line $y = -\frac{3}{5}x$ (graph the line with a solid line) and the region below that line. The graph of the solution set of the system is the intersection of the graphs of the inequalities.

 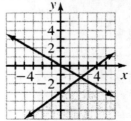

52. First we sketch the graph of $x > 2$ and the graph of $y \le -1$. The graph of $x > 2$ is the region to the right of the vertical line $x = 2$ (graph the line with a dashed line). The graph of $y \le -1$ is the line $y = -1$ (graph the line with a solid line) and the region below that line. The graph of the solution set of the system is the intersection of the graphs of the inequalities.

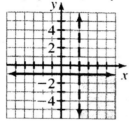

53. a. We are given equations for the upper and lower limits on w and we know that the height is between 58 and 74 inches, inclusive. The system is

$$w \le 3h - 54$$
$$w \ge 3h - 68$$
$$h \ge 58$$
$$h \le 74$$

b. First we sketch the graph of $w \le 3h - 54$, the graph of $w \ge 3h - 68$, the graph of $h \ge 58$, and the graph of $h \le 74$. The graph of $w \le 3h - 54$ is the line $w = 3h - 54$ (graph the line with a solid line) and the region below that line. The graph of $w \ge 3h - 68$ is the line $w = 3h - 68$ (graph the line with a solid line) and the region above that line. The graph of $h \ge 58$ is the vertical line $h = 58$ (graph the line with a solid line) and the region to the right of that line. The graph of $h \le 74$ is the vertical line $h = 74$ (graph the line with a solid line) and the region to the left of that line. The graph of the solution set of the system is the intersection of the graphs of the inequalities.

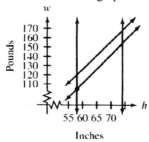

c. Upper limit:
$$w = 3(70) - 54$$
$$= 210 - 54$$
$$= 156$$
Lower limit:
$$w = 3(70) - 68$$
$$= 210 - 68$$
$$= 142$$

The ideal weights are approximately between 142 pounds and 156 pounds, inclusive.

54. Answers may vary. Some possible answers:

Solutions: $(2,-1)$, $(5,1)$, $(10,5)$

Non-solutions: $(0,0)$, $(1,2)$, $(3,2)$

Chapter 6 Test

1. $y = -\dfrac{2}{5}x - 1$

$y = -2x + 7$

To begin, we graph both equations in the same coordinate system.

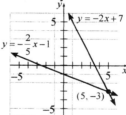

The intersection point is $(5,-3)$. So, the solution is the ordered pair $(5,-3)$.

2. $y = \dfrac{2}{3}x + 4$

$3x + 4y = -2$

To begin, write both equations in slope-intercept form.

$y = \dfrac{2}{3}x + 4$ $3x + 4y = -2$

                                $4y = -3x - 2$

                                $y = -\dfrac{3}{4}x - \dfrac{1}{2}$

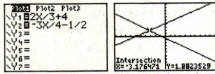

The intersection point is approximately $(-3.176, 1.882)$, so the approximate solution to the system is $(-3.18, 1.88)$.

3. $5x - 2y = 4$
   $$y = 3x - 1$$
   Substitute $3x - 1$ for y in the first equation and solve for x.
   $$5x - 2(3x - 1) = 4$$
   $$5x - 6x + 2 = 4$$
   $$-x + 2 = 4$$
   $$-x = 2$$
   $$x = -2$$
   Substitute -2 for x in the equation $y = 3x - 1$ and solve for y.
   $$y = 3(-2) - 1 = -6 - 1 = -7$$
   The solution is $(-2, -7)$.

4. $3x + 4y = 9$
   $$x - 2y = -7$$
   Solve the second equation for x.
   $$x - 2y = -7$$
   $$x = 2y - 7$$
   Substitute $2y - 7$ for x in the first equation and solve for y.
   $$3(2y - 7) + 4y = 9$$
   $$6y - 21 + 4y = 9$$
   $$10y - 21 = 9$$
   $$10y = 30$$
   $$y = 3$$
   Substitute 3 for y in the equation $x = 2y - 7$ and solve for x.
   $$x = 2(3) - 7 = 6 - 7 = -1$$
   The solution is $(-1, 3)$.

5. $-7x - 2y = -8$ Equation (1)
   $$5x + 4y = -2$$ Equation (2)
   To eliminate the y terms, we multiply both sides of equation (1) by 2, yielding the system
   $$-14x - 4y = -16$$
   $$5x + 4y = -2$$
   The coefficients of the y terms are equal in absolute value and opposite in sign. Add the left sides and the right sides of the equations and solve for x.
   $$-14x - 4y = -16$$
   $$\underline{5x + 4y = -2}$$
   $$-9x = -18$$
   $$x = 2$$
   Substitute 2 for x in equation (2) and solve for y.
   $$5(2) + 4y = -2$$
   $$10 + 4y = -2$$
   $$4y = -12$$
   $$y = -3$$
   The solution is $(2, -3)$.

6. $2x - 5y = -18$
   $$3x + 4y = -4$$
   To eliminate the x terms, we multiply both sides of equation (1) by 3 and both sides of equation (2) by -2, yielding the system
   $$6x - 15y = -54$$
   $$-6x - 8y = 8$$
   The coefficients of the x terms are equal in absolute value and opposite in sign. Add the left sides and the right sides of the equations and solve for y.
   $$6x - 15y = -54$$
   $$\underline{-6x - 8y = 8}$$
   $$-23y = -46$$
   $$y = 2$$
   Substitute 2 for y in equation (2) and solve for x.
   $$3x + 4(2) = -4$$
   $$3x + 8 = -4$$
   $$3x = -12$$
   $$x = -4$$
   The solution is $(-4, 2)$.

7. $3x - 5y = -21$

$x = 2(2 - y)$

Substitute $2(2 - y)$ for x in the first equation and solve for y.

$3(2(2 - y)) - 5y = -21$

$3(4 - 2y) - 5y = -21$

$12 - 6y - 5y = -21$

$12 - 11y = -21$

$-11y = -33$

$y = 3$

Substitute 3 for y in the equation $x = 2(2 - y)$ and solve for x.

$x = 2(2 - 3) = 2(-1) = -2$

The solution is $(-2, 3)$.

8. $2x - 3y = 4$

$-4x + 6y = -8$

To eliminate the x terms, we multiply both sides of equation (1) by 2, yielding the system

$4x - 6y = 8$

$-4x + 6y = -8$

The coefficients of the x terms are equal in absolute value and opposite in sign. Add the left sides and the right sides of the equations and solve for y.

$4x - 6y = 8$

$\underline{-4x + 6y = -8}$

$0 = 0 \quad$ true

The result is a true statement of the form $a = a$. Therefore, the system is dependent. The solution set is the infinite set of ordered pairs that satisfy the equation $2x - 3y = 4$.

9. $x = 2y - 3$

$3x - 6y = 12$

Substitute $2y - 3$ for x in the second equation and solve for y.

$3(2y - 3) - 6y = 12$

$6y - 9 - 6y = 12$

$-9 = 12 \quad$ false

The result is a false statement. Therefore, the system is inconsistent. The solution set is the empty set, $\varnothing$.

10. $4x - 7y = 6$

$-5x + 2y = -21$

To eliminate the x terms, we multiply both sides of equation (1) by 5 and both sides of equation (2) by 4, yielding the system

$20x - 35y = 30$

$-20x + 8y = -84$

The coefficients of the x terms are equal in absolute value and opposite in sign. Add the left sides and the right sides of the equations and solve for y.

$20x - 35y = 30$

$\underline{-20x + 8y = -84}$

$-27y = -54$

$y = 2$

Substitute 2 for y in equation (1) and solve for x.

$4x - 7(2) = 6$

$4x - 14 = 6$

$4x = 20$

$x = 5$

The solution is $(5, 2)$.

11. $y = -1.94x + 8.62$

$y = 1.25x - 2.38$

$1.25x - 2.38 = -1.94x + 8.62$

$3.19x - 2.38 = 8.62$

$3.19x = 11$

$x = \dfrac{11}{3.19} \approx 3.448$

Substitute 3.448 for x in the equation $y = 1.25x - 2.38$ and solve for y.

$y = 1.25(3.448) - 2.38 = 1.93$

The approximate solution is $(3.45, 1.93)$.

12. **a.** The points C and D lie on the graph of $y = ax + b$, so they are solutions to the equation $y = ax + b$.

b. The points D and F lie on the graph of $y = cx + d$, so they are solutions to the equation $y = cx + d$.

c. The point D lies on both graphs, so it is a solution to both equations.

d. The points A, B, and E do not lie on either graph, so they are not solutions to either equation.

13. We can start by writing equations for the two lines.

Blue Line:

Examining the graph, the line crosses the y-axis at the point $(0,1)$, so the y-intercept is $(0,1)$.

The point $(6,2)$ also appears to be on the graph,

so $m = \dfrac{2-1}{6-0} = \dfrac{1}{6}$. The equation of the line is

$y = \dfrac{1}{6}x + 1$.

Red Line:

Examining the graph, the line crosses the y-axis at the point $(0,-3)$, so the y-intercept is $(0,-3)$.

The point $(3,-2)$ also appears to be on the

graph, so $m = \dfrac{-2-(-3)}{3-0} = \dfrac{-2+3}{3} = \dfrac{1}{3}$. The

equation of the line is $y = \dfrac{1}{3}x - 3$.

Using the two equations, we get the system

$y = \dfrac{1}{6}x + 1$

$y = \dfrac{1}{3}x - 3$

Substitute $\dfrac{1}{3}x - 3$ for y in the first equation and

solve for x.

$\dfrac{1}{3}x - 3 = \dfrac{1}{6}x + 1$

$6\left(\dfrac{1}{3}x - 3\right) = 6\left(\dfrac{1}{6}x + 1\right)$

$2x - 18 = x + 6$

$x - 18 = 6$

$x = 24$

Substitute 24 for x in the equation $y = \dfrac{1}{6}x + 1$

and solve for y.

$y = \dfrac{1}{6}(24) + 1 = 4 + 1 = 5$

The solution is $(24,5)$.

14. Answers may vary. One possibility:

$3x - y = 2$

$-2x + 4y = 12$

15. A: This point is the origin so it has coordinates $(0,0)$.

B: This point is the y-intercept for line l_1.
Write the equation for l_1 in slope-intercept form.

$4x + 3y = 33$

$3y = -4x + 33$

$y = -\dfrac{4}{3}x + 11$

The coordinates of the point are $(0,11)$.

C: This point is the intersection of the two lines. Therefore, it is the solution of the system

$y = -\dfrac{4}{3}x + 11$

$y = 2x - 4$

Substitute $2x - 4$ for y in the first equation and solve x.

$2x - 4 = -\dfrac{4}{3}x + 11$

$\dfrac{10}{3}x - 4 = 11$

$\dfrac{10}{3}x = 15$

$x = \dfrac{9}{2}$

Substitute $\dfrac{9}{2}$ for x in the equation

$y = 2x - 4$ and solve for y.

$y = 2\left(\dfrac{9}{2}\right) - 4 = 9 - 4 = 5$

The coordinates of the point are $\left(\dfrac{9}{2},5\right)$.

D: This point is the x-intercept of line l_2. Let
$y = 0$ in the equation and solve for x.

$0 = 2x - 4$

$4 = 2x$

$2 = x$

The coordinates of the point are $(2,0)$.

16. a. Start by creating a scatter diagram of the data.

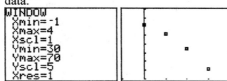

The data appear fairly linear so we use the regression feature of a graphing utility to obtain the line of best fit.

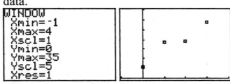

A reasonable model for the data is $p = -8.42t + 62.63$.

b. Start by creating a scatter diagram of the data.

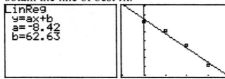

The data appear fairly linear so we use the regression feature of a graphing utility to obtain the line of best fit.

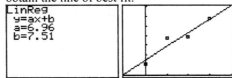

A reasonable model for the data is $p = 6.96t + 7.51$.

c. To estimate when the percentages will be equal, we solve the system

$$p = -8.42t + 62.63$$
$$p = 6.96t + 7.51$$

Substitute $6.96t + 7.51$ for p in the first equation and solve for t.

$$6.96t + 7.51 = -8.42t + 62.63$$
$$15.38t + 7.51 = 62.63$$
$$15.38t = 55.12$$
$$t = \frac{55.12}{15.38} \approx 3.584$$

Substitute 3.584 for t in the equation $p = 6.96t + 7.51$ and solve for p.

$$p = 6.96(3.584) + 7.51 \approx 32.455$$

The approximate solution of the system is $(3.6, 32.5)$.

According to the models, the percentage of time Justice O'Connor voted with liberal justices was the same as the percentage with conservative justices (32.5% each) in 2004.

17. a. Since the number of guns in circulation in the U.S. has increased by about 4.9 million per year, we can model the situation by a linear equation. The slope is 4.9 million guns per year. Since the number of guns in circulation in 2000 was 230 million, the n-intercept is $(0, 230)$. So, a reasonable model is $n = 4.9t + 230$.

b. By the same reasoning as in part (a), the model is $n = 2.9t + 280$.

c.
$$n = 4.9t + 230$$
$$n = 2.9t + 280$$

Substitute $2.9t + 280$ for n in the first equation and solve for t.

$$2.9t + 280 = 4.9t + 230$$
$$-2t + 280 = 230$$
$$-2t = -50$$
$$t = 25$$

Substitute 25 for t in the equation $n = 2.9t + 280$ and solve for n.

$$n = 2.9(25) + 280 = 352.5$$

The solution of the system is $(25, 352.5)$.

According to the models, there will be (on average) one gun per person in the U.S. (352.5 million) in 2025. However, not everyone will own a gun in that year.

18. Let x = the amount invested in the 3% account and y = the amount invested in 7% account. The total amount invested is $7,000 so our first equation is $x + y = 7,000$.

The total interest after one year is $410. Therefore, our second equation is $0.03x + 0.07y = 410$.

The system is

$$x + y = 7,000$$

$$0.03x + 0.07y = 410$$

Solve the first equation for y.

$$x + y = 7,000$$

$$y = -x + 7,000$$

Substitute $-x + 7,000$ for y in the second equation and solve for x.

$$0.03x + 0.07(-x + 7,000) = 410$$

$$0.03x - 0.07x + 490 = 410$$

$$-0.04x + 490 = 410$$

$$-0.04x = -80$$

$$x = 2000$$

Substitute 2,000 for x in the equation $y = -x + 7,000$ and solve for y.

$$y = -(2,000) + 7,000 = 5,000$$

She should invest $2,000 in the 3% acoount and $5000 in the 7% account.

19. $5x - 2y \le 6$

$$-2y \le -5x + 6$$

$$y \ge \frac{5}{2}x - 3$$

The graph of $y \ge \frac{5}{2}x - 3$ is the line $y = \frac{5}{2}x - 3$ and the region above that line. We use a solid line along the border to indicate that the points on the line $y = \frac{5}{2}x - 3$ are solutions to $y \ge \frac{5}{2}x - 3$.

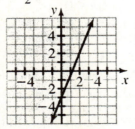

20. $y < -3$

The graph of $y < -3$ is the region below the horizontal line $y = -3$. We graph the line with a dashed line to indicate that points on the line $y = -3$ are not solutions to $y < -3$.

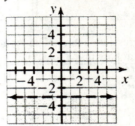

21. First we get y alone on one side of each inequality.

$$y \le -3x + 4 \qquad x - 3y > 6$$

$$-3y > -x + 6$$

$$y < \frac{1}{3}x - 2$$

Next we sketch the graph of $y \le -3x + 4$ and the graph of $y < \frac{1}{3}x - 2$. The graph of $y \le -3x + 4$ is the line $y = -3x + 4$ (graph the line with a solid line) and the region below that line. The graph of $y < \frac{1}{3}x - 2$ is the region below the line $y = \frac{1}{3}x - 2$ (graph the line with a dashed line).

The graph of the solution set of the system is the intersection of the graphs of the inequalities.

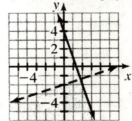

22. First we sketch the graph of $y > 2$ and the graph of $x \geq -3$. The graph of $y > 2$ is the region above the line $y = 2$ (graph the line with a dashed line). The graph of $x \geq -3$ is the vertical line $x = -3$ (graph the line with a solid line) and the region to the right of that line. The graph of the solution set of the system is the intersection of the graphs of the inequalities.

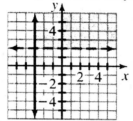

Cumulative Review of Chapters 1 – 6

1.

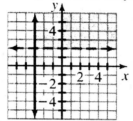

Fahrenheit degrees

2. A cricket chirps 129 times per minute when the temperature is $70°F$.

3.
$$-\frac{26}{27} \cdot \frac{12}{13} = -\frac{2 \cdot 13 \cdot 2 \cdot 2 \cdot 3}{3 \cdot 3 \cdot 3 \cdot 13}$$
$$= -\frac{3 \cdot 13 \cdot 2 \cdot 2 \cdot 2}{3 \cdot 13 \cdot 3 \cdot 3}$$
$$= -\frac{2 \cdot 2 \cdot 2}{3 \cdot 3}$$
$$= -\frac{8}{9}$$

4.
$$\frac{5}{7} - \left(-\frac{3}{5}\right) = \frac{5}{7} + \frac{3}{5}$$
$$= \frac{5}{7} \cdot \frac{5}{5} + \frac{3}{5} \cdot \frac{7}{7}$$
$$= \frac{25}{35} + \frac{21}{35}$$
$$= \frac{46}{35}$$

5. $a^2 - bc + b^2 = (-3)^2 - (-2)(4) + (-2)^2$
$$= 9 - (-8) + 4$$
$$= 9 + 8 + 4$$
$$= 21$$

6. $m = \dfrac{-8 - (-2)}{5 - (-3)} = \dfrac{-8 + 2}{5 + 3} = \dfrac{-6}{8} = -\dfrac{3}{4}$

Since the slope is negative, the line is decreasing.

7. a. The line is decreasing, so $m < 0$. The line crosses the y-axis above the x-axis, so $b > 0$.

b. The line is increasing, so $m > 0$. The line crosses the y-axis below the x-axis, so $b < 0$.

c. The line is horizontal, so $m = 0$. The line crosses the y-axis above the x-axis, so $b > 0$.

d. The line is decreasing, so $m < 0$. The line crosses the y-axis below the x-axis, so $b < 0$.

8. Equation 1:
The input $x = 0$ yields the output $y = 49$, so the y-intercept is $(0, 49)$.
$m = \dfrac{41 - 49}{1 - 0} = \dfrac{-8}{1} = -8$, so the slope is -8. The equation of the line is $y = -8x + 49$.

Equation 2:
The input $x = 0$ yields the output $y = 11$, so the y-intercept is $(0, 11)$.
$m = \dfrac{15 - 11}{1 - 0} = \dfrac{4}{1} = 4$, so the slope is 4. The equation of the line is $y = 4x + 11$.

Equation 3:
$m = \dfrac{37 - 39}{4 - 3} = \dfrac{-2}{1} = -2$, so the slope is -2.
The input $x = 3$ yields the output $y = 39$, the the ordered pair $(3, 39)$ is a solution to the equation. Using this point and the slope, we get
$$y - y_1 = m(x - x_1)$$
$$y - 39 = -2(x - 3)$$
$$y - 39 = -2x + 6$$
$$y = -2x + 45$$
The equation of the line is $y = -2x + 45$.

Equation 4:

$m = \dfrac{20-14}{4-2} = \dfrac{6}{2} = 3$, so the slope is 3.

The input $x = 2$ yields the output $y = 14$, the the ordered pair $(2,14)$ is a solution to the equation. Using this point and the slope, we get

$y - y_1 = m(x - x_1)$

$y - 14 = 3(x - 2)$

$y - 14 = 3x - 6$

$\quad y = 3x + 8$

The equation of the line is $y = 3x + 8$.

9. $2 - 5(4x + 8) = 3(2x - 7) + 3$

$\quad 2 - 20x - 40 = 6x - 21 + 3$

$\quad\quad -20x - 38 = 6x - 18$

$\quad\quad -26x - 38 = -18$

$\quad\quad\quad -26x = 20$

$\quad\quad\quad\quad x = -\dfrac{20}{26} = -\dfrac{10}{13}$

The solution is $-\dfrac{10}{13}$.

10. $-3(2p - w) - (7p + 2w) + 5$

$= -6p + 3w - 7p - 2w + 5$

$= -13p + w + 5$

11. $\dfrac{2}{3}(6w + 9y - 15) = \dfrac{2}{3}(6w) + \dfrac{2}{3}(9y) - \dfrac{2}{3}(15)$

$\quad\quad\quad\quad\quad\quad = 4w + 6y - 10$

12. $\quad\quad \dfrac{3m}{5} - \dfrac{2}{3} = \dfrac{4}{5}$

$\quad 15\left(\dfrac{3m}{5} - \dfrac{2}{3}\right) = 15\left(\dfrac{4}{5}\right)$

$\quad\quad 9m - 10 = 12$

$\quad\quad\quad\quad 9m = 22$

$\quad\quad\quad\quad m = \dfrac{22}{9}$

13. $ax + by = c$

$\quad by = -ax + c$

$\quad \dfrac{by}{b} = \dfrac{-ax + c}{b}$

$\quad y = \dfrac{-ax + c}{b}$ or $y = \dfrac{c - ax}{b}$

14. $6 + 3(4 + x) = 6 + 12 + 3x$

$\quad\quad\quad\quad\quad\quad = 18 + 3x$

15. $4 - \dfrac{x}{3} = 2$

$\quad -\dfrac{x}{3} = -2$

$\quad\quad x = 6$

The solution is 6.

16. $y = 2x - 4$

The equation is in slope-intercept form, so the slope is $m = 2$ and the y-intercept is $(0,-4)$. Starting with the point $(0,-4)$ and writing the slope as $m = \dfrac{2}{1}$, we can obtain a second point on the graph by adding 2 units to the y coordinate and adding 1 unit to the x-coordinate to obtain $(1,-2)$. We plot the two points and connect them with a line, extending the line in both directions.

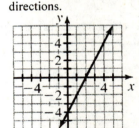

17. $x - 2y = 6$

$\quad -2y = -x + 6$

$\quad\quad y = \dfrac{1}{2}x - 3$

The equation is now in slope-intercept form, so we see that the slope is $m = \dfrac{1}{2}$ and the y-intercept is $(0,-3)$. Starting with the point $(0,-3)$ and using the slope, we can obtain a second point on the graph by adding 1 unit to the y-coordinate and adding 2 units to the x-coordinate to obtain $(2,-2)$. We plot the two points and connect them with a line, extending the line in both directions.

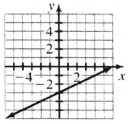

18. $5x + 2y - 12 = 0$

$$5x + 2y = 12$$
$$2y = -5x + 12$$
$$y = -\frac{5}{2}x + 6$$

The equation is now in slope-intercept form, so we see that the slope is $m = -\frac{5}{2}$ and the y-intercept is $(0,6)$. Starting with the point $(0,6)$ and writing the slope as $m = \frac{-5}{2}$, we can obtain a second point on the graph by subtracting 5 units from the y-coordinate and adding 2 units to the x-coordinate to obtain $(2,1)$. We plot the two points and connect them with a line, extending the line in both directions.

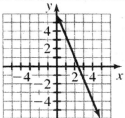

19. $y = -3$

This is the equation of a horizontal line. The y-intercept is $(0,-3)$ and the slope is $m = 0$.

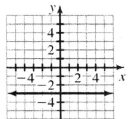

20. $2x - 5y = 10$

To find the x-intercept, we let $y = 0$ and solve for x.

$$2x - 5(0) = 10$$
$$2x = 10$$
$$x = 5$$

The x-intercept is $(5,0)$.

To find the y-intercept, we let $x = 0$ and solve for y.

$$2(0) - 5y = 10$$
$$-5y = 10$$
$$y = -2$$

The y-intercept is $(0,-2)$.

We plot the two intercepts, connect them with a line, and extend the line in both directions.

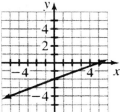

21. $y = -2x + 4$

The equation is in slope-intercept form, so the y-intercept is $(0,4)$.

To find the x-intercept, we let $y = 0$ and solve for x.

$$0 = -2x + 4$$
$$2x = 4$$
$$x = 2$$

The x-intercept is $(2,0)$.

We plot the two intercepts, connect them with a line, and extend the line in both directions.

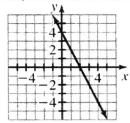

22. The graph of $x = 4$ is a vertical line passing through the point $(4, 0)$.

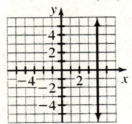

23. Write both equations in $y = mx + b$ form.

$$2x + 5y = 7 \qquad\qquad y = \frac{2}{3}x - 3$$
$$5y = -2x + 7$$
$$y = -\frac{2}{5}x + \frac{7}{5}$$

Since the slopes are different, the lines are not parallel.

24.
$$y - y_1 = m(x - x_1)$$
$$y - (-2) = -\frac{2}{5}(x - 3)$$
$$y + 2 = -\frac{2}{5}x + \frac{6}{5}$$
$$y = -\frac{2}{5}x - \frac{4}{5}$$

Check:
$$y = -\frac{2}{5}x - \frac{4}{5}$$
$$-2 \overset{?}{=} -\frac{2}{5}(3) - \frac{4}{5}$$
$$-2 \overset{?}{=} -\frac{6}{5} - \frac{4}{5}$$
$$-2 \overset{?}{=} -\frac{10}{5}$$
$$-2 \overset{?}{=} -2 \quad \text{true}$$

25. $(-5, 1)$; $(-2, -3)$
$$m = \frac{-3 - 1}{-2 - (-5)} = \frac{-3 - 1}{-2 + 5} = \frac{-4}{3} = -\frac{4}{3}$$
$$y - 1 = -\frac{4}{3}(x - (-5))$$
$$y - 1 = -\frac{4}{3}(x + 5)$$
$$y - 1 = -\frac{4}{3}x - \frac{20}{3}$$
$$y = -\frac{4}{3}x - \frac{17}{3}$$

26. Since the x-coordinates are the same for the two points, the line is vertical and has an undefined slope. The equation is $x = -2$.

27. We locate -3 on the x-axis and move up to the graph. Then we move to the y-axis to find that $y = 2$ when $x = -3$.

28. We locate -1 on the y-axis and move right until we reach the graph. Then we move up to the x-axis to find that $x = 6$ when $y = -1$.

29. The graph crosses the x-axis at the point $(3, 0)$ so the x-intercept is $(3, 0)$.

30. The graph crosses the y-axis at the point $(0, 1)$, so the y-intercept is $(0, 1)$. From problem 29, the x-intercept is $(3, 0)$. Therefore,
$$m = \frac{0 - 1}{3 - 0} = -\frac{1}{3}$$
The equation of the line is $y = -\frac{1}{3}x + 1$.

31.
$$-3(x - 5) > 18$$
$$-3x + 15 > 18$$
$$-3x > 3$$
$$\frac{-3x}{-3} < \frac{3}{-3}$$
$$x < -1$$
Interval notation: $(-\infty, -1)$

32. $y = 2x - 3$

$x + y = 3$

Write both equations in slope-intercept form.

$y = 2x - 3$ $x + y = 3$

                        $y = -x + 3$

Graph both equations in the same coordinate system.

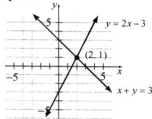

The intersection point is $(2,1)$, so the solution to the system is $(2,1)$.

33. $y = -2.9x + 7.8$

$y = 1.3x - 6.1$

 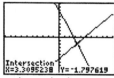

The approximate intersection point is $(3.310, -1.798)$, so the approximate solution to the system is $(3.31, -1.80)$.

34. $3x + 5y = -1$ Equation (1)

$2x - 3y = 12$ Equation (2)

To eliminate the x terms, we multiply both sides of equation (1) by 2 and both sides of equation (2) by -3, yielding the system

$6x + 10y = -2$

$-6x + 9y = -36$

The coefficients of the x terms are equal in absolute value and opposite in sign. Add the left sides and the right sides of the equations and solve for y.

$6x + 10y = -2$

$\underline{-6x + 9y = -36}$

$\qquad 19y = -38$

$\qquad\quad y = -2$

Substitute -2 for y in equation (2) and solve for x.

$2x - 3(-2) = 12$

$2x + 6 = 12$

$2x = 6$

$x = 3$

The solution is $(3, -2)$.

35. $4x - y - 9 = 0$

$y = 5 - 3x$

Substitute $5 - 3x$ for y in the first equation and solve for x.

$4x - (5 - 3x) - 9 = 0$

$4x - 5 + 3x - 9 = 0$

$7x - 14 = 0$

$7x = 14$

$x = 2$

Substitute 2 for x in the equation $y = 5 - 3x$ and solve for y.

$y = 5 - 3(2) = 5 - 6 = -1$

The solution is $(2, -1)$.

36. $y = -2.9x + 97.8$

$y = 3.1x - 45.6$

Rewrite the system in standard form.

$2.9x + y = 97.8$ Equation (1)

$3.1x - y = 45.6$ Equation (2)

The coefficients of the y terms are equal in absolute value and opposite in sign. Add the left sides and the right sides of the equations and solve for x.

$2.9x + y = 97.8$

$\underline{3.1x - y = 45.6}$

$\qquad 6x = 143.4$

$\qquad\quad x = \dfrac{143.4}{6} = 23.9$

Substitute 23.9 for x in equation (1) and solve for y.

$2.9(23.9) + y = 97.8$

$69.31 + y = 97.8$

$y = 28.49$

The solution is $(23.9, 28.49)$.

37. Graphing:
$$2x - 3y = 19$$
$$y = 4x - 13$$
Write both equations in slope-intercept form.

$2x - 3y = 19$ $\qquad$ $y = 4x - 13$

$-3y = -2x + 19$

$y = \dfrac{2}{3}x - \dfrac{19}{3}$

Graph both equations in the same coordinate system.

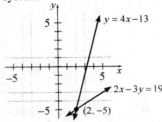

The intersection point is $(2, -5)$, so the solution of the system is $(2, -5)$.

Substitution:
$$2x - 3y = 19$$
$$y = 4x - 13$$
Substitute $4x - 13$ for y in the first equation and solve for x.

$2x - 3(4x - 13) = 19$

$2x - 12x + 39 = 19$

$-10x + 39 = 19$

$-10x = -20$

$x = 2$

Substitute 2 for x in the equation $y = 4x - 13$ and solve for y.

$y = 4(2) - 13 = 8 - 13 = -5$

The solution is $(2, -5)$.

Elimination:
$$2x - 3y = 19$$
$$y = 4x - 13$$
Write the equations in standard form.

$2x - 3y = 19$ $\quad$ Equation (1)

$4x - y = 13$ $\quad$ Equation (2)

To eliminate the y terms, we multiply both sides of equation (2) by -3, yielding the system

$2x - 3y = 19$

$-12x + 3y = -39$

The coefficients of the y terms are equal in

absolute value and opposite in sign. Add the left sides and the right sides of the equations and solve for x.

$2x - 3y = 19$

$\underline{-12x + 3y = -39}$

$-10x = -20$

$x = 2$

Substitute 2 for x in equation (2) and solve for y.

$4(2) - y = 13$

$8 - y = 13$

$-y = 5$

$y = -5$

The solution is $(2, -5)$.

38. We begin by finding the equations for each table.

Table 43:

The input $x = 0$ yields the output $y = 97$, so the y-intercept is $(0, 97)$.

$m = \dfrac{93 - 97}{1 - 0} = \dfrac{-4}{1} = -4$, so the slope is -4.

The first equation is $y = -4x + 97$.

Table 44:

The input $x = 0$ yields the output $y = 7$, so the y-intercept is $(0, 7)$.

$m = \dfrac{9 - 7}{1 - 0} = \dfrac{2}{1} = 2$, so the slope is 2.

The second equation is $y = 2x + 7$.

Now we solve the system
$$y = -4x + 97$$
$$y = 2x + 7$$
Substitute $2x + 7$ for y in the first equation and solve for x.

$2x + 7 = -4x + 97$

$6x + 7 = 97$

$6x = 90$

$x = 15$

Substitute 15 for x in the equation $y = 2x + 7$ and solve for y.

$y = 2(15) + 7 = 30 + 7 = 37$

The solution is $(15, 37)$.

39. The graph of $y < -\dfrac{2}{3}x + 3$ is the region below

the line $y = -\dfrac{2}{3}x + 3$. We use a dashed line

along the border to indicate that the points on the

line $y = -\dfrac{2}{3}x + 3$ are not solutions to

$y < -\dfrac{2}{3}x + 3$.

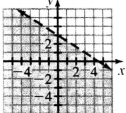

40. $3x - 5y \le 10$

$x > -4$

First we get y alone on one side of the first
inequality.

$3x - 5y \le 10$ $\qquad\qquad x > -4$

$-5y \le -3x + 10$

$y \ge \dfrac{3}{5}x - 2$

Next we sketch the graph of $y \ge \dfrac{3}{5}x - 2$ and the

graph of $x > -4$. The graph of $y \ge \dfrac{3}{5}x - 2$ is the

line $y = \dfrac{3}{5}x - 2$ (graph the line with a solid line)

and the region above that line. The graph of
$x > -4$ is the region to the right of the vertical
line $x = -4$ (graph the line with a dashed line).
The graph of the solution set of the system is the
intersection of the graphs of the inequalities.

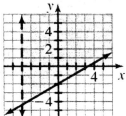

41. a. Answers may vary. If the population
increases, the number of traffic deaths is
likely to rise as will the number of miles
traveled. If the number of miles traveled
rises at a faster rate than the number of
traffice deaths, then the fatality rate will be
decreasing even though the number of
deaths is increasing.

b.

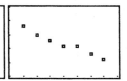

c. Using the regression capability of a graphing
calculator, we get
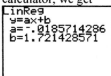
A reasonable model is $r = -0.0186t + 1.72$.

d. The equation in part (c) is in slope-intercept
form, so the r-intercept is $(0, 1.72)$. This
indicates that there were approximately 1.72
deaths per 100 million miles traveled in
1990.

e. Substitute 1.0 for r in the equation
$r = -0.0186t + 1.72$ and solve for t.

$1.0 = -0.0186t + 1.72$

$-0.72 = -0.0186t$

$t = \dfrac{-0.72}{-0.0186} \approx 38.7$

The model predicts that the death rate will
be 1.0 deaths per 100 million miles traveled
in 2029. The Transportation Secretary's plan
is ambitious.

f. For 2008, we have $t = 18$.

$r = -0.0186(18) + 1.72 = 1.3852$

The model predicts that the fatality rate in
2008 will be about 1.39 deaths per 100
million miles traveled.

42. a. Start by creating a scatter diagram of the data.

The data appear fairly linear so we use the regression feature of a graphing utility to obtain the line of best fit.

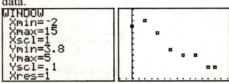

A reasonable model for the data is $r = 0.080t + 1.30$.

b. Start by creating a scatter diagram of the data.

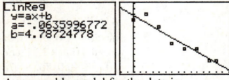

The data appear fairly linear so we use the regression feature of a graphing utility to obtain the line of best fit.

A reasonable model for the data is $r = -0.064t + 4.79$.

c. To estimate when the percentages will be equal, we solve the system
$$r = 0.080t + 1.30$$
$$r = -0.064t + 4.79$$
Substitute $-0.064t + 4.79$ for r in the first equation and solve for t.
$$-0.064t + 4.79 = 0.080t + 1.30$$
$$-0.144t + 4.79 = 1.30$$
$$-0.144t = -3.49$$
$$t = \frac{-3.49}{-0.144} \approx 24.236$$

Substitute 24.236 for t in the equation $r = -0.064t + 4.79$ and solve for r.
$$r = -0.064(24.236) + 4.79 \approx 3.239$$
The approximate solution of the system is $(24.2, 3.2)$.

According to the models, the divorce rates in Japan and the United States will be equal (about 3.2 per 1000 people) in 2014.

43. a. The college enrollment has decreased by 375 students per year, so the slope (average rate of change) is -375 students per year.

b. Since there were 25,700 students at the college in 2000, the E-intercept is $(0, 25700)$. Using the slope from part (a) and the E-intercept, the equation of the line is $E = -375t + 25,700$.

c. Substitute 21,500 for E and solve for t.
$$21,500 = -375t + 25,700$$
$$375t = 4200$$
$$t = \frac{4200}{375} = 11.2$$
Enrollment at the college will reach 21,500 in the year 2011.

d. Solve the following inequality:
$$-375t + 25,700 < 20,000$$
$$-375t < -5,700$$
$$t > \frac{-5,700}{-375} = 15.2$$
Enrollment will be less than 20,000 after 2015.

44. a. Since the number of lines controlled by Bell companies decreased by 6.8 million per year, we can model the situation by a linear equation. The slope is -6.8 million lines per year. Since the number of lines controlled in 2000 was 181.3 million, the n-intercept is $(0, 181.3)$. So, a reasonable model is $n = -6.8t + 181.3$.

b. By the same reasoning as in part (a), the model is $n = 6.8t + 9.4$.

c. Since a line is controlled by a Bell company or some other company, any increase in the number controlled by one group must be matched by an equal decrease in the number controlled by the other group (assuming the number of lines remains constant). Therefore, it makes sense that the slopes (average rate of change) are equal in absolute value, but opposite in sign.

d. $n = -6.8t + 181.3$

$n = 6.8t + 9.4$

Substitute $6.8t + 9.4$ for n in the first equation and solve for t.

$6.8t + 9.4 = -6.8t + 181.3$

$13.6t + 9.4 = 18113$

$13.6t = 171.9$

$$t = \frac{171.9}{13.6} \approx 12.640$$

Substitute 12.640 for t in the equation $n = 6.8t + 9.4$ and solve for n.

$n = 6.8(12.640) + 9.4 = 95.352$

The approximate solution of the system is $(12.64, 95.35)$.

According to the models, there will be an equal number of lines (95.35 million each) controlled by Bell companies and other companies in 2013.

e. After 2013. Since Bell companies no longer have to lease lines to competing companies, it will take longer for the competing companies to gain an equal market share.

45. Let x = the number of quarts of the 16% solution and y = the number of quarts of 28% solution. The total number of quarts is 12, so our first equation is $x + y = 12$.

The total amount of pure acid between the two solutions must be the same as the total amount of pure acid in the mixture, so our second equation is $0.16x + 0.28y = 0.20(12)$.

The system is

$x + y = 12$

$0.16x + 0.28y = 0.20(12)$

Solve the first equation for y.

$x + y = 12$

$y = -x + 12$

Substitute $-x + 12$ for y in the second equation and solve for x.

$0.16x + 0.28(-x + 12) = 0.20(12)$

$0.16x - 0.28x + 3.36 = 2.4$

$-0.12x + 3.36 = 2.4$

$-0.12x = -0.96$

$x = 8$

Substitute 8 for x in the equation $y = -x + 12$ and solve for y.

$y = -(8) + 12 = 4$

The mixture should contain 8 quarts of the 16% acid solution and 4 quarts of the 28% acid solution.

Chapter 7
Polynomials

Homework 7.1

1. $y = 2x^2$

First, we list some solutions in the table. Then we sketch a curve that contains the points corresponding to the solutions.

x	y
-3	$2(-3)^2 = 18$
-2	$2(-2)^2 = 8$
-1	$2(-1)^2 = 2$
0	$2(0)^2 = 0$
1	$2(1)^2 = 2$
2	$2(2)^2 = 8$
3	$2(3)^2 = 18$

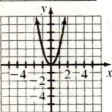

3. $y = -x^2$

First, we list some solutions in the table. Then we sketch a curve that contains the points corresponding to the solutions.

x	y
-3	$-(-3)^2 = -9$
-2	$-(-2)^2 = -4$
-1	$-(-1)^2 = -1$
0	$-(0)^2 = 0$
1	$-(1)^2 = -1$
2	$-(2)^2 = -4$
3	$-(3)^2 = -9$

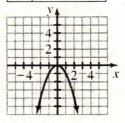

5. $y = x^2 + 1$

First, we list some solutions in the table. Then we sketch a curve that contains the points corresponding to the solutions.

x	y
-3	$(-3)^2 + 1 = 10$
-2	$(-2)^2 + 1 = 5$
-1	$(-1)^2 + 1 = 2$
0	$(0)^2 + 1 = 1$
1	$(1)^2 + 1 = 2$
2	$(2)^2 + 1 = 5$
3	$(3)^2 + 1 = 10$

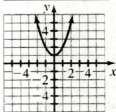

7. $y = 3x^2 - 5$

First, we list some solutions in the table. Then we sketch a curve that contains the points corresponding to the solutions.

x	y
-3	$3(-3)^2 - 5 = 22$
-2	$3(-2)^2 - 5 = 7$
-1	$3(-1)^2 - 5 = -2$
0	$3(0)^2 - 5 = -5$
1	$3(1)^2 - 5 = -2$
2	$3(2)^2 - 5 = 7$
3	$3(3)^2 - 5 = 22$

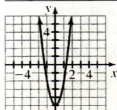

9. $y = x^2 + 4x$

First, we list some solutions in the table. Then we sketch a curve that contains the points corresponding to the solutions.

x	y
-5	$(-5)^2 + 4(-5) = 5$
-4	$(-4)^2 + 4(-4) = 0$
-3	$(-3)^2 + 4(-3) = -3$
-2	$(-2)^2 + 4(-2) = -4$
-1	$(-1)^2 + 4(-1) = -3$
0	$(0)^2 + 4(0) = 0$
1	$(1)^2 + 4(1) = 5$
2	$(2)^2 + 4(2) = 12$
3	$(3)^2 + 4(3) = 21$

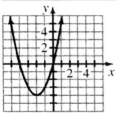

13. $y = x^2 - 4x + 3$

First, we list some solutions in the table. Then we sketch a curve that contains the points corresponding to the solutions.

x	y
-3	$(-3)^2 - 4(-3) + 3 = 24$
-2	$(-2)^2 - 4(-2) + 3 = 15$
-1	$(-1)^2 - 4(-1) + 3 = 8$
0	$(0)^2 - 4(0) + 3 = 3$
1	$(1)^2 - 4(1) + 3 = 0$
2	$(2)^2 - 4(2) + 3 = -1$
3	$(3)^2 - 4(3) + 3 = 0$
4	$(4)^2 - 4(4) + 3 = 3$
5	$(5)^2 - 4(5) + 3 = 8$

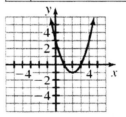

11. $y = -3x^2 - 6x$

First, we list some solutions in the table. Then we sketch a curve that contains the points corresponding to the solutions.

x	y
-3	$-3(-3)^2 - 6(-3) = -9$
-2	$-3(-2)^2 - 6(-2) = 0$
-1	$-3(-1)^2 - 6(-1) = 3$
0	$-3(0)^2 - 6(0) = 0$
1	$-3(1)^2 - 6(1) = -9$
2	$-3(2)^2 - 6(2) = -24$
3	$-3(3)^2 - 6(3) = -45$

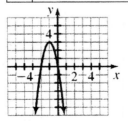

15. $y = -2x^2 + 4x - 2$

First, we list some solutions in the table. Then we sketch a curve that contains the points corresponding to the solutions.

x	y
-3	$-2(-3)^2 + 4(-3) - 2 = -32$
-2	$-2(-2)^2 + 4(-2) - 2 = -18$
-1	$-2(-1)^2 + 4(-1) - 2 = -8$
0	$-2(0)^2 + 4(0) - 2 = -2$
1	$-2(1)^2 + 4(1) - 2 = 0$
2	$-2(2)^2 + 4(2) - 2 = -2$
3	$-2(3)^2 + 4(3) - 2 = -8$

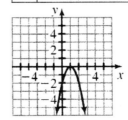

17. The graph contains the point $(2,-5)$, so $y = -5$ when $x = 2$.

19. The graph contains the points $(-2, 3)$ and $(0, 3)$, so $x = -2$ or $x = 0$ when $y = 3$.

21. The graph contains the points $(-1, 4)$, so $x = -1$ when $y = 4$.

23. There are no points on the graph for which the y-coordinate is 5. That is, there is no value of x for which $y = 5$.

25. The graph intersects the x-axis at the points $(-3, 0)$ and $(1, 0)$, so the x-intercepts are $(-3, 0)$ and $(1, 0)$.

27. The parabola opens downward, so the vertex is the maximum point $(-1, 4)$.

29. The graph contains the point $(0, 3)$, so $y = 3$ when $x = 0$.

31. The graph contains the points $(1, 0)$ and $(3, 0)$, so $x = 1$ or $x = 3$ when $y = 0$.

33. There are no points on the graph for which the y-coordinate is -3. That is, there is no value of x for which $y = -3$.

35. The graph intersects the y-axis at the points $(0, 3)$, so the y-intercept is $(0, 3)$.

37. The minimum point is $(2, -1)$.

39. **a.** $y = x^2 - 3$

First, we list some solutions in the table. Then we sketch a curve that contains the points corresponding to the solutions.

x	y
-3	$(-3)^2 - 3 = 6$
-2	$(-2)^2 - 3 = 1$
-1	$(-1)^2 - 3 = -2$
0	$(0)^2 - 3 = -3$
1	$(1)^2 - 3 = -2$
2	$(2)^2 - 3 = 1$
3	$(3)^2 - 3 = 6$

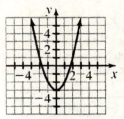

b. Answers may vary. One possibility can be found in part (a).

c. For each solution, the y-coordinate is 3 less than the square of the x-coordinate.

41. The table contains the ordered pair $(1, 3)$, so $y = 3$ when $x = 1$.

43. The table contains the ordered pairs $(1, 3)$ and $(3, 3)$, so $x = 1$ or $x = 3$ when $y = 3$.

45. **a.** The smallest possible value of x^2 is 0. When we square a value, the result cannot be negative.

b. The smallest possible value of $x^2 + 3$ is 3. This is three more the smallest possible value of x^2.

c. The vertex of $y = x^2 + 3$ is $(0, 3)$. Explanations may vary. One possibility follows: The y-coordinate of the vertex will be 3 more than the y-coordinate of the vertex of $y = x^2$. Since the vertex of $y = x^2$ is $(0, 0)$, the vertex of $y = x^2 + 3$ is $(0, 3)$.

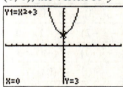

47. **a.** **(i)** The output for $x = 2$ is $y = (2)^2 + 1 = 5$. There is only one output for this input.

(ii) The output for $x = 4$ is $y = (4)^2 + 1 = 17$. There is only one output for this input.

(iii) The output for $x = -3$ is $y = (-3)^2 + 1 = 10$. There is only one output for this input.

b. For $y = x^2 + 1$, one output will originate form any single input. Explanation will vary.

c. Examples will vary. For each case, one output will originate form any single input.

d. Examples and explanations will vary. One output will originate form any single input.

e. For *any* quadratic equation in two variables, one output will originate form any single input. Explanations will vary.

49. $y = -2x - 1$

The slope is $m = -2$, and the y-intercept is $(0, -1)$. We first plot $(0, -1)$. From this point we move 1 unit to the right and 2 units down, where we plot the point $(1, -3)$. We then sketch the line that contains these two points.

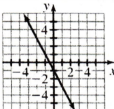

51. $y = -2x^2 - 1$

First, we list some solutions in the table. Then we sketch a curve that contains the points corresponding to the solutions.

x	y
-3	$-2(-3)^2 - 1 = -19$
-2	$-2(-2)^2 - 1 = -9$
-1	$-2(-1)^2 - 1 = -3$
0	$-2(0)^2 - 1 = -1$
1	$-2(1)^2 - 1 = -3$
2	$-2(2)^2 - 1 = -9$
3	$-2(3)^2 - 1 = -19$

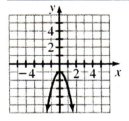

53. a. $y = 2x$

The slope is $m = 2$, and the y-intercept is $(0, 0)$. We first plot $(0, 0)$. From this point we move 1 unit to the right and 2 units up, where we plot the point $(1, 2)$. We then sketch the line that contains these two points.

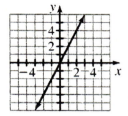

b. $y = x^2$

First, we list some solutions in the table. Then we sketch a curve that contains the points corresponding to the solutions.

x	y
-3	$(-3)^2 = 9$
-2	$(-2)^2 = 4$
-1	$(-1)^2 = 1$
0	$(0)^2 = 0$
1	$(1)^2 = 1$
2	$(2)^2 = 4$
3	$(3)^2 = 9$

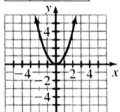

c. The y-intercept is $(0, 0)$ for both curves.

d. The curve $y = x^2$ is steeper. Explanations will vary.

55. $8(3x - 2) = 4(x - 5)$

$$24x - 16 = 4x - 20$$
$$20x - 16 = -20$$
$$20x = -4$$
$$x = \frac{-4}{20} = -\frac{1}{5}$$

This is a linear equation in one variable.

57. $8(3x - 2) - 4(x - 5)$

$$= 24x - 16 - 4x + 20$$
$$= 20x + 4$$

This is a linear expression in one variable.

Homework 7.2

1. A quadratic model would be reasonable for modeling the data since the data points lie in a parabolic shape.

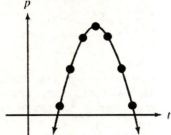

3. We use graphing calculator to create a scattergram of the data.

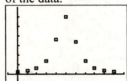

The student is not correct. A quadratic model is not appropriate. Explanations will vary. One possibility: the data do not have a parabolic shape.

5. a.

Yes, the model fits the data well.

b. The year 2005 is represented by $t = 15$. We estimate that the model contains the point (15, 93), so sales in 2005 are approximately 93 million CD albums.

c. We estimate that the model contains the points (6, 340) and (14, 340). Thus, sales were 340 million CD albums in the years $1990 + 6 = 1996$ and $1990 + 14 = 2004$.

d. The parabola opens downward, so the vertex is the maximum point. We estimate it to be the point (9.9, 811). It means that sales were about 811 million CD albums in the year $1990 + 9.9 \approx 2000$. According to the model, this was the largest sales ever. Answers may vary.

e. We estimate that the model intersects the t-axis at the points (4.4, 0) and (15.3, 0), which are the t-intercepts. This means that there were no CD album sales in the years $1990 + 4.4 \approx 1994$ and $1990 + 15.3 \approx 2005$. Model breakdown has occurred for both results.

7. a.

Years since 1990

Yes, the model fits the data well.

b. The year 2001 is represented by $t = 11$. We estimate that the model contains the point (11, 9.9), so there were approximately 9.9 thousand cases in 2001. This is an average of $\dfrac{9.9 \text{ thousand}}{5 \cdot 48} = \dfrac{9900}{240} \approx 41.3$ cases per workday.

c. We estimate that the model contains the points (2, 9) and (10, 9). Thus, 9 thousand cases were heard in the years $1990 + 2 = 1992$ and $1990 + 10 = 2000$.

d. We estimate the model's minimum point to be (6, 7.6). This means that least number of cases was heard in $1990 + 6 = 1996$. In actuality, there were 7.6 thousand cases heard in both 1995 and 1996 – the least number of cases for years between 1994 and 2000, inclusive.

9. a.

Anniversary

Yes, the model fits the data well.

b. We estimate that the model intersects the p-axis at the point (0, 95), which is the p-intercept. This indicates that 95% of married people are still married when they first get married. Model breakdown has occurred.

c. We estimate that the model contains the point (20, 43), so 43% of married people reach their 20th anniversary.

d. We estimate that the model contains the point (17, 50), so 50% of married people reach their 17th anniversary

e. No, the part of the parabola that lies to the right of the vertex will not describe the situation well. Explanations may vary.

11. a. The year 1998 is represented by $t = 3$. We estimate that the model contains the point (3, 157). So, we estimate that there were $157 billion in foreign investment in 1998.

b. From the table, we find that the actual foreign investment in 1998 was $143 billion.

c. Our result from part (a) was an overestimate. The data point for 1998 is below the model's graph. The error in our estimate is $157 − $143 = $14 billion.

13. a.

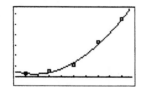

Yes, the model fits the data well.

b. The year 2011 is represented by $t = 16$. We substitute 16 for t in the equation:

$n = t^2 - 3.48t + 5.8$

$n = (16)^2 - 3.48(16) + 5.8 = 206.12$

So, we predict that there will be about 206.1 thousand electric cars in the year 2011.

c. We adjust the WINDOW setting on the calculator and use TRACE:

 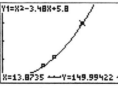

We predict that 150 thousand electric cars will be in use in the year $1995 + 13.88 \approx 2009$.

d. No, the part of the parabola that lies to the left of the vertex will not describe the situation well. Explanations may vary.

15. a.

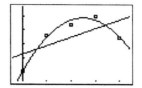

The quadratic model comes closer to the points in the scattergram.

b. Mark Cuban and the Mavericks would like the linear model to be the most accurate for future seasons. Explanations may vary. One possibility follows: The number of win would continue to go up under the linear model.

c.

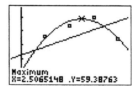

According to the quadratic model, the Mavericks had the greatest number of wins during the year $2000 + 2.51 \approx 2003$ (or the 2002-2003 season), with 59 wins.

d. We adjust the WINDOW setting on the calculator and use TRACE:

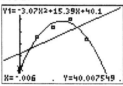

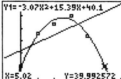

We predict that the Mavericks won 40 games in the years $2000 - 0.006 \approx 2000$ and $2000 + 5.02 \approx 2005$ (or the 1999-2000 season and the 2004-2005 season).

17. If the point (c, d) is below the parabola, then the model overestimates the value of p when $t = c$. Explanations may vary.

19. Answers may vary.

21. Answers may vary. One possibility follows: A *quadratic model* is a quadratic equation in two variables that describes the relationship between two quantities for an authentic situation.

23. a.

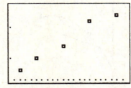

Yes, the model fits the data well.

b. Answers may vary depending on the data points selected to create the model. We use the points $(5, 110)$ and $(20, 117)$ to find an equation of the form $M = mt + b$. We begin by finding the slope:

$$m = \frac{117 - 110}{20 - 5} = \frac{7}{15} \approx 0.47$$

So, the equation of the line is of the form $M = 0.47t + b$. To find b, we substitute the coordinates of the point $(5, 110)$ into the equation and solve for b:

$$M = 0.47t + b$$
$$110 = 0.47(5) + b$$
$$110 = 2.35 + b$$
$$107.65 = b$$

So, the equation is $M = 0.47t + 107.65$.

c. Answers may vary [depending on the equation found in part (b).] The year 2010 is represented by $t = 30$, so we substitute 30 for t in the equation:

$$M = 0.47t + 107.65$$
$$M = 0.47(30) + 107.65 = 121.75$$

We predict that the male birth rate in China in 2010 will be 121.75 male births per 100 female births.

d. Answers may vary [depending on the equation found in part (b)]. We substitute 100 for M in our linear model:

$$M = 0.47t + 107.65$$
$$100 = 0.47t + 107.65$$
$$-7.65 = 0.47t$$
$$t = \frac{-7.65}{0.47} \approx -16.27$$

We predict that the number of male and female births was equal in the year $1980 + (-16.27) \approx 1964$.

25. Answers may vary.

27. $3x - 2y = -4$ Equation (1)

    $4x + 5y = 33$ Equation (2)

We multiply both sides of equation (1) by 5, multiply both sides of equation (2) by 2, and add the results to eliminate the y-variable:

$$15x - 10y = -20$$
$$\underline{8x + 10y = 66}$$
$$23x + 0 = 46$$
$$23x = 46$$
$$x = 2$$

We substitute 2 for x in equation (1) and solve for y:

$$3x - 2y = -4$$
$$3(2) - 2y = -4$$
$$6 - 2y = -4$$
$$-2y = -10$$
$$y = 5$$

The solution is $(2, 5)$. This is a system of two linear equations in two variables.

29.

$$3x - 2y = -4$$
$$-2y = -3x - 4$$
$$\frac{-2y}{-2} = \frac{-3x}{-2} - \frac{4}{-2}$$
$$y = \frac{3}{2}x + 2$$

The slope is $m = \dfrac{3}{2}$, and the y-intercept is $(0, 2)$.

We first plot $(0, 2)$. From this point we move 2 units to the right and 3 units up, where we plot the point $(2, 5)$. We then sketch the line that contains these two points.

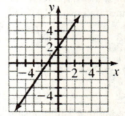

This is a linear equation in two variables.

Homework 7.3

1. $3x^2 - 4x + 2$

The term $3x^2$ has degree 2, which is larger than the degrees of the other terms. So $3x^2 - 4x + 2$ is a quadratic (or second-degree) polynomial in one variable.

3. $-7x^3 - 9x - 4$

The term $-7x^3$ has degree 3, which is larger than the degrees of the other terms. So $-7x^3 - 9x - 4$ is a cubic (or third-degree) polynomial in one variable.

5. $3p^5q^2 - 5p^3q^3 + 7pq^4$

The term $3p^5q^2$ has degree $5 + 2 = 7$, which is larger than the degrees of the other terms. So $3p^5q^2 - 5p^3q^3 + 7pq^4$ is a seventh-degree polynomial in two variables.

7. $2x + 4x = 6x$

9. $-4x - 9x = -13x$

11. $3t^2 + 5t^2 = 8t^2$

13. $-8a^4b^3 - 3a^4b^3 = -11a^4b^3$

15. $4x^2 + x^2 = 4x^2 + 1x^2 = 5x^2$

17. $7x^2 - 3x$

The terms $7x^2$ and $-3x$ are not like terms, so they cannot be combined.

19. $5b^3 - 8b^3 = -3b^3$

21. $-x^6 + 7x^6 = 6x^6$

23. $2t^3w^5 + 4t^5w^3$

The terms $2t^3w^5$ and $4t^5w^3$ are not like terms, so they cannot be combined.

25. $-2.5p^4 + 9.9p^4 = 7.4p^4$

27. $8x^2 + 2x - 3x^2 - 5x = 8x^2 - 3x^2 + 2x - 5x$
$$= 5x^2 - 3x$$

29. $9x - 4x^2 + 5x - 2 + 3x^2 - 6$
$$= -4x^2 + 3x^2 + 9x + 5x - 2 - 6$$
$$= -x^2 + 14x - 8$$

31. $5x + 8x^3 - x^2 + 4x^3 + 2x - x^3$
$$= 8x^3 + 4x^3 - x^3 - x^2 + 5x + 2x$$
$$= 11x^3 - x^2 + 7x$$

33. $20.3t^2 - 5.4t - 45.1t^3 - 3.6t + 93.8t^2$
$$= -45.1t^3 + 20.3t^2 + 93.8t^2 - 5.4t - 3.6t$$
$$= -45.1t^3 + 114.1t^2 - 9t$$

35. $(5x^2 - 4x - 2) + (-9x^2 - 3x + 8)$
$$= 5x^2 - 9x^2 - 4x - 3x - 2 + 8$$
$$= -4x^2 - 7x + 6$$

37. $(4x^2 - 3x + 2) + (-4x^2 + 3x - 2)$
$$= 4x^2 - 4x^2 - 3x + 3x + 2 - 2$$
$$= 0$$

39. $(4x^3 - 7x^2 + 2x - 9) + (-7x^3 - 3x^2 + 5x - 2)$
$$= 4x^3 - 7x^3 - 7x^2 - 3x^2 + 2x + 5x - 9 - 2$$
$$= -3x^3 - 10x^2 + 7x - 11$$

41. $(5a^3 - a) + (8a^3 - 3a^2 - 1)$
$$= 5a^3 + 8a^3 - 3a^2 - a - 1$$
$$= 13a^3 - 3a^2 - a - 1$$

43. $(5a^2 - 3ab + 7b^2) + (-4a^2 - 6ab + 2b^2)$
$$= 5a^2 - 4a^2 - 3ab - 6ab + 7b^2 + 2b^2$$
$$= a^2 - 9ab + 9b^2$$

45. $(14.1x^3 - 7.9x^2 - 4.8x + 31.9) + (-8.2x^3 + 28.8x^2 - 9.5x + 32.2)$
$$= 14.1x^3 - 8.2x^3 - 7.9x^2 + 28.8x^2 - 4.8x - 9.5x + 31.9 + 32.2$$
$$= 5.9x^3 + 20.9x^2 - 14.3x + 64.1$$

47. $(-3x + 8) - (4x - 3) = -3x + 8 - 4x + 3$
$$= -3x - 4x + 8 + 3$$
$$= -7x + 11$$

49. $(2x^2 + 5x - 1) - (4x^2 + 9x - 7)$
$= (2x^2 + 5x - 1) - 4x^2 - 9x + 7)$
$= 2x^2 - 4x^2 + 5x - 9x - 1 + 7$
$= -2x^2 - 4x + 6$

51. $(6y^3 - 2y^2 - 4y + 5) - (5y^3 - 8y^2 - 5y + 3)$
$= 6y^3 - 2y^2 - 4y + 5 - 5y^3 + 8y^2 + 5y - 3$
$= 6y^3 - 5y^3 - 2y^2 + 8y^2 - 4y + 5y + 5 - 3$
$= y^3 + 6y^2 + y + 2$

53. $(5x^3 - 9x^2 + 2x - 4) - (5x^3 - 9x^2 + 2x - 4)$
$= 5x^3 - 9x^2 + 2x - 4 - 5x^3 + 9x^2 - 2x + 4$
$= 5x^3 - 5x^3 - 9x^2 + 9x^2 + 2x - 2x - 4 + 4$
$= 0$

55. $(3x^2 + 7xy - 2y^2) - (5x^2 - 4xy - 3y^2)$
$= 3x^2 + 7xy - 2y^2 - 5x^2 + 4xy + 3y^2$
$= 3x^2 - 5x^2 + 7xy + 4xy - 2y^2 + 3y^2$
$= -2x^2 + 11xy + y^2$

57. $(2.54x^2 + 6.29x - 7.99) - (-4.21x^2 - 8.45x + 9.29)$
$= 2.54x^2 + 6.29x - 7.99 + 4.21x^2 + 8.45x - 9.29$
$= 2.54x^2 + 4.21x^2 + 6.29x + 8.45x - 7.99 - 9.29$
$= 6.75x^2 + 14.74x - 17.28$

59. $(2x^2 + 5) + (3x + 9) = 2x^2 + 5 + 3x + 9$
$\qquad\qquad\qquad\quad = 2x^2 + 3x + 5 + 9$
$\qquad\qquad\qquad\quad = 2x^2 + 3x + 14$

61. $(5x^3 - 5x + 8) - (8x^2 - 3x - 3)$
$= 5x^3 - 5x + 8 - 8x^2 + 3x + 3$
$= 5x^3 - 8x^2 - 5x + 3x + 8 + 3$
$= 5x^3 - 8x^2 - 2x + 11$

63. $(2x^2 + 6x - 1) - (3x^2 - 5x + 4)$
$= 2x^2 + 6x - 1 - 3x^2 + 5x - 4$
$= 2x^2 - 3x^2 + 6x + 5x - 1 - 4$
$= -x^2 + 11x - 5$

65. $(6x^3 - 2x) - (4x^2 - 7x + 2) + (3x^3 - x^2)$
$= 6x^3 - 2x - 4x^2 + 7x - 2 + 3x^3 - x^2$
$= 6x^3 + 3x^3 - 4x^2 - x^2 - 2x + 7x - 2$
$= 9x^3 - 5x^2 + 5x - 2$

67. a. $(13.28t + 440.09) + (3.42t + 468.14)$
$= 13.28t + 440.09 + 3.42t + 468.14$
$= 13.28t + 3.42t + 440.09 + 468.14$
$= 16.70t + 908.23$
This expression represents the total number (in thousands) of women and men who have earned bachelor's degrees in the year that is t years since 1980.

b. Note that $t = 31$ represents the year 2011. Substitute 31 for t in our expression:
$16.70(31) + 908.23 = 1425.93$
This means that there will be a total of about 1425.93 thousand, or 1.43 million, women and men who will earn a bachelor's degree in 2011.

c. $(13.28t + 440.09) - (3.42t + 468.14)$
$= 13.28t + 440.09 - 3.42t - 468.14$
$= 13.28t - 3.42t + 440.09 - 468.14$
$= 9.86t - 28.05$
This expression represents the difference in the number (in thousands) of bachelor's degrees earned by women and men in the year that is t years since 1980.

d. Substitute 31 for t in our expression:
$9.86(31) - 28.05 = 277.61$
This means that women will earn 277.61 thousand more bachelor's degrees than men in 2011.

69. a.

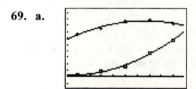

Both models fit the respective data well.

b. $(0.155t^2 - 1.72t + 5.28) + (-0.138t^2 + 3.22t - 1.38)$
$= 0.155t^2 - 1.72t + 5.28 - 0.138t^2 + 3.22t - 1.38$
$= 0.155t^2 - 0.138t^2 - 1.72t + 3.22t + 5.28 - 1.38$
$= 0.017t^2 + 1.5t + 3.9$
This expression represents the percentage of vehicles that are either crossover utility vehicles or SUVs at t years since 1990.

c. Note that $t = 18$ represents the year 2008. Substitute 18 for t in our expression:
$$0.017(18)^2 + 1.5(18) + 3.9 \approx 36.41$$
This means that the percentage of U.S. vehicle sales that are either crossover utility vehicles or SUVs will be about 36.4% in 2008.

d. $(0.155t^2 - 1.72t + 5.28) - (-0.138t^2 + 3.22t - 1.38)$
$$= 0.155t^2 - 1.72t + 5.28 + 0.138t^2 - 3.22t + 1.38$$
$$= 0.155t^2 + 0.138t^2 - 1.72t - 3.22t + 5.28 + 1.38$$
$$= 0.293t^2 - 4.94t + 6.66$$
This expression represents the difference of the percentages of crossover utility vehicles and SUVs at t years since 1990.

e. Substitute 18 for t in our expression:
$$0.293(18)^2 - 4.94(18) + 6.66 \approx 12.67$$
This means that the percentage of U.S. vehicle sales that are crossover utility vehicles will be 12.7 percentage points more than the percentage of U.S. vehicle sales that are SUVs in 2008.

71. Answers may vary. One possibility follows:
The student failed to distribute the subtraction throughout the entire expression.
$$(5x^2 + 3x + 7) - (3x^2 + 2x + 1)$$
$$= 5x^2 + 3x + 7 - 3x^2 - 2x - 1$$
$$= 5x^2 - 3x^2 + 3x - 2x + 7 - 1$$
$$= 2x^2 + x + 6$$

73. Answers may vary. One possibility follows:
The two polynomials $5x^3 - 3x^2 + 2x - 1$ and $x^3 - x^2 - x - 1$ are each of degree 3 and have a sum of $6x^3 - 4x^2 + x - 2$.

75. Answers may vary. One possibility follows:
The two polynomials $2x^3 + 2x^2 - 4x + 3$ and $-2x^3 + x^2 - x + 1$ are each of degree 3 and have a sum of $3x^2 - 5x + 4$.

77. $2x^3 + 5x^3 = (2 + 5)x^3 = 7x^3$

79. $x^2 - 3x + 5x^2 = x^2 + 5x^2 - 3x$
$$= 6x^2 - 3x$$

81. $4x^2 - (x^2 + x) = 4x^2 - x^2 - x$
$$= 3x^2 - x$$

83. a.

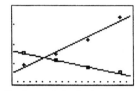

Both models fit the respective data well.

b. $(-0.14t + 10.29) + (0.32t + 4.70)$
$$= -0.14t + 10.29 + 0.32t + 4.70$$
$$= -0.14t + 0.32t + 10.29 + 4.70$$
$$= 0.18t + 14.99$$
This expression represents the total petroleum consumed (in millions of barrels per day) in the United States at t years since 1980.

c. Note that $t = 31$ represents the year 2011. Substitute 31 for t in our expression:
$$0.18(31) + 14.99 = 20.57$$
This means that the United States will consume 20.57 million barrels of petroleum per day in the year 2011.

d. $A = -0.14t + 10.29$ Equation (1)
$A = 0.32t + 4.70$ Equation (2)
We begin by substituting $-0.14t + 10.29$ for A in equation (2):
$$A = 0.32t + 4.70$$
$$-0.14t + 10.29 = 0.32t + 4.70$$
$$-0.46t + 10.29 = 4.70$$
$$-0.46t = -5.59$$
$$t = \frac{-5.59}{-0.46} \approx 12.15$$
The United States consumed the same amount of domestic petroleum as imported petroleum in the year $1980 + 12.15 \approx 1992$.

85. $3x^2 - 1 - 2x^2 = 3x^2 - 2x^2 - 1$
$$= x^2 - 1$$
This is a quadratic (or second-degree) polynomial in one variable.

87. $y = 3x^2 - 1 - 2x^2$

$\quad\quad = x^2 - 1$

First, we list some solutions in the table. Then we sketch a curve that contains the points corresponding to the solutions.

x	y
-3	$(-3)^2 - 1 = 8$
-2	$(-2)^2 - 1 = 3$
-1	$(-1)^2 - 1 = 0$
0	$(0)^2 - 1 = -1$
1	$(1)^2 - 1 = 0$
2	$(2)^2 - 1 = 3$
3	$(3)^2 - 1 = 8$

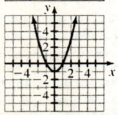

This is a quadratic equation in two variables.

Homework 7.4

1. $x^4 x^3 = x^{4+3} = x^7$

3. $w^8 w = w^8 w^1 = w^{8+1} = w^9$

5. $-5x^4 \left(-6x^3\right) = -5(-6)\left(x^4 x^3\right) = 30x^7$

7. $4p^2 t\left(-9p^3 t^2\right) = 4(-9)\left(p^2 p^3\right)\left(t^1 t^2\right) = -36p^5 t^3$

9. $\dfrac{4}{5}x^3\left(-\dfrac{7}{2}x^2\right) = \dfrac{4}{5}\left(-\dfrac{7}{2}\right)\left(x^3 x^2\right) = -\dfrac{14}{5}x^5$

11. $3w(w - 2) = 3w \cdot w - 3w \cdot 2$

$\quad\quad\quad\quad\quad = 3w^2 - 6w$

13. $2.8x(9.4x - 7.3) = 2.8x \cdot 9.4x - 2.8x \cdot 7.3$

$\quad\quad\quad\quad\quad\quad\quad = 26.32x^2 - 20.44x$

15. $-4x(2x^2 + 3) = -4x \cdot 2x^2 + (-4x) \cdot 3$

$\quad\quad\quad\quad\quad\quad = -8x^3 - 12x$

17. $2mn^2(3m^2 + 5n) = 2mn^2 \cdot 3m^2 + 2mn^2 \cdot 5n$

$\quad\quad\quad\quad\quad\quad\quad = 6m^3 n^2 + 10mn^3$

19. $2x(3x^2 - 2x + 7) = 2x \cdot 3x^2 + 2x \cdot (-2x) + 2x \cdot 7$

$\quad\quad\quad\quad\quad\quad\quad = 6x^3 - 4x^2 + 14x$

21. $-3t^2\left(2t^2 + 4t - 2\right)$

$\quad = -3t^2 \cdot 2t^2 + \left(-3t^2\right) \cdot 4t + \left(-3t^2\right) \cdot (-2)$

$\quad = -6t^4 - 12t^3 + 6t^2$

23. $2xy\left(3x^2 - 4xy + 5y^2\right)$

$\quad = 2xy \cdot 3x^2 - 2xy \cdot 4xy + 2xy \cdot 5y^2$

$\quad = 6x^3 y - 8x^2 y^2 + 10xy^3$

25. $(x + 2)(x + 4) = x \cdot x + x \cdot 4 + 2 \cdot x + 2 \cdot 4$

$\quad\quad\quad\quad\quad\quad = x^2 + 4x + 2x + 8$

$\quad\quad\quad\quad\quad\quad = x^2 + 6x + 8$

27. $(x - 2)(x + 5) = x \cdot x + x \cdot 5 + (-2) \cdot x + (-2) \cdot 5$

$\quad\quad\quad\quad\quad\quad = x^2 + 5x - 2x - 10$

$\quad\quad\quad\quad\quad\quad = x^2 + 3x - 10$

29. $(a - 3)(a - 2) = a \cdot a + a(-2) + (-3)a + (-3)(-2)$

$\quad\quad\quad\quad\quad\quad = a^2 - 2a - 3a + 6$

$\quad\quad\quad\quad\quad\quad = a^2 - 5a + 6$

31. $(x + 6)(x - 6) = x \cdot x + x(-6) + 6 \cdot x + 6(-6)$

$\quad\quad\quad\quad\quad\quad = x^2 - 6x + 6x - 36$

$\quad\quad\quad\quad\quad\quad = x^2 - 36$

33. $(x - 5.3)(x - 9.2)$

$\quad = x \cdot x + x(-9.2) + (-5.3)x + (-5.3)(-9.2)$

$\quad = x^2 - 9.2x - 5.3x + 48.76$

$\quad = x^2 - 14.5x + 48.76$

35. $(5y - 2)(3y + 4)$

$\quad = 5y \cdot 3y + 5y \cdot 4 + (-2) \cdot 3y + (-2) \cdot 4$

$\quad = 15y^2 + 20y - 6y - 8$

$\quad = 15y^2 + 14y - 8$

37. $(2x + 4)(2x + 4) = 2x \cdot 2x + 2x \cdot 4 + 4 \cdot 2x + 4 \cdot 4$

$\quad\quad\quad\quad\quad\quad = 4x^2 + 8x + 8x + 16$

$\quad\quad\quad\quad\quad\quad = 4x^2 + 16x + 16$

39. $(3x-1)(3x-1)$

$= 3x \cdot 3x + 3x(-1) + (-1) \cdot 3x + (-1)(-1)$

$= 9x^2 - 3x - 3x + 1$

$= 9x^2 - 6x + 1$

41. $(3x-5y)(4x+y)$

$= 3x \cdot 4x + 3x \cdot y + (-5y) \cdot 4x + (-5y) \cdot y$

$= 12x^2 + 3xy - 20xy - 5y^2$

$= 12x^2 - 17xy - 5y^2$

43. $(2a-3b)(3a-2b)$

$= 2a \cdot 3a + 2a(-2b) + (-3b) \cdot 3a + (-3b)(-2b)$

$= 6a^2 - 4ab - 9ab + 6b^2$

$= 6a^2 - 13ab + 6b^2$

45. $(3x-4)(3x+4)$

$= 3x \cdot 3x + 3x \cdot 4 + (-4) \cdot 3x + (-4) \cdot 4$

$= 9x^2 + 12x - 12x - 16$

$= 9x^2 - 16$

47. $(9x+4y)(9x-4y)$

$= 9x \cdot 9x + 9x(-4y) + 4y \cdot 9x + 4y(-4y)$

$= 81x^2 - 36xy + 36xy - 16y^2$

$= 81x^2 - 16y^2$

49. $(2.5x+9.1)(4.6x-7.7)$

$= 2.5x \cdot 4.6x + 2.5x(-7.7) + 9.1 \cdot 4.6x + 9.1(-7.7)$

$= 11.5x^2 - 19.25x + 41.86x - 70.07$

$= 11.5x^2 + 22.61x - 70.07$

51. $(0.37x+20.45)(-1.7x+50.8)$

$= 0.37x(-1.7x) + 0.37x \cdot 50.8 + 20.45(-1.7x) +$ $20.45 \cdot 50.8$

$= -0.629x^2 + 18.796x - 34.765x + 1038.86$

$= -0.629x^2 - 15.969x + 1038.86$

53. $(x+6)(x^2-3) = x \cdot x^2 + x \cdot (-3) + 6 \cdot x^2 + 6 \cdot (-3)$

$\qquad = x^3 - 3x + 6x^2 - 18$

$\qquad = x^3 + 6x^2 - 3x - 18$

55. $(2t^2-5)(3t-2)$

$= 2t^2 \cdot 3t + 2t^2(-2) + (-5) \cdot 3t + (-5)(-2)$

$= 6t^3 - 4t^2 - 15t + 10$

57. $(3a^2+5b^2)(2a^2-3b^2)$

$= 3a^2 \cdot 2a^2 + 3a^2(-3b^2) + 5b^2 \cdot 2a^2 + 5b^2(-3b^2)$

$= 6a^4 - 9a^2b^2 + 10a^2b^2 - 15b^4$

$= 6a^4 + a^2b^2 - 15b^4$

59. $(x+2)(x^2+3x+5)$

$= x \cdot x^2 + x \cdot 3x + x \cdot 5 + 2 \cdot x^2 + 2 \cdot 3x + 2 \cdot 5$

$= x^3 + 3x^2 + 5x + 2x^2 + 6x + 10$

$= x^3 + 3x^2 + 2x^2 + 5x + 6x + 10$

$= x^3 + 5x^2 + 11x + 10$

61. $(x+2)(x^2-2x+4)$

$= x \cdot x^2 + x(-2x) + x \cdot 4 + 2 \cdot x^2 + 2(-2x) + 2 \cdot 4$

$= x^3 - 2x^2 + 4x + 2x^2 - 4x + 8$

$= x^3 - 2x^2 + 2x^2 + 4x - 4x + 8$

$= x^3 + 8$

63. $(2b^2-3b+2)(b-4)$

$= 2b^2 \cdot b + 2b^2(-4) + (-3b) \cdot b + (-3b)(-4) +$ $2 \cdot b + 2(-4)$

$= 2b^3 - 8b^2 - 3b^2 + 12b + 2b - 8$

$= 2b^3 - 11b^2 + 14b - 8$

65. $(2x^2+3)(3x^2-x+4)$

$= 2x^2 \cdot 3x^2 + 2x^2(-x) + 2x^2 \cdot 4 + 3 \cdot 3x^2 +$ $3(-x) + 3 \cdot 4$

$= 6x^4 - 2x^3 + 8x^2 + 9x^2 - 3x + 12$

$= 6x^4 - 2x^3 + 17x^2 - 3x + 12$

67. $(4w^2-2w+1)(3w^2-2)$

$= 4w^2 \cdot 3w^2 + 4w^2(-2) + (-2w) \cdot 3w^2 +$ $(-2w)(-2) + 1 \cdot 3w^2 + 1(-2)$

$= 12w^4 - 8w^2 - 6w^3 + 4w + 3w^2 - 2$

$= 12w^4 - 6w^3 - 8w^2 + 3w^2 + 4w - 2$

$= 12w^4 - 6w^3 - 5w^2 + 4w - 2$

69. $(2x^2+4x-1)(3x^2-x+2)$

$= 2x^2 \cdot 3x^2 + 2x^2(-x) + 2x^2 \cdot 2 + 4x \cdot 3x^2 +$ $4x(-x) + 4x \cdot 2 + (-1) \cdot 3x^2 + (-1)(-x) + (-1) \cdot 2$

$= 6x^4 - 2x^3 + 4x^2 + 12x^3 - 4x^2 + 8x - 3x^2 + x - 2$

$= 6x^4 - 2x^3 + 12x^3 + 4x^2 - 4x^2 - 3x^2 + 8x + x - 2$

$= 6x^4 + 10x^3 - 3x^2 + 9x - 2$

71. $(a+b+c)(a+b-c)$

$= a \cdot a + a \cdot b + a(-c) + b \cdot a + b \cdot b + b(-c) + c \cdot a +$
$c \cdot b + c(-c)$

$= a^2 + ab - ac + ab + b^2 - bc + ac + bc - c^2$

$= a^2 + b^2 - c^2 + ab + ab - ac + ac - bc + bc$

$= a^2 + b^2 - c^2 + 2ab$

73. a. $\underbrace{(1.5t + 30.4)}_{\substack{\text{dollars} \\ \text{subscription}}}$ $\underbrace{(-0.8t + 66.4)}_{\text{millions of subscriptions}}$

We use the fact that $\dfrac{\text{subscriptions}}{\text{subscriptions}} = 1$ to

simplify the expression:

$\dfrac{\text{dollars}}{\text{subscription}} \cdot \text{millions of subscriptions}$

$= \text{dollars} \cdot \text{millions}$

$= \text{millions of dollars}$

Thus, the units of the expression are millions of dollars.

b. $(1.5t + 30.4)(-0.8t + 66.4)$

$= 1.5t(-0.8t) + 1.5t \cdot 66.4 + 30.4(-0.8t) +$
$30.4 \cdot 66.4$

$= -1.2t^2 + 99.6t - 24.32t + 2018.56$

$= -1.2t^2 + 75.28t + 2018.56$

This expression represents the total monthly cost (in millions of dollars) of cable TV in the United States at t years since 2000.

c. Substitute 11 for t in the expression:
$-1.2t^2 + 75.28t + 2018.56$

$= -1.2(11)^2 + 75.28(11) + 2018.56$

$= 2701.44$

In the year 2011, the total monthly cost of cable TV in the United States will be about $2701 million, or about $2.7 billion.

75. a. $\underbrace{(1.15t + 19)}_{\substack{\text{thousands of dollars} \\ \text{teacher}}}$ $\underbrace{(0.044t + 1.97)}_{\text{millions of teachers}}$

We use the fact that $\dfrac{\text{teachers}}{\text{teachers}} = 1$ to simplify

the expression:

$\dfrac{\text{thousands of dollars}}{\text{teacher}} \cdot \text{millions of teachers}$

$= \text{thousands of dollars} \cdot \text{millions}$

$= \text{billions of dollars}$

Thus, the units of the expression are billions of dollars.

b. $(1.15t + 19)(0.044t + 1.97)$

$= 1.15t \cdot 0.044t + 1.15t \cdot 1.97 + 19 \cdot 0.044t +$
$19 \cdot 1.97$

$= 0.0506t^2 + 2.2655t + 0.836t + 37.43$

$= 0.0506t^2 + 3.1015t + 37.43$

$\approx 0.05t^2 + 3.10t + 37.43$

This expression represents the total money (in billions of dollars) paid for teacher salaries in the year that is t years since 1980.

c. Substitute 30 for t in the expression:
$0.05t^2 + 3.10t + 37.43$

$= 0.05(30)^2 + 3.10(30) + 37.43$

$= 175.43$

In the year 2010, the total money paid for teacher salaries will be $175.43 billion.

77. Answers may vary. One possibility follows: The student subtracted the rather than multiply. The correct result is:

$6x(-4x) = 6(-4) \cdot x \cdot x = -24x^2$.

79. a. (i) $(2x + 3)(4x + 5)$

$= 2x \cdot 4x + 2x \cdot 5 + 3 \cdot 4x + 3 \cdot 5$

$= 8x^2 + 10x + 12x + 15$

$= 8x^2 + 22x + 15$

This is a quadratic polynomial.

(ii) $(3x - 7)(5x + 2)$

$= 3x \cdot 5x + 3x \cdot 2 + (-7) \cdot 5x + (-7) \cdot 2$

$= 15x^2 + 6x - 35x - 14$

$= 15x^2 - 29x - 14$

This is a quadratic polynomial.

b. Examples will vary. The result will be a quadratic polynomial.

c. In general, the product of two linear polynomials will be a quadratic polynomial. Explanations will vary. One possibility follows: Each linear polynomial will be of degree one. Multiplying these will result in a polynomial that is degree two.

81. a. $(x + 4)(x + 7) = x \cdot x + x \cdot 7 + 4 \cdot x + 4 \cdot 7$

$= x^2 + 7x + 4x + 28$

$= x^2 + 11x + 28$

b. $(x + 7)(x + 4) = x \cdot x + x \cdot 4 + 7 \cdot x + 7 \cdot 4$

$= x^2 + 4x + 7x + 28$

$= x^2 + 11x + 28$

c. Answers may vary. One possibility follows: The fact that $(x+4)(x+7) = (x+7)(x+4)$ is a result of the commutative property of multiplication.

83. The expressions $(x-5)(x+2)$, $x(x-3)-10$, $x^2 - 3x - 10$, and $(x+2)(x-5)$ are all equivalent.

85. $(3x-5)(2x^2 - 4x + 2)$
$= 3x \cdot 2x^2 + 3x(-4x) + 3x \cdot 2 + (-5) \cdot 2x^2 + (-5)(-4x) + (-5) \cdot 2$
$= 6x^3 - 12x^2 + 6x - 10x^2 + 20x - 10$
$= 6x^3 - 12x^2 - 10x^2 + 6x + 20x - 10$
$= 6x^3 - 22x^2 + 26x - 10$

87. $(3x-5) - (2x^2 - 4x + 2) = 3x - 5 - 2x^2 + 4x - 2$
$= -2x^2 + 3x + 4x - 5 - 2$
$= -2x^2 + 7x - 7$

89. $y = 3x(x-2)$
$y = 3x \cdot x + 3x(-2)$
$y = 3x^2 - 6x$
The equation is quadratic. The graph will be a parabola.

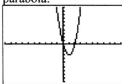

91. $y = (x-3)(x-4)$
$y = x \cdot x + x(-4) + (-3)x + (-3)(-4)$
$y = x^2 - 4x - 3x + 12$
$y = x^2 - 7x + 12$
The equation is quadratic. The graph will be a parabola.

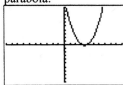

93. $y = (x+2) - (3x+5)$
$y = x + 2 - 3x - 5$
$y = x - 3x + 2 - 5$
$y = -2x - 3$
The equation is linear. The graph will be a line.

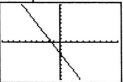

95. $y = (2x+1)(5x-2)$
$y = 2x \cdot 5x + 2x(-2) + 1 \cdot 5x + 1(-2)$
$y = 10x^2 - 4x + 5x - 2$
$y = 10x^2 + x - 2$
The equation is quadratic. The graph will be a parabola.

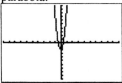

97. $2w - 5 = 7w + 5$
$2w - 5 - 7w = 7w + 5 - 7w$
$-5w - 5 = 5$
$-5w - 5 + 5 = 5 + 5$
$-5w = 10$
$\dfrac{-5w}{-5} = \dfrac{10}{-5}$
$w = -2$
This is a linear equation in one variable.

99. $(2w-5)(7w+5)$
$= 2w \cdot 7w + 2w \cdot 5 + (-5) \cdot 7w + (-5) \cdot 5$
$= 14w^2 + 10w - 35w - 25$
$= 14x^2 - 25w - 25$
This is a quadratic (or second-degree) polynomial in one variable.

Homework 7.5

1. $(xy)^8 = x^8 y^8$

3. $(6x)^2 = 6^2 x^2 = 36x^2$

5. $(4x)^3 = 4^3 x^3 = 64x^3$

7. $(-8x)^2 = (-8)^2 x^2 = 64x^2$

9. $(-3x)^3 = (-3)^3 x^3 = -27x^3$

11. $(-a)^5 = (-1a)^5 = (-1)^5 a^5 = -1a^5 = -a^5$

13. $(x+5)^2 = x^2 + 2 \cdot x \cdot 5 + 5^2$
$= x^2 + 10x + 25$

15. $(x-4)^2 = x^2 - 2 \cdot x \cdot 4 + 4^2$
$= x^2 - 8x + 16$

17. $(2x+3)^2 = (2x)^2 + 2 \cdot 2x \cdot 3 + 3^2$
$= 4x^2 + 12x + 9$

19. $(5y-2)^2 = (5y)^2 - 2 \cdot 5y \cdot 2 + 2^2$
$= 25y^2 - 20y + 4$

21. $(3x+6)^2 = (3x)^2 + 2 \cdot 3x \cdot 6 + 6^2$
$= 9x^2 + 36x + 36$

23. $(9x-2)^2 = (9x)^2 - 2 \cdot 9x \cdot 2 + 2^2$
$= 81x^2 - 36x + 4$

25. $(2a+5b)^2 = (2a)^2 + 2 \cdot 2a \cdot 5b + (5b)^2$
$= 4a^2 + 20ab + 25b^2$

27. $(8x-3y)^2 = (8x)^2 - 2 \cdot 8x \cdot 3y + (3y)^2$
$= 64x^2 - 48xy + 9y^2$

29. $y = (x+6)^2$
$y = x^2 + 2 \cdot x \cdot 6 + 6^2$
$y = x^2 + 12x + 36$

31. $y = (x-3)^2 + 1$
$y = x^2 - 2 \cdot x \cdot 3 + 3^2 + 1$
$y = x^2 - 6x + 9 + 1$
$y = x^2 - 6x + 10$

33. $y = 2(x+4)^2 - 3$
$y = 2(x^2 + 2 \cdot x \cdot 4 + 4^2) - 3$
$y = 2(x^2 + 8x + 16) - 3$
$y = 2x^2 + 16x + 32 - 3$
$y = 2x^2 + 16x + 29$

35. $y = -3(x-1)^2 - 2$
$y = -3(x^2 - 2 \cdot x \cdot 1 + 1^2) - 2$
$y = -3(x^2 - 2x + 1) - 2$
$y = -3x^2 + 6x - 3 - 2$
$y = -3x^2 + 6x - 5$

37. $(x+4)(x-4) = x^2 - 4^2 = x^2 - 16$

39. $(t-7)(t+7) = t^2 - 7^2 = t^2 - 49$

41. $(x+1)(x-1) = x^2 - 1^2 = x^2 - 1$

43. $(3a+7)(3a-7) = (3a)^2 - 7^2 = 9a^2 - 49$

45. $(2x-3)(2x+3) = (2x)^2 - 3^2 = 4x^2 - 9$

47. $(3x+6y)(3x-6y) = (3x)^2 - (6y)^2$
$= 9x^2 - 36y^2$

49. $(3t-7w)(3t+7w) = (3t)^2 - (7w)^2$
$= 9t^2 - 49w^2$

51. $(x+11)(x-11) = x^2 - 11^2 = x^2 - 121$

53. $(4x^2 - 5x) - (2x^3 - 8x) = 4x^2 - 5x - 2x^3 + 8x$
$= -2x^3 + 4x^2 - 5x + 8x$
$= -2x^3 + 4x^2 + 3x$

55. $(6m-2n)^2 = (6m)^2 - 2 \cdot 6m \cdot 2n + (2n)^2$
$= 36m^2 - 24mn + 4n^2$

57. $5t(-2t^2) = 5(-2)t^1 \cdot t^2 = -10t^{1+2} = -10t^3$

59. $(3x+4)(x^2 - x + 2)$
$= 3x \cdot x^2 + 3x(-x) + 3x \cdot 2 + 4 \cdot x^2 + 4(-x) + 4 \cdot 2$
$= 3x^3 - 3x^2 + 6x + 4x^2 - 4x + 8$
$= 3x^3 - 3x^2 + 4x^2 + 6x - 4x + 8$
$= 3x^3 + x^2 + 2x + 8$

61. $(2t-3p)(2t+3p) = (2t)^2 - (3p)^2$
$$= 4t^2 - 9p^2$$

63. $2xy^2(4x^2 - 8x - 5)$
$$= 2xy^2 \cdot 4x^2 + 2xy^2 \cdot (-8x) + 2xy^2 \cdot (-5)$$
$$= 8x^3y^2 - 16x^2y^2 - 10xy^2$$

65. $(-6x^2 - 4x + 5) + (-2x^2 + 3x - 8)$
$$= -6x^2 + (-2x^2) - 4x + 3x + 5 - 8$$
$$= -8x^2 - x - 3$$

67. $(x+3)(x-7) = x \cdot x + x \cdot (-7) + 3 \cdot x + 3 \cdot (-7)$
$$= x^2 - 7x + 3x - 21$$
$$= x^2 - 4x - 21$$

69. $(4w-8)^2 = (4w)^2 - 2 \cdot 4w \cdot 8 + 8^2$
$$= 16w^2 - 64w + 64$$

71. $(x^2 + 2x - 5)(x - 3)$
$$= x^2 \cdot x + x^2(-3) + 2x \cdot x + 2x(-3) + (-5)x +$$
$$(-5)(-3)$$
$$= x^3 - 3x^2 + 2x^2 - 6x - 5x + 15$$
$$= x^3 - x^2 - 11x + 15$$

73. $(3x - 7y)(2x + 3y)$
$$= 3x \cdot 2x + 3x \cdot 3y + (-7y)(2x) + (-7y)(3y)$$
$$= 6x^2 + 9xy - 14xy - 21y^2$$
$$= 6x^2 - 5xy - 21y^2$$

75. $(6x-7)(6x+7) = (6x)^2 - 7^2$
$$= 36x^2 - 49$$

77. $(2t+7)^2 = (2t)^2 + 2 \cdot 2t \cdot 7 + 7^2$
$$= 4t^2 + 28t + 49$$

79. Answers may vary. One possibility follows: The student failed to raise the 4 to the second power. The correct answer is:
$$(4x)^2 = 4^2 x^2 = 16x^2$$

81. Answers may vary. One possibility follows: The student has left out the middle term. The correct answer is:
$$(x+7)^2 = x^2 + 2 \cdot x \cdot 7 + 7^2$$
$$= x^2 + 14x + 49$$

83. Answers may vary. One possibility follows:

Since the y-values differ, the expression $(x-5)^2$ is not equivalent to $x^2 - 5^2$.

Since the y-values differ, the expression $(x-5)^2$ is not equivalent to $x^2 + 5^2$.

The correct simplification is:
$$(x-5)^2 = x^2 - 2 \cdot x \cdot 5 + 5^2$$
$$= x^2 - 10x + 25$$

Since the y-values are always the same, the two expressions are equivalent.

85. a. Answers may vary. One possibility follows:

Since the y-values differ, the expression $(x+4)^2$ is not equivalent to $x^2 + 4^2$.

b. $(x+4)^2 = x^2 + 2 \cdot x \cdot 4 + 4^2$
$$= x^2 + 8x + 16$$

c. Answers may vary. One possibility follows:

Since the y-values are always the same, the two expressions are equivalent.

87. If $A = 2$ and $B = 3$, then we obtain the statement
$$(A+B)^2 = A^2 + B^2$$
$$(2+3)^2 = 2^2 + 3^2$$
$$5^2 = 4+9$$
$$25 = 13$$
Explanations may vary. One possibility follows: Since the last line is not true, this shows that $(A+B)^2 = A^2 + B^2$ is not true for all values of A and B. In other words, $(A+B)^2 \neq A^2 + B^2$.

A true statement is $(A+B)^2 = A^2 + 2AB + B^2$.

89. The expressions $(x-2)^2$, $x(x-4)+4$, and $x^2 - 4x + 4$ are all equivalent.

91. Answers may vary. One possibility follows: Since $(x+7)^2$ is equivalent to $(x+7)(x+7)$, the $14x$ term comes from the "outer" and "inner" multiplications.

93. a. (i) $2^5 = 32, 2^4 = 16, 2^3 = 8, 2^2 = 4, 2^1 = 2$

    (ii) Answers may vary. One possibility follows: Each result is half of the previous result. For example,
$$16 = \frac{1}{2} \cdot 32, \; 8 = \frac{1}{2} \cdot 16, \text{ and so on.}$$

    (iii) $2^0 = \frac{1}{2} \cdot 2 = 1$

  b. $3^4 = 81, 3^3 = 27, 3^2 = 9, 3^1 = 3, 3^0 = 1$

  c. If b is a nonzero real number, then $b^0 = 1$.

95. $y = 2x(5x-3)$
$$y = 2x \cdot 5x - 2x \cdot 3$$
$$y = 10x^2 - 6x$$
This is a quadratic equation in two variables. The graph will be a parabola.

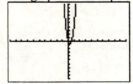

97. $y = (4x-3)(6x-5)$
$$y = 4x \cdot 6x + 4x(-5) + (-3) \cdot 6x + (-3)(-5)$$
$$y = 24x^2 - 20x - 18x + 15$$
$$y = 24x^2 - 38x + 15$$
This is a quadratic equation in two variables. The graph will be a parabola.

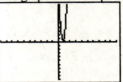

99. $y = x^2 - (x-3)^2$
$$y = x^2 - (x^2 - 2 \cdot x \cdot 3 + 3^2)$$
$$y = x^2 - (x^2 - 6x + 9)$$
$$y = x^2 - x^2 + 6x - 9$$
$$y = 6x - 9$$
This is a linear equation in two variables. The graph will be a line.

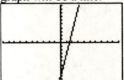

101. $2x - 5y = 15$ Equation (1)
    $y = 3x - 16$ Equation (2)
From equation (2), we substitute $3x - 16$ for y in equation (1) and solve for x:
$$2x - 5y = 15$$
$$2x - 5(3x - 16) = 15$$
$$2x - 15x + 80 = 15$$
$$-13x + 80 = 15$$
$$-13x = -65$$
$$x = 5$$
We substitute 5 for x in equation (2) to find y:
$$y = 3x - 16$$
$$y = 3(5) - 16 = -1$$
The solution is $(5, -1)$. This is a system of two linear equations in two variables.

103. $2x - 5y = 15$

$$-5y = -2x + 15$$

$$\frac{-5y}{-5} = \frac{-2x}{-5} + \frac{15}{-5}$$

$$y = \frac{2}{5}x - 3$$

The slope is $m = \frac{2}{5}$, and the y-intercept is $(0, -3)$.

We first plot $(0, -3)$. From this point we move 5 units to the right and 2 units up, where we plot the point $(5, -1)$. We then sketch the line that contains these two points.

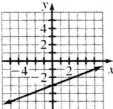

This is a linear equation in two variables.

Homework 7.6

1. $x^3 x^5 = x^{3+5} = x^8$

3. $r^5 r = r^5 r^1 = r^{5+1} = r^6$

5. $5x^4 \left(3x^5\right) = (5 \cdot 3)\left(x^4 x^5\right) = 15x^9$

7. $\left(-4b^3\right)\left(-8b^5\right) = -4(-8)\left(b^3 b^5\right) = 32b^8$

9. $\left(6a^2 b^5\right)\left(9a^4 b^3\right) = (6 \cdot 9)\left(a^2 a^4\right)\left(b^5 b^3\right) = 54a^6 b^8$

11. $(rt)^7 = r^7 t^7$

13. $(8x)^2 = 8^2 x^2 = 64x^2$

15. $(2xy)^5 = 2^5 x^5 y^5 = 32x^5 y^5$

17. $(-2a)^4 = (-2)^4 a^4 = 16a^4$

19. $(9xy)^0 = 1$

21. $\dfrac{a^5}{a^2} = a^{5-2} = a^3$

23. $\dfrac{6x^7}{3x^3} = \dfrac{6}{3} \cdot \dfrac{x^7}{x^3} = 2 \cdot x^{7-3} = 2x^4$

25. $\dfrac{15x^6 y^8}{12x^3 y} = \dfrac{15}{12} \cdot \dfrac{x^6}{x^3} \cdot \dfrac{y^8}{y^1} = \dfrac{5}{4} \cdot x^{6-3} \cdot y^{8-1} = \dfrac{5x^3 y^7}{4}$

27. $\left(\dfrac{t}{w}\right)^7 = \dfrac{t^7}{w^7}$

29. $\left(\dfrac{3}{t}\right)^3 = \dfrac{3^3}{t^3} = \dfrac{27}{t^3}$

31. $\left(\dfrac{x}{3}\right)^0 = 1$

33. $\left(r^2\right)^4 = r^{2 \cdot 4} = r^8$

35. $\left(x^4\right)^9 = x^{4 \cdot 9} = x^{36}$

37. $\left(6x^3\right)^2 = 6^2 \left(x^3\right)^2 = 36x^{3 \cdot 2} = 36x^6$

39. $\left(-t^3\right)^4 = \left(-1 \cdot t^3\right)^4 = (-1)^4 \left(t^3\right)^4 = 1 \cdot t^{3 \cdot 4} = t^{12}$

41. $\left(2a^2 a^7\right)^3 = \left(2a^{2+7}\right)^3 = \left(2a^9\right)^3 = 2^3 \left(a^9\right)^3 = 8a^{27}$

43. $\left(x^2 y^3\right)^4 x^5 y^8 = \left(x^2\right)^4 \left(y^3\right)^4 x^5 y^8$

$$= x^8 y^{12} x^5 y^8$$

$$= x^{8+5} y^{12+8}$$

$$= x^{13} y^{20}$$

45. $5x^4 \left(3x^6\right)^2 = 5x^4 \cdot 3^2 \left(x^6\right)^2$

$$= 5x^4 \cdot 9x^{12}$$

$$= 5 \cdot 9x^{4+12}$$

$$= 45x^{16}$$

47. $-3c^6 \left(c^4\right)^5 = -3c^6 \cdot c^{20} = -3c^{6+20} = -3c^{26}$

49. $\left(xy^3\right)^5 (xy)^4 = x^5 \left(y^3\right)^5 x^4 y^4$

$$= x^5 y^{15} x^4 y^4$$

$$= x^{5+4} y^{15+4}$$

$$= x^9 y^{19}$$

51. $\dfrac{10t^5t^7}{8t^4} = \dfrac{10t^{5+7}}{8t^4} = \dfrac{10t^{12}}{8t^4} = \dfrac{10}{8} \cdot t^{12-4} = \dfrac{5}{4} \cdot t^8 = \dfrac{5t^8}{4}$

53. $\dfrac{18x^{10}}{24x^4x^6} = \dfrac{18}{24} \cdot \dfrac{x^{10}}{x^{4+6}} = \dfrac{3}{4} \cdot \dfrac{x^{10}}{x^{10}} = \dfrac{3}{4} \cdot x^{10-10} = \dfrac{3x^0}{4} = \dfrac{3}{4}$

55. $\left(\dfrac{y}{2x}\right)^3 = \dfrac{y^3}{(2x)^3} = \dfrac{y^3}{2^3 x^3} = \dfrac{y^3}{8x^3}$

57. $\left(\dfrac{x^2}{y^5}\right)^4 = \dfrac{\left(x^2\right)^4}{\left(y^5\right)^4} = \dfrac{x^8}{y^{20}}$

59. $\left(\dfrac{r^6}{6}\right)^2 = \dfrac{\left(r^6\right)^2}{6^2} = \dfrac{r^{12}}{36}$

61. $\left(\dfrac{2a^4}{3b^2}\right)^3 = \dfrac{\left(2a^4\right)^3}{\left(3b^2\right)^3} = \dfrac{2^3\left(a^4\right)^3}{3^3\left(b^2\right)^3} = \dfrac{8a^{12}}{27b^6}$

63. $\left(\dfrac{3x^4}{5y^7}\right)^0 = 1$

65. $\left(\dfrac{2a^6b}{3c^5}\right)^3 = \dfrac{\left(2a^6b\right)^3}{\left(3c^5\right)^3} = \dfrac{2^3\left(a^6\right)^3 b^3}{3^3\left(c^5\right)^3} = \dfrac{8a^{18}b^3}{27c^{15}}$

67. $\dfrac{\left(x^4y\right)^4}{x^5} = \dfrac{\left(x^4\right)^4 y^4}{x^5} = \dfrac{x^{16}y^4}{x^5} = x^{16-5}y^4 = x^{11}y^4$

69. $\dfrac{\left(w^3\right)^4}{(2w)^5} = \dfrac{w^{12}}{2^5w^5} = \dfrac{w^{12}}{32w^5} = \dfrac{w^{12-5}}{32} = \dfrac{w^7}{32}$

71. $\dfrac{\left(4x^5y^8\right)^2}{8x^8y^9} = \dfrac{4^2\left(x^5\right)^2\left(y^8\right)^2}{8x^8y^9}$

$\qquad = \dfrac{16x^{10}y^{16}}{8x^8y^9}$

$\qquad = \dfrac{16}{8} \cdot x^{10-8} \cdot y^{16-9}$

$\qquad = 2x^2y^7$

73. We substitute 5 for t in the equation $d = 16t^2$.
$\quad d = 16(5)^2 = 400$
The sky diver will have fallen 400 feet.

75. We substitute 0.57 for r and 12.5 for v in the equation $P = 0.8r^2v^3$.
$\quad P = 0.8(0.57)^2(12.5)^3 = 507.65625$
The windmill can generate approximately 507.66 watts of power.

77. Answers may vary. One possibility follows: The student should have added the exponents instead of multiplying them. The correct simplification is: $x^3x^5 = x^{3+5} = x^8$.

79. Answers may vary. One possibility follows: When raising the product to the power, the student failed to raise both factors to the power. The correct simplification is:
$\left(5x^3\right)^2 = 5^2\left(x^3\right)^2 = 25x^6$.

81. Answers may vary. One possibility follows: When raising the product to the fourth power, the student should have raised the factor 2 to the 4th power instead of multiplying it by 4. The correct simplification is:
$\left(2x^2\right)^4 = 2^4\left(x^2\right)^4 = 16x^8$.

83. $x^3x^2 = x^{3+2} = x^5$

85. $x^3 + x^2$ can be simplified no further.

87. $2x^4 + 3x^4 = 5x^4$

89. $\left(2x^4\right)\left(3x^4\right) = (2 \cdot 3)x^4x^4 = 6x^{4+4} = 6x^8$

91. $(3x)^2 = 3^2x^2 = 9x^2$

93. $(3+x)^2 = 3^2 + 2 \cdot 3 \cdot x + x^2$
$\qquad = 9 + 6x + x^2$
$\qquad = x^2 + 6x + 9$

95. $\dfrac{2}{3}x - \dfrac{5}{6} = \dfrac{1}{2}x$

$\quad 6\left(\dfrac{2}{3}x - \dfrac{5}{6}\right) = 6\left(\dfrac{1}{2}x\right)$

$\qquad 4x - 5 = 3x$

$\qquad\quad x - 5 = 0$

$\qquad\qquad x = 5$

This is a linear equation in one variable.

97. $\dfrac{2}{3}x - \dfrac{5}{6} - \dfrac{1}{2}x = \dfrac{2}{3}x - \dfrac{1}{2}x - \dfrac{5}{6}$

$\qquad\qquad\;\; = \dfrac{4}{6}x - \dfrac{3}{6}x - \dfrac{5}{6}$

$\qquad\qquad\;\; = \dfrac{1}{6}x - \dfrac{5}{6}$

This is a linear (or first-degree) polynomial in one variable.

Homework 7.7

1. $6^{-2} = \dfrac{1}{6^2} = \dfrac{1}{36}$

3. $x^{-4} = \dfrac{1}{x^4}$

5. $b^{-1} = \dfrac{1}{b^1} = \dfrac{1}{b}$

7. $\dfrac{1}{2^{-4}} = 2^4 = 16$

9. $\dfrac{1}{w^{-2}} = w^2$

11. $\dfrac{x^{-3}}{y^5} = x^{-3} \cdot \dfrac{1}{y^5} = \dfrac{1}{x^3} \cdot \dfrac{1}{y^5} = \dfrac{1}{x^3 y^5}$

13. $\dfrac{a^{-4}}{b^{-2}} = a^{-4} \cdot \dfrac{1}{b^{-2}} = \dfrac{1}{a^4} \cdot b^2 = \dfrac{b^2}{a^4}$

15. $\dfrac{2a^3 b^{-5}}{5c^{-8}} = \dfrac{2a^3}{5} \cdot b^{-5} \cdot \dfrac{1}{c^{-8}} = \dfrac{2a^3}{5} \cdot \dfrac{1}{b^5} \cdot c^8 = \dfrac{2a^3 c^8}{5b^5}$

17. $\dfrac{4x^{-9}}{-6y^4 w^{-1}} = \dfrac{4}{-6y^4} \cdot x^{-9} \cdot \dfrac{1}{w^{-1}}$

$\qquad\qquad\;\; = -\dfrac{2}{3y^4} \cdot \dfrac{1}{x^9} \cdot w^1$

$\qquad\qquad\;\; = -\dfrac{2w}{3x^9 y^4}$

19. $\left(x^{-2}\right)^7 = x^{-2(7)} = x^{-14} = \dfrac{1}{x^{14}}$

21. $\left(t^{-4}\right)^{-3} = t^{(-4)(-3)} = t^{12}$

23. $\left(6t^{-4}\right)\left(5t^2\right) = 6 \cdot 5 \cdot t^{-4} t^2 = 30t^{-4+2} = 30t^{-2} = \dfrac{30}{t^2}$

25. $\left(-4x^{-1}\right)\left(3x^{-8}\right) = -4 \cdot 3 \cdot x^{-1} x^{-8} = -12x^{-9} = -\dfrac{12}{x^9}$

27. $\left(-4x^3 y^{-7}\right)\left(-x^{-5} y^4\right) = (-4)(-1)x^3 x^{-5} y^{-7} y^4$

$\qquad\qquad\qquad\qquad\quad = 4x^{-2} y^{-3}$

$\qquad\qquad\qquad\qquad\quad = \dfrac{4}{x^2 y^3}$

29. $\dfrac{x^2}{x^6} = x^{2-6} = x^{-4} = \dfrac{1}{x^4}$

31. $\dfrac{a^{-3}}{a^5} = a^{-3-5} = a^{-8} = \dfrac{1}{a^8}$

33. $\dfrac{a^3}{a^{-2}} = a^{3-(-2)} = a^{3+2} = a^5$

35. $\dfrac{7x^{-3}}{4x^{-9}} = \dfrac{7x^{-3-(-9)}}{4} = \dfrac{7x^{-3+9}}{4} = \dfrac{7x^6}{4}$

37. $\dfrac{2^{-1}}{2^4} = 2^{-1-4} = 2^{-5} = \dfrac{1}{2^5} = \dfrac{1}{32}$

39. $\dfrac{5^{-6}}{5^{-4}} = 5^{-6-(-4)} = 5^{-6+4} = 5^{-2} = \dfrac{1}{5^2} = \dfrac{1}{25}$

41. $\dfrac{3^4 w^{-8}}{3^2 w^{-3}} = 3^{4-2} w^{-8-(-3)} = 3^2 w^{-5} = \dfrac{3^2}{w^5} = \dfrac{9}{w^5}$

43. $\left(8c^{-3}\right)^2 = 8^2 \left(c^{-3}\right)^2 = 64c^{-6} = \dfrac{64}{c^6}$

45. $\left(2x^{-1}\right)^{-5} = 2^{-5}\left(x^{-1}\right)^{-5} = \dfrac{1}{2^5} \cdot x^5 = \dfrac{x^5}{32}$

47. $\left(x^{-2} y^5\right)^{-6} = \left(x^{-2}\right)^{-6}\left(y^5\right)^{-6} = x^{12} y^{-30} = \dfrac{x^{12}}{y^{30}}$

49. $\left(2a^{-6} b\right)^{-3} = 2^{-3}\left(a^{-6}\right)^{-3} b^{-3}$

$\qquad\qquad\qquad = 2^{-3} a^{18} b^{-3}$

$\qquad\qquad\qquad = \dfrac{a^{18}}{2^3 b^3}$

$\qquad\qquad\qquad = \dfrac{a^{18}}{8b^3}$

51. $\left(ab^2\right)^3\left(a^3\right)^{-2} = a^3\left(b^2\right)^3\left(a^3\right)^{-2}$
$$= a^3b^6a^{-6}$$
$$= a^{3-6}b^6$$
$$= a^{-3}b^6$$
$$= \frac{b^6}{a^3}$$

53. $\dfrac{1}{(xy)^{-3}} = (xy)^3 = x^3y^3$

55. $\left(\dfrac{x^{-3}}{y^2}\right)^4 = \dfrac{\left(x^{-3}\right)^4}{\left(y^2\right)^4} = \dfrac{x^{-12}}{y^8} = \dfrac{1}{x^{12}y^8}$

57. $\left(\dfrac{2}{c^{-7}}\right)^{-3} = \dfrac{2^{-3}}{\left(c^{-7}\right)^{-3}} = \dfrac{2^{-3}}{c^{21}} = \dfrac{1}{2^3c^{21}} = \dfrac{1}{8c^{21}}$

59. $\left(\dfrac{3r^{-5}}{t^9}\right)^3 = \dfrac{3^3\left(r^{-5}\right)^3}{\left(t^9\right)^3} = \dfrac{27r^{-15}}{t^{27}} = \dfrac{27}{r^{15}t^{27}}$

61. $\left(\dfrac{8a^{-3}b}{c^{-5}d^4}\right)^{-2} = \dfrac{8^{-2}\left(a^{-3}\right)^{-2}b^{-2}}{\left(c^{-5}\right)^{-2}\left(d^4\right)^{-2}}$
$$= \dfrac{8^{-2}a^6b^{-2}}{c^{10}d^{-8}}$$
$$= \dfrac{a^6d^8}{8^2b^2c^{10}}$$
$$= \dfrac{a^6d^8}{64b^2c^{10}}$$

63. $\dfrac{\left(2a^{-2}b\right)^{-3}}{\left(3cd^{-3}\right)^2} = \dfrac{2^{-3}\left(a^{-2}\right)^{-3}b^{-3}}{3^2c^2\left(d^{-3}\right)^2}$
$$= \dfrac{2^{-3}a^6b^{-3}}{3^2c^2d^{-6}}$$
$$= \dfrac{a^6d^6}{2^33^2b^3c^2}$$
$$= \dfrac{a^6d^6}{8\cdot9b^3c^2}$$
$$= \dfrac{a^6d^6}{72b^3c^2}$$

65. $\dfrac{6b^{-3}c^4}{8b^2c^{-3}} = \dfrac{6b^{-3-2}c^{4-(-3)}}{8} = \dfrac{6c^7}{8b^5} = \dfrac{3c^7}{4b^5}$

67. a. $s = dt^{-1}$
$$s = \dfrac{d}{t}$$

    b. $s = \dfrac{d}{t} = \dfrac{186}{3} = 62$

This result means that an object that travels 186 miles in 3 hours at a constant speed is traveling at a speed of 62 miles per hour.

69. a. $L = 5760d^{-2}$
$$L = \dfrac{5760}{d^2}$$

    b. $L = \dfrac{5760}{d^2} = \dfrac{5760}{8^2} = \dfrac{5760}{84} = 90$

This result means that the sound level is 90 decibels at a distance of 8 yards from the amplifier.

71. For 4.9×10^4 the decimal point must move four places to the right. Thus, the standard decimal is 49,000

73. For 8.59×10^{-3} the decimal point must move three places to the left. Thus, the standard decimal is 0.00859.

75. For 2.95×10^{-4} the decimal point must move four places to the left. Thus, the standard decimal is 0.000295.

77. For -4.512×10^8 the decimal point must move eight places to the right. Thus, the standard decimal is $-451,200,000$.

79. For 45,700,000 the decimal point needs to be moved seven places to the left so that the new number is between 1 and 10. Thus, the scientific notation is 4.57×10^7 .

81. For 0.0000659 the decimal point needs to be moved five places to the right so that the new number is between 1 and 10. Thus, the scientific notation is 6.59×10^{-5} .

83. For $-5,987,000,000,000$ the decimal point needs to be moved twelve places to the left so that the absolute value of the new number is between 1 and 10. Thus, the scientific notation is -5.987×10^{12}.

85. For 0.000001 the decimal point needs to be moved six places to the right so that the new number is between 1 and 10. Thus, the scientific notation is 1×10^{-6}.

87. $4.8\,\text{E}-8 = 0.000000048$
$1.7\,\text{E}-6 = 0.0000017$
$3.58\,\text{E}\,7 = 35,800,000$
$1.28\,\text{E}\,9 = 1,280,000,000$

89. $4 \times 10^{11} = 400,000,000,000$

91. $4.66 \times 10^{10} = 46,600,000,000$

93. $6.3 \times 10^{-4} = 0.00063$

95. $3,720,000 = 3.72 \times 10^{6}$

97. $27,000,000 = 2.7 \times 10^{7}$

99. $0.0000075 = 7.5 \times 10^{-6}$

101. Answers may vary. One possibility follows: The student should not move the 5 to the denominator. The correct answer is:
$$\frac{5x^{-3}y^2}{w} = \frac{5y^2}{x^3 w}.$$

103. Answers may vary. One possibility follows: The student failed to subtract the powers correctly. The correct answer is:
$$\frac{x^6}{x^{-4}} = x^{6-(-4)} = x^{6+4} = x^{10}.$$

105. **a.** $1^5 = 1$

     b. $1^0 = 1$

     c. $1^{-3} = \frac{1}{1^3} = \frac{1}{1} = 1$

     d. $1^n = 1$, where n is an integer.

107. $-3x^{-3} + \left(-3x^{-3}\right) = -6x^{-3} = -\dfrac{6}{x^3}$

109. $-3x^{-3}\left(-3x^{-3}\right) = (-3)(-3)x^{-3}x^{-3} = 9x^{-6} = \dfrac{9}{x^6}$

111. $\left(-3x^{-3}\right)^3 = (-3)^3\left(x^{-3}\right)^3 = -27x^{-9} = -\dfrac{27}{x^9}$

113. $\left(-3x^3\right)^{-3} = (-3)^{-3}\left(x^3\right)^{-3} = \dfrac{1}{(-3)^3\left(x^3\right)^3} = -\dfrac{1}{27x^9}$

115. **a.** (i) $\dfrac{x^4}{x^{-3}} = x^{4-(-3)} = x^{4+3} = x^7$

     (ii) $\dfrac{x^4}{x^{-3}} = x^4 \cdot \dfrac{1}{x^{-3}} = x^4 \cdot x^3 = x^{4+3} = x^7$

     b. The results are the same.

117 – 121. Answers may vary.

Chapter 7 Review Exercises

1. $y = -2x^2$

First, we list some solutions in the table. Then we sketch a curve that contains the points corresponding to the solutions.

x	y
-3	$-2(-3)^2 = -18$
-2	$-2(-2)^2 = -8$
-1	$-2(-1)^2 = -2$
0	$-2(0)^2 = 0$
1	$-2(1)^2 = -2$
2	$-2(2)^2 = -8$
3	$-2(3)^2 = -18$

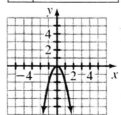

2. $y = 3x^2 - 4$

First, we list some solutions in the table. Then we sketch a curve that contains the points corresponding to the solutions.

x	y
-3	$3(-3)^2 - 4 = 23$
-2	$3(-2)^2 - 4 = 8$
-1	$3(-1)^2 - 4 = -1$
0	$3(0)^2 - 4 = -4$
1	$3(1)^2 - 4 = -1$
2	$3(2)^2 - 4 = 8$
3	$3(3)^2 - 4 = 23$

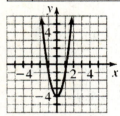

3. $y = 2x^2 - 8x + 4$

First, we list some solutions in the table. Then we sketch a curve that contains the points corresponding to the solutions.

x	y
-3	$2(-3)^2 - 8(-3) + 4 = 46$
-2	$2(-2)^2 - 8(-2) + 4 = 28$
-1	$2(-1)^2 - 8(-1) + 4 = 14$
0	$2(0)^2 - 8(0) + 4 = 4$
1	$2(1)^2 - 8(1) + 4 = -2$
2	$2(2)^2 - 8(2) + 4 = -4$
3	$2(3)^2 - 8(3) + 4 = -2$
4	$2(4)^2 - 8(4) + 4 = 4$
5	$2(5)^2 - 8(5) + 4 = 14$

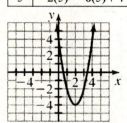

4. The table contains the ordered pair $(3, 2)$, so $y = 2$ when $x = 3$.

5. The table contains the ordered pair $(2, 1)$, so $y = 1$ when $x = 2$.

6. The table contains the ordered pairs $(1, 2)$ and $(3, 2)$, so $x = 1$ or $x = 3$ when $y = 2$.

7. The table contains the ordered pair $(2, 1)$, so $x = 2$ when $y = 1$.

8. a.

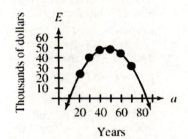

Yes, the model fits the data well.

b. We estimate that the model contains the point $(19, 21)$, so the average annual expenditure for 19-year-old Americans is approximately $21 thousand.

c. We estimate that the model contains the points $(27, 350)$ and $(67, 35)$. Thus, the average annual expenditure for 27-year-old Americans and 67-year-old Americans is $35 thousand.

d. The parabola opens downward, so the vertex is the maximum point. We estimate it to be the point $(47, 50)$. It means that the average annual expenditure for 47-year-old Americans is $50 thousand. According to the model, this is the largest annual expenditure for any age.

e. We estimate that the model intersects the a-axis at the points $(10, 0)$ and $(84, 0)$, which are the a-intercepts. This means that both 10-year-old Americans and 84-year-old Americans do not spend any money. Model breakdown has occurred for both results.

9. a.

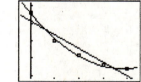

The quadratic model comes closer to the points in the scattergram.

b. The quadratic model predicts that Americans' confidence in executives will eventually increase. Explanations may vary. The quadratic model has a minimum point.

c.

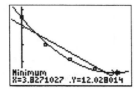

The minimum point is approximately (3.83, 12.03). It means that, in the year $2000 + 3.83 \approx 2004$, approximately 12.0% of Americans had confidence in executives at major corporations, which was the lowest percentage in any year.

d. We adjust the WINDOW setting on the calculator and use TRACE:

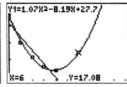

We predict that about 17.1% of Americans will have faith in executives at major corporations in the year 2006.

e. We use TRACE:

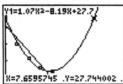

We predict that 27.7% of Americans will again have faith in executives at major corporations in the year $2000 + 7.66 \approx 2008$.

10. $(-4x^3 + 7x^2 - x) + (-5x^3 - 9x^2 + 3)$
$= -4x^3 - 5x^3 + 7x^2 - 9x^2 - x + 3$
$= -9x^3 - 2x^2 - x + 3$

11. $(6x^3 - 2x^2 + 5) - (8x^3 - 4x^2 + 3x)$
$= 6x^3 - 2x^2 + 5 - 8x^3 + 4x^2 - 3x$
$= 6x^3 - 8x^3 - 2x^2 + 4x^2 - 3x + 5$
$= -2x^3 + 2x^2 - 3x + 5$

12. $4x - 9x^2 - 6x + 2 - 7x^3 + x^2 - 8x$
$= -7x^3 - 9x^2 + x^2 + 4x - 6x - 8x + 2$
$= -7x^3 - 8x^2 - 10x + 2$

13. a. $(4.4t + 7.4) + (-4.7t + 86.8)$
$= 4.4t - 4.7t + 7.4 + 86.8$
$= -0.3t + 94.2$
This expression represents the percentage of airplanes that are regional jet, turboprops, or large jets.

b. Substitute 10 for t in our expression:
$-0.3(10) + 94.2 = 91.2$
Note that $t = 10$ represents the year 2010. The above result means that 91.2% of airplanes will be regional jets, turboprops, or large jets in the year 2010.

c. $(4.4t + 7.4) - (-4.7t + 86.8)$
$= 4.4t + 7.4 + 4.7t - 86.8$
$= 4.4t + 4.7t + 7.4 - 86.8$
$= 9.1t - 79.4$
This expression represents the difference in of the percentage of airplanes that are regional jets and the percentage of airplanes that are turboprops or large jets.

d. Substitute 10 for t in our expression:
$9.1(10) - 79.4 = 11.6$
This means that, in 2010, the percentage of airplanes that are regional jets will be 11.6 percentage points more than the percentage of airplanes that are turboprops or large jets.

14. $-3x^2\left(7x^5\right) = -3(7)\left(x^2 x^5\right) = -21x^7$

15. $5x^3(2x^2 - 7x + 4) = 5x^3 \cdot 2x^2 + 5x^3(-7x) + 5x^3 \cdot 4$
$= 10x^5 - 35x^4 + 20x^3$

16. $(w - 3)(w - 9)$
$= w \cdot w + w \cdot (-9) + (-3) \cdot w + (-3)(-9)$
$= w^2 - 9w - 3w + 27$
$= w^2 - 12w + 27$

17. $(2a + 5b)(3a - 8b)$
$= 2a \cdot 3a + 2a(-8b) + 5b \cdot 3a + 5b(-8b)$
$= 6a^2 - 16ab + 15ab - 40b^2$
$= 6a^2 - ab - 40b^2$

18. $(3x^2 - 4)(5x + 6)$
$= 3x^2 \cdot 5x + 3x^2 \cdot 6 + (-4) \cdot 5x + (-4) \cdot 6$
$= 15x^3 + 18x^2 - 20x - 24$

19. $(x+4)(x^2-3x+5)$
$= x \cdot x^2 + x(-3x) + x \cdot 5 + 4 \cdot x^2 + 4(-3x) + 4 \cdot 5$
$= x^3 - 3x^2 + 5x + 4x^2 - 12x + 20$
$= x^3 - 3x^2 + 4x^2 + 5x - 12x + 20$
$= x^3 + x^2 - 7x + 20$

20. $(4b^2 - b + 3)(2b - 7)$
$= 4b^2 \cdot 2b + 4b^2(-7) + (-b) \cdot 2b + (-b)(-7) +$
$3 \cdot 2b + 3(-7)$
$= 8b^3 - 28b^2 - 2b^2 + 7b + 6b - 21$
$= 8b^3 - 30b^2 + 13b - 21$

21. $(-2t)^3 = (-2)^3 t^3 = -8t^3$

22. $(x+7)^2 = x^2 + 2 \cdot x \cdot 7 + 7^2$
$= x^2 + 14x + 49$

23. $(x-4)^2 = x^2 - 2 \cdot x \cdot 4 + 4^2$
$= x^2 - 8x + 16$

24. $(2p+5)^2 = (2p)^2 + 2 \cdot (2p) \cdot 5 + 5^2$
$= 4p^2 + 20p + 25$

25. $-5(c+2)^2 = -5(c^2 + 2 \cdot c \cdot 2 + 2^2)$
$= -5(c^2 + 4c + 4)$
$= -5c^2 - 20c - 20$

26. $(x+6)(x-6) = x^2 - 6^2 = x^2 - 36$

27. $(4m-7n)(4m+7n) = (4m)^2 - (7n)^2$
$= 16m^2 - 49n^2$

28. $(3p-2t)^2 = (3p)^2 - 2 \cdot (3p) \cdot (2t) + (2t)^2$
$= 9p^2 - 12pt + 4t^2$

29. $(2a^2 - a + 3)(a^2 + 2a - 1)$
$= 2a^2(a^2) + 2a^2(2a) + 2a^2(-1) + (-a)(a^2) +$
$(-a)(2a) + (-a)(-1) + 3(a^2) + 3(2a) + 3(-1)$
$= 2a^4 + 4a^3 - 2a^2 - a^3 - 2a^2 + a + 3a^2 + 6a - 3$
$= 2a^4 + 4a^3 - a^3 - 2a^2 - 2a^2 + 3a^2 + a + 6a - 3$
$= 2a^4 + 3a^3 - a^2 + 7a - 3$

30. The product of a linear polynomial and a quadratic polynomial is a cubic polynomial. Examples may vary.

31. $y = (x-5)^2 + 3$
$y = x^2 - 2 \cdot x \cdot 5 + 5^2 + 3$
$y = x^2 - 10x + 25 + 3$
$y = x^2 - 10x + 28$

32. $y = -2(x+3)^2 - 6$
$y = -2(x^2 + 2 \cdot x \cdot 3 + 3^2) - 6$
$y = -2(x^2 + 6x + 9) - 6$
$y = -2x^2 - 12x - 18 - 6$
$y = -2x^2 - 12x - 24$

33. $\left(-x^5\right)^2 = (-1)^2 \left(x^5\right)^2 = 1x^{5 \cdot 2} = x^{10}$

34. $(2x^3)(6x^4) = 2 \cdot 6x^3 x^4 = 12x^{3+4} = 12x^7$

35. $\left(8a^2 b^3\right)\left(-5a^4 b^9\right) = 8 \cdot (-5)\left(a^2 a^4\right)\left(b^3 b^9\right)$
$= -40a^6 b^{12}$

36. $\dfrac{8x^4 y^8}{16x^3 y^5} = \dfrac{8}{16} \cdot x^{4-3} \cdot y^{8-5} = \dfrac{1}{2} x^1 y^3 = \dfrac{xy^3}{2}$

37. $\left(\dfrac{x}{2}\right)^3 = \dfrac{x^3}{2^3} = \dfrac{x^3}{8}$

38. $\left(2x^9 y^3\right)^5 = 2^5 \left(x^9\right)^5 \left(y^3\right)^5 = 32x^{45} y^{15}$

39. $3x^6 \left(5x^4\right)^2 = 3x^6 \cdot 5^2 \left(x^4\right)^2$
$= 3x^6 \cdot 25x^8$
$= 3 \cdot 25x^{6+8}$
$= 75x^{14}$

40. $\dfrac{15c^2 c^7}{10c^4} = \dfrac{15c^{2+7}}{10c^4} = \dfrac{15c^9}{10c^4} = \dfrac{15}{10} \cdot c^{9-4} = \dfrac{3}{2} \cdot c^5 = \dfrac{3c^5}{2}$

41. $\left(\dfrac{a^4}{9}\right)^2 = \dfrac{\left(a^4\right)^2}{9^2} = \dfrac{a^8}{81}$

42. $\left(\dfrac{-9x^5}{5y^7}\right)^0 = 1$

43. $\dfrac{\left(3x^5 y^4\right)^2}{6x^7 y^3} = \dfrac{3^2\left(x^5\right)^2\left(y^4\right)^2}{6x^7 y^3}$

$= \dfrac{9x^{10} y^8}{6x^7 y^3}$

$= \dfrac{9}{6}\cdot x^{10-7}\cdot y^{8-3}$

$= \dfrac{3}{2}x^3 y^5$

$= \dfrac{3x^3 y^5}{2}$

44. $\left(\dfrac{3x^5}{4x^2}\right)^3 = \left(\dfrac{3x^{5-2}}{4}\right)^3 = \left(\dfrac{3x^3}{4}\right)^3 = \dfrac{3^3\left(x^3\right)^3}{4^3} = \dfrac{27x^9}{64}$

45. $4^{-3} = \dfrac{1}{4^3} = \dfrac{1}{64}$

46. $\dfrac{1}{x^{-5}} = x^5$

47. $\dfrac{r^{-7}}{r^3} = r^{-7-3} = r^{-10} = \dfrac{1}{r^{10}}$

48. $\dfrac{t^{-5}}{t^{-2}} = t^{-5-(-2)} = t^{-5+2} = t^{-3} = \dfrac{1}{t^3}$

49. $\dfrac{x^{-5} y^2}{w^{-7}} = x^{-5}\cdot y^2\cdot\dfrac{1}{w^{-7}} = \dfrac{1}{x^5}\cdot y^2\cdot w^7 = \dfrac{y^2 w^7}{x^5}$

50. $\left(c^6\right)^{-4} = c^{6(-4)} = c^{-24} = \dfrac{1}{c^{24}}$

51. $\left(-4x^3\right)\left(5x^{-9}\right) = -4\cdot 5\cdot x^3 x^{-9} = -20x^{-6} = -\dfrac{20}{x^6}$

52. $\left(2x^{-3}\right)^{-5} = 2^{-5}\left(x^{-3}\right)^{-5} = 2^{-5} x^{15} = \dfrac{x^{15}}{2^5} = \dfrac{x^{15}}{32}$

53. $\left(3x^{-4} y^6\right)^{-3} = 3^{-3}\left(x^{-4}\right)^{-3}\left(y^6\right)^{-3}$

$= 3^{-3} x^{12} y^{-18}$

$= \dfrac{x^{12}}{3^3 y^{18}}$

$= \dfrac{x^{12}}{27 y^{18}}$

54. $\dfrac{6a^{-2} b^3}{8a^{-5} b^6} = \dfrac{6}{8}\cdot\dfrac{a^{-2}}{a^{-5}}\cdot\dfrac{b^3}{b^6}$

$= \dfrac{3}{4}\cdot a^{-2-(-5)}\cdot b^{3-6}$

$= \dfrac{3}{4}\cdot a^3\cdot b^{-3}$

$= \dfrac{3}{4}\cdot a^3\cdot\dfrac{1}{b^3}$

$= \dfrac{3a^3}{4b^3}$

55. $\dfrac{\left(3a^{-2} b\right)^2}{\left(2ab^{-3}\right)^3} = \dfrac{3^2\left(a^{-2}\right)^2 b^2}{2^3 a^3\left(b^{-3}\right)^3}$

$= \dfrac{9a^{-4} b^2}{8a^3 b^{-9}}$

$= \dfrac{9a^{-4-3} b^{2-(-9)}}{8}$

$= \dfrac{9a^{-7} b^{11}}{8}$

$= \dfrac{9b^{11}}{8a^7}$

56. $\left(\dfrac{x^3 y^{-2}}{w^{-4}}\right)^{-7} = \dfrac{\left(x^3\right)^{-7}\left(y^{-2}\right)^{-7}}{\left(w^{-4}\right)^{-7}} = \dfrac{x^{-21} y^{14}}{w^{28}} = \dfrac{y^{14}}{x^{21} w^{28}}$

57. Answers may vary. One possibility follows: The student failed to subtract the powers correctly. The correct answer is:

$\dfrac{x^9}{x^{-6}} = x^{9-(-6)} = x^{9+6} = x^{15}$.

58. For 5.832×10^8 the decimal point must move eight places to the right. Thus, the standard decimal is $583,200,000$.

59. For 3.17×10^{-4} the decimal point must move four places to the left. Thus, the standard decimal is 0.000317.

60. For $74,200,000$ the decimal point needs to be moved seven places to the left so that the new number is between 1 and 10. Thus, the scientific notation is 7.42×10^7.

61. For 0.00008 the decimal point needs to be moved five places to the right so that the new number is between 1 and 10. Thus, the scientific notation is 8×10^{-5}.

62. For 1,426,000,000 the decimal point needs to be moved nine places to the left so that the new number is between 1 and 10. Thus, the scientific notation is 1.426×10^{9}.

Chapter 7 Test

1. $y = -2x^2 + 4x + 1$

First, we list some solutions in the table. Then we sketch a curve that contains the points corresponding to the solutions.

x	y
-3	$-2(-3)^2 + 4(-3) + 1 = -29$
-2	$-2(-2)^2 + 4(-2) + 1 = -15$
-1	$-2(-1)^2 + 4(-1) + 1 = -5$
0	$-2(0)^2 + 4(0) + 1 = 1$
1	$-2(1)^2 + 4(1) + 1 = 3$
2	$-2(2)^2 + 4(2) + 1 = 1$
3	$-2(3)^2 + 4(3) + 1 = -5$

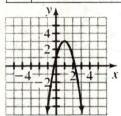

2. The graph contains the point $(-5, 5)$, so $y = 5$ when $x = -5$.

3. The graph contains the points $(-3, -3)$ and $(-1, -3)$, so $x = -3$ or $x = -1$ when $y = -3$.

4. The graph contains the point $(-2, -4)$, so $x = -2$ when $y = -4$.

5. There are no points on the graph for which the y-coordinate is -5. That is, there is no value of x for which $y = -5$.

6. The graph intersects the x-axis at the points $(-4, 0)$ and $(0, 0)$, so the x-intercepts are $(-4, 0)$ and $(0, 0)$.

7. The minimum point is $(-2, -4)$.

8. a.

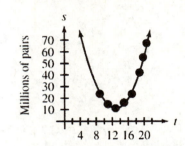

Years since 1980

Yes, the model fits the data well.

b. The year 1996 is represented by $t = 16$. We estimate that the model contains the point $(16, 19)$, so sales in 1996 were approximately 19 million pairs.

c. We estimate that the model contains the points $(8, 30)$ and $(18, 30)$. Thus, sales were 30 million pairs the years $1980 + 8 = 1988$ and $1980 + 18 = 1998$.

d. The parabola opens upward, so the vertex is the minimum point. We estimate it to be the point $(13, 11)$. It means that sales were 11 million pairs in the year $1980 + 13 = 1993$. According to the model, this is the lowest sales ever.

9. $(2x^3 - 4x^2 + 7x) - (6x^3 - 3x^2 + 9x)$
$= 2x^3 - 4x^2 + 7x - 6x^3 + 3x^2 - 9x$
$= 2x^3 - 6x^3 - 4x^2 + 3x^2 + 7x - 9x$
$= -4x^3 - x^2 - 2x$

10. $-5x^2(3x^2 - 8x + 2)$
$= -5x^2 \cdot 3x^2 + (-5x^2)(-8x) + (-5x^2) \cdot 2$
$= -15x^4 + 40x^3 - 10x^2$

11. $(3p - 2)(4p + 6)$
$= 3p \cdot 4p + 3p \cdot 6 + (-2) \cdot 4p + (-2) \cdot 6$
$= 12p^2 + 18p - 8p - 12$
$= 12p^2 + 10p - 12$

12. $(2k-5)(3k^2-4k-2)$
$= 2k \cdot 3k^2 + 2k(-4k) + 2k(-2) + (-5) \cdot 3k^2 +$
$(-5)(-4k) + (-5)(-2)$
$= 6k^3 - 8k^2 - 4k - 15k^2 + 20k + 10$
$= 6k^3 - 8k^2 - 15k^2 - 4k + 20k + 10$
$= 6k^3 - 23k^2 + 16k + 10$

13. $(x-6)^2 = x^2 - 2 \cdot x \cdot 6 + 6^2$
$\qquad = x^2 - 12x + 36$

14. $(4a+7b)^2 = (4a)^2 + 2(4a)(7b) + (7b)^2$
$\qquad\qquad = 16a^2 + 56ab + 49b^2$

15. $(3w^2 - 2w + 3)(w^2 + w - 2)$
$= 3w^2(w^2) + 3w^2(w) + 3w^2(-2) + (-2w)(w^2) +$
$(-2w)(w) + (-2w)(-2) + 3(w^2) + 3(w) + 3(-2)$
$= 3w^4 + 3w^3 - 6w^2 - 2w^3 - 2w^2 + 4w + 3w^2 +$
$3w - 6$
$= 3w^4 + 3w^3 - 2w^3 - 6w^2 - 2w^2 + 3w^2 + 4w +$
$3w - 6$
$= 3w^4 + w^3 - 5w^2 + 7w - 6$

16. a. $(-2.17t + 97.74) + (2.02t - 4.84)$
$= -2.17t + 2.02t + 97.74 - 4.84)$
$= -0.15t + 92.90$
This expression represents the total viewing share for broadcast and ad-supported cable television at t years since 1980.

b. Substitute 31 for t: $-0.15(31) + 92.90 = 88.25$
This means that the total viewing share for broadcast and ad-supported cable television will be 88.25% in the year $1980 + 31 = 2011$.

c. $(-2.17t + 97.74) - (2.02t - 4.84)$
$= -2.17t + 97.74 - 2.02t + 4.84$
$= -2.17t - 2.02t + 97.74 + 4.84$
$= -4.19t + 102.58$
This expression represents the difference in viewing share for broadcast and ad-supported cable television at t years since 1980.

d. Substitute 31 for t:
$-4.19(31) + 102.58 = -27.31$
This means that broadcast television will have 27.31 less percentage points of viewing share than ad-supported cable television in the year $1980 + 31 = 2011$.

17. $y = -3(x-1)^2 + 5$
$y = -3(x^2 - 2 \cdot x \cdot 1 + 1^2) + 5$
$y = -3(x^2 - 2x + 1) + 5$
$y = -3x^2 + 6x - 3 + 5$
$y = -3x^2 + 6x + 2$

18. Answers may vary. One possibility follows: The student failed to find the middle term in the simplified expression. The correct answer is:
$(x+4)^2 = x^2 + 2 \cdot x \cdot 4 + 4^2$
$\qquad\qquad = x^2 + 8x + 16$

19. $\dfrac{6x^7y^4}{8x^3y^9} = \dfrac{6}{8} \cdot x^{7-3} \cdot y^{4-9} = \dfrac{3}{4}x^4y^{-5} = \dfrac{3x^4}{4y^5}$

20. $\left(4a^3b^5\right)^3 a^6 b = 4^3\left(a^3\right)^3\left(b^5\right)^3 a^6 b$
$\qquad\qquad = 64a^9 b^{15} a^6 b$
$\qquad\qquad = 64a^{9+6}b^{15+1}$
$\qquad\qquad = 64a^{15}b^{16}$

21. $\left(\dfrac{x^3}{y^4}\right)^6 = \dfrac{\left(x^3\right)^6}{\left(y^4\right)^6} = \dfrac{x^{18}}{y^{24}}$

22. $\left(7x^{-3}\right)^{-2} = 7^{-2}\left(x^{-3}\right)^{-2} = 7^{-2}x^6 = \dfrac{x^6}{7^2} = \dfrac{x^6}{49}$

23. $\left(\dfrac{x^2y^{-6}}{w^{-3}}\right)^4 = \dfrac{\left(x^2\right)^4\left(y^{-6}\right)^4}{\left(w^{-3}\right)^4} = \dfrac{x^8 y^{-24}}{w^{-12}} = \dfrac{x^8 w^{12}}{y^{24}}$

24. $\dfrac{2p^{-5}t^2}{4p^{-2}t^{-3}} = \dfrac{1}{2} \cdot p^{-5-(-2)} \cdot t^{2-(-3)} = \dfrac{1}{2} \cdot p^{-3} \cdot t^5 = \dfrac{t^5}{2p^3}$

25. For 0.000468 the decimal point needs to be moved four places to the right so that the new number is between 1 and 10. Thus, the scientific notation is 4.68×10^{-4}.

26. For 1.65×10^7 the decimal point must move seven places to the right. Thus, the standard decimal is 16,500,000.

Chapter 8
Factoring Polynomials and Solving Polynomial Equations

Homework 8.1

1. $x^2 + 5x + 6$

We are looking for two factors of $c = 6$ whose sum is $b = 5$. Since c is positive the two factors have the same sign, and since b is positive the factors are both positive.

Factors	1,6	2,3
Sum	7	5

$x^2 + 5x + 6 = (x + 2)(x + 3)$

3. $t^2 + 9t + 20$

We are looking for two factors of $c = 20$ whose sum is $b = 9$. Since c is positive the two factors have the same sign, and since b is positive the factors are both positive.

Factors	1, 20	2, 10	4, 5
Sum	21	12	9

$t^2 + 9t + 20 = (t + 4)(t + 5)$

5. $x^2 + 8x + 16$

We are looking for two factors of $c = 16$ whose sum is $b = 8$. Since c is positive the two factors have the same sign, and since b is positive the factors are both positive.

Factors	1, 16	2, 8	4, 4
Sum	17	10	8

$x^2 + 8x + 16 = (x + 4)(x + 4) = (x + 4)^2$

7. $x^2 - 2x - 8$

We are looking for two factors of $c = -8$ whose sum is $b = -2$. Since c is negative, the two factors will have opposite signs. Since b is also negative, the factor with the larger absolute value will be negative.

Factors	−8,1	−4,2
Sum	−7	−2

$x^2 - 2x - 8 = (x - 4)(x + 2)$

9. $a^2 - 6a - 16$

We are looking for two factors of $c = -16$ whose sum is $b = -6$. Since c is negative, the two factors will have opposite signs. Since b is also negative, the factor with the larger absolute value will be negative.

Factors	−16,1	−8,2	−4,4
Sum	−15	−6	0

$a^2 - 6a - 16 = (a - 8)(a + 2)$

11. $x^2 + 5x - 24$

We are looking for two factors of $c = -24$ whose sum is $b = 5$. Since c is negative the two factors have opposite signs, and since b is positive the factor with the larger absolute value must be positive.

Factors	−1,24	−2,12	−3 , 8	−4,6
Sum	23	10	5	2

$x^2 + 5x - 24 = (x - 3)(x + 8)$

13. $x^2 + 8x - 12$

We are looking for two factors of $c = -12$ whose sum is $b = 8$. Since c is negative the two factors have opposite signs, and since b is positive the factor with the larger absolute value must be positive.

Factors	−1,12	−2,6	−3,4
Sum	11	4	1

Since none of the possibilities work, we can say that $x^2 + 8x - 12$ is prime.

15. $3t - 28 + t^2 = t^2 + 3t - 28$

We are looking for two factors of $c = -28$ whose sum is $b = 3$. Since c is negative the two factors have opposite signs, and since b is positive the factor with the larger absolute value must be positive.

Factors	−1,28	−2,14	−4,7
Sum	27	12	3

$3t - 28 + t^2 = t^2 + 3t - 28$
$$= (t - 4)(t + 7)$$

17. $x^2 - 10x + 16$

We are looking for two factors of $c = 16$ whose sum is $b = -10$. Since c is positive the two factors have the same sign, and since b is negative the factors are both negative.

Factors	$-1, -16$	$-2, -8$	$-4, -4$
Sum	-17	-10	-8

$x^2 - 10x + 16 = (x - 8)(x - 2)$

19. $24 - 11x + x^2 = x^2 - 11x + 24$

We are looking for two factors of $c = 24$ whose sum is $b = -11$. Since c is positive the two factors have the same sign, and since b is negative the factors are both negative.

Factors	$-1, -24$	$-2, -12$	$-3, -8$	$-4, -6$
Sum	-25	-14	-11	-10

$24 - 11x + x^2 = x^2 - 11x + 24$
$= (x - 8)(x - 3)$

21. $x^2 - 3x + 10$

We are looking for two factors of $c = 10$ whose sum is $b = -3$. Since c is positive the two factors have the same sign, and since b is negative the factors are both negative.

Factors	$-1, -10$	$-2, -5$
Sum	-11	-7

Since none of the possibilities work, we can say that $x^2 - 3x + 10$ is prime.

23. $r^2 - 10r + 25$

We are looking for two factors of $c = 25$ whose sum is $b = -10$. Since c is positive the two factors have the same sign, and since b is negative the factors are both negative.

Factors	$-1, -25$	$-5, -5$
Sum	-26	-10

$r^2 - 10r + 25 = (r - 5)(r - 5) = (r - 5)^2$

25. $x^2 + 36 - 12x = x^2 - 12x + 36$

We are looking for two factors of $c = 36$ whose sum is $b = -12$. Since c is positive the two factors have the same sign, and since b is negative the factors are both negative.

Factors	$-1, -36$	$-2, -18$	$-3, -12$	$-4, -9$	$-6, -6$
Sum	-37	-20	-15	-13	-12

$x^2 + 36 - 12x = x^2 - 12x + 36$
$= (x - 6)(x - 6)$
$= (x - 6)^2$

27. $x^2 + 10xy + 9y^2 = x^2 + (10y)x + 9y^2$

We need two monomials whose product is $9y^2$ and whose sum is $10y$. Since $9y^2$ has a positive coefficient, the coefficients of the monomials must be the same sign. Since the coefficient of $10y$ is positive, both coefficients must be positive.

$1y \cdot 9y = 9y^2$ and $1y + 9y = 10y$

Therefore, $x^2 + 10xy + 9y^2 = (x + y)(x + 9y)$.

29. $m^2 - mn - 6n^2 = m^2 - (n)m - 6n^2$

We need two monomials whose product is $-6n^2$ and whose sum is $-n$. Since $-6n^2$ has a negative coefficient, the coefficients of the monomials must be opposite signs. Since the coefficient of $-n$ is negative, the coefficient with the larger absolute value will be negative.

$-3n \cdot 2n = -6n^2$ and $-3n + 2n = -n$

Therefore, $m^2 - mn - 6n^2 = (m - 3n)(m + 2n)$.

31. $a^2 - 7ab + 6b^2 = a^2 - (7b)a + 6b^2$

We need two monomials whose product is $6b^2$ and whose sum is $-7b$. Since $6b^2$ has a positive coefficient, the coefficients of the monomials must be the same sign. Since the coefficient of $-7b$ is negative, both coefficients must be negative.

$(-1b)(-6b) = 6b^2$ and $(-1b) + (-6b) = -7b$

Therefore, $a^2 - 7ab + 6b^2 = (a - 6b)(a - b)$.

33. $x^2 - 25 = x^2 - 5^2$
$= (x - 5)(x + 5)$

35. $x^2 - 81 = x^2 - 9^2$
$$= (x-9)(x+9)$$

37. $t^2 - 1 = t^2 - 1^2$
$$= (t-1)(t+1)$$

39. $x^2 + 36$

We are looking for two factors of $c = 36$ whose sum is $b = 0$. Since 36 is positive, the two factors have the same sign. However, there are no factors of 36 whose sum is 0. Therefore, $x^2 + 36$ is prime.

41. $4x^2 - 25 = (2x)^2 - 5^2$
$$= (2x-5)(2x+5)$$

43. $81r^2 - 1 = (9r)^2 - 1^2$
$$= (9r-1)(9r+1)$$

45. $36x^2 + 49$

The binomial $36x^2 + 49$ is the sum of two squares. Since the sum of two squares does not factor over the integers, $36x^2 + 49$ is prime. (see #39)

47. $49p^2 - 100q^2 = (7p)^2 - (10q)^2$
$$= (7p-10q)(7p+10q)$$

49. $64m^2 - 9n^2 = (8m)^2 - (3n)^2$
$$= (8m-3n)(8m+3n)$$

51. $x^2 - 3x - 18$

We are looking for two factors of $c = -18$ whose sum is $b = -3$. Since c is negative, the two factors will have opposite signs. Since b is also negative, the factor with the larger absolute value will be negative.

Factors	$-18,1$	$-9,2$	$-6,3$
Sum	-17	-7	-3

$x^2 - 3x - 18 = (x-6)(x+3)$

53. $x^2 + 14x + 49$

We are looking for two factors of $c = 49$ whose sum is $b = 14$. Since c is positive the two factors have the same sign, and since b is positive the factors are both positive.

Factors	1, 49	7, 7
Sum	50	14

$x^2 + 14x + 49 = (x+7)(x+7)$
$$= (x+7)^2$$

55. $a^2 - 4 = a^2 - 2^2$
$$= (a-2)(a+2)$$

57. $x^2 + 4x + 12$

We are looking for two factors of $c = 12$ whose sum is $b = 4$. Since c is positive the two factors have the same sign, and since b is positive the factors are both positive.

Factors	1, 12	2, 6	3, 4
Sum	13	8	7

Since none of the possibilities work, we can say that $x^2 + 4x + 12$ is prime.

59. $x^2 - 8x + 12$

We are looking for two factors of $c = 12$ whose sum is $b = -8$. Since c is positive the two factors have the same sign, and since b is negative the factors are both negative.

Factors	$-1,-12$	$-2,-6$	$-3,-4$
Sum	-13	-8	-7

$x^2 - 8x + 12 = (x-6)(x-2)$

61. $-2w - 48 + w^2 = w^2 - 2w - 48$

We are looking for two factors of $c = -48$ whose sum is $b = -2$. Since c is negative, the two factors will have opposite signs. Since b is also negative, the factor with the larger absolute value will be negative.

Factors	$-48,1$	$-24,2$	$-16,3$	$-12,4$	$-8,6$
Sum	-47	-22	-13	-8	-2

$-2w - 48 + w^2 = w^2 - 2w + 48$
$$= (w-8)(w+6)$$

63. $x^2 - 8x + 16$

We are looking for two factors of $c = 16$ whose sum is $b = -8$. Since c is positive the two factors have the same sign, and since b is negative the factors are both negative.

Factors	$-1, -16$	$-2, -8$	$-4, -4$
Sum	-17	-10	-8

$$x^2 - 8x + 16 = (x-4)(x-4)$$
$$= (x-4)^2$$

65. $w^2 + 49$

The binomial $w^2 + 49$ is the sum of two squares. Since the sum of two squares does not factor over the integers, $w^2 + 49$ is prime. (see #39)

67. $m^2 - 6mn - 27n^2 = m^2 - (6n)m - 27n^2$

We need to monomials whose product is $-27n^2$ and whose sum is $-6n$. Since $-27n^2$ has a negative coefficient, the coefficients of the monomials must have opposite signs. Since the coefficient of $-6n$ is negative, the coefficient with the larger absolute value must be negative. $(-9n)(3n) = -27n^2$ and $-9n + 3n = -6n$
Therefore,
$$m^2 - 6mn - 27n^2 = (m-9n)(m+3n).$$

69. $32 - 18x + x^2 = x^2 - 18x + 32$

We are looking for two factors of $c = 32$ whose sum is $b = -18$. Since c is positive the two factors have the same sign, and since b is negative the factors are both negative.

Factors	$-1, -32$	$-2, -16$	$-4, -8$
Sum	-33	-18	-12

$$32 - 18x + x^2 = x^2 - 18x + 32$$
$$= (x-16)(x-2)$$

71. $100p^2 - 9t^2 = (10p)^2 - (3t)^2$
$$= (10p - 3t)(10p + 3t)$$

73. $p^2 + 12p + 36$

We are looking for two factors of $c = 36$ whose sum is $b = 12$. Since c is positive the two factors have the same sign, and since b is positive the factors are both positive.

Factors	1, 36	2, 18	3, 12	4, 9	6, 6
Sum	37	20	15	13	12

$$p^2 + 12p + 36 = (p+6)(p+6)$$
$$= (p+6)^2$$

75. The student is incorrectly factoring $A^2 + B^2$ as $A^2 + B^2 = (A+B)(A+B)$, but

$(A+B)(A+B) = A^2 + 2AB + B^2$. The binomial $x^2 + 9$ is the sum of two squares. Since the sum of two squares does not factor over the integers, $x^2 + 9$ is prime.

77. $(x-3)(x+7) = x^2 + 7x - 3x - 21$
$$= x^2 + 4x - 21$$
$(x+7)(x-3) = x^2 - 3x + 7x - 21$
$$= x^2 + 4x - 21$$
Therefore,
$$(x-3)(x+7) = (x+7)(x-3) = x^2 + 4x - 21.$$

79. $x^2 - 5x - 24$

We are looking for two factors of $c = -24$ whose sum is $b = -5$. Since c is negative, the two factors will have opposite signs. Since b is also negative, the factor with the larger absolute value will be negative.

Factors	$-24, 1$	$-12, 2$	$-8, 3$	$-6, 4$
Sum	-23	-10	-5	-2

$$x^2 - 5x - 24 = (x-8)(x+3)$$
$$(x-8)(x+3) = x^2 + 3x - 8x - 24$$
$$= x^2 - 5x - 24$$
Multiplying the factors yields the original expression.

81. Answers may vary.

83. $x^2 + kx + 12$

We are looking for two factors of $c = 12$ whose sum is $b = k$. Since c is positive the two factors have the same sign.

Factors	1, 12	2, 6	3, 4	−1,−12	−2,−6	−3,−4
Sum, k	13	8	7	−13	−8	−7

The possible values for k are -13, -8, -7, 7, 8, and 13.

85. Answers may vary.

87. $(x-9)(x+2) = x \cdot x + 2 \cdot x + (-9) \cdot x + (-9) \cdot 2$

$\qquad = x^2 + 2x - 9x - 18$

$\qquad = x^2 - 7x - 18$

89. $x^2 - 15x + 50$

We are looking for two factors of $c = 50$ whose sum is $b = -15$. Since c is positive the two factors have the same sign, and since b is negative the factors are both negative.

Factors	−1,−50	−2,−25	−5,−10
Sum	−51	−27	−15

$x^2 - 15x + 50 = (x-10)(x-5)$

91. $(3x-7)(3x+7) = 3x \cdot 3x + 7 \cdot 3x + (-7) \cdot 3x + (-7) \cdot 7$

$\qquad = 9x^2 + 21x - 21x - 49$

$\qquad = 9x^2 - 49$

93. $25x^2 - 36 = (5x)^2 - 6^2$

$\qquad = (5x-6)(5x+6)$

95. $(5p+7w) - (2p-4w) = 5p + 7w - 2p + 4w$

$\qquad = 5p - 2p + 7w + 4w$

$\qquad = 3p + 11w$

Description: *linear (first degree) polynomial in two variables*

97. $(5p+7w)(2p-4w)$

$= (5p)(2p) + (5p)(-4w) + (7w)(2p) + (7w)(-4w)$

$= 10p^2 - 20pw + 14pw - 28w^2$

$= 10p^2 - 6pw - 28w^2$

Description: *factored quadratic (second degree) polynomial in two variables*

99. $p^2 - 11pw + 18w^2 = p^2 - (11w)p + 18w^2$

We need to monomials whose product is $18w^2$ and whose sum is $-11w$. Since $18w^2$ has a positive coefficient, the coefficients of the monomials must be the same sign. Since the coefficient of $-11w$ is negative, both coefficients must be negative.

$(-2w)(-9w) = 18w^2$ and $(-2w) + (-9w) = -11w$

Therefore, $p^2 - 11pw + 18w^2 = (p-9w)(p-2w)$.

Description: *quadratic (second degree) polynomial in two variables*

Homework 8.2

1. The GCF is 2, so we get

$6x + 8 = 2 \cdot 3x + 2 \cdot 4$

$\qquad = 2(3x+4)$

3. The GCF is $5w$, so we get

$20w^2 + 35w = 5w \cdot 4w + 5w \cdot 7$

$\qquad = 5w(4w+7)$

5. The GCF is $6x$, so we get

$12x^2 - 30x = 6x \cdot 2x - 6x \cdot 5$

$\qquad = 6x(2x-5)$

7. The GCF is $3ab$, so we get

$6a^2b - 9ab = 3ab \cdot 2a - 3ab \cdot 3$

$\qquad = 3ab(2a-3)$

9. The GCF is $4x^2y^2$, so we get

$8x^3y^2 + 12x^2y^3 = 4x^2y^2 \cdot 2x + 4x^2y^2 \cdot 3y$

$\qquad = 4x^2y^2(2x+3y)$

11. The GCF is 5, so we get

$15x^3 - 10x - 30 = 5 \cdot 3x^3 - 5 \cdot 2x - 5 \cdot 6$

$\qquad = 5(3x^3 - 2x - 6)$

13. The GCF is $4t$, so we get

$12t^4 + 8t^3 - 16t = 4t \cdot 3t^3 + 4t \cdot 2t^2 - 4t \cdot 4$

$\qquad = 4t(3t^3 + 2t^2 - 4)$

15. The GCF is $5ab$, so we get
$$10a^4b - 15a^3b + 25ab = 5ab \cdot 2a^3 - 5ab \cdot 3a^2 + 5ab \cdot 5$$
$$= 5ab(2a^3 - 3a^2 + 5)$$

17. $2x^2 - 18 = 2(x^2 - 9)$
$$= 2(x^2 - 3^2)$$
$$= 2(x-3)(x+3)$$

19. $3m^2 + 21m + 30 = 3(m^2 + 7m + 10)$
$$= 3(m+2)(m+5)$$

21. $2x^2 - 18x + 36 = 2(x^2 - 9x + 18)$
$$= 2(x-6)(x-3)$$

23. $3x^3 - 27x = 3x(x^2 - 9)$
$$= 3x(x^2 - 3^2)$$
$$= 3x(x-3)(x+3)$$

25. $4r^3 - 16r^2 - 20r = 4r(r^2 - 4r - 5)$
$$= 4r(r-5)(r+1)$$

27. $6x^4 - 24x^2 = 6x^2(x^2 - 4)$
$$= 6x^2(x^2 - 2^2)$$
$$= 6x^2(x-2)(x+2)$$

29. $8m^4n - 18m^2n = 2m^2n(4m^2 - 9)$
$$= 2m^2n((2m)^2 - 3^2)$$
$$= 2m^2n(2m-3)(2m+3)$$

31. $5x^4 + 10x^3 - 120x^2 = 5x^2(x^2 + 2x - 24)$
$$= 5x^2(x-4)(x+6)$$

33. $8x - 2x^3 = -2x^3 + 8x$
$$= -2x(x^2 - 4)$$
$$= -2x(x^2 - 2^2)$$
$$= -2x(x-2)(x+2)$$

35. $36t^2 + 32t + 4t^3 = 4t^3 + 36t^2 + 32t$
$$= 4t(t^2 + 9t + 8)$$
$$= 4t(t+1)(t+8)$$

37. $-12x^3 + 27x = -3x(4x^2 - 9)$
$$= -3x((2x)^2 - 3^2)$$
$$= -3x(2x-3)(2x+3)$$

39. $-3x^3 - 18x^2 + 48x = -3x(x^2 + 6x - 16)$
$$= -3x(x-2)(x+8)$$

41. $6a^4b + 36a^3b + 54a^2b = 6a^2b(a^2 + 6a + 9)$
$$= 6a^2b(a+3)(a+3)$$
$$= 6a^2b(a+3)^2$$

43. The GCF is $(x-3)$, so we get
$$5x^2(x-3) + 2(x-3) = (5x^2 + 2)(x-3)$$

45. The GCF is $(2x+5)$, so we get
$$6x^2(2x+5) - 7(2x+5) = (6x^2 - 7)(2x+5)$$

47. $2p^3 + 6p^2 + 5p + 15 = 2p^2(p+3) + 5(p+3)$
$$= (2p^2 + 5)(p+3)$$

49. $6x^3 - 2x^2 + 21x - 7 = 2x^2(3x-1) + 7(3x-1)$
$$= (2x^2 + 7)(3x-1)$$

51. $15w^3 + 5w^2 - 6w - 2 = 5w^2(3w+1) - 2(3w+1)$
$$= (5w^2 - 2)(3w+1)$$

53. $16x^3 - 12x^2 - 36x + 27 = 4x^2(4x-3) - 9(4x-3)$
$$= (4x^2 - 9)(4x-3)$$
$$= ((2x)^2 - 3^2)(4x-3)$$
$$= (2x-3)(2x+3)(4x-3)$$

55. $2b^3 - 5b^2 - 18b + 45 = b^2(2b-5) - 9(2b-5)$
$$= (b^2 - 9)(2b-5)$$
$$= (b^2 - 3^2)(2b-5)$$
$$= (b-3)(b+3)(2b-5)$$

57. $x^3 - x^2 - x + 1 = x^2(x-1) - 1(x-1)$
$$= (x^2 - 1)(x-1)$$
$$= (x^2 - 1^2)(x-1)$$
$$= (x-1)(x+1)(x-1)$$
$$= (x-1)^2(x+1)$$

59. $3x + 3y + ax + ay = 3(x+y) + a(x+y)$
$$= (3+a)(x+y)$$

61. $2xy - 8x + 3y - 12 = 2x(y-4) + 3(y-4)$
$$= (2x+3)(y-4)$$

63. $81x^2 - 25 = (9x)^2 - 5^2$
$$= (9x-5)(9x+5)$$

65. $w^2 - 10w + 16$
We are looking for two factors of $c = 16$ whose sum is $b = -10$. Since c is positive, the two factors will have the same sign. Since b is negative, the factors will both be negative.
$(-8)(-2) = 16$ and $(-8) + (-2) = -10$
Therefore, $w^2 - 10w + 16 = (w-8)(w-2)$.

67. $24 - 10x + x^2 = x^2 - 10x + 24$
We are looking for two factors of $c = 24$ whose sum is $b = -10$. Since c is positive, the two factors will have the same sign. Since b is negative, the factors will both be negative.
$(-6)(-4) = 24$ and $(-6) + (-4) = -10$
Therefore, $w^2 - 10w + 24 = (w-6)(w-4)$.

69. $20a^2b - 15ab^3 = 5ab(4a) - 5ab(3b^2)$
$$= 5ab(4a - 3b^2)$$

71. $3r^2 + 30r + 75 = 3 \cdot r^2 + 3 \cdot 10r + 3 \cdot 25$
$$= 3(r^2 + 10r + 25)$$
$$= 3(r+5)(r+5)$$
$$= 3(r+5)^2$$

73. $64x^3 - 49x = x(64x^2 - 49)$
$$= x((8x)^2 - 7^2)$$
$$= x(8x-7)(8x+7)$$

75. $-m^2 + 6m - 9 = -1 \cdot m^2 + (-1)(-6m) + (-1)(9)$
$$= -1(m^2 - 6m + 9)$$
$$= -1(m-3)(m-3)$$
$$= -(m-3)^2$$

77. $x^3 + 9x^2 - 4x - 36 = x^2(x+9) - 4(x+9)$
$$= (x^2 - 4)(x+9)$$
$$= (x^2 - 2^2)(x+9)$$
$$= (x-2)(x+2)(x+9)$$

79. $2m^3n - 10m^2n^2 + 12mn^3$
$$= 2mn \cdot m^2 - 2mn \cdot 5mn + 2mn \cdot 6n^2$$
$$= 2mn(m^2 - 5mn + 6n^2)$$
$$= 2mn(m-3n)(m-2n)$$

81. The factorization is not complete.
$$6x^3 + 8x^2 + 15x + 20 = 2x^2(3x+4) + 5(3x+4)$$
$$= (2x^2 + 5)(3x+4)$$

83. The factorization is not complete.
$$4x^3 + 28x^2 + 40x = 4x(x^2 + 7x + 10)$$
$$= 4x(x+2)(x+5)$$

85. The student should factor out the GCF first.
$$4x^2 - 100 = 4(x^2 - 25)$$
$$= 4(x^2 - 5^2)$$
$$= 4(x-5)(x+5)$$

87. Answers may vary.

89. The factorization is not complete.
$$2x^2 + 10x + 12 = 2\left(x^2 + 5x + 6\right)$$
$$= 2(x+2)(x+3)$$

91. $2x(x-3)(x+4) = 2x\left(x^2 + 4x - 3x - 12\right)$
$$= 2x\left(x^2 + x - 12\right)$$
$$= 2x^3 + 2x^2 - 24x$$

93. $5x^2 - 40x + 80 = 5\left(x^2 - 8x + 16\right)$
$$= 5(x-4)(x-4)$$
$$= 5(x-4)^2$$

95. $6x^3 - 9x^2 - 4x + 6 = 3x^2(2x-3) - 2(2x-3)$
$$= \left(3x^2 - 2\right)(2x-3)$$

97. $(x-3)\left(x^2 + 5\right) = x^3 + 5x - 3x^2 - 15$
$$= x^3 - 3x^2 + 5x - 15$$

99. The equation $y = -4x + 1$ is in slope-intercept form, so the slope is $m = -4$ and the y-intercept is $(0,1)$. We first plot $(0,1)$. From this point we move 1 unit to the right and 4 units down, plotting the point $(1,-3)$. We then draw the line that connects the two points, extending in both directions.

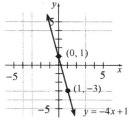

Description: *linear equation in two variables*

101. $-4x + 1 = 2x - 5$
$$-6x + 1 = -5$$
$$-6x = -6$$
$$x = 1$$
The solution is 1.
Description: *linear equation in one variable*

103. $y = -4x + 1$
$y = 2x - 5$
Substitute $2x - 5$ for y in the first equation and solve for x.
$$2x - 5 = -4x + 1$$
$$6x - 5 = 1$$
$$6x = 6$$
$$x = 1$$
Substitute 1 for x in the equation $y = 2x - 5$ and solve for y.
$$y = 2(1) - 5 = 2 - 5 = -3$$
The solution is $(1,-3)$.
Description: *system of linear equations in two variables*

Homework 8.3

1. *ac* Method:
$a \cdot c = 2 \cdot 3 = 6$
We are looking for two factors of 6 whose sum is $b = 7$. Since the product is positive, the factors will have the same sign. The sum is positive so the factors will be positive.

factor 1	factor 2	sum	
1	6	7	← okay
2	3	5	

$2x^2 + 7x + 3 = 2x^2 + x + 6x + 3$
$$= x(2x+1) + 3(2x+1)$$
$$= (x+3)(2x+1)$$

Trial and Error Method:
First note that there are no common factors and that $a = 2$, $b = 7$, and $c = 3$. Since c is positive the signs of our factors will be the same. Since b is positive, the signs in our factors will be positive. We will consider factorizations with this form:
$$(\underline{\quad}x + \underline{\quad})(\underline{\quad}x + \underline{\quad})$$
Since $a = 2$ can be factored as $1 \cdot 2$, we have the following form:
$$(x + \underline{\quad})(2x + \underline{\quad})$$
$|c| = |3| = 3$ can be factored as $1 \cdot 3$. Since the original expression had no common factors, the binomials we select cannot have a common factor.
$$(x+3)(2x+1) \rightarrow 2x^2 + 7x + 3$$
The correct factorization is
$$2x^2 + 7x + 3 = (x+3)(2x+1).$$

3. *ac* Method:

$a \cdot c = 5 \cdot 2 = 10$

We are looking for two factors of 10 whose sum is $b = 11$. Since the product is positive, the factors will have the same sign. The sum is positive so the factors will be positive.

factor 1	factor 2	sum
1	10	11 ← okay
2	5	7

$5x^2 + 11x + 2 = 5x^2 + 1x + 10x + 2$
$$= x(5x+1) + 2(5x+1)$$
$$= (x+2)(5x+1)$$

Trial and Error Method:

First note that there are no common factors and that $a = 5$, $b = 11$, and $c = 2$. Since c is positive the signs of our factors will be the same. Since b is positive, the signs in our factors will be positive. We will consider factorizations with this form:

$(\underline{}x + \underline{})(\underline{}x + \underline{})$

Since $a = 5$ can be factored as $1 \cdot 5$, we have the following form:

$(x + \underline{})(5x + \underline{})$

$|c| = |2| = 2$ can be factored as $1 \cdot 2$.

$(x+2)(5x+1) \rightarrow 5x^2 + 11x + 2$

The correct factorization is

$5x^2 + 11x + 2 = (x+2)(5x+1)$.

5. *ac* Method:

$a \cdot c = 3 \cdot 4 = 12$

We are looking for two factors of 12 whose sum is 8. Since the product is positive, the factors will have the same sign. The sum is positive so the factors will be positive.

factor 1	factor 2	sum
1	12	13
2	6	8 ← okay
3	4	7

$3x^2 + 8x + 4 = 3x^2 + 2x + 6x + 4$
$$= x(3x+2) + 2(3x+2)$$
$$= (x+2)(3x+2)$$

Trial and Error Method:

First note that there are no common factors and that $a = 3$, $b = 8$, and $c = 4$. Since c is positive the signs of our factors will be the same. Since b is positive, the signs in our factors will be positive. We will consider factorizations with this form:

$(\underline{}x + \underline{})(\underline{}x + \underline{})$

Since $a = 3$ can be factored as $1 \cdot 3$, we have the following form:

$(x + \underline{})(3x + \underline{})$

$|c| = |4| = 4$ can be factored as $1 \cdot 4$ and $2 \cdot 2$.

$(x+1)(3x+4) \rightarrow 3x^2 + 7x + 4$
$(x+4)(3x+1) \rightarrow 3x^2 + 13x + 4$
$(x+2)(3x+2) \rightarrow 3x^2 + 8x + 4$

The correct factorization is

$3x^2 + 8x + 4 = (x+2)(3x+2)$

7. *ac* Method:

$a \cdot c = 2 \cdot (-6) = -12$

We are looking for two factors of -12 whose sum is 1. Since the product is negative, the factors will have opposite signs. The sum is positive so the factor with the larger absolute value will be positive.

factor 1	factor 2	sum
-1	12	1
-2	6	4
-3	4	1 ← okay

$2t^2 + t - 6 = 2t^2 - 3t + 4t - 6$
$$= t(2t-3) + 2(2t-3)$$
$$= (t+2)(2t-3)$$

Trial and Error Method:

First note that there are no common factors and that $a = 2$, $b = 1$, and $c = -6$. Since c is negative, the signs in our factors will be opposites. We will consider factorizations with this form:

$(\underline{}t + \underline{})(\underline{}t - \underline{})$

If our choice results in a middle term with the wrong sign, we simply switch the signs of the factors.

Since $a = 2$ can be factored as $1 \cdot 2$, we have the following form:

$(t + \underline{})(2t - \underline{})$

$|c| = |-6| = 6$ can be factored as $1 \cdot 6$ and $2 \cdot 3$.

Since the original expression had no common factors, the binomials we select cannot have a common factor.

$$(t+6)(2t-1) \rightarrow 2t^2 + 11t - 6$$

$$(t+2)(2t-3) \rightarrow 2t^2 + t - 6$$

The correct factorization is

$$2t^2 + t - 6 = (t+2)(2t-3).$$

9. *ac* Method:

$a \cdot c = 6 \cdot 6 = 36$

We are looking for two factors of 36 whose sum is -13. Since the product is positive, the factors will have the same sign. The sum is negative so the factors will be negative.

factor 1	factor 2	sum
−1	−36	−37
−2	−18	−20
−3	−12	−15
−4	−9	−13← okay
−6	−6	−12

$$6x^2 - 13x + 6 = 6x^2 - 4x - 9x + 6$$
$$= 2x(3x-2) - 3(3x-2)$$
$$= (2x-3)(3x-2)$$

Trial and Error Method:

First note that there are no common factors and that $a = 6$, $b = -13$, and $c = 6$. Since c is positive the signs of our factors will be the same. Since b is negative, the signs in our factors will be negative. We will consider factorizations with this form:

$$(\_\_\, x - \_\_)(\_\_\, x - \_\_)$$

Since $a = 6$ can be factored as $1 \cdot 6$ or $2 \cdot 3$, we have the following forms:

$$(x - \_\_)(6x - \_\_)$$
$$(2x - \_\_)(3x - \_\_)$$

$|c| = |6| = 6$ can be factored as $1 \cdot 6$ and $2 \cdot 3$. Since the original expression had no common factors, the binomials we select cannot have a common factor.

$$(x-6)(6x-1) \rightarrow 6x^2 - 37x + 6$$
$$(2x-3)(3x-2) \rightarrow 6x^2 - 13x + 6$$

The correct factorization is

$$6x^2 - 13x + 6 = (2x-3)(3x-2).$$

11. *ac* Method:

$a \cdot c = 4 \cdot 25 = 100$

We are looking for two factors of 100 whose sum is 20. Since the product is positive, the factors will have the same sign. The sum is positive so the factors will both be positive. Since we are adding

two positive numbers to get 20, neither factor can exceed 20.

factor 1	factor 2	sum
5	20	25
10	10	20 ← okay

$$4x^2 + 20x + 25 = 4x^2 + 10x + 10x + 25$$
$$= 2x(2x+5) + 5(2x+5)$$
$$= (2x+5)(2x+5)$$
$$= (2x+5)^2$$

Trial and Error Method:

First note that there are no common factors and that $a = 4$, $b = 20$, and $c = 25$. Since c is positive, the signs in our factors will be the same. Since b is also positive, the signs of our factors will be positive. We will consider factorizations with this form:

$$(\_\_\, x + \_\_)(\_\_\, x + \_\_)$$

Since $a = 4$ can be factored as $1 \cdot 4$ and $2 \cdot 2$, we have the following forms:

$$(x + \_\_)(4x + \_\_)$$
$$(2x + \_\_)(2x + \_\_)$$

$|c| = |25| = 25$ can be factored as $1 \cdot 25$ and $5 \cdot 5$. Therefore, we have

$$(x+1)(4x+25) \rightarrow 4x^2 + 29x + 25$$
$$(x+25)(4x+1) \rightarrow 4x^2 + 101x + 25$$
$$(2x+1)(2x+25) \rightarrow 4x^2 + 52x + 25$$
$$(2x+5)(2x+5) \rightarrow 4x^2 + 20x + 25$$

The correct factorization is

$$4x^2 + 20x + 25 = (2x+5)(2x+5)$$
$$= (2x+5)^2$$

13. *ac* Method:

$a \cdot c = 6 \cdot 6 = 36$

We are looking for two factors of 36 whose sum is -37. Since the product is positive, the factors will have the same sign. The sum is negative so the factors will be negative.

factor 1	factor 2	sum
−1	−36	−37 ← okay
−2	−18	−20
−3	−12	−15
−4	−9	−13
−6	−6	−12

271

$$6x^2 - 37x + 6 = 6x^2 - x - 36x + 6$$
$$= x(6x - 1) - 6(6x - 1)$$
$$= (x - 6)(6x - 1)$$

Trial and Error Method:
First note that there are no common factors and that $a = 6$, $b = -37$, and $c = 6$. Since c is positive the signs of our factors will be the same. Since b is negative, the signs in our factors will be negative. We will consider factorizations with this form:

$$(\underline{}x - \underline{})(\underline{}x - \underline{})$$

Since $a = 6$ can be factored as $1 \cdot 6$ or $2 \cdot 3$, we have the following forms:

$$(x - \underline{})(6x - \underline{})$$
$$(2x - \underline{})(3x - \underline{})$$

$|c| = |6| = 6$ can be factored as $1 \cdot 6$ and $2 \cdot 3$. Since the original expression had no common factors, the binomials we select cannot have a common factor.

$$(x - 6)(6x - 1) \rightarrow 6x^2 - 37x + 6$$
$$(2x - 3)(3x - 2) \rightarrow 6x^2 - 13x + 6$$

The correct factorization is
$$6x^2 - 37x + 6 = (x - 6)(6x - 1).$$

15. *ac* **Method:**
$$a \cdot c = 2 \cdot 4 = 8$$
We need two factors of 8 whose sum is 5. Since the product is positive, the factors will have the same sign. The sum is positive so the factors will both be positive. Since we are adding two positive numbers to get 5, neither factor can exceed 5.

factor 1	factor 2	sum
2	4	6

The only remaining possibility for the factors does not yield the correct sum so the expression is prime.

Trial and Error Method:
First note that there are no common factors and that $a = 2$, $b = 5$, and $c = 4$. Since c is positive, the signs in our factors will be the same. Since b is also positive, the signs of our factors will be positive. We will consider factorizations with this form:

$$(\underline{}r + \underline{})(\underline{}r + \underline{})$$

Since $a = 2$ can be factored as $1 \cdot 2$, we have the following form:

$$(r + \underline{})(2r + \underline{})$$

$|c| = |4| = 4$ can be factored as $1 \cdot 4$ and $2 \cdot 2$. Since the original expression had no common factors, the

binomials we select cannot have a common factor.
$$(r + 4)(2r + 1) \rightarrow 2r^2 + 9r + 4$$
The only remaining possibility does not result in the correct middle term. Therefore, the expression is prime.

17. *ac* **Method:**
$$a \cdot c = 18 \cdot -4 = -72$$
We are looking for two factors of -72 whose sum is 21. Since the product is negative, the factors will have opposite signs. The sum is positive so the factor with the larger absolute value will be positive.

factor 1	factor 2	sum
-1	72	71
-2	36	34
-3	24	21 ← okay
-4	18	14
-6	12	6
-8	9	1

$$18x^2 + 21x - 4 = 18x^2 - 3x + 24x - 4$$
$$= 3x(6x - 1) + 4(6x - 1)$$
$$= (3x + 4)(6x - 1)$$

Trial and Error Method:
First note that there are no common factors and that $a = 18$, $b = 21$, and $c = -4$. Since c is negative, the signs in our factors will be opposites. We will consider factorizations with this form:

$$(\underline{}x + \underline{})(\underline{}x - \underline{})$$

If our choice results in a middle term with the wrong sign, we simply switch the signs of the factors.
Since $a = 18$ can be factored as $1 \cdot 18$, $2 \cdot 9$, and $3 \cdot 6$, we have the following forms:

$$(x + \underline{})(18x - \underline{})$$
$$(2x + \underline{})(9x - \underline{})$$
$$(3x + \underline{})(6x - \underline{})$$

$|c| = |-4| = 4$ can be factored as $1 \cdot 4$ and $2 \cdot 2$.
Since the original expression had no common factors, the binomials we select cannot have a common factor.

$$(x + 4)(18x - 1) \rightarrow 18x^2 + 71x - 4$$
$$(2x + 1)(9x - 4) \rightarrow 18x^2 + x - 4$$
$$(3x + 4)(6x - 1) \rightarrow 18x^2 + 21x - 4$$

The correct factorization is
$$18x^2 + 21x - 4 = (3x + 4)(6x - 1)$$

19. <u>*ac* Method:</u>

$a \cdot c = 3 \cdot 24 = 72$

We are looking for two factors of 72 whose sum is -22. Since the product is positive, the factors will have the same sign. The sum is negative so the factors will be negative.

factor 1	factor 2	sum
-1	-72	-73
-2	-36	-38
-3	-24	-27
-4	-18	$-22 \leftarrow$ okay
-6	-12	-18
-8	-9	-17

$$3m^2 - 22m + 24 = 3m^2 - 4m - 18m + 24$$
$$= m(3m-4) - 6(3m-4)$$
$$= (m-6)(3m-4)$$

<u>Trial and Error Method:</u>

First note that there are no common factors and that $a = 3$, $b = -22$, and $c = 24$. Since c is positive the signs of our factors will be the same. Since b is negative, the signs in our factors will be negative. We will consider factorizations with this form:

$$(\underline{\quad}m - \underline{\quad})(\underline{\quad}m - \underline{\quad})$$

Since $a = 3$ can be factored as $1 \cdot 3$, we have the following form:

$$(m - \underline{\quad})(3m - \underline{\quad})$$

$|c| = |24| = 24$ can be factored as $1 \cdot 24$, $2 \cdot 12$, $3 \cdot 8$, and $4 \cdot 6$. Since the original expression had no common factors, the binomials we select cannot have a common factor.

$$(m-24)(3m-1) \rightarrow 3m^2 - 73m + 24$$
$$(m-12)(3m-2) \rightarrow 3m^2 - 38m + 24$$
$$(m-3)(3m-8) \rightarrow 3m^2 - 17m + 24$$
$$(m-6)(3m-4) \rightarrow 3m^2 - 22m + 24$$

The correct factorization is

$$3m^2 - 22m + 24 = (m-6)(3m-4).$$

21. <u>*ac* Method:</u>

$a \cdot c = 2 \cdot 40 = 80$

We are looking for two factors of 80 whose sum is -21. Since the product is positive, the factors will have the same sign. The sum is negative so the factors will be negative.

factor 1	factor 2	sum
-1	-80	-81
-2	-40	-42
-4	-20	-24
-5	-16	$-21 \leftarrow$ okay
-8	-10	-18

$$2x^2 - 21x + 40 = 2x^2 - 5x - 16x + 40$$
$$= x(2x-5) - 8(2x-5)$$
$$= (x-8)(2x-5)$$

<u>Trial and Error Method:</u>

First note that there are no common factors and that $a = 2$, $b = -21$, and $c = 40$. Since c is positive the signs of our factors will be the same. Since b is negative, the signs in our factors will be negative. We will consider factorizations with this form:

$$(\underline{\quad}x - \underline{\quad})(\underline{\quad}x - \underline{\quad})$$

Since $a = 2$ can be factored as $1 \cdot 2$, we have the following form:

$$(x - \underline{\quad})(2x - \underline{\quad})$$

$|c| = |40| = 40$ can be factored as $1 \cdot 40$, $2 \cdot 20$, $4 \cdot 10$, and $5 \cdot 8$. Since the original expression had no common factors, the binomials we select cannot have a common factor.

$$(x-40)(2x-1) \rightarrow 2x^2 - 81x + 40$$
$$(x-8)(2x-5) \rightarrow 2x^2 - 21x + 40$$

The correct factorization is

$$2x^2 - 21x + 40 = (x-8)(2x-5).$$

23. <u>*ac* Method:</u>

$a \cdot c = 2 \cdot 3 = 6$

We are looking for two factors of 6 whose sum is 5. Since the product is positive, the factors will have the same sign. The sum is positive so the factors will both be positive. Since we are adding two positive numbers to get 5, neither factor can exceed 5.

factor 1	factor 2	sum
2	3	$5 \leftarrow$ okay

$$2a^2 + 5ab + 3b^2 = 2a^2 + 2ab + 3ab + 3b^2$$
$$= 2a(a+b) + 3b(a+b)$$
$$= (2a+3b)(a+b)$$

Trial and Error Method:
First note that there are no common factors and
that $a = 2$, $b = 5$, and $c = 3$. Since c is positive,
the signs in our factors will be the same. Since b is
also positive, the signs of our factors will be
positive. We will consider factorizations with this
form:

$$(\underline{\quad}a + \underline{\quad}b)(\underline{\quad}a + \underline{\quad}b)$$

Since $a = 2$ can be factored as $1 \cdot 2$, we have the
following form:

$$(a + \underline{\quad}b)(2a + \underline{\quad}b)$$

$|c| = |3| = 3$ can be factored as $1 \cdot 3$.

$$(a + b)(2a + 3b) \rightarrow 2a^2 + 5ab + 3b^2$$

$$(a + 3b)(2a + b) \rightarrow 2a^2 + 7ab + 3b^2$$

The correct factorization is

$$2a^2 + 5ab + 3b^2 = (a + b)(2a + 3b).$$

25. *ac* Method:

$$a \cdot c = 5 \cdot (-8) = -40$$

We are looking for two factors of -40 whose sum
is 18. Since the product is negative, the factors will
have opposite signs. The sum is positive so the
factor with the larger absolute value will be
positive.

factor 1	factor 2	sum
−1	40	39
−2	20	18 ← okay
−4	10	6
−5	8	3

$$5x^2 + 18xy - 8y^2 = 5x^2 - 2xy + 20xy - 8y^2$$
$$= x(5x - 2y) + 4y(5x - 2)$$
$$= (x + 4y)(5x - 2y)$$

Trial and Error Method:
First note that there are no common factors and
that $a = 5$, $b = 18$, and $c = -8$. Since c is
negative, the signs in our factors will be opposites.
We will consider factorizations with this form:

$$(\underline{\quad}x + \underline{\quad}y)(\underline{\quad}x - \underline{\quad}y)$$

Since $a = 5$ can be factored as $1 \cdot 5$, we have the
following form:

$$(x + \underline{\quad}y)(5x - \underline{\quad}y)$$

If our choice results in a middle term with the
wrong sign, we simply switch the signs of the
factors.

$|c| = |-8| = 8$ can be factored as $1 \cdot 8$ and $2 \cdot 4$.

$$(x + y)(5x - 8y) \rightarrow 5x^2 - 3xy - 8y^2$$
$$(x + 8y)(5x - y) \rightarrow 5x^2 + 39xy - 8y^2$$
$$(x + 2y)(5x - 4y) \rightarrow 5x^2 + 6xy - 8y^2$$
$$(x + 4y)(5x - 2y) \rightarrow 5x^2 + 18xy - 8y^2$$

The correct factorization is

$$5x^2 + 18xy - 8y^2 = (x + 4y)(5x - 2y).$$

27. *ac* Method:
Start by factoring out the GCF, 3.

$$6b^2 - 15bc + 6c^2 = 3(2b^2 - 5bc + 2c^2)$$

Now factor the expression in parentheses.

$$a \cdot c = 2 \cdot 2 = 4$$

We are looking for two factors of 4 whose sum
is -5. Since the product is positive, the factors will
have the same sign. The sum is negative so the
factors will be negative.

factor 1	factor 2	sum
−1	−4	−5 ← okay
−2	−2	−4

$$2b^2 - 5bc + 2c^2 = 2b^2 - bc - 4bc + 2c^2$$
$$= b(2b - c) - 2c(2b - c)$$
$$= (b - 2c)(2b - c)$$

Therefore,

$$6b^2 - 15bc + 6c^2 = 3(b - 2c)(2b - c)$$

Trial and Error Method:
First factor out the GCF, 3.

$$6b^2 - 15bc + 6c^2 = 3(2b^2 - 5bc + 2c^2)$$

Now factor the expression in parentheses. Note that
$a = 2$, $b = -5$, and $c = 2$. Since c is positive the
signs of our factors will be the same. Since b is
negative, the signs in our factors will be negative.
We will consider factorizations with this form:

$$(\underline{\quad}b - \underline{\quad}c)(\underline{\quad}b - \underline{\quad}c)$$

Since $a = 2$ can be factored as $1 \cdot 2$, we have the
following form:

$$(b - \underline{\quad}c)(2b - \underline{\quad}c)$$

$|c| = |2| = 2$ can be factored as $1 \cdot 2$. Since the
parenthetical expression had no common factors,
the binomials we select cannot have a common
factor.

$$(b - 2c)(2b - c) \rightarrow 2b^2 - 5bc + 2c^2$$

The correct factorization is

$$6b^2 - 15bc + 6c^2 = 3(b - 2c)(2b - c).$$

29. $4x^2 + 26x + 30$

Start by factoring out the GCF, 2.

$4x^2 + 26x + 30 = 2(2x^2 + 13x + 15)$

The leading coefficient of the expression inside the parentheses is not 1, so we try factoring by grouping.

$a \cdot c = 2 \cdot 15 = 30$

We need two numbers whose product is 30 and whose sum is 13. The product and sum are both positive, so the numbers are both positive. Since $10 \cdot 3 = 30$ and $10 + 3 = 13$, we get

$$4x^2 + 26x + 30 = 2(2x^2 + 13x + 15)$$
$$= 2(2x^2 + 3x + 10x + 15)$$
$$= 2(x(2x + 3) + 5(2x + 3))$$
$$= 2(x + 5)(2x + 3)$$

31. $20a^2 - 40a + 15$

Start by factoring out the GCF, 5.

$20a^2 - 40a + 15 = 5(4a^2 - 8a + 3)$

The leading coefficient of the expression inside the parentheses is not 1, so we try factoring by grouping.

$a \cdot c = 4 \cdot 3 = 12$

We need two numbers whose product is 12 and whose sum is -8. The product is positive but the sum is negative, so the two numbers must be negative. Since $(-2)(-6) = 12$ and

$(-2) + (-6) = -8$, we get

$$20a^2 - 40a + 15 = 5(4a^2 - 8a + 3)$$
$$= 5(4a^2 - 2a - 6a + 3)$$
$$= 5(2a(2a - 1) - 3(2a - 1))$$
$$= 5(2a - 3)(2a - 1)$$

33. $24x^2 + 15x - 9$

Start by factoring out the GCF, 3.

$24x^2 + 15x - 9 = 3(8x^2 + 5x - 3)$

The leading coefficient of the expression inside the parentheses is not 1, so we try factoring by grouping.

$a \cdot c = 8 \cdot (-3) = -24$

We need two numbers whose product is -24 and whose sum is 5. The product is negative, so the numbers have opposite signs. The sum is positive, so the number with the larger absolute value must

be positive. Since $(8)(-3) = -24$ and $8 + (-3) = 5$, we get

$$24x^2 + 15x - 9 = 3(8x^2 + 5x - 3)$$
$$= 3(8x^2 + 8x - 3x - 3)$$
$$= 3(8x(x + 1) - 3(x + 1))$$
$$= 3(8x - 3)(x + 1)$$

35. $-20x^2 + 22x + 12$

Start by factoring out the GCF, -2.

$-20x^2 + 22x + 12 = -2(10x^2 - 11x - 6)$

The leading coefficient of the expression inside the parentheses is not 1, so we try factoring by grouping.

$a \cdot c = 10(-6) = -60$

We need two numbers whose product is -60 and whose sum is -11. The product is negative, so the numbers have opposite signs. The sum is negative, so the number with the larger absolute value must be negative. Since $-15(4) = -60$ and

$-15 + 4 = -11$, we get

$$-20x^2 + 22x + 12 = -2(10x^2 - 11x - 6)$$
$$= -2(10x^2 - 15x + 4x - 6)$$
$$= -2(5x(2x - 3) + 2(2x - 3))$$
$$= -2(5x + 2)(2x - 3)$$

37. $4w^4 - 6w^3 - 12w^2$

Start by factoring out the GCF, $2w^2$.

$4w^4 - 6w^3 - 12w^2 = 2w^2(2w^2 - 3w - 6)$

The leading coefficient of the expression inside the parentheses is not 1, so we try factoring by grouping.

$a \cdot c = 2(-6) = -12$

We need two numbers whose product is -12 and whose sum is -3. The product is negative, so the numbers have opposite signs. The sum is negative, so the number with the larger absolute value must be negative.

Factors	1,−12	2,−6	3,−4
Sum	−11	−4	−1

Since none of the possibilities work, the expression inside the parentheses is prime. Therefore,

$$4w^4 - 6w^3 - 12w^2 = 2w^2(2w^2 - 3w - 6).$$

39. $10x^4 - 5x^3 - 50x^2$

Start by factoring out the GCF, $5x^2$.

$$10x^4 - 5x^3 - 50x^2 = 5x^2(2x^2 - x - 10)$$

The leading coefficient of the expression inside the parentheses is not 1, so we try factoring by grouping.

$$a \cdot c = 2(-10) = -20$$

We need two numbers whose product is -20 and whose sum is -1. The product is negative, so the numbers have opposite signs. The sum is negative, so the number with the larger absolute value must be negative.

Since $4(-5) = -20$ and $4 + (-5) = -1$, we get

$$10x^4 - 5x^3 - 50x^2 = 5x^2(2x^2 - x - 10)$$
$$= 5x^2(2x^2 + 4x - 5x - 10)$$
$$= 5x^2(2x(x+2) - 5(x+2))$$
$$= 5x^2(2x - 5)(x + 2)$$

41. $6a^2 - 34ab - 12b^2$

Start by factoring out the GCF, 2.

$$6a^2 - 34ab - 12b^2 = 2(3a^2 - 17ab - 6b^2)$$

The leading coefficient of the expression inside the parentheses is not 1, so we try factoring by grouping.

$$a \cdot c = 3(-6) = -18$$

We need two numbers whose product is -18 and whose sum is -17. The product is negative, so the numbers have opposite signs. The sum is negative, so the number with the larger absolute value must be negative.

Since $-18(1) = -18$ and $-18 + 1 = -17$, we get

$$6a^2 - 34ab - 12b^2 = 2(3a^2 - 17ab - 6b^2)$$
$$= 2(3a^2 - 18ab + 1ab - 6b^2)$$
$$= 2(3a(a - 6b) + b(a - 6b))$$
$$= 2(3a + b)(a - 6b)$$

43. $12r^3 + 40r^2w + 32rw^2$

Start by factoring out the GCF, $4r$.

$$12r^3 + 40r^2w + 32rw^2 = 4r(3r^2 + 10rw + 8w^2)$$

The leading coefficient of the expression inside the parentheses is not 1, so we try factoring by grouping.

$$a \cdot c = 3 \cdot 8 = 24$$

We need two numbers whose product is 24 and whose sum is 10. The product and sum are positive, so the numbers are both positive.

Since $4 \cdot 6 = 24$ and $4 + 6 = 10$, we get

$$12r^3 + 40r^2w + 32rw^2$$
$$= 4r(3r^2 + 10rw + 8w^2)$$
$$= 4r(3r^2 + 4rw + 6rw + 8w^2)$$
$$= 4r(r(3r + 4w) + 2w(3r + 4w))$$
$$= 4r(r + 2w)(3r + 4w)$$

45. $x^2 - 6x - 27$

We are looking for two factors of $c = -27$ whose sum is $b = -6$. Since c is negative, the two factors will have opposite signs. Since b is negative, the factor with the larger absolute value will be negative.

$$(3)(-9) = -27 \text{ and } 3 + (-9) = -6$$

Therefore, $x^2 - 6x - 27 = (x - 9)(x + 3)$.

47. $-48x^2 + 40x = -8x \cdot 6x + (-8x) \cdot (-5)$
$$= -8x(6x - 5)$$

49. $5a^3 + 2a^2 - 15a - 6 = a^2(5a + 2) - 3(5a + 2)$
$$= (a^2 - 3)(5a + 2)$$

51. $x^2 + 9$

Since the sum of two squares is prime over the integers, $x^2 + 9$ is prime.

53. $4x^2 - 12x + 9$

The leading coefficient is not 1, so we try factoring by grouping.

$$a \cdot c = 4 \cdot 9 = 36$$

We need two numbers whose product is 36 and whose sum is -12. The product is positive, so the numbers have the same sign. The sum is negative, so the numbers are both negative.

Since $(-6)(-6) = 36$ and $-6 + (-6) = -12$, we get

$$4x^2 - 12x + 9 = 4x^2 - 6x - 6x + 9$$
$$= 2x(2x - 3) - 3(2x - 3)$$
$$= (2x - 3)(2x - 3)$$
$$= (2x - 3)^2$$

55. $-17p^2 + 17 = -17(p^2 - 1)$

$$= -17(p^2 - 1^2)$$

$$= -17(p-1)(p+1)$$

57. $24 + 10x + x^2 = x^2 + 10x + 24$

We are looking for two factors of $c = 24$ whose sum is $b = 10$. Since c is positive the two factors will have the same sign. Since b is positive, the factors will both be positive.

$4 \cdot 6 = 24$ and $4 + 6 = 10$

Therefore, $24 + 10x + x^2 = x^2 + 10x + 24$

$$= (x+6)(x+4).$$

59. $b^2 - 3bc - 28c^2$

We are looking for two factors of $c = -28$ whose sum is $b = -3$. Since c is negative, the two factors will have opposite signs. Since b is negative, the factor with the larger absolute value will be negative.

$(-7)(4) = -28$ and $-7 + 4 = -3$

Therefore, $b^2 - 3bc - 28c^2 = (b + 4c)(b - 7c)$.

61. $8t^2 - 10t + 3$

The leading coefficient is not 1, so we try factoring by grouping.

$a \cdot c = 8 \cdot 3 = 24$

We need two numbers whose product is 24 and whose sum is -10. The product is positive, so the numbers have the same sign. The sum is negative, so the numbers are both negative.

Since $(-6)(-4) = 24$ and $-6 + (-4) = -10$, we get

$8t^2 - 10t + 3 = 8t^2 - 4t - 6t + 3$

$$= 4t(2t - 1) - 3(2t - 1)$$

$$= (4t - 3)(2t - 1)$$

63. $7x^4 - 28x^2 = 7x^2(x^2 - 4)$

$$= 7x^2(x^2 - 2^2)$$

$$= 7x^2(x - 2)(x + 2)$$

65. $12p^3 - 4p^2 - 27p + 9 = 4p^2(3p-1) - 9(3p-1)$

$$= (4p^2 - 9)(3p - 1)$$

$$= ((2p)^2 - 3^2)(3p - 1)$$

$$= (2p - 3)(2p + 3)(3p - 1)$$

67. $x^2 - 6x + 12$

We are looking for two factors of $c = 12$ whose sum is $b = -6$. Since c is positive, the two factors will have the same sign. Since b is negative, the factors will both be negative.

Factors	$-1, -12$	$-2, -6$	$-3, -4$
Sum	-13	-8	-7

Since none of the possibilities work, we can say that $x^2 - 6x + 12$ is prime.

69. $3x^4 - 21x^3 + 30x^2 = 3x^2(x^2 - 7x + 10)$

To factor the expression in parentheses, we are looking for two factors of $c = 10$ whose sum is $b = -7$. Since c is positive, the two factors will have the same sign. Since b is negative, the factors will both be negative.

$(-2)(-5) = 10$ and $(-2) + (-5) = -7$

Therefore, $3x^4 - 21x^3 + 30x^2 = 3x^2(x^2 - 7x + 10)$

$$= 3x^2(x - 2)(x - 5).$$

71. $20x^2 + 16x^4 - 42x^3 = 16x^4 - 42x^3 + 20x^2$

$$= 2x^2(8x^2 - 21x + 10)$$

The leading coefficient of the expression inside the parentheses is not 1, so we try factoring by grouping.

$a \cdot c = 8 \cdot 10 = 80$

We need two numbers whose product is 80 and whose sum is -21. The product is positive but the sum is negative, so the two numbers must be negative. Since $(-5)(-16) = 80$ and

$(-5) + (-16) = -21$, we get

$20x^2 + 16x^4 - 42x^3 = 16x^4 - 42x^3 + 20x^2$

$$= 2x^2(8x^2 - 21x + 10)$$

$$= 2x^2(8x^2 - 16x - 5x + 10)$$

$$= 2x^2(8x(x - 2) - 5(x - 2))$$

$$= 2x^2(8x - 5)(x - 2)$$

73. $36a^2 - 49b^2 = (6a)^2 - (7b)^2$
$$= (6a - 7b)(6a + 7b)$$

75. $-2x^2y + 8xy + 24y = -2y(x^2 - 4x - 12)$

To factor the expression in parentheses, we are looking for two factors of $c = -12$ whose sum is $b = -4$. Since c is negative, the two factors will have opposite signs. Since b is negative, the factor with the larger absolute value will be negative.
$$(2)(-6) = -12 \quad \text{and} \quad 2 + (-6) = -4$$
Therefore,
$$-2x^2y + 8xy + 24y = -2y(x^2 - 4x - 12)$$
$$= -2y(x - 6)(x + 2).$$

77. $4y^3 - 9y^2 - 9y = y(4y^2 - 9y - 9)$

The leading coefficient of the expression inside the parentheses is not 1, so we try factoring by grouping.
$$a \cdot c = 4(-9) = -36$$
We need two numbers whose product is -36 and whose sum is -9. The product is negative, so the two numbers will have opposite signs. The sum is negative, so the number with the larger absolute value must be negative.
Since $(3)(-12) = -36$ and $3 + (-12) = -9$, we get
$$4y^3 - 9y^2 - 9y = y(4y^2 - 9y - 9)$$
$$= y(4y^2 + 3y - 12y - 9)$$
$$= y(y(4y + 3) - 3(4y + 3))$$
$$= y(y - 3)(4y + 3)$$

79. The student is not correct. If we multiply the right side of the equation, we see that the middle term is not correct. The student disregarded the coefficient of x^2
$$2x^2 + 7x + 10 = (2x + 5)(x + 1)$$

81. The student's work is incorrect because they have not completely factored. In each factor of the student's answer, there is a common factor that can be pulled out.
$$8x^2 + 28x + 12 = 4(2x^2 + 7x + 3)$$
$$= 4(x + 3)(2x + 1)$$

83. Answers may vary.

85. $3x^2 + 16x - 12$
The leading coefficient is not 1, so we try factoring by grouping.
$$a \cdot c = 3(-12) = -36$$
We need two numbers whose product is -36 and whose sum is 16. The product is negative, so the two numbers will have opposite signs. The sum is positive, so the number with the larger absolute value will be positive
Since $(-2)(18) = -36$ and $-2 + (18) = 16$, we get
$$3x^2 + 16x - 12 = 3x^2 - 2x + 18x - 12$$
$$= x(3x - 2) + 6(3x - 2)$$
$$= (x + 6)(3x - 2)$$

87. $(4x - 7)(3x - 1) = 12x^2 - 4x - 21x + 7$
$$= 12x^2 - 25x + 7$$

89. $(x - 3)(2x^2 + 3x - 5)$
$$= x(2x^2 + 3x - 5) - 3(2x^2 + 3x - 5)$$
$$= 2x^3 + 3x^2 - 5x - 6x^2 - 9x + 15$$
$$= 2x^3 - 3x^2 - 14x + 15$$

91. $6x^3 + 10x^2 - 4x = 2x(3x^2 + 5x - 2)$

The leading coefficient of the expression inside the parentheses is not 1, so we try factoring by grouping.
$$a \cdot c = 3(-2) = -6$$
We need two numbers whose product is -6 and whose sum is 5. The product is negative, so the two numbers will have opposite signs. The sum is positive, so the number with the larger absolute value must be positive.
Since $(-1)(6) = -6$ and $-1 + 6 = 5$, we get
$$6x^3 + 10x^2 - 4x = 2x(3x^2 + 5x - 2)$$
$$= 2x(3x^2 - 1x + 6x - 2)$$
$$= 2x(x(3x - 1) + 2(3x - 1))$$
$$= 2x(x + 2)(3x - 1)$$

93. $y = x^2 - 3$

First we list some solutions of $y = x^2 - 3$. Then, we sketch a curve that contains the points corresponding to the solutions.

x	y	(x, y)
-3	$(-3)^2 - 3 = 6$	$(-3, 6)$
-1	$(-1)^2 - 3 = -2$	$(-1, -2)$
0	$(0)^2 - 3 = -3$	$(0, -3)$
1	$(1)^2 - 3 = -2$	$(1, -2)$
3	$(3)^2 - 3 = 6$	$(3, 6)$

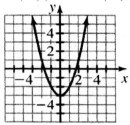

Description: *quadratic equation in two variables*

95. $x^2 - 2x - 3$

We are looking for two factors of $c = -3$ whose sum is $b = -2$. Since c is negative, the two factors will have opposite signs. Since b is negative, the factor with the larger absolute value will be negative.

$(1)(-3) = -3$ and $1 + (-3) = -2$

Therefore, $x^2 - 2x - 3 = (x - 3)(x + 1)$.

Description: *quadratic (or second degree) polynomial in one variable*

97. $x^2 - 2x - 3 = (-5)^2 - 2(-5) - 3$
$$= 25 - 2(-5) - 3$$
$$= 25 + 10 - 3$$
$$= 32$$

Description: *quadratic (or second degree) polynomial in one variable*

Homework 8.4

1. $x^3 + 27 = x^3 + 3^3$
$$= (x + 3)(x^2 - 3x + 3^2)$$
$$= (x + 3)(x^2 - 3x + 9)$$

3. $x^3 + 125 = x^3 + 5^3$
$$= (x + 5)(x^2 - 5x + 5^2)$$
$$= (x + 5)(x^2 - 5x + 25)$$

5. $x^3 - 8 = x^3 - 2^3$
$$= (x - 2)(x^2 + 2x + 2^2)$$
$$= (x - 2)(x^2 + 2x + 4)$$

7. $x^3 - 1 = x^3 - 1^3$
$$= (x - 1)(x^2 + 1x + 1^2)$$
$$= (x - 1)(x^2 + x + 1)$$

9. $8t^3 + 27 = (2t)^3 + 3^3$
$$= (2t + 3)((2t)^2 - 3(2t) + 3^2)$$
$$= (2t + 3)(4t^2 - 6t + 9)$$

11. $27x^3 - 8 = (3x)^3 - 2^3$
$$= (3x - 2)((3x)^2 + 2(3x) + 2^2)$$
$$= (3x - 2)(9x^2 + 6x + 4)$$

13. $5x^3 + 40 = 5(x^3 + 8)$
$$= 5(x^3 + 2^3)$$
$$= 5(x + 2)(x^2 - 2x + 2^2)$$
$$= 5(x + 2)(x^2 - 2x + 4)$$

15. $2x^3 - 54 = 2(x^3 - 27)$
$$= 2(x^3 - 3^3)$$
$$= 2(x - 3)(x^2 + 3x + 3^2)$$
$$= 2(x - 3)(x^2 + 3x + 9)$$

17. $8x^3 + 27y^3 = (2x)^3 + (3y)^3$
$$= (2x + 3y)((2x)^2 - (2x)(3y) + (3y)^2)$$
$$= (2x + 3y)(4x^2 - 6xy + 9y^2)$$

19. $64a^3 - 27b^3 = (4a)^3 - (3b)^3$

$$= (4a - 3b)\left((4a)^2 + (4a)(3b) + (3b)^2\right)$$

$$= (4a - 3b)\left(16a^2 + 12ab + 9b^2\right)$$

21. $x^2 - 64 = x^2 - 8^2$

$$= (x - 8)(x + 8)$$

23. $m^2 + 11m + 28$

The leading coefficient is 1 so we need two integers whose product is $c = 28$ and whose sum is $b = 11$. Since $4 \cdot 7 = 28$ and $4 + 7 = 11$, we get $m^2 + 11m + 28 = (m + 4)(m + 7)$.

25. $2t^2 + 2t - 24 = 2\left(t^2 + t - 12\right)$

The leading coefficient inside the parentheses is 1 so we need two integers whose product is $c = -12$ and whose sum is $b = 1$. Since $4 \cdot (-3) = -12$ and $4 + (-3) = 1$, we get

$$2t^2 + 2t - 24 = 2(t - 3)(t + 4).$$

27. $x^2 + 49$

Since the sum of two squares is prime over the integers, $x^2 + 49$ is prime.

29. $-3x^2 + 24x - 45 = -3\left(x^2 - 8x + 15\right)$

The leading coefficient inside the parentheses is 1 so we need two integers whose product is $c = 15$ and whose sum is $b = -8$. Since $(-5) \cdot (-3) = 15$ and $(-5) + (-3) = -8$, we get

$$-3x^2 + 24x - 45 = -3(x - 5)(x - 3).$$

31. $1 + 15p^2 - 8p = 15p^2 - 8p + 1$

The leading coefficient is not 1, so we try factoring by grouping. $a \cdot c = 15 \cdot 1 = 15$, so we need two numbers whose product is 15 and whose sum is $b = -8$. Since $(-5)(-3) = 15$ and $(-5) + (-3) = -8$, we get

$$15p^2 - 8p + 1 = 15p^2 - 5p - 3p + 1$$

$$= 5p(3p - 1) - 1(3p - 1)$$

$$= (5p - 1)(3p - 1)$$

33. $x^2 - 2x + 1$

The leading coefficient is 1 so we need two integers whose product is $c = 1$ and whose sum is $b = -2$. Since $(-1)(-1) = 1$ and $(-1) + (-1) = -2$, we get

$$x^2 - 2x + 1 = (x - 1)(x - 1)$$

$$= (x - 1)^2$$

35. $24r^2 + 4r - 4 = 4\left(6r^2 + r - 1\right)$

The leading coefficient inside the parentheses is not 1, so we try factoring by grouping. $a \cdot c = 6(-1) = -6$, so we need two numbers whose product is -6 and whose sum is $b = 1$. Since $(3)(-2) = -6$ and $3 + (-2) = 1$, we get

$$24r^2 + 4r - 4 = 4\left(6r^2 + r - 1\right)$$

$$= 4\left(6r^2 + 3r - 2r - 1\right)$$

$$= 4\left(3r(2r + 1) - 1(2r + 1)\right)$$

$$= 4(3r - 1)(2r + 1)$$

37. $-4ab^3 + 6a^2b^2 = \left(-2ab^2\right) \cdot 2b + \left(-2ab^2\right)(-3a)$

$$= -2ab^2(2b - 3a)$$

39. $a^2 - ab - 20b^2$

We need two monomials whose product is $-20b^2$ and whose sum is $-b$. Since $-20b^2$ has a negative coefficient, the coefficients of the monomials must have opposite signs. Since the coefficient of $-b$ is negative, the coefficient with the larger absolute value must be negative.

$(-5b)(4b) = -20b^2$ and $(-5b) + (4b) = -b$

Therefore, $a^2 - ab - 20b^2 = (a - 5b)(a + 4b)$.

41. $8x^3 - 20x^2 - 2x + 5 = 4x^2(2x - 5) - 1(2x - 5)$

$$= \left(4x^2 - 1\right)(2x - 5)$$

$$= \left((2x)^2 - 1^2\right)(2x - 5)$$

$$= (2x - 1)(2x + 1)(2x - 5)$$

43. $-12x^4 - 4x^3 = -4x^3 \cdot 3x + \left(-4x^3\right) \cdot 1$

$$= -4x^3(3x + 1)$$

45. $15a^4 + 25a^3 + 10a^2 = 5a^2\left(3a^2 + 5a + 2\right)$

The leading coefficient inside the parentheses is not 1, so we try factoring by grouping. $a \cdot c = 3 \cdot 2 = 6$, so we need two numbers whose product is 6 and whose sum is $b = 5$. Since $(3)(2) = 6$ and $3 + 2 = 5$, we get

$$15a^4 + 25a^3 + 10a^2 = 5a^2\left(3a^2 + 5a + 2\right)$$
$$= 5a^2\left(3a^2 + 3a + 2a + 2\right)$$
$$= 5a^2\left(3a(a+1) + 2(a+1)\right)$$
$$= 5a^2\left(3a + 2\right)(a+1)$$

47. $24 - 14x + x^2 = x^2 - 14x + 24$

The leading coefficient is 1 so we need two integers whose product is $c = 24$ and whose sum is $b = -14$. Since $-12 \cdot (-2) = 24$ and

$-12 + (-2) = -14$, we get

$$24 - 14x + x^2 = x^2 - 14x + 24$$
$$= (x - 12)(x - 2)$$

49. $2w^4 + 4w^3 - 8w^2 = 2w^2\left(w^2 + 2w - 4\right)$

The leading coefficient inside the parentheses is 1 so we need two integers whose product is $c = -4$ and whose sum is $b = 2$. Since the product is negative, the integers have opposite signs. Since the sum is positive, the integer with the larger absolute value must be positive.

Factors	$-1, 4$	$-2, 2$
Sum	3	0

Since none of the possibilities work, the expression inside the parentheses is prime.

$$2w^4 + 4w^3 - 8w^2 = 2w^2\left(w^2 + 2w - 4\right)$$

51. $12x^4 - 27x^2 = 3x^2\left(4x^2 - 9\right)$
$$= 3x^2\left((2x)^2 - 3^2\right)$$
$$= 3x^2\left(2x - 3\right)(2x + 3)$$

53. $x^2 + 10x + 25$

The leading coefficient is 1 so we need two integers whose product is $c = 25$ and whose sum is $b = 10$. Since $5 \cdot 5 = 25$ and $5 + 5 = 10$, we get

$$x^2 + 10x + 25 = (x + 5)(x + 5)$$
$$= (x + 5)^2$$

55. $2x^2 + x - 21$

The leading coefficient is not 1, so we try factoring by grouping. $a \cdot c = 2(-21) = -42$, so we need two numbers whose product is -42 and whose sum is $b = 1$. Since $7(-6) = -42$ and $7 + (-6) = 1$, we get

$$2x^2 + x - 21 = 2x^2 + 7x - 6x - 21$$
$$= x(2x + 7) - 3(2x + 7)$$
$$= (x - 3)(2x + 7)$$

57. $m^3 - 13m^2n + 36mn^2 = m\left(m^2 - 13mn + 36n^2\right)$

For the expression in parentheses, we need two monomials whose product is $36n^2$ and whose sum is $-13n$. Since $36n^2$ has a positive coefficient, the coefficients of the monomials must have the same sign. Since the coefficient of $-13n$ is negative, the coefficients must both be negative. $(-9n)(-4n) = 36n^2$ and $-9n + (-4n) = -13n$ Therefore,

$$m^3 - 13m^2n + 36mn^2 = m\left(m^2 - 13mn + 36n^2\right)$$
$$= m(m - 9n)(m - 4n)$$

59. $4x - 5 + 2x^2 = 2x^2 + 4x - 5$

The leading coefficient is not 1, so we try factoring by grouping. $a \cdot c = 2(-5) = -10$, so we need two integers whose product is -10 and whose sum is $b = 4$. Since the product is negative, the integers will have opposite signs. Since the sum is positive, the integer with the larger absolute value will be positive.

Factors	$-1, 10$	$-2, 5$
Sum	9	3

Since none of the possibilities work, the expression is prime.

61. $100x^2 - 9y^2 = (10x)^2 - (3y)^2$
$$= (10x - 3y)(10x + 3y)$$

63. $4x^2 + 12x + 9$

The leading coefficient is not 1, so we try factoring by grouping. $a \cdot c = 4 \cdot 9 = 36$, so we need two numbers whose product is 36 and whose sum is $b = 12$. Since $6 \cdot 6 = 36$ and $6 + 6 = 12$, we get

$4x^2 + 12x + 9 = 4x^2 + 6x + 6x + 9$
$= 2x(2x + 3) + 3(2x + 3)$
$= (2x + 3)(2x + 3)$
$= (2x + 3)^2$

65. $18x^3 + 27x^2 - 8x - 12 = 9x^2(2x + 3) - 4(2x + 3)$
$= (9x^2 - 4)(2x + 3)$
$= ((3x)^2 - 2^2)(2x + 3)$
$= (3x - 2)(3x + 2)(2x + 3)$

67. $3a^3 - 10a^2b + 8ab^2 = a(3a^2 - 10ab + 8b^2)$
$= a(3a^2 - 6ab - 4ab + 8b^2)$
$= a(3a(a - 2b) - 4b(a - 2b))$
$= a(3a - 4b)(a - 2b)$

69. $x^2 - 9x - 20$

The leading coefficient is 1 so we need two integers whose product is $c = -20$ and whose sum is $b = -9$. Since the product is negative, the integers will have opposite signs. Since the sum is also negative, the integer with the larger absolute value will be negative.

Factors	1, −20	2, −10	4, −5
Sum	−19	−8	−1

Since none of the possibilities work, the expression is prime.

71. $x^3 - 1000 = x^3 - 10^3$
$= (x - 10)(x^2 + 10x + 10^2)$
$= (x - 10)(x^2 + 10x + 100)$

73. $64a^3 + 27 = (4a)^3 + 3^3$
$= (4a + 3)((4a)^2 - 3(4a) + 3^2)$
$= (4a + 3)(16a^2 - 12a + 9)$

75. $3x^2 + 24 = 3 \cdot x^2 + 3 \cdot 8$
$= 3(x^2 + 8)$

77. Answers may vary.

79. Answers may vary.

81. The coefficient on x in the second factor is incorrect.

$x^3 - 27 = x^3 - 3^3$
$= (x - 3)(x^2 + 3x + 3^2)$
$= (x - 3)(x^2 + 3x + 9)$

83. No; the middle term in the second factor should be AB.

85. In terms of correctness, it does not matter if the GCF is factored out in the first step or the last step. However, factoring it out in the first step often makes the coefficients easier to work with in the resulting expression.

87. $(5x - 7)(5x + 7) = 5x \cdot 5x + 5x \cdot 7 - 7 \cdot 5x - 7 \cdot 7$
$= 25x^2 + 35x - 35x - 49$
$= 25x^2 - 49$

89. $81x^2 - 16 = (9x)^2 - 4^2$
$= (9x - 4)(9x + 4)$

91. $3x^3 + 9x^2 - 12x = 3x(x^2 + 3x - 4)$
$= 3x(x - 1)(x + 4)$

93. $-2x(7x^2 - 5x + 1) = -2x(7x^2) - 2x(-5x) - 2x(1)$
$= -14x^3 + 10x^2 - 2x$

95. The equation $y = 4x - 5$ is in slope-intercept form, so the slope is $m = 4$ and the y-intercept is $(0, -5)$. We first plot $(0, -5)$. From this point we move 1 unit to the right and 4 units up, plotting the point $(1, -1)$. We then draw the line that connects the two points, extending in both directions.

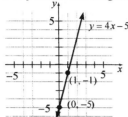

Description: *linear equation in two variables*

97. $4x - 5 = 3x + 2$

$x - 5 = 2$

$x = 7$

The solution is 7.

Description: *linear equation in one variable*

99. $(4x - 5)(3x + 2) = 4x \cdot 3x + 4x \cdot 2 - 5 \cdot 3x - 5 \cdot 2$

$= 12x^2 + 8x - 15x - 10$

$= 12x^2 - 7x - 10$

Description: *quadratic (or second degree) polynomial in one variable*

Homework 8.5

1. $(x - 2)(x - 7) = 0$

$x - 2 = 0$ or $x - 7 = 0$

$x = 2$ $x = 7$

The solutions are 2 and 7.

3. $x^2 + 7x + 10 = 0$

$(x + 2)(x + 5) = 0$

$x + 2 = 0$ or $x + 5 = 0$

$x = -2$ $x = -5$

The solutions are -2 and -5.

5. $w^2 + 3w - 28 = 0$

$(w - 4)(w + 7) = 0$

$w - 4 = 0$ or $w + 7 = 0$

$w = 4$ $w = -7$

The solutions are -7 and 4.

7. $x^2 - 14x + 49 = 0$

$(x - 7)(x - 7) = 0$

$x - 7 = 0$

$x = 7$

The solution is 7.

9. $r^2 + 3r = 4$

$r^2 + 3r - 4 = 0$

$(r - 1)(r + 4) = 0$

$r - 1 = 0$ or $r + 4 = 0$

$r = 1$ $r = -4$

The solutions are -4 and 1.

11. $5x = x^2 + 6$

$x^2 - 5x + 6 = 0$

$(x - 3)(x - 2) = 0$

$x - 3 = 0$ or $x - 2 = 0$

$x = 3$ $x = 2$

The solutions are 2 and 3.

13. $k^2 - 64 = 0$

$(k - 8)(k + 8) = 0$

$k - 8 = 0$ or $k + 8 = 0$

$k = 8$ $k = -8$

The solutions are -8 and 8.

15. $36x^2 - 49 = 0$

$(6x - 7)(6x + 7) = 0$

$6x - 7 = 0$ or $6x + 7 = 0$

$6x = 7$ $6x = -7$

$x = \dfrac{7}{6}$ $x = -\dfrac{7}{6}$

The solutions are $-\dfrac{7}{6}$ and $\dfrac{7}{6}$.

17.
$$x^2 = 49$$
$$x^2 - 49 = 0$$
$$(x-7)(x+7) = 0$$
$$x - 7 = 0 \quad \text{or} \quad x + 7 = 0$$
$$x = 7 \qquad x = -7$$
The solutions are -7 and 7.

19.
$$w^2 = w$$
$$w^2 - w = 0$$
$$w(w-1) = 0$$
$$w = 0 \quad \text{or} \quad w - 1 = 0$$
$$w = 1$$
The solutions are 0 and 1.

21. $(3x-7)(2x+5) = 0$
$$3x - 7 = 0 \quad \text{or} \quad 2x + 5 = 0$$
$$3x = 7 \qquad 2x = -5$$
$$x = \frac{7}{3} \qquad x = -\frac{5}{2}$$
The solutions are $-\frac{5}{2}$ and $\frac{7}{3}$.

23.
$$4x^2 - 8x + 3 = 0$$
$$(2x-3)(2x-1) = 0$$
$$2x - 3 = 0 \quad \text{or} \quad 2x - 1 = 0$$
$$2x = 3 \qquad 2x = 1$$
$$x = \frac{3}{2} \qquad x = \frac{1}{2}$$
The solutions are $\frac{1}{2}$ and $\frac{3}{2}$.

25.
$$10k^2 + 3k - 4 = 0$$
$$(5x+4)(2x-1) = 0$$
$$5x + 4 = 0 \quad \text{or} \quad 2x - 1 = 0$$
$$5x = -4 \qquad 2x = 1$$
$$x = -\frac{4}{5} \qquad x = \frac{1}{2}$$
The solutions are $-\frac{4}{5}$ and $\frac{1}{2}$.

27.
$$25x^2 + 20x + 4 = 0$$
$$(5x+2)(5x+2) = 0$$
$$5x + 2 = 0$$
$$5x = -2$$
$$x = -\frac{2}{5}$$
The solution is $-\frac{2}{5}$.

29.
$$x^2 - 2x = 12 - 6x$$
$$x^2 + 4x - 12 = 0$$
$$(x+6)(x-2) = 0$$
$$x + 6 = 0 \quad \text{or} \quad x - 2 = 0$$
$$x = -6 \qquad x = 2$$
The solutions are -6 and 2.

31.
$$3t^2 - 4t = 6t - 8$$
$$3t^2 - 10t + 8 = 0$$
$$(3t-4)(t-2) = 0$$
$$3t - 4 = 0 \quad \text{or} \quad t - 2 = 0$$
$$3t = 4 \qquad t = 2$$
$$t = \frac{4}{3}$$
The solutions are $\frac{4}{3}$ and 2.

33.
$$(x-2)(x-5) = 4$$
$$x^2 - 5x - 2x + 10 = 4$$
$$x^2 - 7x + 10 = 4$$
$$x^2 - 7x + 6 = 0$$
$$(x-6)(x-1) = 0$$
$$x - 6 = 0 \quad \text{or} \quad x - 1 = 0$$
$$x = 6 \qquad x = 1$$
The solutions are 1 and 6.

35.
$$4x(x+1) = 15$$
$$4x^2 + 4x = 15$$
$$4x^2 + 4x - 15 = 0$$
$$(2x+5)(2x-3) = 0$$
$$2x+5 = 0 \quad \text{or} \quad 2x-3 = 0$$
$$2x = -5 \qquad\qquad 2x = 3$$
$$x = -\frac{5}{2} \qquad\qquad x = \frac{3}{2}$$

The solutions are $-\frac{5}{2}$ and $\frac{3}{2}$.

37.
$$(2w-3)(w-2) = 3$$
$$2w^2 - 4w - 3w + 6 = 3$$
$$2w^2 - 7w + 6 = 3$$
$$2w^2 - 7w + 3 = 0$$
$$(2w-1)(w-3) = 0$$
$$2w-1 = 0 \quad \text{or} \quad w-3 = 0$$
$$2w = 1 \qquad\qquad w = 3$$
$$w = \frac{1}{2}$$

The solutions are $\frac{1}{2}$ and 3.

39.
$$(x-2)^2 - 3x = -6$$
$$x^2 - 4x + 4 - 3x = -6$$
$$x^2 - 7x + 4 = -6$$
$$x^2 - 7x + 10 = 0$$
$$(x-5)(x-2) = 0$$
$$x-5 = 0 \quad \text{or} \quad x-2 = 0$$
$$x = 5 \qquad\qquad x = 2$$

The solutions are 2 and 5.

41.
$$\frac{1}{4}x^2 + 2x + 3 = 0$$
$$4\left(\frac{1}{4}x^2 + 2x + 3\right) = 4 \cdot 0$$
$$x^2 + 8x + 12 = 0$$
$$(x+2)(x+6) = 0$$
$$x+2 = 0 \quad \text{or} \quad x+6 = 0$$
$$x = -2 \qquad\qquad x = -6$$

The solutions are -2 and -6.

43.
$$\frac{3}{8}m^2 + m - 2 = 0$$
$$8\left(\frac{3}{8}m^2 + m - 2\right) = 8 \cdot 0$$
$$3m^2 + 8m - 16 = 0$$
$$(3m-4)(m+4) = 0$$
$$3m-4 = 0 \quad \text{or} \quad m+4 = 0$$
$$3m = 4 \qquad\qquad m = -4$$
$$m = \frac{4}{3}$$

The solutions are -4 and $\frac{4}{3}$.

45.
$$-\frac{1}{3}x^2 + \frac{1}{3}x + 10 = 6$$
$$-3\left(-\frac{1}{3}x^2 + \frac{1}{3}x + 10\right) = -3(6)$$
$$x^2 - x - 30 = -18$$
$$x^2 - x - 12 = 0$$
$$(x-4)(x+3) = 0$$
$$x-4 = 0 \quad \text{or} \quad x+3 = 0$$
$$x = 4 \qquad\qquad x = -3$$

The solutions are -3 and 4.

47. $x^2 - 6x + 5 = -3$

The graphs of $y = x^2 - 6x + 5$ and $y = -3$ intersect only at the points $(2,-3)$ and $(4,-3)$, which have x-coordinates 2 and 4. So, 2 and 4 are the solutions of $x^2 - 6x + 5 = -3$.

49. $x^2 - 6x + 5 = -4$

The graphs of $y = x^2 - 6x + 5$ and $y = -4$ intersect only at the point $(3,-4)$, which has x-coordinate 3. So, 3 is the solution of $x^2 - 6x + 5 = -4$.

51. $-x^2 - 4x - 1 = 2$

The graphs of $y = -x^2 - 4x - 1$ and $y = 2$ intersect only at the points $(-3,2)$ and $(-1,2)$, which have x-coordinates -3 and -1. So, -3 and -1 are the solutions of $-x^2 - 4x - 1 = 2$.

53. $-x^2 - 4x - 1 = 4$

The graphs of $y = -x^2 - 4x - 1$ and $y = 4$ never intersect. Therefore, the equation $-x^2 - 4x - 1 = 4$ has no real solution.

55. $x^2 - 5x = -2x + 3$

Graph the equations $y = x^2 - 5x$ and $y = -2x + 3$, then find the intersection points of the two graphs. The x-coordinates of the intersection points, if any, are the solutions to the equation.

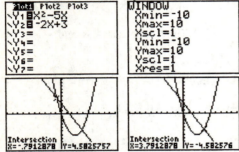

The approximate intersection points are $(-0.79, 4.58)$ and $(3.79, -4.58)$, so the approximate solutions to the equation are -0.79 and 3.79.

57. $-x^2 - 2x + 4 = -x - 5$

Graph the equations $y = -x^2 - 2x + 4$ and $y = -x - 5$, then find the intersection points of the two graphs. The x-coordinates of the intersection points, if any, are the solutions to the equation.

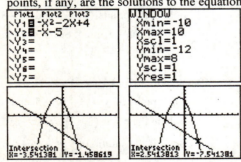

The approximate intersection points are $(-3.54, -1.46)$ and $(2.54, -7.54)$, so the approximate solutions to the equation are -3.54 and 2.54.

59. $y = 2x^2 - 8x + 10$

From the table, the inputs $x = 1$ and $x = 3$ yield the output $y = 4$. Therefore, 1 and 3 are solutions to the equation $2x^2 - 8x + 10 = 4$.

61. $y = 2x^2 - 8x + 10$

From the table, the input $x = 2$ yield the output $y = 2$. Therefore, 2 is a solution to the equation $2x^2 - 8x + 10 = 2$.

63. $y = x^2 - 3x - 28$

Substitute 0 for y and solve for x.

$x^2 - 3x - 28 = 0$

$(x - 7)(x + 4) = 0$

$x - 7 = 0$ or $x + 4 = 0$

$x = 7$ $\qquad$ $x = -4$

So, the x-intercepts are $(7, 0)$ and $(-4, 0)$.

65. $y = x^2 - 8x + 16$

Substitute 0 for y and solve for x.

$x^2 - 8x + 16 = 0$

$(x - 4)(x - 4) = 0$

$x - 4 = 0$

$x = 4$

The x-intercept is $(4, 0)$.

67. $y = 2x^2 - x - 3$

Substitute 0 for y and solve for x.

$2x^2 - x - 3 = 0$

$(2x - 3)(x + 1) = 0$

$2x - 3 = 0$ or $x + 1 = 0$

$2x = 3$ $\qquad$ $x = -1$

$x = \dfrac{3}{2}$

So, the x-intercepts are $(-1, 0)$ and $\left(\dfrac{3}{2}, 0\right)$.

69. $2x^3 - 5x^2 - 18x + 45 = 0$

$x^2(2x - 5) - 9(2x - 5) = 0$

$(x^2 - 9)(2x - 5) = 0$

$(x - 3)(x + 3)(2x - 5) = 0$

$x - 3 = 0$ or $x + 3 = 0$ or $2x - 5 = 0$

$x = 3$ $\qquad$ $x = -3$ $\qquad$ $2x = 5$

$x = \dfrac{5}{2}$

The solutions are -3, 3, and $\dfrac{5}{2}$.

71. $3x^3 - 4x^2 - 12x + 16 = 0$

$x^2(3x-4) - 4(3x-4) = 0$

$(x^2 - 4)(3x-4) = 0$

$(x-2)(x+2)(3x-4) = 0$

$x-2=0$ or $x+2=0$ or $3x-4=0$

$x=2$ $\qquad x=-2$ $\qquad 3x=4$

$\qquad\qquad\qquad\qquad\qquad x=\dfrac{4}{3}$

The solutions are -2, 2, and $\dfrac{4}{3}$.

73. $18y^3 - 27y^2 = 8y - 12$

$18y^3 - 27y^2 - 8y + 12 = 0$

$9y^2(2y-3) - 4(2y-3) = 0$

$(9y^2 - 4)(2y-3) = 0$

$(3y-2)(3y+2)(2y-3) = 0$

$3y-2=0$ or $3y+2=0$ or $2y-3=0$

$3y=2$ $\qquad 3y=-2$ $\qquad 2y=3$

$y=\dfrac{2}{3}$ $\qquad y=-\dfrac{2}{3}$ $\qquad y=\dfrac{3}{2}$

The solutions are $-\dfrac{2}{3}$, $\dfrac{2}{3}$, and $\dfrac{3}{2}$.

75. $5x^2 + 10x - 40 = 0$

$5(x^2 + 2x - 8) = 0$

$5(x-2)(x+4) = 0$

$x-2=0$ or $x+4=0$

$x=2$ $\qquad x=-4$

The solutions are -4 and 2.

77. $2x^3 - 50x = 0$

$2x(x^2 - 25) = 0$

$2x(x-5)(x+5) = 0$

$2x=0$ or $x-5=0$ or $x+5=0$

$x=0$ $\qquad x=5$ $\qquad x=-5$

The solutions are -5, 0, and 5.

79. $3p^3 + 15p^2 = -18p$

$3p^3 + 15p^2 + 18p = 0$

$3p(p^2 + 5p + 6) = 0$

$3p(p+2)(p+3) = 0$

$3p=0$ or $p+2=0$ or $p+3=0$

$p=0$ $\qquad p=-2$ $\qquad p=-3$

The solutions are -3, -2, and 0.

81. $2x^3 = 4x^2 + 16x$

$2x^3 - 4x^2 - 16x = 0$

$2x(x^2 - 2x - 8) = 0$

$2x(x-4)(x+2) = 0$

$2x=0$ or $x-4=0$ or $x+2=0$

$x=0$ $\qquad x=4$ $\qquad x=-2$

The solutions are -2, 0, and 4.

83. The student divided by x which caused the loss of a solution. The student should have moved all the terms to one side of the equation and then factored that side.

$4x^2 = 8x$

$4x^2 - 8x = 0$

$4x(x-2) = 0$

$4x=0$ or $x-2=0$

$x=0$ $\qquad x=2$

The solutions are 0 and 2.

85. The equation $x=2$ is incorrect. The student should have written $2=0$ which has no real solution.

$2x^2 - 26x + 80 = 0$

$2(x^2 - 13x + 40) = 0$

$2(x-8)(x-5) = 0$

$2=0$ or $x-8=0$ or $x-5=0$

false $\qquad\quad x=8$ $\qquad x=5$

The solutions are 5 and 8.

87. Answers may vary.

89. a. $y = x^2 - 6x + 8$

$$3 = x^2 - 6x + 8$$

$$x^2 - 6x + 5 = 0$$

$$(x-5)(x-1) = 0$$

$$x - 5 = 0 \quad \text{or} \quad x - 1 = 0$$

$$x = 5 \qquad\qquad x = 1$$

The solutions are 1 and 5.

b. $-1 = x^2 - 6x + 8$

$$x^2 - 6x + 9 = 0$$

$$(x-3)(x-3) = 0$$

$$x - 3 = 0$$

$$x = 3$$

The solution is 3.

c. Since the input $x = 3$ yields a single output, the vertex lies on the line $x = 3$. Thus, the vertex is $(3, -1)$. Two additional points are $(1, 3)$ and $(5, 3)$. From the equation $y = x^2 - 6x + 8$, we know the y-intercept is $(0, 8)$. We then sketch a curve that contains these points.

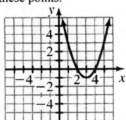

91. Answers may vary.

93. $x^2 + 3x = 7x + 12$

$$x^2 - 4x - 12 = 0$$

$$(x-6)(x+2) = 0$$

$$x - 6 = 0 \quad \text{or} \quad x + 2 = 0$$

$$x = 6 \qquad\qquad x = -2$$

The solutions are -2 and 6.

95. $x + 3 = 7x + 12$

$$-6x + 3 = 12$$

$$-6x = 9$$

$$x = -\frac{3}{2}$$

The solution is $-\frac{3}{2}$.

97. $A = P + PRT$

$$A = P(1 + RT)$$

$$\frac{A}{1 + RT} = \frac{P(1 + RT)}{(1 + RT)}$$

$$P = \frac{A}{1 + RT}$$

99. $x^2 - 4x - 5 = (x-5)(x+1)$

Description: *quadratic (or second-degree) polynomial in one variable*

101. $x^2 - 4x - 5 = 0$

$$(x-5)(x+1) = 0$$

$$x - 5 = 0 \quad \text{or} \quad x + 1 = 0$$

$$x = 5 \qquad\qquad x = -1$$

The solutions are -1 and 5.

Description: *quadratic equation in one variable*

103. $y = x^2 - 4x - 5$

First we list some solutions of $y = x^2 - 4x - 5$.

Then, we sketch a curve that contains the points corresponding to the solutions.

x	y	(x, y)
-2	$(-2)^2 - 4(-2) - 5 = 7$	$(-2, 7)$
0	$(0)^2 - 4(0) - 5 = -5$	$(0, -5)$
2	$(2)^2 - 4(2) - 5 = -9$	$(2, -9)$
4	$(4)^2 - 4(4) - 5 = -5$	$(4, -5)$
6	$(6)^2 - 4(6) - 5 = 7$	$(6, 7)$

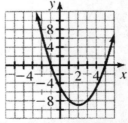

Description: *quadratic equation in two variables*

Homework 8.6

1. a. $n = t^2 + 4t + 22$

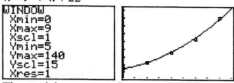

The model appears to fit the data well.

b. For 2010, we have $t = 15$.

$$n = (15)^2 + 4(15) + 22$$
$$= 225 + 60 + 22$$
$$= 307$$

The model predicts that there will be approximately 307 thousand special-education students with Autism in 2010. Some research will show that model breakdown has occurred for the 1988 estimate.

c. $\qquad 43 = t^2 + 4t + 22$

$$t^2 + 4t - 21 = 0$$
$$(t+7)(t-3) = 0$$
$$t + 7 = 0 \quad \text{or} \quad t - 3 = 0$$
$$t = -7 \qquad\qquad t = 3$$

The model predicts that there were 43 thousand special education students with Autism in 1988 and 1998.

3. a.

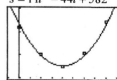

The curvature in the graph suggests that the data would be better fit by a quadratic model. The average rate of change is not constant.

b. $s = 11t^2 - 44t + 582$

The model appears to fit the data well.

c. $\qquad 637 = 11t^2 - 44t + 582$

$$0 = 11t^2 - 44t - 55$$
$$0 = (11t - 55)(t + 1)$$
$$11t - 55 = 0 \quad \text{or} \quad t + 1 = 0$$
$$11t = 55 \qquad\qquad t = -1$$
$$t = 5$$

The model predicts that total spending by travelers in the U.S. was $637 billion in 1999 and 2005.

d.

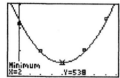

The vertex means that total spending by travelers in the U.S. in 2002 was $538 billion, the lowest of any year.

5. a.

The curvature in the graph suggests that the data would be better fit by a quadratic model since the average rate of change is not constant.

b.

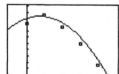

The model appears to fit the data fairly well.

c. For 2004, we have $t = 5$.

$$r = -\frac{1}{3}(5)^2 + \frac{1}{2}(5) + 9 \approx 3.167$$

The model predicts that the revenue for U.S. Airways was approximately $3.2 billion in 2004.

d. To find the t-intercepts, set $r = 0$ and solve for t.

$$0 = -\frac{1}{3}t^2 + \frac{1}{2}t + 9$$

$$-6 \cdot 0 = -6\left(-\frac{1}{3}t^2 + \frac{1}{2}t + 9\right)$$

$$0 = 2t^2 - 3t - 54$$

$$0 = (t - 6)(2t + 9)$$

$$t - 6 = 0 \quad \text{or} \quad 2t + 9 = 0$$

$$t = 6 \qquad\qquad 2t = -9$$

$$t = -\frac{9}{2}$$

The t-intercepts are $(-4.5, 0)$ and $(6, 0)$. The intercepts indicate that U.S. Airways' revenue was zero dollars in 1994 and 2005. Since this is not realistic, model breakdown has occurred.

7. a.

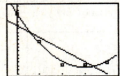

The quadratic model appears to be a better fit to the data.

b. Since economists felt that sales of Firestone tires would continue to decline, the linear model (with its constant negative average rate of change) better fits this viewpoint. The economists would likely feel the linear model would best describe future sales.

c. After an initial decline in tire sales, Firestone experienced an increase in sales. This would indicate that the graph of sales should initial be decreasing, then 'turn around' so that it is increasing. Firestone would likely feel the quadratic model better predicts future sales.

d.

$$\frac{4}{5}t^2 - \frac{24}{5}t + 23 = 19$$

$$\frac{4}{5}t^2 - \frac{24}{5}t + 4 = 0$$

$$\frac{5}{4}\left(\frac{4}{5}t^2 - \frac{24}{5}t + 4\right) = \frac{5}{4} \cdot 0$$

$$t^2 - 6 + 5 = 0$$

$$(t - 5)(t - 1) = 0$$

$$t - 5 = 0 \quad \text{or} \quad t - 1 = 0$$

$$t = 5 \qquad\qquad t = 1$$

The quadratic model predicts that sales of Firestone tires were 19 million tires in 2000 and 2004.

9.

$$t^2 - 3t + 5 = 23$$

$$t^2 - 3t - 18 = 0$$

$$(t - 6)(t + 3) = 0$$

$$t - 6 = 0 \quad \text{or} \quad t + 3 = 0$$

$$t = 6 \qquad\qquad t = -3$$

The model predicts that the company's profit was $23 thousand in 2002 and will be again in 2011.

11.

$$2t^2 - 13t + 25 = 10$$

$$2t^2 - 13t + 15 = 0$$

$$(2t - 3)(t - 5) = 0$$

$$2t - 3 = 0 \quad \text{or} \quad t - 5 = 0$$

$$2t = 3 \qquad\qquad t = 5$$

$$t = \frac{3}{2}$$

The model predicts that the company's revenue was $10 million in 2007 and will be again in 2010.

13. a.

$$-16t^2 + 64t + 3 = 3$$

$$-16t^2 + 64t = 0$$

$$t^2 - 4t = 0$$

$$t(t - 4) = 0$$

$$t = 0 \quad \text{or} \quad t - 4 = 0$$

$$t = 4$$

The ball will be at a height of 3 feet after 0 seconds and again after 4 seconds.

There are two such times because the ball is at a height of 3 feet when it is hit and again on its way back down to the ground.

b.

$$-16t^2 + 64t + 3 = 51$$

$$-16t^2 + 64t - 48 = 0$$

$$t^2 - 4t + 3 = 0$$

$$(t - 3)(t - 1) = 0$$

$$t - 3 = 0 \quad \text{or} \quad t - 1 = 0$$

$$t = 3 \qquad\qquad t = 1$$

One second after the ball is hit, and again 3 seconds after it is hit, the ball is at a height of 51 feet. It makes sense that there are two times, because the ball reaches 51 feet both on its way up and on its way down.

c.
$$-16t^2 + 64t + 3 = 67$$
$$-16t^2 + 64t - 64 = 0$$
$$t^2 - 4t + 4 = 0$$
$$(t-2)(t-2) = 0$$
$$t - 2 = 0$$
$$t = 2$$

Two seconds after the ball is hit, the ball is at a height of 67 feet. It makes sense that there is only one time if the ball reaches a height of 67 feet only at the top of its climb.

15. Let L = the length of the table (in feet) and W = the width (in feet).
Since the length is 7 feet more than the width, our first equation is $L = 7 + W$. Because the area is
$$A = LW$$
$$60 = LW$$
we get the following system:
$$L = 7 + W$$
$$60 = LW$$
Substitute $7 + W$ for L in the second equation and solve for W.
$$60 = (7 + W)W$$
$$60 = 7W + W^2$$
$$0 = W^2 + 7W - 60$$
$$0 = (W - 5)(W + 12)$$
$$W - 5 = 0 \quad \text{or} \quad W + 12 = 0$$
$$W = 5 \qquad\qquad W = -12$$
The width must be positive, so discard the negative solution. Substitute 5 for W in the equation $L = 7 + W$ and solve for L.
$$L = 7 + 5 = 12$$
The table has a length of 12 feet and a width of 5 feet.

17. Let L = the length of the office (in feet) and W = the width (in feet).
Since the length is twice the width, our first equation is $L = 2W$. Because the area is
$$A = LW$$
$$98 = LW$$
we get the following system:
$$L = 2W$$
$$98 = LW$$
Substitute $2W$ for L in the second equation and solve for W.

$$98 = (2W)W$$
$$98 = 2W^2$$
$$49 = W^2$$
$$0 = W^2 - 49$$
$$0 = (W - 7)(W + 7)$$
$$W - 7 = 0 \quad \text{or} \quad W + 7 = 0$$
$$W = 7 \qquad\qquad W = -7$$
The width must be positive, so discard the negative solution. Substitute 7 for W in the equation $L = 2W$ and solve for L.
$$L = 2(7) = 14$$
The office has a length of 14 feet and a width of 7 feet.

19. Answers may vary. A linear model is appropriate if the average rate of change is roughly constant. A quadratic model may be appropriate if the rate of change switches from increasing to decreasing, or from decreasing to increasing. In either case, a scatter diagram can be helpful in determining which model is appropriate.

21. a.

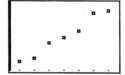

The graph appears to be more linear than parabolic in shape. Therefore, a linear model would better fit the data.

b.
```
LinReg
y=ax+b
a=121.0714286
b=422.7142857
```
Using the regression feature of a graphing calculator, an equation of a model is $T = 121.07t + 422.71$.

c. Substitute 2800 for T in the equation from part (b) and solve for t.
$$121.07t + 422.71 = 2800$$
$$121.07t = 2377.29$$
$$t = \frac{2377.29}{121.07} \approx 19.64$$
The model predicts that the average tax refund will be $2800 in 2010.

23. a. Since per person consumption of seafood has increased by about 0.14 pound per year, a linear model is appropriate. The slope is 0.14 pound per year. Since the number of pounds consumed per capita was 10.3 in 1960, the c-intercept is $(0,10.3)$. So, a reasonable model is $c = 0.14t + 10.3$.

b. Substitute 17.4 for c in our model and solve for t.
$$0.14t + 10.3 = 17.4$$
$$0.14t = 7.1$$
$$t = \frac{7.1}{0.14} \approx 50.71$$

The model predicts that the U.S. annual per person consumption of seafood will be 17.4 pounds in 2011.

25. Let c = average cost of campus housing and t = the number of years after 1999. The average cost has increased linearly with slope
$$m = \frac{5475 - 4340}{2004 - 1999} = \frac{1135}{5} = 227.$$

Since the average cost was \$4340 in 1999, the c-intercept is $(0,4340)$. So, a reasonable model is
$$c = 227t + 4340.$$
For 2011, we have $t = 12$.
$$c = 227(12) + 4340 = 7064$$

The model predicts that the average cost of campus housing will be \$7064 for the school year ending in 2011.

27. – 33. Answers may vary.

Chapter 8 Review Exercises

1. $x^2 + 9x + 20$
The leading coefficient is 1 so we need two integers whose product is $c = 20$ and whose sum is $b = 9$. Since $4 \cdot 5 = 20$ and $4 + 5 = 9$, we get
$$x^2 + 9x + 20 = (x+4)(x+5)$$

2. $6x^2 - 2x - 8 = 2(3x^2 - x - 4)$
The leading coefficient of the expression inside the parentheses is not 1, so we try factoring by grouping. $a \cdot c = 3(-4) = -12$, so we need two numbers whose product is -12 and whose sum is

-1. Since $3(-4) = -12$ and $3 + (-4) = -1$, we get
$$6x^2 - 2x - 8 = 2(3x^2 - x - 4)$$
$$= 2(3x^2 + 3x - 4x - 4)$$
$$= 2(3x(x+1) - 4(x+1))$$
$$= 2(3x - 4)(x+1)$$

3. $81x^2 - 49 = (9x)^2 - 7^2$
$$= (9x - 7)(9x + 7)$$

4. $x^2 + 14x + 49$
The leading coefficient is 1 so we need two integers whose product is $c = 49$ and whose sum is $b = 14$. Since $7 \cdot 7 = 49$ and $7 + 7 = 14$, we get
$$x^2 + 14x + 49 = (x+7)(x+7)$$
$$= (x+7)^2$$

5. $-18t^4 - 33t^3 + 30t^2 = -3t^2(6t^2 + 11t - 10)$
The leading coefficient of the expression inside the parentheses is not 1, so we try factoring by grouping. $a \cdot c = 6(-10) = -60$, so we need two numbers whose product is -60 and whose sum is 11. Since $15(-4) = -60$ and $15 + (-4) = 11$, we get
$$-18t^4 - 33t^3 + 30t^2 = -3t^2(6t^2 + 11t - 10)$$
$$= -3t^2(6t^2 + 15t - 4t - 10)$$
$$= -3t^2(3t(2t+5) - 2(2t+5))$$
$$= -3t^2(3t - 2)(2t + 5)$$

6. $p^2 - 3pq - 54q^2$
The leading coefficient is 1, so we need two monomials whose product is $-54q^2$ and whose sum is $-3q$. Since $(6q)(-9q) = -54q^2$ and $6q + (-9q) = -3q$, we get
$$p^2 - 3pq - 54q^2 = (p - 9q)(p + 6q).$$

7. $32 - 12x + x^2 = x^2 - 12x + 32$

The leading coefficient is 1 so we need two integers whose product is $c = 32$ and whose sum is $b = -12$. Since $-8 \cdot (-4) = 32$ and $-4 + (-8) = -12$, we get

$32 - 12x + x^2 = x^2 - 12x + 32$
$= (x - 8)(x - 4)$

8. $-9x^2 + 4 = -1(9x^2 - 4)$
$= -1((3x)^2 - 2^2)$
$= -(3x - 2)(3x + 2)$

9. $4w^2 + 25 = (2w)^2 + 5^2$

Since the sum of two squares is prime over the integers, the expression $4w^2 + 25$ is prime.

10. $20m^2n - 45mn^3 = 5mn \cdot 4m - 5mn \cdot 9n^2$
$= 5mn(4m - 9n^2)$

11. $16x^2 + 14x + 2x^3 = 2x^3 + 16x^2 + 14x$
$= 2x(x^2 + 8x + 7)$

The leading coefficient of the expression in parentheses is 1, so we need two integers whose product is $c = 7$ and whose sum is $b = 8$. Since $7 \cdot 1 = 7$ and $7 + 1 = 8$, we get

$16x^2 + 14x + 2x^3 = 2x^3 + 16x^2 + 14x$
$= 2x(x^2 + 8x + 7)$
$= 2x(x + 1)(x + 7)$.

12. $16x^3 - 32x^2 + 16x = 16x(x^2 - 2x + 1)$

The leading coefficient of the expression in parentheses is 1, so we need two integers whose product is $c = 1$ and whose sum is $b = -2$. Since $(-1)(-1) = 1$ and $(-1) + (-1) = -2$, we get

$16x^3 - 32x^2 + 16x = 16x(x^2 - 2x + 1)$
$= 16x(x - 1)(x - 1)$
$= 16(x - 1)^2$.

13. $24x^3 - 32x^2 = 8x^2 \cdot 3x - 8x^2 \cdot 4$
$= 8x^2(3x - 4)$

14. $5x^4y - 35x^3y + 60x^2y = 5x^2y(x^2 - 7x + 12)$

The leading coefficient of the expression in parentheses is 1, so we need two integers whose product is $c = 12$ and whose sum is $b = -7$. Since $(-3)(-4) = 12$ and $(-3) + (-4) = -7$, we get

$5x^4y - 35x^3y + 60x^2y = 5x^2y(x^2 - 7x + 12)$
$= 5x^2y(x - 3)(x - 4)$.

15. $-m^2 - 2m + 35 = -1(m^2 + 2m - 35)$

The leading coefficient of the expression in parentheses is 1, so we need two integers whose product is $c = -35$ and whose sum is $b = 2$. Since $7(-5) = -35$ and $7 + (-5) = 2$, we get

$-m^2 - 2m + 35 = -1(m^2 + 2m - 35)$
$= -(m - 5)(m + 7)$.

16. $4r^3 - 10r^2 + 6r - 15 = 2r^2(2r - 5) + 3(2r - 5)$
$= (2r^2 + 3)(2r - 5)$

17. $2p^2 + 7p + 3$

The leading coefficient is not 1, so we try factoring by grouping. $a \cdot c = 2(3) = 6$, so we need two numbers whose product is 6 and whose sum is 7. Since $1(6) = 6$ and $1 + 6 = 7$, we get

$2p^2 + 7p + 3 = 2p^2 + 1p + 6p + 3$
$= p(2p + 1) + 3(2p + 1)$
$= (p + 3)(2p + 1)$.

18. $x^2 - 9x + 20$

The leading coefficient is 1 so we need two integers whose product is $c = 20$ and whose sum is $b = -9$. Since $-5 \cdot (-4) = 20$ and $-5 + (-4) = -9$, we get

$x^2 - 9x + 20 = (x - 5)(x - 4)$.

19. $6t^2 + 11ty - 10y^2$

The leading coefficient is not 1, so we try factoring by grouping. $a \cdot c = 6\left(-10y^2\right) = -60y^2$, so we need two monomials whose product is $-60y^2$ and whose sum is $11y$. Since $-4y\left(15y\right) = -60y^2$ and $-4y + 15y = 11y$, we get

$$6t^2 + 11ty - 10y^2 = 6t^2 - 4ty + 15ty - 10y^2$$
$$= 2t\left(3t - 2y\right) + 5y\left(3t - 2y\right)$$
$$= \left(2t + 5y\right)\left(3t - 2y\right)$$

20. $2x^3 - 50x = 2x \cdot x^2 - 2x \cdot 25$
$$= 2x\left(x^2 - 25\right)$$
$$= 2x\left(x^2 - 5^2\right)$$
$$= 2x\left(x - 5\right)\left(x + 5\right)$$

21. $x^2 - 10x + 25$

The leading coefficient is 1 so we need two integers whose product is $c = 25$ and whose sum is $b = -10$. Since $-5 \cdot \left(-5\right) = 25$ and $-5 + \left(-5\right) = -10$, we get

$$x^2 - 10x + 25 = \left(x - 5\right)\left(x - 5\right)$$
$$= \left(x - 5\right)^2$$

22. $p^2 - 81 = p^2 - 9^2$
$$= \left(p - 9\right)\left(p + 9\right)$$

23. $8w^2 - 12w + 3$

The leading coefficient is not 1, so we try factoring by grouping. $a \cdot c = 8\left(3\right) = 24$, so we need two numbers whose product is 24 and whose sum is -12. Since the product is positive and the sum is negative, the two numbers must be negative.

factor 1	factor 2	sum
-1	-24	-25
-2	-12	-14
-3	-8	-11
-4	-6	-10

Since none of the possibilities work, we can say that the expression is prime.

24. $x^2 + 4 = x^2 + 2^2$

Since the sum of two squares is not factorable over the integers, the expression $x^2 + 4$ is prime.

25. $4x^3 - 4x^2 - 9x + 9 = 4x^2\left(x - 1\right) - 9\left(x - 1\right)$
$$= \left(4x^2 - 9\right)\left(x - 1\right)$$
$$= \left(\left(2x\right)^2 - 3^2\right)\left(x - 1\right)$$
$$= \left(2x - 3\right)\left(2x + 3\right)\left(x - 1\right)$$

26. $4x^2 + 20x + 25$

The leading coefficient is not 1, so we try factoring by grouping. $a \cdot c = 4\left(25\right) = 100$, so we need two numbers whose product is 100 and whose sum is 20. Since $10 \cdot 10 = 100$ and $10 + 10 = 20$, we get

$$4x^2 + 10x + 10x + 25 = 2x\left(2x + 5\right) + 5\left(2x + 5\right)$$
$$= \left(2x + 5\right)\left(2x + 5\right)$$
$$= \left(2x + 5\right)^2$$

27. $12w^3 - 50w^2 + 8w = 2w\left(6w^2 - 25w + 4\right)$

The leading coefficient of the expression inside the parentheses is not 1, so we try factoring by grouping. $a \cdot c = 6 \cdot 4 = 24$, so we need two numbers whose product is 24 and whose sum is -25. Since $\left(-1\right)\left(-24\right) = 24$ and $\left(-1\right) + \left(-24\right) = -25$, we get

$$12w^3 - 50w^2 + 8w = 2w\left(6w^2 - 25w + 4\right)$$
$$= 2w\left(6w^2 - w - 24w + 4\right)$$
$$= 2w\left(w\left(6w - 1\right) - 4\left(6w - 1\right)\right)$$
$$= 2w\left(w - 4\right)\left(6w - 1\right)$$

28. $49a^2 - 9b^2 = \left(7a\right)^2 - \left(3b\right)^2$
$$= \left(7a - 3b\right)\left(7a + 3b\right)$$

29. $x^2 - 7x - 12$

The leading coefficient is 1 so we need two integers whose product is $c = -12$ and whose sum is $b = -7$. Since the product is negative, the integers will have opposite signs. Since the sum is also negative, the factor with the larger absolute value will be negative.

Factors	$1, -12$	$2, -6$	$3, -4$
Sum	-11	-4	-1

Since none of the possibilities work, the expression $x^2 - 7x - 12$ is prime.

30. $x^3 + 3x^2 - 4x - 12 = x^2(x+3) - 4(x+3)$
$$= (x^2 - 4)(x+3)$$
$$= (x-2)(x+2)(x+3)$$

31. $r^3 + 8 = r^3 + 2^3$
$$= (r+2)(r^2 - 2\cdot r + 2^2)$$
$$= (r+2)(r^2 - 2r + 4)$$

32. $8t^3 - 27 = (2t)^3 - 3^3$
$$= (2t-3)((2t)^2 + 3\cdot 2t + 3^2)$$
$$= (2t-3)(4t^2 + 6t + 9)$$

33. $x^2 + 25 = x^2 + 5^2$
The student tried to factor the sum of two squares as $a^2 + b^2 = (a+b)(a+b)$. However, the sum of two squares is not factorable over the integers, so the expression is prime.
Note: $(a+b)(a+b) = a^2 + 2ab + b^2$ not $a^2 + b^2$

34. The student factored out the GCF, but did not factor completely since the expression inside parentheses can be factored further.
$$5x^3 + 35x^2 + 60x = 5x(x^2 + 7x + 12)$$
$$= 5x(x+3)(x+4)$$

35. For $x^2 + kx + 24$ to be factorable, we need two integers whose product is 24 and whose sum is k. Since 24 is positive, the two integers must have the same sign.

Factors	Sum $= k$
−1,−24	−25
−2,−12	−14
−3,−8	−11
−4,−6	−10
4,6	10
3,8	11
2,12	14
1,24	25

36. $x^2 + 9x + 14 = 0$
$$(x+2)(x+7) = 0$$
$$x+2 = 0 \quad \text{or} \quad x+7 = 0$$
$$x = -2 \qquad\qquad x = -7$$
The solutions are -2 and -7.

37. $\qquad 2x^3 + 16x^2 = -24x$
$$2x^3 + 16x^2 + 24x = 0$$
$$2x(x^2 + 8x + 12) = 0$$
$$2x(x+2)(x+6) = 0$$
$$2x = 0 \quad \text{or} \quad x+2 = 0 \quad \text{or} \quad x+6 = 0$$
$$x = 0 \qquad\quad x = -2 \qquad\quad x = -6$$
The solutions are -6, -2, and 0.

38. $(m-3)(m+2) = -4$
$$m^2 - m - 6 = -4$$
$$m^2 - m - 2 = 0$$
$$(m-2)(m+1) = 0$$
$$m-2 = 0 \quad \text{or} \quad m+1 = 0$$
$$m = 2 \qquad\qquad m = -1$$
The solutions are -1 and 2.

39. $\qquad t^2 - 6t + 9 = 0$
$$(t-3)(t-3) = 0$$
$$t-3 = 0$$
$$t = 3$$
The solution is 3.

40. $\qquad x^2 - 3x = 5x - 15$
$$x^2 - 8x + 15 = 0$$
$$(x-5)(x-3) = 0$$
$$x-5 = 0 \quad \text{or} \quad x-3 = 0$$
$$x = 5 \qquad\qquad x = 3$$
The solutions are 3 and 5.

41. $\qquad 25x^2 - 81 = 0$
$$(5x)^2 - 9^2 = 0$$
$$(5x-9)(5x+9) = 0$$
$$5x-9 = 0 \quad \text{or} \quad 5x+9 = 0$$
$$5x = 9 \qquad\qquad 5x = -9$$
$$x = \frac{9}{5} \qquad\qquad x = -\frac{9}{5}$$
The solutions are $-\frac{9}{5}$ and $\frac{9}{5}$.

42.
$$2x^3 - 7x^2 - 2x + 7 = 0$$
$$x^2(2x - 7) - 1(2x - 7) = 0$$
$$(x^2 - 1)(2x - 7) = 0$$
$$(x - 1)(x + 1)(2x - 7) = 0$$
$$x - 1 = 0 \quad \text{or} \quad x + 1 = 0 \quad \text{or} \quad 2x - 7 = 0$$
$$x = 1 \qquad x = -1 \qquad 2x = 7$$
$$x = \frac{7}{2}$$

The solutions are -1, 1, and $\dfrac{7}{2}$.

43.
$$6x^2 + x - 2 = 0$$
$$(3x + 2)(2x - 1) = 0$$
$$3x + 2 = 0 \quad \text{or} \quad 2x - 1 = 0$$
$$3x = -2 \qquad 2x = 1$$
$$x = -\frac{2}{3} \qquad x = \frac{1}{2}$$

The solutions are $-\dfrac{2}{3}$ and $\dfrac{1}{2}$.

44.
$$3x^2 = 15x$$
$$3x^2 - 15x = 0$$
$$3x(x - 5) = 0$$
$$3x = 0 \quad \text{or} \quad x - 5 = 0$$
$$x = 0 \qquad x = 5$$

The solutions are 0 and 5.

45.
$$8r^2 - 18r + 9 = 0$$
$$(4r - 3)(2r - 3) = 0$$
$$4r - 3 = 0 \quad \text{or} \quad 2r - 3 = 0$$
$$4r = 3 \qquad 2r = 3$$
$$r = \frac{3}{4} \qquad r = \frac{3}{2}$$

The solutions are $\dfrac{3}{4}$ and $\dfrac{3}{2}$.

46.
$$a^2 = 2a + 35$$
$$a^2 - 2a - 35 = 0$$
$$(a - 7)(a + 5) = 0$$
$$a - 7 = 0 \quad \text{or} \quad a + 5 = 0$$
$$a = 7 \qquad a = -5$$

The solutions are -5 and 7.

47.
$$x(x - 4) = 12$$
$$x^2 - 4x = 12$$
$$x^2 - 4x - 12 = 0$$
$$(x - 6)(x + 2) = 0$$
$$x - 6 = 0 \quad \text{or} \quad x + 2 = 0$$
$$x = 6 \qquad x = -2$$

The solutions are -2 and 6.

48.
$$\frac{1}{3}x^2 - \frac{1}{3}x - 4 = 6$$
$$3\left(\frac{1}{3}x^2 - \frac{1}{3}x - 4\right) = 3(6)$$
$$x^2 - x - 12 = 18$$
$$x^2 - x - 30 = 0$$
$$(x - 6)(x + 5) = 0$$
$$x - 6 = 0 \quad \text{or} \quad x + 5 = 0$$
$$x = 6 \qquad x = -5$$

The solutions are -5 and 6.

49.
$$3x^3 - 2x^2 = 27x - 18$$
$$3x^3 - 2x^2 - 27x + 18 = 0$$
$$x^2(3x - 2) - 9(3x - 2) = 0$$
$$(x^2 - 9)(3x - 2) = 0$$
$$(x - 3)(x + 3)(3x - 2) = 0$$
$$x - 3 = 0 \quad \text{or} \quad x + 3 = 0 \quad \text{or} \quad 3x - 2 = 0$$
$$x = 3 \qquad x = -3 \qquad 3x = 2$$
$$x = \frac{2}{3}$$

The solutions are -3, $\dfrac{2}{3}$, and 3.

50.
$$a^2 = 4$$
$$a^2 - 4 = 0$$
$$a^2 - 2^2 = 0$$
$$(a - 2)(a + 2) = 0$$
$$a - 2 = 0 \quad \text{or} \quad a + 2 = 0$$
$$a = 2 \qquad a = -2$$

The solutions are -2 and 2.

51. $5p^2 + 20p - 60 = 0$

$\frac{1}{5}\left(5p^2 + 20p - 60\right) = \frac{1}{5}(0)$

$p^2 + 4p - 12 = 0$

$(p-2)(p+6) = 0$

$p - 2 = 0$ or $p + 6 = 0$

$p = 2$ $\qquad\;$ $p = -6$

The solutions are −6 and 2.

52. $2x^2 + 4x - 5 = 2x + 3$

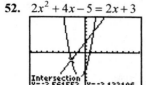 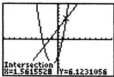

The approximate intersection points are $(-2.56, -2.12)$ and $(1.56, 6.12)$, which have *x*-coordinates of -2.56 and 1.56, respectively. Therefore, the approximate solutions to the equation are -2.56 and 1.56.

53. $y = -3x^2 + 6x + 20$

From the table, the input $x = 1$ yields the output $y = 23$. Therefore, 1 is the solution to the equation $-3x^2 + 6x + 20 = 23$.

54. $y = -3x^2 + 6x + 20$

From the table, the inputs $x = -2$ and $x = 4$ yield the output $y = -4$. Therefore, −2 and 4 are solutions to the equation $-3x^2 + 6x + 20 = -4$.

55. From the table we see the maximum value for $y = -3x^2 + 6x + 20$ is $y = 23$. There are no input values such that the output is 27. Therefore, the equation $-3x^2 + 6x + 20 = 27$ has no real solutions.

56. $y = -3x^2 + 6x + 20$

From the table, the inputs $x = 0$ and $x = 2$ yield the output $y = 20$. Therefore, 0 and 2 are solutions to the equation $-3x^2 + 6x + 20 = 20$.

57. $y = x^2 - 49$

$0 = x^2 - 49$

$0 = (x-7)(x+7)$

$x - 7 = 0$ or $x + 7 = 0$

$x = 7$ $\qquad\;$ $x = -7$

The solutions are −7 and 7, so the *x*-intercepts are $(-7, 0)$ and $(7, 0)$.

58. $y = 8x^2 - 14x - 15$

$0 = 8x^2 - 14x - 15$

$0 = (4x + 3)(2x - 5)$

$4x + 3 = 0$ $\quad$ or $\quad 2x - 5 = 0$

$4x = -3$ $\qquad\qquad\;\; 2x = 5$

$x = -\dfrac{3}{4}$ $\qquad\quad\; x = \dfrac{5}{2}$

The solutions are $-\dfrac{3}{4}$ and $\dfrac{5}{2}$, so the *x*-intercepts are $\left(-\dfrac{3}{4}, 0\right)$ and $\left(\dfrac{5}{2}, 0\right)$.

59. Answers may vary. One example:

$(x-3)(x+6) = 0$

$x^2 + 3x - 18 = 0$

60. Answers may vary. One example:

$x(x+2)(x-1) = 0$

$x^3 + x^2 - 2x = 0$

61. a.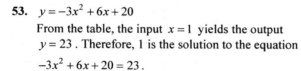

The curvature in the graph suggests that the data would be better fit by a quadratic model since the average rate of change is not constant.

b.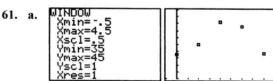

The model appears to fit the data reasonably well.

c. For 2004, we have $t = 5$.

$r = -(5)^2 + 4(5) + 39 = 34$

The model predicts that the number of asthma episodes per 1000 people was 34 in 2004.

d. $27 = -t^2 + 4t + 39$

$\quad 0 = t^2 - 4t - 12$

$\quad 0 = (t - 6)(t + 2)$

$\quad t - 6 = 0 \quad \text{or} \quad t + 2 = 0$

$\qquad t = 6 \qquad\qquad t = -2$

The annual asthma episode rate was 27 per 1000 people in 1997 and 2005.

e.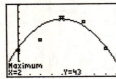

The vertex is $(2, 43)$. This means that in 2001, the annual asthma episode rate was 43 episodes per 1000 people (the largest rate ever, according to the model).

62. Let L = the length of the banner (in feet) and W = the width (in feet).

Since the length is 4 feet more than twice the width, our first equation is $L = 4 + 2W$. Because the area is

$A = LW$

$30 = LW$

we get the following system:

$L = 4 + 2W$

$30 = LW$

Substitute $4 + 2W$ for L in the second equation and solve for W.

$30 = (4 + 2W)W$

$30 = 4W + 2W^2$

$0 = 2W^2 + 4W - 30$

$0 = W^2 + 2W - 15$

$0 = (W - 3)(W + 5)$

$W - 3 = 0 \quad \text{or} \quad W + 5 = 0$

$\quad W = 3 \qquad\qquad W = -5$

The width must be positive, so discard the negative solution. Substitute 3 for W in the equation $L = 4 + 2W$ and solve for L.

$L = 4 + 2(3) = 10$

The banner has a length of 10 feet and a width of 3 feet.

Chapter 8 Test

1. $x^2 - 3x - 40$

The leading coefficient is 1 so we need two integers whose product is $c = -40$ and whose sum is $b = -3$. Since $-8 \cdot (5) = -40$ and $-8 + 5 = -3$, we get $x^2 - 3x - 40 = (x - 8)(x + 5)$.

2. $24 + x^2 - 10x = x^2 - 10x + 24$

The leading coefficient is 1 so we need two integers whose product is $c = 24$ and whose sum is $b = -10$. Since $-6(-4) = 24$ and $-6 + (-4) = -10$, we get

$24 + x^2 - 10x = x^2 - 10x + 24$

$\qquad\qquad\qquad\quad = (x - 6)(x - 4)$.

3. $8m^2n^3 - 10m^3n = 2m^2n \cdot 4n^2 - 2m^2n \cdot 5m$

$\qquad\qquad\qquad\quad = 2m^2n\left(4n^2 - 5m\right)$

4. $p^2 - 14pq + 40q^2$

The leading coefficient is 1, so we need two monomials whose product is $40q^2$ and whose sum is $-14q$. Since $(-10q)(-4q) = 40q^2$ and $-10q + (-4q) = -14q$, we get

$p^2 - 14pq + 40q^2 = (p - 10q)(p - 4q)$

5. $25p^2 - 36y^2 = (5p)^2 - (6y)^2$

$\qquad\qquad\qquad = (5p - 6y)(5p + 6y)$

6. $3x^4y - 21x^3y + 36x^2y = 3x^2y\left(x^2 - 7x + 12\right)$

The leading coefficient of the expression in parentheses is 1, so we need two integers whose product is $c = 12$ and whose sum is $b = -7$. Since $(-3)(-4) = 12$ and $(-3) + (-4) = -7$, we get

$3x^4y - 21x^3y + 36x^2y = 3x^2y\left(x^2 - 7x + 12\right)$

$\qquad\qquad\qquad\qquad\quad = 3x^2y(x - 4)(x - 3)$.

7. $8x^3 + 20x^2 - 18x - 45 = 4x^2(2x + 5) - 9(2x + 5)$

$\qquad\qquad\qquad\qquad\quad = \left(4x^2 - 9\right)(2x + 5)$

$\qquad\qquad\qquad\qquad\quad = (2x - 3)(2x + 3)(2x + 5)$

8. $8x^2 - 26x + 15$

The leading coefficient is not 1, so we try factoring by grouping. $a \cdot c = 8(15) = 120$, so we need two numbers whose product is 120 and whose sum is -26. Since the product is positive and the sum is negative, the two numbers must be negative. Since $-6(-20) = 120$ and $-6 + (-20) = -26$, we get

$$8x^2 - 26x + 15 = 8x^2 - 6x - 20x + 15$$
$$= 2x(4x - 3) - 5(4x - 3)$$
$$= (2x - 5)(4x - 3)$$

9. $64x^3 - 1 = (4x)^3 - 1^3$

$$= (4x - 1)\left((4x)^2 + 4x(1) + 1^2\right)$$
$$= (4x - 1)(16x^2 + 4x + 1)$$

10. $(x - 5)(x + 2) = x^2 - 3x - 10$
$$= (x + 2)(x - 5)$$

11. The student did not factor completely.
$$5x^3 + 3x^2 - 20x - 12 = x^2(5x + 3) - 4(5x + 3)$$
$$= (x^2 - 4)(5x + 3)$$
$$= (x - 2)(x + 2)(5x + 3)$$

12. $x^2 - 13x + 36 = 0$
$$(x - 4)(x - 9) = 0$$
$$x - 4 = 0 \quad \text{or} \quad x - 9 = 0$$
$$x = 4 \qquad\qquad x = 9$$
The solutions are 4 and 9.

13. $49x^2 - 9 = 0$
$$(7x)^2 - 3^2 = 0$$
$$(7x - 3)(7x + 3) = 0$$
$$7x - 3 = 0 \quad \text{or} \quad 7x + 3 = 0$$
$$7x = 3 \qquad\qquad 7x = -3$$
$$x = \frac{3}{7} \qquad\qquad x = -\frac{3}{7}$$
The solutions are $-\frac{3}{7}$ and $\frac{3}{7}$.

14. $t(t + 14) = 2(t - 18)$
$$t^2 + 14t = 2t - 36$$
$$t^2 + 12t + 36 = 0$$
$$(t + 6)(t + 6) = 0$$
$$t + 6 = 0$$
$$t = -6$$
The solution is -6.

15. $\frac{1}{4}p^2 - \frac{1}{2}p - 6 = 0$
$$4\left(\frac{1}{4}p^2 - \frac{1}{2}p - 6\right) = \frac{1}{4}(0)$$
$$p^2 - 2p - 24 = 0$$
$$(p - 6)(p + 4) = 0$$
$$p - 6 = 0 \quad \text{or} \quad p + 4 = 0$$
$$p = 6 \qquad\qquad p = -4$$
The solutions are -4 and 6.

16. $3x^3 - 12x = 8 - 2x^2$
$$3x^3 + 2x^2 - 12x - 8 = 0$$
$$x^2(3x + 2) - 4(3x + 2) = 0$$
$$(x^2 - 4)(3x + 2) = 0$$
$$(x - 2)(x + 2)(3x + 2) = 0$$
$$x - 2 = 0 \quad \text{or} \quad x + 2 = 0 \quad \text{or} \quad 3x + 2 = 0$$
$$x = 2 \qquad\qquad x = -2 \qquad\qquad 3x = -2$$
$$x = -\frac{2}{3}$$
The solutions are -2, $-\frac{2}{3}$, and 2.

17. $2x^3 = 8x^2 + 10x$
$$2x^3 - 8x^2 - 10x = 0$$
$$2x(x^2 - 4x - 5) = 0$$
$$2x(x - 5)(x + 1) = 0$$
$$2x = 0 \quad \text{or} \quad x - 5 = 0 \quad \text{or} \quad x + 1 = 0$$
$$x = 0 \qquad\qquad x = 5 \qquad\qquad x = -1$$
The solutions are -1, 0, and 5.

18. $x^2 + 6x + 7 = 2$

The graphs of $y = x^2 + 6x + 7$ and $y = 2$ intersect only at the points $(-5, 2)$ and $(-1, 2)$, which have x-coordinates -5 and -1. So, -5 and -1 are the solutions of $x^2 + 6x + 7 = 2$.

19. $x^2 + 6x + 7 = -1$

The graphs of $y = x^2 + 6x + 7$ and $y = -1$ intersect only at the points $(-4, -1)$ and $(-2, -1)$, which have x-coordinates -4 and -2. So, -4 and -2 are the solutions of $x^2 + 6x + 7 = -1$.

20. $x^2 + 6x + 7 = -2$

The graphs of $y = x^2 + 6x + 7$ and $y = -2$ intersect only at the point $(-3, -2)$, which has x-coordinate -3. So, -3 is the solution of $x^2 + 6x + 7 = -2$.

21. $x^2 + 6x + 7 = -4$

The graphs of $y = x^2 + 6x + 7$ and $y = -4$ do not intersect. So, there are no real solutions to the equation $x^2 + 6x + 7 = -4$.

22. $y = 10x^2 - 11x - 6$

$0 = 10x^2 - 11x - 6$

$0 = (5x + 2)(2x - 3)$

$5x + 2 = 0 \quad$ or $\quad 2x - 3 = 0$

$5x = -2 \qquad\qquad 2x = 3$

$x = -\dfrac{2}{5} \qquad\qquad x = \dfrac{3}{2}$

The solutions are $-\dfrac{2}{5}$ and $\dfrac{3}{2}$.

23. Answers may vary. One example:

$(x + 3)(x - 8) = 0$

$x^2 - 5x - 24 = 0$

24. a.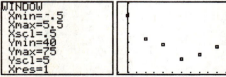

The curvature in the graph suggests that the data would be better fit by a quadratic model since the average rate of change is not constant.

b.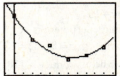

The model appears to fit the data fairly well.

c. For 2007, we have $t = 7$.

$p = 2(7)^2 - 13(7) + 69 = 76$

The model predicts that 76% of Americans in 2007 thought the environment should have received top priority.

d. $63 = 2t^2 - 13t + 69$

$0 = 2t^2 - 13t + 6$

$0 = (2t - 1)(t - 6)$

$2t - 1 = 0 \quad$ or $\quad t - 6 = 0$

$2t = 1 \qquad\qquad t = 6$

$t = 0.5$

The solutions are 0.5 and 6.

The percent of Americans who thought that the environment should be given top priority was 63% in 2001 and 2006.

e.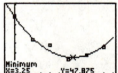

The vertex is $(3.25, 47.875)$. This means that in 2003, about 48% of Americans thought that the environment should have been given top priority, the lowest percentage in any year, according to the model (the actual percentage in 2003 was 47%).

25. $r = t^2 - 4t + 34$

$66 = t^2 - 4t + 34$

$0 = t^2 - 4t - 32$

$0 = (t - 8)(t + 4)$

$t - 8 = 0 \quad$ or $\quad t + 4 = 0$

$t = 8 \qquad\qquad t = -4$

The solutions are -4 and 8.

The model predicts that revenue was $66 million in 2001 and will be again in 2013.

Cumulative Review of Chapters 1 – 8

1. The independent variable is age a in years. The dependent variable is the percentage p who work.

2. If the data point lies above the model, the model is underestimating the value of p because the model predicts a lower value for p.

3. $-2 - 4 = -6$
   The change in temperature is $-6°F$.

4. $m = \dfrac{3 - (-1)}{5 - (-3)} = \dfrac{3+1}{5+3} = \dfrac{4}{8} = \dfrac{1}{2}$

   Since the slope is positive, the line is increasing.

5. The equation is linear with slope $m = -4 = \dfrac{-4}{1}$.

   Therefore, the value of y will decrease 4 units for a 1 unit increase in x.

6. $y - y_1 = m(x - x_1)$

   $y - (-6) = \dfrac{2}{3}(x - (-2))$

   $y + 6 = \dfrac{2}{3}(x + 2)$

   $y + 6 = \dfrac{2}{3}x + \dfrac{4}{3}$

   $y = \dfrac{2}{3}x - \dfrac{14}{3}$

7. Begin by finding the slope.
   $m = \dfrac{-3 - 7}{2 - (-4)} = \dfrac{-10}{6} = -\dfrac{5}{3}$

   Using the slope and the point $(2, -3)$, we get

   $y - y_1 = m(x - x_1)$

   $y - (-3) = -\dfrac{5}{3}(x - 2)$

   $y + 3 = -\dfrac{5}{3}x + \dfrac{10}{3}$

   $y = -\dfrac{5}{3}x + \dfrac{1}{3}$

8. From the graph we see that the y-intercept is $(0, 1)$ and the x-intercept is $(-2, 0)$. The slope is then

   $m = \dfrac{0 - 1}{-2 - 0} = \dfrac{-1}{-2} = \dfrac{1}{2}$

   The equation of the line is $y = \dfrac{1}{2}x + 1$.

9. $3(x + 5) = 9$; $3(x + 5) = 9$

   $\qquad\qquad\qquad 3x + 15 = 9$

   $\qquad\qquad\qquad\quad 3x = -6$

   $\qquad\qquad\qquad\qquad x = -2$

   The solution is -2.

10. $7 - 4\left(\dfrac{x}{2}\right)$; $7 - 4\left(\dfrac{x}{2}\right) = 7 - 2x$

11. $2x - 3y = 7$

    $\qquad y = 4x - 9$

    Substitute $4x - 9$ for y in the first equation and solve for x.

    $2x - 3(4x - 9) = 7$

    $2x - 12x + 27 = 7$

    $-10x + 27 = 7$

    $-10x = -20$

    $x = 2$

    Substitute 2 for x in the equation $y = 4x - 9$ and solve for y.

    $y = 4(2) - 9 = -1$

    The solution is $(2, -1)$.

12. $3x + 4y = 4$ Equation (1)

    $7x - 5y = 38$ Equation (2)

    To eliminate the variable y, multiply both sides of equation (1) by 5 and both sides of equation (2) by 4, yielding

    $15x + 20y = 20$

    $28x - 20y = 152$

    The coefficients of y are equal in absolute value, but opposite in sign. Add the left sides and right sides of the equations and solve for x.

    $15x + 20y = 20$

    $\underline{28x - 20y = 152}$

    $\qquad 43x = 172$

    $\qquad\quad x = 4$

    Substitute 4 for x in equation (1) and solve for y.

$$3x + 4y = 4$$
$$3(4) + 4y = 4$$
$$12 + 4y = 4$$
$$4y = -8$$
$$y = -2$$

The solution is $(4, -2)$.

13. $4x - 5y = -22$
     $x + 2y = 1$

Graphing by hand:
Begin by writing each equation in slope-intercept form.

$$4x - 5y = -22 \qquad\qquad x + 2y = 1$$
$$-5y = -4x - 22 \qquad\qquad 2y = -x + 1$$
$$y = \frac{4}{5}x + \frac{22}{5} \qquad\qquad y = -\frac{1}{2}x + \frac{1}{2}$$

Next we graph both equations in the same coordinate system.

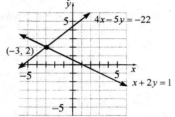

The intersection point is $(-3, 2)$. So, the ordered pair $(-3, 2)$ is the solution.

Substitution:
$$4x - 5y = -22$$
$$x + 2y = 1$$

Solve the second equation for x.
$$x + 2y = 1$$
$$x = -2y + 1$$

Substitute $-2y + 1$ for x in the first equation and solve for y.
$$4(-2y + 1) - 5y = -22$$
$$-8y + 4 - 5y = -22$$
$$-13y + 4 = -22$$
$$-13y = -26$$
$$y = 2$$

Substitute 2 for y in the equation $x = -2y + 1$ and solve for x.
$$x = -2(2) + 1 = -3$$

The solution is $(-3, 2)$.

Elimination:
$$4x - 5y = -22 \qquad \text{Equation (1)}$$
$$x + 2y = 1 \qquad \text{Equation (2)}$$

To eliminate the x terms, we multiply both sides of equation (2) by -4, yielding the system
$$4x - 5y = -22$$
$$-4x - 8y = -4$$

The coefficients of the x terms are equal in absolute value and opposite in sign. Add the left sides and the right sides of the equations and solve for y.

$$\begin{array}{r} 4x - 5y = -22 \\ \underline{-4x - 8y = -4} \\ -13y = -26 \\ y = 2 \end{array}$$

Substitute 2 for y in equation (2) and solve for x.
$$x + 2(2) = 1$$
$$x + 4 = 1$$
$$x = -3$$

The solution is $(-3, 2)$.

Preference may vary.

14. We can start by writing equations for the two lines.
     Red Line:
     Examining the graph, it appears that the y-intercept is $(0, 3)$ and the point $(3, 2)$ is on the graph.

$$m = \frac{2 - 3}{3 - 0} = \frac{-1}{3} = -\frac{1}{3}$$

Therefore, the equation of the line is
$$y = -\frac{1}{3}x + 3 .$$

Blue Line:
Examining the graph, it appears that the y-intercept is $(0, -3)$ and the point $(3, -2)$ is on the graph.

$$m = \frac{-2 - (-3)}{3 - 0} = \frac{1}{3}$$

Therefore, the equation of the line is
$$y = \frac{1}{3}x - 3 .$$

Using the two equations, we get the system
$$y = -\frac{1}{3}x + 3$$
$$y = \frac{1}{3}x - 3$$

Substitute $\frac{1}{3}x - 3$ for y in the first equation and solve for x.

$$\frac{1}{3}x - 3 = -\frac{1}{3}x + 3$$

$$\frac{2}{3}x - 3 = 3$$

$$\frac{2}{3}x = 6$$

$$x = 9$$

Substitute 9 for x in the equation $y = \frac{1}{3}x - 3$ and

solve for y.

$$y = \frac{1}{3}(9) - 3 = 0$$

The solution is $(9, 0)$.

15. $5(-2) - (-6)^2 + 4 = 5(-2) - (36) + 4$
$$= -10 - 36 + 4$$
$$= -46 + 4$$
$$= -42$$

16. $9 - (6-8)^3 + 4 \div (-2) = 9 - (-2)^3 + 4 \div (-2)$
$$= 9 - (-8) + 4 \div (-2)$$
$$= 9 - (-8) + (-2)$$
$$= 9 + 8 - 2$$
$$= 17 - 2$$
$$= 15$$

17. $b^2 - 4ac = (-2)^2 - 4(-4)(5)$
$$= 4 - 4(-4)(5)$$
$$= 4 - (-80)$$
$$= 4 + 80$$
$$= 84$$

18. $\dfrac{c + a^2}{a - b^2} = \dfrac{5 + (-4)^2}{(-4) - (-2)^2}$
$$= \frac{5 + 16}{-4 - 4}$$
$$= \frac{21}{-8}$$
$$= -\frac{21}{8}$$

19. The line $x = 4$ intersects the graph at $(4, -5)$. So, when $x = 4$, $y = -5$.

20. The line $y = 4$ intersects the graph at the point $(1, 4)$. So, when $y = 4$, $x = 1$.

21. The line $y = 3$ intersects that graph at the points $(0, 3)$ and $(2, 3)$. So, when $y = 3$, $x = 0$ or $x = 2$.

22. The line $y = 5$ does not intersect the graph. So, there are no values for x such that $y = 5$.

23. The graph crosses the x-axis when $x = -1$ and $x = 3$. So, the x-intercepts are $(-1, 0)$ and $(3, 0)$.

24. The maximum point is at the vertex, $(1, 4)$.

25. The graph of $x = 2$ is a vertical line passing through the point $(2, 0)$.

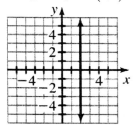

26. $y = 2x^2$

First we list some solutions of $y = 2x^2$. Then, we sketch a curve that contains the points corresponding to the solutions.

x	y	(x, y)
-2	$2(-2)^2 = 8$	$(-2, 8)$
-1	$2(-1)^2 = 2$	$(-1, 2)$
0	$2(0)^2 = 0$	$(0, 0)$
1	$2(1)^2 = 2$	$(1, 2)$
2	$2(2)^2 = 8$	$(2, 8)$

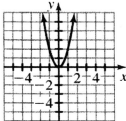

Description: *quadratic equation in two variables*

27. $4m^2 - 49n^2 = (2m)^2 - (7n)^2$
$$= (2m - 7n)(2m + 7n)$$
Description: *quadratic (or second degree) polynomial in two variables*

28. $(3p - 7q)^2 = (3p)^2 - 2(3p)(7q) + (7q)^2$
$$= 9p^2 - 42pq + 49q^2$$
Description: *quadratic (or second degree) polynomial in two variables*

29. $w^2 + 5w - 14$
The leading coefficient is 1 so we need two integers whose product is $c = -14$ and whose sum is $b = 5$. Since $7(-2) = -14$ and $7 + (-2) = 5$, we get $w^2 + 5w - 14 = (w - 2)(w + 7)$.
Description: *quadratic (or second degree) polynomial in one variable*

30. $5x^2 + 18x - 8 = 0$
$$(5x - 2)(x + 4) = 0$$
$$5x - 2 = 0 \quad \text{or} \quad x + 4 = 0$$
$$5x = 2 \qquad\qquad x = -4$$
$$x = \frac{2}{5}$$
The solutions are -4 and $\frac{2}{5}$.
Description: *quadratic equation in one variable*

31. $x^2 = 2x + 35$
$$x^2 - 2x - 35 = 0$$
$$(x - 7)(x + 5) = 0$$
$$x - 7 = 0 \quad \text{or} \quad x + 5 = 0$$
$$x = 7 \qquad\qquad x = -5$$
The solutions are -5 and 7.
Description: *quadratic equation in one variable*

32. $6x^3y + x^2y - 15xy = xy(6x^2 + x - 15)$
$$= xy(3x + 5)(2x - 3)$$
Description: *fourth-degree polynomial in two variables*

33. $(8p - 3)(2p + 3) = 16p^2 + 24p - 6p - 9$
$$= 16p^2 + 18p - 9$$
Description: *quadratic (or second degree) polynomial in one variable*

34. $-4xy(5x^2 + 2xy - 3y^2)$
$$= -4xy(5x^2) - 4xy(2xy) - 4xy(-3y^2)$$
$$= -20x^3y - 8x^2y^2 + 12xy^3$$
Description: *fourth-degree polynomial in two variables*

35. $a^2 - 3ab - 40b^2$
The leading coefficient is 1, so we need two monomials whose product is $-40b^2$ and whose sum is $-3b$. Since $(-8b)(5b) = -40b^2$ and $-8b + 5b = -3b$, we get $a^2 - 3ab - 40b^2 = (a - 8b)(a + 5b)$.
Description: *quadratic (or second degree) polynomial in two variables*

36. $5(t - 2) = 4 - 7t$
$$5t - 10 = 4 - 7t$$
$$12t - 10 = 4$$
$$12t = 14$$
$$t = \frac{14}{12} = \frac{7}{6}$$
The solution is $\frac{7}{6}$.
Description: *linear equation in one variable*

37. $(3x^2 - 5x) + (-7x^2 - 2x + 9)$
$$= 3x^2 - 5x - 7x^2 - 2x + 9$$
$$= 3x^2 - 7x^2 - 5x - 2x + 9$$
$$= -4x^2 - 7x + 9$$
Description: *quadratic (or second degree) polynomial in one variable*

38. $3x^2 - 33x + 54 = 3(x^2 - 11x + 18)$
$$= 3(x - 9)(x - 2)$$
Description: *quadratic (or second degree) polynomial in one variable*

39. $2x - 4y = 8$
x-intercept: $2x - 4(0) = 8$
$$2x = 8$$
$$x = 4$$
The x-intercept is $(4, 0)$.

y-intercept: $2(0) - 4y = 8$

$$-4y = 8$$

$$y = -2$$

The y-intercept is $(0, -2)$.

We plot the intercepts and connect them with a line, extending the line in both directions.

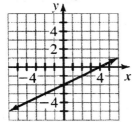

Description: *linear equation in two variables*

40. $(2m - 5)(3m^2 - 2m + 4)$

$$= 2m(3m^2 - 2m + 4) - 5(3m^2 - 2m + 4)$$

$$= 6m^3 - 4m^2 + 8m - 15m^2 + 10m - 20$$

$$= 6m^3 - 19m^2 + 18m - 20$$

Description: *cubic (or third degree) polynomial in one varible*

41. $(x^2 - x) - (2x^3 - 4x^2 + 5x)$

$$= x^2 - x - 2x^3 + 4x^2 - 5x$$

$$= -2x^3 + 5x^2 - 6x$$

Description: *cubic (or third degree) polynomial in one varible*

42. $4r^3 + 8r^2 - 9r - 18 = 4r^2(r + 2) - 9(r + 2)$

$$= (4r^2 - 9)(r + 2)$$

$$= ((2r)^2 - 3^2)(r + 2)$$

$$= (2r - 3)(2r + 3)(r + 2)$$

Description: *cubic (or third degree) polynomial in one varible*

43. $8x^3 - 27 = (2x)^3 - 3^3$

$$= (2x - 3)((2x)^2 + (3)(2x) + 3^2)$$

$$= (2x - 3)(4x^2 + 6x + 9)$$

Description: *cubic (or third degree) polynomial in one varible*

44. $S = 2\pi r^2 + rh$

$$S - 2\pi r^2 = rh$$

$$\frac{S - 2\pi r^2}{r} = \frac{rh}{r}$$

$$\frac{S - 2\pi r^2}{r} = h \quad \text{or} \quad h = \frac{S - 2\pi r^2}{r}$$

45. $2(x - 1) \le 5(x + 2)$

$$2x - 2 \le 5x + 10$$

$$-3x - 2 \le 10$$

$$-3x \le 12$$

$$\frac{-3x}{-3} \ge \frac{12}{-3}$$

$$x \ge -4$$

Interval: $[-4, \infty)$

46. $4x - 5y \ge 20$

First we get y by itself on one side of the inequality.

$$4x - 5y \ge 20$$

$$-5y \ge -4x + 20$$

$$y \le \frac{4}{5}x - 4$$

The graph of $y \le \frac{4}{5}x - 4$ is the line $y = \frac{4}{5}x - 4$ and the region below that line. We use a solid line along the border to indicate that the points on the line are solutions to $4x - 5y \ge 20$.

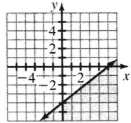

47. $y > -\dfrac{1}{2}x + 2$

$y < 3$

First we sketch the graph of $y > -\dfrac{1}{2}x + 2$ and the

graph of $y < 3$. The graph of $y > -\dfrac{1}{2}x + 2$ is the

region above the line $y = -\dfrac{1}{2}x + 2$ (graph the line

with a dashed line). The graph of $y < 3$ is the

region below the horizontal line $y = 3$ (graph the

line with a dashed line). The graph of the solution

set of the system is the intersection of the graphs of

the inequalities.

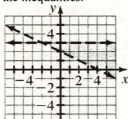

48. $(x-2)(x+4) = (x+4)(x-2)$

$\qquad = x(x+2) - 8$

$\qquad = x^2 + 2x - 8$

49. $y = -3(x-2)^2$

$y = -3(x^2 - 4x + 4)$

$y = -3x^2 + 12x - 12$

50. $y = x^2 + 4x - 21$

$0 = x^2 + 4x - 21$

$0 = (x+7)(x-3)$

$x + 7 = 0 \quad$ or $\quad x - 3 = 0$

$\quad x = -7 \qquad\qquad x = 3$

The solutions are -7 and 3.

Therefore, the x-intercepts are $(-7,0)$ and $(3,0)$.

51. $2x - 5y = 20$

$2x - 5(0) = 20$

$2x = 20$

$x = 10$

The solution is 10.

Therefore, the x-intercept is $(10,0)$.

52. $\left(-2ab^2\right)^3 = (-2)^3\, a^3 \left(b^2\right)^3$

$\qquad\qquad = -8a^3 b^6$

53. $\dfrac{2^2\, r^{-2}}{2^5\, r^{-7}} = 2^{2-5}\, r^{-2-(-7)}$

$\qquad = 2^{-3}\, r^5$

$\qquad = \dfrac{r^5}{2^3}$

$\qquad = \dfrac{r^5}{8}$

54. $\left(4x^{-8}y^5\right)\left(2x^3 y^{-2}\right) = 4 \cdot 2x^{-8+3}\, y^{5+(-2)}$

$\qquad\qquad\qquad = 8x^{-5}y^3$

$\qquad\qquad\qquad = \dfrac{8y^3}{x^5}$

55. $\left(\dfrac{x^2 y^{-6}}{w^{-4}}\right)^3 = \left(\dfrac{x^2 w^4}{y^6}\right)^3$

$\qquad\qquad = \dfrac{\left(x^2\right)^3 \left(w^4\right)^3}{\left(y^6\right)^3}$

$\qquad\qquad = \dfrac{x^6 w^{12}}{y^{18}}$

56. Let x = the number of \$20 tickets and y = the
number of \$35 tickets.

There are a total of 6000 tickets, so our first
equation is $x + y = 6000$.

The total revenue is \$147,000. The revenue for
each ticket type is obtained by multiplying the
ticket price by the number of tickets sold at that
price. Therefore, our second equation is
$20x + 35y = 147,000$

The system is

$\qquad x + y = 6000 \qquad$ Equation (1)

$20x + 35y = 147,000 \qquad$ Equation (2)

To eliminate the x terms, multiply both sides of
equation (1) by -20, yielding the system
$-20x - 20y = -120,000$

$\quad 20x + 35y = 147,000$

Add the left sides and right sides and solve for y.

$$-20x - 20y = -120,000$$
$$\underline{20x + 35y = 147,000}$$
$$15y = 27,000$$
$$y = 1800$$

Substitute 1800 for y in equation (1) and solve for x.

$$x + (1800) = 6000$$
$$x = 4200$$

The theater should sell 4200 of the $20 tickets and $1800 of the $35 tickets.

57. **a.**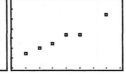

b. Using the regression feature of a graphing calculator, we obtain the equation $B = 0.063t + 10.13$.

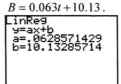

c. The slope is 0.063; Jones' best time in the 100-meter run increases by 0.063 second each year.

d. For the year 2003, we have $t = 13$.

$$B = 0.063(13) + 10.13 \approx 10.95$$

The model predicts her best time would have been about 10.95 seconds.

e. The smallest change in Jones' best time was 0 seconds. This occurred between 2001 and 2002.

f. Answers may vary. With the use of performance enhancing drugs, we would expect her running times to decrease. However, her times have been slowly increasing.

58. **a.** Since the life expectancy for Okinawans has increased by 0.10 year each year, we can model the situation by a linear equation. The slope is 0.10 year per year. Since the life expectancy in 1995 was 77.2 years, the L-intercept is $(0, 77.2)$. So, a reasonable model is $L = 0.10t + 77.2$.

b. By the same reasoning as in part (a), the model is $L = 0.20t + 76.7$.

c. $L = 0.10t + 77.2$

$L = 0.20t + 76.7$

Substitute $0.20t + 76.7$ for L in the first equation and solve for t.

$$0.20t + 76.7 = 0.10t + 77.2$$
$$0.10t + 76.7 = 77.2$$
$$0.10t = 0.5$$
$$t = 5$$

Substitute 5 for t in the equation $L = 0.20t + 76.7$ and solve for L.

$$L = 0.20(5) + 76.7 = 77.7$$

The solution of the system is $(5, 77.7)$.

According to the models, the life expectancies were the same (77.7 years) in 2000.

59. **a.** The model appears to fit the data well.

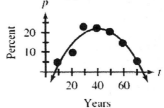

b. The graph appears to cross the t-axis at $t = 7$ and $t = 74$. So, the t-intercepts are $(7, 0)$ and $(74, 0)$. This would indicate that no children aged 7 or younger, nor adults aged 74 or older, visit online trading sites (model breakdown).

c. The vertex is approximately $(40.7, 22.0)$; this means that 22% of 41-year-old adults visit online trading sites, the highest percentage for any age group, according to the model.

d. Based on the graph, approximately 12.9% of 19-year old Americans visit online trading sites.

e. Based on the graph, roughly 18% of 26-year-old and 55-year-old Americans visit online trading sites.

60.
$$p = t^2 - 2t + 8$$
$$32 = t^2 - 2t + 8$$
$$0 = t^2 - 2t - 24$$
$$0 = (t - 6)(t + 4)$$
$$t - 6 = 0 \quad \text{or} \quad t + 4 = 0$$
$$t = 6 \qquad t = -4$$
The solutions are -4 and 6.
The model predicts that the profit was \$32 million in 2001 and will be again in 2011.

61. Let L = the length of the rug (in feet) and
W = the width (in feet).
Since the length is 8 feet more than the width, our first equation is $L = 8 + W$. Because the area is
$$A = LW$$
$$84 = LW$$
we get the following system:
$$L = 8 + W$$
$$84 = LW$$
Substitute $8 + W$ for L in the second equation and solve for W.
$$84 = (8 + W)W$$
$$84 = 8W + W^2$$
$$0 = W^2 + 8W - 84$$
$$0 = (W - 6)(W + 14)$$
$$W - 6 = 0 \quad \text{or} \quad W + 14 = 0$$
$$W = 6 \qquad W = -14$$
The width must be positive, so discard the negative solution. Substitute 6 for W in the equation
$L = 8 + W$ and solve for L.
$$L = 8 + 6 = 14$$
The rug has a length of 14 feet and a width of 6 feet.

Chapter 9
Solving Quadratic Equations

Homework 9.1

1. $\sqrt{4} = 2$, because $2^2 = 4$.

3. $\sqrt{81} = 9$, because $9^2 = 81$.

5. $\sqrt{121} = 11$, because $11^2 = 121$.

7. $\sqrt{144} = 12$, because $12^2 = 144$.

9. $\sqrt{16} = 4$, because $4^2 = 16$. Therefore, $-\sqrt{16} = -1\left(\sqrt{16}\right) = -1(4) = -4$.

11. $\sqrt{81} = 9$, because $9^2 = 81$. Therefore, $-\sqrt{81} = -1\left(\sqrt{81}\right) = -1(9) = -9$.

13. $\sqrt{-9}$ is not a real number, because the radicand -9 is negative.

15. $-\sqrt{-25}$ is not a real number, because the radicand -25 is negative.

17. The number 30 is not a perfect square, so $\sqrt{30}$ is irrational.
$\sqrt{30} \approx 5.48$

19. The number 78 is not a perfect square, so $\sqrt{78}$ is irrational.
$\sqrt{78} \approx 8.83$

21. The number 196 is a perfect square ($196 = 14^2$), so $\sqrt{196}$ is rational.
$\sqrt{196} = 14$

23. $\sqrt{20} = \sqrt{4 \cdot 5} = \sqrt{4}\sqrt{5} = 2\sqrt{5}$

25. $\sqrt{45} = \sqrt{9 \cdot 5} = \sqrt{9}\sqrt{5} = 3\sqrt{5}$

27. $\sqrt{27} = \sqrt{9 \cdot 3} = \sqrt{9}\sqrt{3} = 3\sqrt{3}$

29. $\sqrt{50} = \sqrt{25 \cdot 2} = \sqrt{25}\sqrt{2} = 5\sqrt{2}$

31. $\sqrt{300} = \sqrt{100 \cdot 3} = \sqrt{100}\sqrt{3} = 10\sqrt{3}$

33. $-\sqrt{98} = -\sqrt{49 \cdot 2} = -\sqrt{49}\sqrt{2} = -7\sqrt{2}$

35. $4\sqrt{72} = 4\sqrt{36 \cdot 2}$
$= 4\sqrt{36}\sqrt{2}$
$= 4 \cdot 6\sqrt{2}$
$= 24\sqrt{2}$

37. $3\sqrt{120} = 3\sqrt{4 \cdot 30}$
$= 3\sqrt{4}\sqrt{30}$
$= 3 \cdot 2\sqrt{30}$
$= 6\sqrt{30}$

39. $\sqrt{9x} = \sqrt{9 \cdot x} = \sqrt{9}\sqrt{x} = 3\sqrt{x}$

41. $\sqrt{64t} = \sqrt{64 \cdot t} = \sqrt{64}\sqrt{t} = 8\sqrt{t}$

43. $\sqrt{81x^2} = \sqrt{81 \cdot x^2} = \sqrt{81}\sqrt{x^2} = 9x$

45. $\sqrt{225t^2w^2} = \sqrt{225 \cdot t^2 \cdot w^2}$
$= \sqrt{225}\sqrt{t^2}\sqrt{w^2}$
$= 15tw$

47. $\sqrt{5x^2} = \sqrt{5 \cdot x^2} = \sqrt{5}\sqrt{x^2} = \left(\sqrt{5}\right)x = x\sqrt{5}$

49. $7\sqrt{39x^2y} = 7\sqrt{39 \cdot x^2 \cdot y}$
$= 7\sqrt{x^2 \cdot 39y}$
$= 7\sqrt{x^2}\sqrt{39y}$
$= 7x\sqrt{39y}$

51. $\sqrt{12p} = \sqrt{4 \cdot 3 \cdot p}$
$= \sqrt{4 \cdot 3p}$
$= \sqrt{4}\sqrt{3p}$
$= 2\sqrt{3p}$

53.
$$2\sqrt{63x} = 2\sqrt{9 \cdot 7 \cdot x}$$
$$= 2\sqrt{9 \cdot 7x}$$
$$= 2\sqrt{9}\sqrt{7x}$$
$$= 2 \cdot 3\sqrt{7x}$$
$$= 6\sqrt{7x}$$

55.
$$\sqrt{60a^2b^2} = \sqrt{4 \cdot 15 \cdot a^2 \cdot b^2}$$
$$= \sqrt{4 \cdot a^2 \cdot b^2 \cdot 15}$$
$$= \sqrt{4}\sqrt{a^2}\sqrt{b^2}\sqrt{15}$$
$$= 2ab\sqrt{15}$$

57.
$$3\sqrt{125xy^2} = 3\sqrt{25 \cdot 5 \cdot x \cdot y^2}$$
$$= 3\sqrt{25 \cdot y^2 \cdot 5x}$$
$$= 3\sqrt{25}\sqrt{y^2}\sqrt{5x}$$
$$= 3 \cdot 5y\sqrt{5x}$$
$$= 15y\sqrt{5x}$$

59. Since the motorist was traveling on asphalt, the drag factor is $F = 0.75$. The length of the skid mark is given as 210 feet, so $D = 210$.
$$S = \sqrt{30FD}$$
$$= \sqrt{30(0.75)(210)}$$
$$= \sqrt{4725}$$
$$= \sqrt{7 \cdot 5 \cdot 5 \cdot 3 \cdot 3 \cdot 3}$$
$$= \sqrt{5 \cdot 5 \cdot 3 \cdot 3 \cdot 3 \cdot 7}$$
$$= \sqrt{5 \cdot 5}\sqrt{3 \cdot 3}\sqrt{3 \cdot 7}$$
$$= 5 \cdot 3 \cdot \sqrt{21}$$
$$= 15\sqrt{21} \approx 68.7$$
The motorist was traveling at about 68.7 miles per hour before braking.

61. No, the student is not correct. $\sqrt{-25}$ is not a real number since the radicand -25 is negative. Also note that $(-5)^2 = 25 \neq -25$.

63. Since $16 < 22 < 25$, we get
$$\sqrt{16} < \sqrt{22} < \sqrt{25}$$
$$4 < \sqrt{22} < 5$$
So, $\sqrt{22}$ lies between 4 and 5.

65. Since $64 < 71 < 81$, we get
$$\sqrt{64} < \sqrt{71} < \sqrt{81}$$
$$8 < \sqrt{71} < 9$$
So, $\sqrt{71}$ lies between 8 and 9.

67. a. i. 2
   ii. 5
   iii. 8

   b. Each of the numbers 2, 5, and 8 is larger than its principle square root.

   c. i. $\sqrt{0.2}$
   ii. $\sqrt{0.5}$
   iii. $\sqrt{0.8}$

   d. Each of the numbers 0.2, 0.5, and 0.8 is smaller than its principle square root.

   e. Taking the principle square root of a number greater than 1 will result in a smaller result. Taking the principle square root of a number between 0 and 1 (not including 0 or 1) will result in a larger result.

69. No; $\sqrt{10} = \sqrt{2 \cdot 5} = \sqrt{2}\sqrt{5}$
Therefore, $\sqrt{10}$ is $\sqrt{2}$ times as big as $\sqrt{5}$.

71. $y = \sqrt{x}$

First we list some solutions of $y = \sqrt{x}$. Then we sketch a curve that contains the points corresponding to solutions.

x	$y = \sqrt{x}$	(x, y)
0	$\sqrt{0} = 0$	$(0,0)$
1	$\sqrt{1} = 1$	$(1,1)$
4	$\sqrt{4} = 2$	$(4,2)$
9	$\sqrt{9} = 3$	$(9,3)$

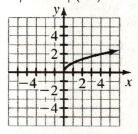

73. $(7x)^2 = 7^2 \cdot x^2 = 49x^2$

75. $\sqrt{49x^2} = \sqrt{49 \cdot x^2} = \sqrt{49}\sqrt{x^2} = 7x$

77. $y = -2x^2$

First we list some solutions of $y = -2x^2$. Then we sketch a curve that contains the points corresponding to solutions.

x	$y = -2x^2$	(x, y)
-2	$-2(-2)^2 = -8$	$(-2, -8)$
-1	$-2(-1)^2 = -2$	$(-1, -2)$
0	$-2(0)^2 = 0$	$(0, 0)$
1	$-2(1)^2 = -2$	$(1, -2)$
2	$-2(2)^2 = -8$	$(2, -8)$

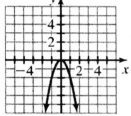

Description: *quadratic equation in two variables*

79. $\qquad x^2 = 6x - 8$

$x^2 - 6x + 8 = 0$

$(x-4)(x-2) = 0$

$x - 4 = 0 \quad \text{or} \quad x - 2 = 0$

$\qquad x = 4 \qquad\qquad x = 2$

The solutions are 2 and 4.
Description: *quadratic equation in one variable*

81. $\sqrt{68x} = \sqrt{4 \cdot 17 \cdot x}$

$\qquad = \sqrt{4 \cdot 17x}$

$\qquad = \sqrt{4}\sqrt{17x}$

$\qquad = 2\sqrt{17x}$

Description: *radical expression in one variable*

Homework 9.2

1. $\sqrt{\dfrac{25}{36}} = \dfrac{\sqrt{25}}{\sqrt{36}} = \dfrac{5}{6}$

3. $\sqrt{\dfrac{121}{x^2}} = \dfrac{\sqrt{121}}{\sqrt{x^2}} = \dfrac{11}{x}$

5. $\sqrt{\dfrac{7x}{25}} = \dfrac{\sqrt{7x}}{\sqrt{25}} = \dfrac{\sqrt{7x}}{5}$

7. $\sqrt{\dfrac{19}{a^2}} = \dfrac{\sqrt{19}}{\sqrt{a^2}} = \dfrac{\sqrt{19}}{a}$

9. $\sqrt{\dfrac{5}{x^2 y^2}} = \dfrac{\sqrt{5}}{\sqrt{x^2 y^2}} = \dfrac{\sqrt{5}}{\sqrt{x^2}\sqrt{y^2}} = \dfrac{\sqrt{5}}{xy}$

11. $-\sqrt{\dfrac{8}{49}} = -\dfrac{\sqrt{8}}{\sqrt{49}} = -\dfrac{\sqrt{4 \cdot 2}}{7} = -\dfrac{\sqrt{4}\sqrt{2}}{7} = -\dfrac{2\sqrt{2}}{7}$

13. $\sqrt{\dfrac{20}{81}} = \dfrac{\sqrt{20}}{\sqrt{81}} = \dfrac{\sqrt{4 \cdot 5}}{9} = \dfrac{\sqrt{4}\sqrt{5}}{9} = \dfrac{2\sqrt{5}}{9}$

15. $\sqrt{\dfrac{75w}{36}} = \dfrac{\sqrt{75w}}{\sqrt{36}} = \dfrac{\sqrt{25 \cdot 3w}}{6} = \dfrac{\sqrt{25}\sqrt{3w}}{6} = \dfrac{5\sqrt{3w}}{6}$

17. $\sqrt{\dfrac{4a}{b^2}} = \dfrac{\sqrt{4a}}{\sqrt{b^2}} = \dfrac{\sqrt{4}\sqrt{a}}{b} = \dfrac{2\sqrt{a}}{b}$

19. $\sqrt{\dfrac{80}{x^2}} = \dfrac{\sqrt{80}}{\sqrt{x^2}} = \dfrac{\sqrt{16 \cdot 5}}{x} = \dfrac{\sqrt{16}\sqrt{5}}{x} = \dfrac{4\sqrt{5}}{x}$

21. $\sqrt{\dfrac{7r^2 t}{81}} = \dfrac{\sqrt{7r^2 t}}{\sqrt{81}} = \dfrac{\sqrt{r^2 \cdot 7t}}{9} = \dfrac{\sqrt{r^2}\sqrt{7t}}{9} = \dfrac{r\sqrt{7t}}{9}$

23. $\dfrac{2}{\sqrt{3}} = \dfrac{2}{\sqrt{3}} \cdot \dfrac{\sqrt{3}}{\sqrt{3}} = \dfrac{2\sqrt{3}}{\sqrt{9}} = \dfrac{2\sqrt{3}}{3}$

25. $\dfrac{6}{\sqrt{5}} = \dfrac{6}{\sqrt{5}} \cdot \dfrac{\sqrt{5}}{\sqrt{5}} = \dfrac{6\sqrt{5}}{\sqrt{25}} = \dfrac{6\sqrt{5}}{5}$

27. $\dfrac{a}{\sqrt{13}} = \dfrac{a}{\sqrt{13}} \cdot \dfrac{\sqrt{13}}{\sqrt{13}} = \dfrac{a\sqrt{13}}{\sqrt{169}} = \dfrac{a\sqrt{13}}{13}$

29. $\dfrac{7}{\sqrt{x}} = \dfrac{7}{\sqrt{x}} \cdot \dfrac{\sqrt{x}}{\sqrt{x}} = \dfrac{7\sqrt{x}}{\sqrt{x^2}} = \dfrac{7\sqrt{x}}{x}$

31. $\sqrt{\dfrac{2}{7}} = \dfrac{\sqrt{2}}{\sqrt{7}} = \dfrac{\sqrt{2}}{\sqrt{7}} \cdot \dfrac{\sqrt{7}}{\sqrt{7}} = \dfrac{\sqrt{14}}{\sqrt{49}} = \dfrac{\sqrt{14}}{7}$

33. $\sqrt{\dfrac{11}{2}} = \dfrac{\sqrt{11}}{\sqrt{2}} = \dfrac{\sqrt{11}}{\sqrt{2}} \cdot \dfrac{\sqrt{2}}{\sqrt{2}} = \dfrac{\sqrt{22}}{\sqrt{4}} = \dfrac{\sqrt{22}}{2}$

35. $\sqrt{\dfrac{7}{p}} = \dfrac{\sqrt{7}}{\sqrt{p}} = \dfrac{\sqrt{7}}{\sqrt{p}} \cdot \dfrac{\sqrt{p}}{\sqrt{p}} = \dfrac{\sqrt{7p}}{\sqrt{p^2}} = \dfrac{\sqrt{7p}}{p}$

37. $\sqrt{\dfrac{3}{8}} = \dfrac{\sqrt{3}}{\sqrt{8}} = \dfrac{\sqrt{3}}{2\sqrt{2}} = \dfrac{\sqrt{3}}{2\sqrt{2}} \cdot \dfrac{\sqrt{2}}{\sqrt{2}}$

$= \dfrac{\sqrt{6}}{2\sqrt{4}} = \dfrac{\sqrt{6}}{2 \cdot 2} = \dfrac{\sqrt{6}}{4}$

39. $\sqrt{\dfrac{3x}{50}} = \dfrac{\sqrt{3x}}{\sqrt{50}} = \dfrac{\sqrt{3x}}{5\sqrt{2}} = \dfrac{\sqrt{3x}}{5\sqrt{2}} \cdot \dfrac{\sqrt{2}}{\sqrt{2}}$

$= \dfrac{\sqrt{6x}}{5\sqrt{4}} = \dfrac{\sqrt{6x}}{5 \cdot 2} = \dfrac{\sqrt{6x}}{10}$

41. $\sqrt{\dfrac{5w^2}{12}} = \dfrac{\sqrt{5w^2}}{\sqrt{12}} = \dfrac{\sqrt{w^2}\sqrt{5}}{\sqrt{4}\sqrt{3}} = \dfrac{w\sqrt{5}}{2\sqrt{3}}$

$= \dfrac{w\sqrt{5}}{2\sqrt{3}} \cdot \dfrac{\sqrt{3}}{\sqrt{3}} = \dfrac{w\sqrt{15}}{2\sqrt{9}} = \dfrac{w\sqrt{15}}{2 \cdot 3}$

$= \dfrac{w\sqrt{15}}{6}$

43. $\sqrt{\dfrac{7x^2y}{5}} = \dfrac{\sqrt{7x^2y}}{\sqrt{5}} = \dfrac{\sqrt{x^2}\sqrt{7y}}{\sqrt{5}} = \dfrac{x\sqrt{7y}}{\sqrt{5}}$

$= \dfrac{x\sqrt{7y}}{\sqrt{5}} \cdot \dfrac{\sqrt{5}}{\sqrt{5}} = \dfrac{x\sqrt{35y}}{\sqrt{25}}$

$= \dfrac{x\sqrt{35y}}{5}$

45. $\dfrac{9 + 3\sqrt{2}}{6} = \dfrac{3\left(3 + \sqrt{2}\right)}{3 \cdot 2} = \dfrac{3}{3} \cdot \dfrac{3 + \sqrt{2}}{2} = \dfrac{3 + \sqrt{2}}{2}$

47. $\dfrac{8 - 4\sqrt{7}}{4} = \dfrac{4\left(2 - \sqrt{7}\right)}{4 \cdot 1} = \dfrac{4}{4} \cdot \dfrac{2 - \sqrt{7}}{1} = 2 - \sqrt{7}$

49. $\dfrac{8 + 12\sqrt{13}}{6} = \dfrac{2\left(4 + 6\sqrt{13}\right)}{2 \cdot 3}$

$= \dfrac{2}{2} \cdot \dfrac{4 + 6\sqrt{13}}{3}$

$= \dfrac{4 + 6\sqrt{13}}{3}$

51. $\dfrac{4 + \sqrt{12}}{8} = \dfrac{4 + 2\sqrt{3}}{8}$

$= \dfrac{2\left(2 + \sqrt{3}\right)}{2 \cdot 4}$

$= \dfrac{2}{2} \cdot \dfrac{2 + \sqrt{3}}{4}$

$= \dfrac{2 + \sqrt{3}}{4}$

53. $\dfrac{10 - \sqrt{50}}{20} = \dfrac{10 - 5\sqrt{2}}{20}$

$= \dfrac{5\left(2 - \sqrt{2}\right)}{5 \cdot 4}$

$= \dfrac{5}{5} \cdot \dfrac{2 - \sqrt{2}}{4}$

$= \dfrac{2 - \sqrt{2}}{4}$

55. $\dfrac{9 - \sqrt{45}}{6} = \dfrac{9 - 3\sqrt{5}}{6}$

$= \dfrac{3\left(3 - \sqrt{5}\right)}{3 \cdot 2}$

$= \dfrac{3}{3} \cdot \dfrac{3 - \sqrt{5}}{2}$

$= \dfrac{3 - \sqrt{5}}{2}$

57. a. $T = \sqrt{\dfrac{h}{16}} = \dfrac{\sqrt{h}}{\sqrt{16}} = \dfrac{\sqrt{h}}{4}$

b. $T = \dfrac{\sqrt{1450}}{4} \approx 9.5$

It will take the baseball about 9.5 seconds to reach the ground.

59. In the second line the student is squaring the expression instead of multiplying by $1 = \dfrac{\sqrt{3}}{\sqrt{3}}$.

$\dfrac{5}{\sqrt{3}} = \dfrac{5}{\sqrt{3}} \cdot \dfrac{\sqrt{3}}{\sqrt{3}} = \dfrac{5\sqrt{3}}{\sqrt{9}} = \dfrac{5\sqrt{3}}{3}$

61. The work is correct, but it may be easier to simplify the denominator first.

$$\frac{3}{\sqrt{20}}=\frac{3}{2\sqrt5}=\frac{3}{2\sqrt5}\cdot\frac{\sqrt5}{\sqrt5}=\frac{3\sqrt5}{2\sqrt{25}}=\frac{3\sqrt5}{2\cdot5}=\frac{3\sqrt5}{10}$$

63. $12+8x=4\cdot3+4\cdot2x$
$$=4(3+2x)$$

65. $12+8\sqrt3=4\cdot3+4\cdot2\sqrt3$
$$=4(3+2\sqrt3)$$

67. $14-21x=7\cdot2-7\cdot3x$
$$=7(2-3x)$$

69. $14-21\sqrt7=7\cdot2-7\cdot3\sqrt7$
$$=7(2-3\sqrt7)$$

71. $x^2-4x=0$
$x(x-4)=0$
$x=0$ or $x-4=0$
$x=4$
The solutions are 0 and 4.
Description: *quadratic equation in one variable*

73. $x^2-4x=x\cdot x-4\cdot x$
$$=x(x-4)$$
Description: *quadratic (or second degree) polynomial in one variable*

75. $y=x^2-4x$

First we list some solutions of $y=x^2-4x$. Then we sketch a curve that contains the points corresponding to solutions.

x	$y=x^2-4x$	(x,y)
-1	$(-1)^2-4(-1)=5$	$(-1,5)$
0	$0^2-4(0)=0$	$(0,0)$
2	$2^2-4(2)=-4$	$(2,-4)$
4	$4^2-4(4)=0$	$(4,0)$
5	$5^2-4(5)=5$	$(5,5)$

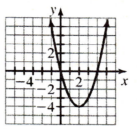

Description: *quadratic equation in two variables*

Homework 9.3

1. $x^2=4$
$x=\pm\sqrt4$
$x=\pm2$

3. $x^2=196$
$x=\pm\sqrt{196}$
$x=\pm14$

5. $x^2=0$
$x=\pm\sqrt0$
$x=0$

7. $x^2=15$
$x=\pm\sqrt{15}$

9. $x^2=20$
$x=\pm\sqrt{20}$
$x=\pm\sqrt{4\cdot5}$
$x=\pm\sqrt4\cdot\sqrt5$
$x=\pm2\sqrt5$

11. $x^2=27$
$x=\pm\sqrt{27}$
$x=\pm\sqrt{9\cdot3}$
$x=\pm\sqrt9\cdot\sqrt3$
$x=\pm3\sqrt3$

13. $x^2 = -49$

Since the square of a real number is nonnegative, we conclude that $x^2 = -49$ has no real number solution.

15. $x^2 - 28 = 0$

$$x^2 = 28$$
$$x = \pm\sqrt{28}$$
$$x = \pm\sqrt{4 \cdot 7}$$
$$x = \pm\sqrt{4} \cdot \sqrt{7}$$
$$x = \pm 2\sqrt{7}$$

17. $x^2 + 17 = 0$

$$x^2 = -17$$

Since the square of a real number is nonnegative, we conclude that $x^2 + 17 = 0$ has no real number solution.

19. $4x^2 = 5$

$$x^2 = \frac{5}{4}$$
$$x = \pm\sqrt{\frac{5}{4}}$$
$$x = \pm\frac{\sqrt{5}}{\sqrt{4}}$$
$$x = \pm\frac{\sqrt{5}}{2}$$

21. $5x^2 = 7$

$$x^2 = \frac{7}{5}$$
$$x = \pm\sqrt{\frac{7}{5}}$$
$$x = \pm\frac{\sqrt{7}}{\sqrt{5}} \cdot \frac{\sqrt{5}}{\sqrt{5}}$$
$$x = \pm\frac{\sqrt{35}}{5}$$

23. $8m^2 = 5$

$$m^2 = \frac{5}{8}$$
$$m = \pm\sqrt{\frac{5}{8}}$$
$$m = \pm\frac{\sqrt{5}}{\sqrt{8}} \cdot \frac{\sqrt{2}}{\sqrt{2}}$$
$$m = \pm\frac{\sqrt{10}}{\sqrt{16}}$$
$$m = \pm\frac{\sqrt{10}}{4}$$

25. $2x^2 + 4 = 7$

$$2x^2 = 3$$
$$x^2 = \frac{3}{2}$$
$$x = \pm\sqrt{\frac{3}{2}}$$
$$x = \pm\frac{\sqrt{3}}{\sqrt{2}} \cdot \frac{\sqrt{2}}{\sqrt{2}}$$
$$x = \pm\frac{\sqrt{6}}{2}$$

27. $5x^2 - 3 = 11$

$$5x^2 = 14$$
$$x^2 = \frac{14}{5}$$
$$x = \pm\sqrt{\frac{14}{5}}$$
$$x = \pm\frac{\sqrt{14}}{\sqrt{5}} \cdot \frac{\sqrt{5}}{\sqrt{5}}$$
$$x = \pm\frac{\sqrt{70}}{5}$$

29. $(x+2)^2 = 16$

$$x + 2 = \pm\sqrt{16}$$
$$x + 2 = \pm 4$$
$$x + 2 = -4 \quad \text{or} \quad x + 2 = 4$$
$$x = -6 \qquad\qquad x = 2$$

The solutions are -6 and 2.

31. $(p-3)^2 = 36$

$p-3 = \pm\sqrt{36}$

$p-3 = \pm 6$

$p-3 = -6 \quad \text{or} \quad p-3 = 6$

$p = -3 \qquad\qquad p = 9$

The solutions are -3 and 9.

33. $(x-7)^2 = 13$

$x-7 = \pm\sqrt{13}$

$x = 7 \pm \sqrt{13}$

The solutions are $7 - \sqrt{13}$ and $7 + \sqrt{13}$.

35. $(x+3)^2 = -16$

Since the square of a real number is nonnegative, we conclude that $(x+3)^2 = -16$ has no real number solution.

37. $(x+2)^2 = 18$

$x+2 = \pm\sqrt{18}$

$x+2 = \pm 3\sqrt{2}$

$x = -2 \pm 3\sqrt{2}$

The solutions are $-2 - 3\sqrt{2}$ and $-2 + 3\sqrt{2}$.

39. $(r-6)^2 = 24$

$r-6 = \pm\sqrt{24}$

$r-6 = \pm 2\sqrt{6}$

$r = 6 \pm 2\sqrt{6}$

The solutions are $6 - 2\sqrt{6}$ and $6 + 2\sqrt{6}$.

41. $(x-5)^2 = 0$

$x-5 = \pm\sqrt{0}$

$x-5 = 0$

$x = 5$

The solution is 5.

43.

$x^2 = 4x + 12$

$x^2 - 4x - 12 = 0$

$(x-6)(x+2) = 0$

$x-6 = 0 \quad \text{or} \quad x+2 = 0$

$x = 6 \qquad\qquad x = -2$

The solutions are -2 and 6.

45. $y^2 - 81 = 0$

$y^2 = 81$

$y = \pm\sqrt{81}$

$y = \pm 9$

The solutions are -9 and 9.

47. $(x-2)^2 = 24$

$x-2 = \pm\sqrt{24}$

$x-2 = \pm 2\sqrt{6}$

$x = 2 \pm 2\sqrt{6}$

The solutions are $2 - 2\sqrt{6}$ and $2 + 2\sqrt{6}$.

49. $3x^2 + 4 = 15$

$3x^2 = 11$

$x^2 = \dfrac{11}{3}$

$x = \pm\sqrt{\dfrac{11}{3}}$

$x = \pm\dfrac{\sqrt{11}}{\sqrt{3}} \cdot \dfrac{\sqrt{3}}{\sqrt{3}}$

$x = \pm\dfrac{\sqrt{33}}{3}$

The solutions are $-\dfrac{\sqrt{33}}{3}$ and $\dfrac{\sqrt{33}}{3}$.

51. $2x^2 - 15 = -7x$

$2x^2 + 7x - 15 = 0$

$(2x-3)(x+5) = 0$

$2x-3 = 0 \quad \text{or} \quad x+5 = 0$

$2x = 3 \qquad\qquad x = -5$

$x = \dfrac{3}{2}$

The solutions are -5 and $\dfrac{3}{2}$.

53. $(t-1)^2 = 1$

$t-1 = \pm\sqrt{1}$

$t-1 = \pm 1$

$t-1 = -1 \quad \text{or} \quad t-1 = 1$

$t = 0 \qquad\qquad t = 2$

The solutions are 0 and 2.

55. Since the lengths of the legs are given, we find the length of the hypotenuse. Substitute $a = 4$ and $b = 5$ into $a^2 + b^2 = c^2$ and solve for c.

$$4^2 + 5^2 = c^2$$
$$16 + 25 = c^2$$
$$41 = c^2$$
$$c = \sqrt{41} \quad \text{(disregard the negative)}$$

The length of the hypotenuse is $\sqrt{41}$ units (about 6.40 units).

57. The length of the hypotenuse is 9 and the length of one of the legs is 4. We substitute $a = 4$ and $c = 9$ into $a^2 + b^2 = c^2$ and solve for b.

$$4^2 + b^2 = 9^2$$
$$16 + b^2 = 81$$
$$b^2 = 65$$
$$b = \sqrt{65} \quad \text{(disregard the negative)}$$

The length of the other leg is $\sqrt{65}$ units (about 8.06 units).

59. The length of the hypotenuse is 12 and the length of one of the legs is 10. We substitute $a = 10$ and $c = 12$ into $a^2 + b^2 = c^2$ and solve for b.

$$10^2 + b^2 = 12^2$$
$$100 + b^2 = 144$$
$$b^2 = 44$$
$$b = \sqrt{44} = 2\sqrt{11} \quad \text{(disregard the negative)}$$

The length of the other side is $2\sqrt{11}$ units (about 6.63 units).

61.
$$a^2 + b^2 = c^2$$
$$6^2 + 8^2 = c^2$$
$$36 + 64 = c^2$$
$$100 = c^2$$
$$c = \sqrt{100} = 10$$

63.
$$a^2 + b^2 = c^2$$
$$2^2 + 3^2 = c^2$$
$$4 + 9 = c^2$$
$$13 = c^2$$
$$c = \sqrt{13}$$

65.
$$a^2 + b^2 = c^2$$
$$4^2 + 8^2 = c^2$$
$$16 + 64 = c^2$$
$$80 = c^2$$
$$c = \sqrt{80} = 4\sqrt{5}$$

67.
$$a^2 + b^2 = c^2$$
$$6^2 + b^2 = 10^2$$
$$36 + b^2 = 100$$
$$b^2 = 64$$
$$b = \sqrt{64} = 8$$

69.
$$a^2 + b^2 = c^2$$
$$2^2 + b^2 = 5^2$$
$$4 + b^2 = 25$$
$$b^2 = 21$$
$$b = \sqrt{21}$$

71.
$$a^2 + b^2 = c^2$$
$$a^2 + 2^2 = 7^2$$
$$a^2 + 4 = 49$$
$$a^2 = 45$$
$$a = \sqrt{45} = 3\sqrt{5}$$

73. First we define c to be the straight line distance (in miles) between the student's home and school. Then we draw a diagram that describes the situation.

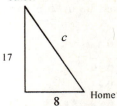

The triangle is a right triangle with legs of length 8 miles and 17 miles.

$$a^2 + b^2 = c^2$$
$$8^2 + 17^2 = c^2$$
$$64 + 289 = c^2$$
$$353 = c^2$$
$$c = \sqrt{353} \approx 18.8 \text{ miles}$$

The length of the trip would be about 18.8 miles.

75. First we define h to be the height (in feet) that the 12-foot ladder can reach. Then we draw a diagram that describes the situation.

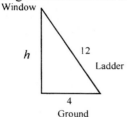

The triangle is a right triangle with one leg measuring 4 feet and a hypotenuse of 12 feet.

$$a^2 + b^2 = c^2$$
$$4^2 + b^2 = 12^2$$
$$16 + b^2 = 144$$
$$b^2 = 128$$
$$L = \sqrt{128} = 8\sqrt{2} \approx 11.3$$

The 12-foot ladder is long enough to reach the bottom of the window.

77.
$$a^2 + b^2 = c^2$$
$$12^2 + b^2 = 19^2$$
$$144 + b^2 = 361$$
$$b^2 = 217$$
$$b = \sqrt{217} \approx 14.7$$

The width of the screen is about 14.7 inches.

79.
$$a^2 + b^2 = c^2$$
$$24^2 + b^2 = 37^2$$
$$576 + b^2 = 1369$$
$$b^2 = 793$$
$$b = \sqrt{793} \approx 28.2$$

The length of the painting is about 28.2 inches.

81. Let d be the distance from point A to point B (in miles).
$$a^2 + b^2 = c^2$$
$$2^2 + 2.5^2 = d^2$$
$$4 + 6.25 = d^2$$
$$10.25 = c^2$$
$$d = \sqrt{10.25} \approx 3.2$$

The distance across the lake is about 3.2 miles.

83. 100 yards = 300 feet
53 yards 1 foot = 160 feet.
Let d be the distance the player runs (in feet).
$$a^2 + b^2 = c^2$$
$$160^2 + 300^2 = d^2$$
$$25,600 + 90,000 = d^2$$
$$115,600 = d^2$$
$$d = \sqrt{115,600} = 340$$

The player would run 340 feet (113 yards 1 foot).

85. First we define d to be the distance between Sioux Falls and Madison (in miles). Then we draw a diagram that describes the situation.

The triangle is a right triangle with a hypotenuse of length 514 miles and one leg of length 266 miles.
$$a^2 + b^2 = c^2$$
$$a^2 + 266^2 = 514^2$$
$$a^2 + 70,756 = 264,196$$
$$a^2 = 193,440$$
$$a = \sqrt{193,440} \approx 439.8$$

The distance between Sioux Falls and Madison is about 440 miles.

87. a.
$$x^2 - 36 = 0$$
$$(x-6)(x+6) = 0$$
$$x - 6 = 0 \quad \text{or} \quad x + 6 = 0$$
$$x = 6 \qquad\qquad x = -6$$
The solutions are −6 and 6.

b. $x^2 - 36 = 0$
$$x^2 = 36$$
$$x = \pm\sqrt{36}$$
$$x = \pm 6$$
The solutions are −6 and 6 .

c. The solutions are the same.

d. Answers may vary.

89. a. Yes; $(x-2)^2 = 7$

$$x-2 = \pm\sqrt{7}$$
$$x = 2 \pm \sqrt{7}$$

The solutions are $2-\sqrt{7}$ and $2+\sqrt{7}$.

b. No, $(x-2)^2 = 7$

$$x^2 - 4x + 4 = 7$$
$$x^2 - 4x - 3 = 0$$

The expression on the left side of the equation is prime, so the equation cannot be solved by factoring.

c. No; to solve by factoring, the expression being factored must be factorable over the set of integers.

91. Answers may vary.

93. a. $x^2 = 36$

$$x = \pm\sqrt{36}$$
$$x = \pm 6$$

The solutions are -6 and 6.

b. $\sqrt{36} = \sqrt{6^2} = 6$

c. Answers may vary;
In part (b) we are looking for the principle square root of 36, which is 6.

In part (a), we are looking for real numbers whose square is 36. Since $(-6)^2 = 36$ and $6^2 = 36$, there are two solutions to the equation.

95. a. Answers may vary. One possibility:

b. A: $a^2 + b^2 = c^2$

$$2^2 + 8^2 = c^2$$
$$4 + 64 = c^2$$
$$68 = c^2$$
$$c = \sqrt{68} \approx 8.25 \text{ in}$$

B: $a^2 + b^2 = c^2$

$$4^2 + 4^2 = c^2$$
$$16 + 16 = c^2$$
$$32 = c^2$$
$$c = \sqrt{32} = 4\sqrt{2} \approx 5.66 \text{ in}$$

C: $a^2 + b^2 = c^2$

$$1^2 + 16^2 = c^2$$
$$1 + 256 = c^2$$
$$257 = c^2$$
$$c = \sqrt{257} \approx 16.03 \text{ in}$$

c. The square had the smallest diagonal length.

d. *W*, *L*, and *D* are variables. The length and width can be varied to achieve the same area. The length of the diagonal varies depending on the values for the width and length of the rectangle.

e. *A* is a constant. The problem specifies that the area be 16 square inches.

97. $(3x^2 - 2)(4x^2 + x - 5)$

$$= 3x^2(4x^2 + x - 5) - 2(4x^2 + x - 5)$$
$$= 12x^4 + 3x^3 - 15x^2 - 8x^2 - 2x + 10$$
$$= 12x^4 + 3x^3 - 23x^2 - 2x + 10$$

99. $3x^4 - 12x^3 - 63x^2 = 3x^2(x^2 - 4x - 21)$

$$= 3x^2(x-7)(x+3)$$

101. $3x + 5y = 15$

x-intercept: $3x + 5(0) = 15$

$$3x = 15$$

$$x = 5$$

The x-intercept is $(5, 0)$.

y-intercept: $3(0) + 5y = 15$

$$5y = 15$$

$$y = 3$$

The y-intercept is $(0, 3)$.

Plot the intercepts and connect with a straight line, extending the line in both directions.

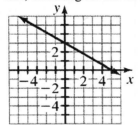

Homework 9.4

1. $x^2 + 6x + c$

We divide $b = 6$ by 2 and then square the result:

$$\left(\frac{6}{2}\right)^2 = 3^2 = 9 = c$$

The expression is $x^2 + 6x + 9$, with factored form $(x + k)^2$ where $k = \frac{b}{2} = \frac{6}{2} = 3$. Therefore,

$$x^2 + 6x + 9 = (x + 3)^2.$$

3. $x^2 + 14x + c$

We divide $b = 14$ by 2 and then square the result:

$$\left(\frac{14}{2}\right)^2 = 7^2 = 49 = c$$

The expression is $x^2 + 14x + 49$, with factored form $(x + k)^2$ where $k = \frac{b}{2} = \frac{14}{2} = 7$. Therefore,

$$x^2 + 14x + 49 = (x + 7)^2.$$

5. $x^2 + 2x + c$

We divide $b = 2$ by 2 and then square the result:

$$\left(\frac{2}{2}\right)^2 = 1^2 = 1 = c$$

The expression is $x^2 + 2x + 1$, with factored form $(x + k)^2$ where $k = \frac{b}{2} = \frac{2}{2} = 1$. Therefore,

$$x^2 + 2x + 1 = (x + 1)^2.$$

7. $x^2 - 8x + c$

We divide $b = -8$ by 2 and then square the result:

$$\left(\frac{-8}{2}\right)^2 = (-4)^2 = 16 = c$$

The expression is $x^2 - 8x + 16$, with factored form $(x + k)^2$ where $k = \frac{b}{2} = \frac{-8}{2} = -4$. Therefore, $x^2 - 8x + 16 = (x - 4)^2$.

9. $x^2 - 10x + c$

We divide $b = -10$ by 2 and then square the result:

$$\left(\frac{-10}{2}\right)^2 = (-5)^2 = 25 = c$$

The expression is $x^2 - 10x + 25$, with factored form $(x + k)^2$ where $k = \frac{b}{2} = \frac{-10}{2} - 5$. Therefore, $x^2 - 10x + 25 = (x - 5)^2$.

11. $x^2 - 20x + c$

We divide $b = -20$ by 2 and then square the result:

$$\left(\frac{-20}{2}\right)^2 = (-10)^2 = 100 = c$$

The expression is $x^2 - 20x + 100$, with factored form $(x + k)^2$ where $k = \frac{-20}{2} = -10$. Therefore,

$$x^2 - 20x + 100 = (x - 10)^2.$$

13. $x^2 + 6x = 5$

Since $\left(\dfrac{6}{2}\right)^2 = 3^2 = 9$, we add 9 to both sides of

the equation so that the left side is a perfect
square trinomial.

$x^2 + 6x + 9 = 5 + 9$

$\quad (x+3)^2 = 14$

$\qquad x + 3 = \pm\sqrt{14}$

$\qquad\quad x = -3 \pm \sqrt{14}$

The solutions are $-3 - \sqrt{14}$ and $-3 + \sqrt{14}$.

15. $x^2 + 14x = -20$

Since $\left(\dfrac{14}{2}\right)^2 = 7^2 = 49$, we add 49 to both sides

of the equation so that the left side is a perfect
square trinomial.

$x^2 + 14x + 49 = -20 + 49$

$\quad (x+7)^2 = 29$

$\qquad x + 7 = \pm\sqrt{29}$

$\qquad\quad x = -7 \pm \sqrt{29}$

The solutions are $-7 - \sqrt{29}$ and $-7 + \sqrt{29}$.

17. $x^2 - 10x = -4$

Since $\left(\dfrac{-10}{2}\right)^2 = (-5)^2 = 25$, we add 25 to both

sides of the equation so that the left side is a
perfect square trinomial.

$x^2 - 10x + 25 = -4 + 25$

$\quad (x-5)^2 = 21$

$\qquad x - 5 = \pm\sqrt{21}$

$\qquad\quad x = 5 \pm \sqrt{21}$

The solutions are $5 - \sqrt{21}$ and $5 + \sqrt{21}$.

19. $t^2 - 2t = 5$

Since $\left(\dfrac{-2}{2}\right)^2 = (-1)^2 = 1$, we add 1 to both sides

of the equation so that the left side is a perfect
square trinomial.

$t^2 - 2t + 1 = 5 + 1$

$\quad (t-1)^2 = 6$

$\qquad t - 1 = \pm\sqrt{6}$

$\qquad\quad t = 1 \pm \sqrt{6}$

The solutions are $1 - \sqrt{6}$ and $1 + \sqrt{6}$.

21. $x^2 + 12x = -4$

Since $\left(\dfrac{12}{2}\right)^2 = 6^2 = 36$, we add 36 to both sides

of the equation so that the left side is a perfect
square trinomial.

$x^2 + 12x + 36 = -4 + 36$

$\quad (x+6)^2 = 32$

$\qquad x + 6 = \pm\sqrt{32}$

$\qquad\quad x = -6 \pm 4\sqrt{2}$

The solutions are $-6 - 4\sqrt{2}$ and $-6 + 4\sqrt{2}$.

23. $x^2 - 4x = 14$

Since $\left(\dfrac{-4}{2}\right)^2 = (-2)^2 = 4$, we add 4 to both

sides of the equation so that the left side is a
perfect square trinomial.

$x^2 - 4x + 4 = 14 + 4$

$\quad (x-2)^2 = 18$

$\qquad x - 2 = \pm\sqrt{18}$

$\qquad x - 2 = \pm 3\sqrt{2}$

$\qquad\quad x = 2 \pm 3\sqrt{2}$

The solutions are $2 - 3\sqrt{2}$ and $2 + 3\sqrt{2}$.

25. $x^2 - 8x = 8$

Since $\left(\dfrac{-8}{2}\right)^2 = (-4)^2 = 16$, we add 16 to both

sides of the equation so that the left side is a
perfect square trinomial.

$x^2 - 8x + 16 = 8 + 16$

$\quad (x-4)^2 = 24$

$\qquad x - 4 = \pm\sqrt{24}$

$\qquad x - 4 = \pm 2\sqrt{6}$

$\qquad\quad x = 4 \pm 2\sqrt{6}$

The solutions are $4 - 2\sqrt{6}$ and $4 + 2\sqrt{6}$.

27. $x^2 - 16x = -70$

Since $\left(\dfrac{-16}{2}\right)^2 = (-8)^2 = 64$, we add 64 to both

sides of the equation so that the left side is a perfect square trinomial.

$x^2 - 16x + 64 = -70 + 64$

$(x-8)^2 = -6$

Since the square of a real number is always nonnegative, the equation has no real number solution.

29. $y^2 + 20y = -40$

Since $\left(\dfrac{20}{2}\right)^2 = 10^2 = 100$, we add 100 to both

sides of the equation so that the left side is a perfect square trinomial.

$y^2 + 20y + 100 = -40 + 100$

$(y+10)^2 = 60$

$y + 10 = \pm\sqrt{60}$

$y + 10 = \pm 2\sqrt{15}$

$y = -10 \pm 2\sqrt{15}$

The solutions are $-10 - 2\sqrt{15}$ and $-10 + 2\sqrt{15}$.

31. $x^2 + 6x + 1 = 0$

$x^2 + 6x = -1$

Since $\left(\dfrac{6}{2}\right)^2 = 3^2 = 9$, we add 9 to both sides of

the equation so that the left side is a perfect square trinomial.

$x^2 + 6x + 9 = -1 + 9$

$(x+3)^2 = 8$

$x + 3 = \pm\sqrt{8}$

$x + 3 = \pm 2\sqrt{2}$

$x = -3 \pm 2\sqrt{2}$

The solutions are $-3 - 2\sqrt{2}$ and $-3 + 2\sqrt{2}$.

33. $x^2 - 4x + 1 = 0$

$x^2 - 4x = -1$

Since $\left(\dfrac{-4}{2}\right)^2 = (-2)^2 = 4$, we add 4 to both sides

of the equation so that the left side is a perfect square trinomial.

$x^2 - 4x + 4 = -1 + 4$

$(x-2)^2 = 3$

$x - 2 = \pm\sqrt{3}$

$x = 2 \pm \sqrt{3}$

The solutions are $2 - \sqrt{3}$ and $2 + \sqrt{3}$.

35. $x^2 - 10x - 7 = 0$

$x^2 - 10x = 7$

Since $\left(\dfrac{-10}{2}\right)^2 = (-5)^2 = 25$, we add 25 to both

sides of the equation so that the left side is a perfect square trinomial.

$x^2 - 10x + 25 = 7 + 25$

$(x-5)^2 = 32$

$x - 5 = \pm\sqrt{32}$

$x - 5 = \pm 4\sqrt{2}$

$x = 5 \pm 4\sqrt{2}$

The solutions are $5 - 4\sqrt{2}$ and $5 + 4\sqrt{2}$.

37. $y^2 + 20y + 120 = 0$

$y^2 + 20y = -120$

Since $\left(\dfrac{20}{2}\right)^2 = 10^2 = 100$, we add 100 to both

sides of the equation so that the left side is a perfect square trinomial.

$y^2 + 20y + 100 = -120 + 100$

$(y+10)^2 = -20$

Since the square of a real number is always nonnegative, the equation has no real number solution.

39. $x^2 + 8x + 4 = 0$

$x^2 + 8x = -4$

Since $\left(\dfrac{8}{2}\right)^2 = 4^2 = 16$, we add 16 to both sides

of the equation so that the left side is a perfect square trinomial.

$x^2 + 8x + 16 = -4 + 16$

$(x + 4)^2 = 12$

$x + 4 = \pm\sqrt{12}$

$x + 4 = \pm2\sqrt{3}$

$x = -4 \pm 2\sqrt{3}$

The solutions are $-4 - 2\sqrt{3}$ and $-4 + 2\sqrt{3}$.

41. $2x^2 - 16x = 6$

$x^2 - 8x = 3$

Since $\left(\dfrac{-8}{2}\right)^2 = (-4)^2 = 16$, we add 16 to both

sides of the equation so that the left side is a perfect square trinomial.

$x^2 - 8x + 16 = 3 + 16$

$(x - 4)^2 = 19$

$x - 4 = \pm\sqrt{19}$

$x = 4 \pm \sqrt{19}$

The solutions are $4 - \sqrt{19}$ and $4 + \sqrt{19}$.

43. $5x^2 + 10x = 35$

$x^2 + 2x = 7$

Since $\left(\dfrac{2}{2}\right)^2 = 1^2 = 1$, we add 1 to both sides of

the equation so that the left side is a perfect square trinomial.

$x^2 + 2x + 1 = 7 + 1$

$(x + 1)^2 = 8$

$x + 1 = \pm\sqrt{8}$

$x + 1 = \pm2\sqrt{2}$

$x = -1 \pm 2\sqrt{2}$

The solutions are $-1 - 2\sqrt{2}$ and $-1 + 2\sqrt{2}$.

45. $3w^2 - 30w = -21$

$w^2 - 10w = -7$

Since $\left(\dfrac{-10}{2}\right)^2 = (-5)^2 = 25$, we add 25 to both

sides of the equation so that the left side is a perfect square trinomial.

$w^2 - 10w + 25 = -7 + 25$

$(w - 5)^2 = 18$

$w - 5 = \pm\sqrt{18}$

$w - 5 = \pm3\sqrt{2}$

$w = 5 \pm 3\sqrt{2}$

The solutions are $5 - 3\sqrt{2}$ and $5 + 3\sqrt{2}$.

47. $6x^2 + 12x - 6 = 0$

$6x^2 + 12x = 6$

$x^2 + 2x = 1$

Since $\left(\dfrac{2}{2}\right)^2 = 1^2 = 1$, we add 1 to both sides of

the equation so that the left side is a perfect square trinomial.

$x^2 + 2x + 1 = 1 + 1$

$(x + 1)^2 = 2$

$x + 1 = \pm\sqrt{2}$

$x = -1 \pm \sqrt{2}$

The solutions are $-1 - \sqrt{2}$ and $-1 + \sqrt{2}$.

49. $4x^2 - 24x + 4 = 0$

$4x^2 - 24x = -4$

$x^2 - 6x = -1$

Since $\left(\dfrac{-6}{2}\right)^2 = (-3)^2 = 9$, we add 9 to both sides

of the equation so that the left side is a perfect square trinomial.

$x^2 - 6x + 9 = -1 + 9$

$(x - 3)^2 = 8$

$x - 3 = \pm\sqrt{8}$

$x - 3 = \pm2\sqrt{2}$

$x = 3 \pm 2\sqrt{2}$

The solutions are $3 - 2\sqrt{2}$ and $3 + 2\sqrt{2}$.

51. $5x^2 + 20x - 20 = 0$

$$5x^2 + 20x = 20$$

$$x^2 + 4x = 4$$

Since $\left(\dfrac{4}{2}\right)^2 = 2^2 = 4$, we add 4 to both sides of

the equation so that the left side is a perfect square trinomial.

$$x^2 + 4x + 4 = 4 + 4$$

$$(x + 2)^2 = 8$$

$$x + 2 = \pm\sqrt{8}$$

$$x + 2 = \pm 2\sqrt{2}$$

$$x = -2 \pm 2\sqrt{2}$$

The solutions are $-2 - 2\sqrt{2}$ and $-2 + 2\sqrt{2}$.

53. $x^2 - 9 = 0$

$$x^2 = 9$$

$$x = \pm\sqrt{9}$$

$$x = \pm 3$$

The solutions are -3 and 3.

55. $r^2 = 11r - 30$

$$r^2 - 11r + 30 = 0$$

$$(r - 6)(r - 5) = 0$$

$$r - 6 = 0 \quad \text{or} \quad r - 5 = 0$$

$$r = 6 \qquad\qquad r = 5$$

The solutions are 5 and 6.

57. $(x - 5)^2 = 32$

$$x - 5 = \pm\sqrt{32}$$

$$x - 5 = \pm 4\sqrt{2}$$

$$x = 5 \pm 4\sqrt{2}$$

The solutions are $5 - 4\sqrt{2}$ and $5 + 4\sqrt{2}$.

59. $3x^2 + 5x = 12$

$$3x^2 + 5x - 12 = 0$$

$$(3x - 4)(x + 3) = 0$$

$$3x - 4 = 0 \quad \text{or} \quad x + 3 = 0$$

$$3x = 4 \qquad\qquad x = -3$$

$$x = \dfrac{4}{3}$$

The solutions are -3 and $\dfrac{4}{3}$.

61. $x^2 = 13$

$$x = \pm\sqrt{13}$$

The solutions are $-\sqrt{13}$ and $\sqrt{13}$.

63. $t^2 - 6t - 3 = 0$

$$t^2 - 6t = 3$$

Since $\left(\dfrac{-6}{2}\right)^2 = (-3)^2 = 9$, we add 9 to both sides

of the equation so that the left side is a perfect square trinomial.

$$t^2 - 6t + 9 = 3 + 9$$

$$(t - 3)^2 = 12$$

$$t - 3 = \pm\sqrt{12}$$

$$t - 3 = \pm 2\sqrt{3}$$

$$t = 3 \pm 2\sqrt{3}$$

The solutions are $3 - 2\sqrt{3}$ and $3 + 2\sqrt{3}$.

65. a. $x^2 - 6x + 8 = 0$

$$(x - 4)(x - 2) = 0$$

$$x - 4 = 0 \quad \text{or} \quad x - 2 = 0$$

$$x = 4 \qquad\qquad x = 2$$

The solutions are 2 and 4.

b. $x^2 - 6x + 8 = 0$

$$x^2 - 6x = -8$$

Since $\left(\dfrac{-6}{2}\right)^2 = (-3)^2 = 9$, we add 9 to both

sides of the equation so that the left side is a perfect square trinomial.

$$x^2 - 6x + 9 = -8 + 9$$

$$(x - 3)^2 = 1$$

$$x - 3 = \pm\sqrt{1}$$

$$x - 3 = \pm 1$$

$$x - 3 = -1 \quad \text{or} \quad x - 3 = 1$$

$$x = 2 \qquad\qquad x = 4$$

The solutions are 2 and 4.

c. Answers may vary.

67. a. Yes;

$x^2 + 4x = 7$

Since $\left(\dfrac{4}{2}\right)^2 = 2^2 = 4$, we add 4 to both sides

of the equation so that the left side is a perfect square trinomial.

$x^2 + 4x + 4 = 7 + 4$

$(x+2)^2 = 11$

$x + 2 = \pm\sqrt{11}$

$x = -2 \pm \sqrt{11}$

The solutions are $-2 - \sqrt{11}$ and $-2 + \sqrt{11}$.

b. No; the solutions involve radicals, so it is not possible to factor the express $x^2 + 4x - 7$ over the set of integers.

c. A quadratic equation of the form $x^2 + bx + c = 0$ can be solved by factoring if $b^2 - 4c$ is a perfect square. (see section 9.5)

69. Answers may vary.

71. $(x+7)^2 = x^2 + 2(7)x + 7^2$

$= x^2 + 14x + 49$

73. $x^2 - 16x + 64 = x^2 - 2(8)x + 8^2$

$= (x-8)^2$

75. $(x-4)^2 = x^2 - 2(4)(x) + 4^2$

$= x^2 - 8x + 16$

Description: *quadratic (or second degree) polynomial in one variable*

77. $(x-4)^2 = 3$

$x - 4 = \pm\sqrt{3}$

$x = 4 \pm \sqrt{3}$

The solutions are $4 - \sqrt{3}$ and $4 + \sqrt{3}$.

Description: *quadratic equation in one variable*

79. $p^2 - 8pq + 16q^2 = p^2 - 2(4q)p + (4q)^2$

$= (p-4q)^2$

Description: *quadratic (or second degree) polynomial in two variables*

Homework 9.5

1. $2x^2 + 5x + 3 = 0$

$a = 2, b = 5, c = 3$

$x = \dfrac{-5 \pm \sqrt{5^2 - 4(2)(3)}}{2(2)}$

$= \dfrac{-5 \pm \sqrt{25 - 24}}{4}$

$= \dfrac{-5 \pm \sqrt{1}}{4}$

$= \dfrac{-5 \pm 1}{4}$

$x = \dfrac{-5 - 1}{4}$ or $x = \dfrac{-5 + 1}{4}$

$= \dfrac{-6}{4}$ $= \dfrac{-4}{4}$

$= -\dfrac{3}{2}$ $= -1$

The solutions are $-\dfrac{3}{2}$ and -1.

3. $4x^2 + 7x + 2 = 0$

$a = 4, b = 7, c = 2$

$x = \dfrac{-7 \pm \sqrt{7^2 - 4(4)(2)}}{2(4)}$

$= \dfrac{-7 \pm \sqrt{49 - 32}}{8}$

$= \dfrac{-7 \pm \sqrt{17}}{8}$

The solutions are $\dfrac{-7 \pm \sqrt{17}}{8}$.

5. $x^2 + 3x - 5 = 0$

$a = 1, b = 3, c = -5$

$x = \dfrac{-3 \pm \sqrt{3^2 - 4(1)(-5)}}{2(1)}$

$= \dfrac{-3 \pm \sqrt{9 + 20}}{2}$

$= \dfrac{-3 \pm \sqrt{29}}{2}$

The solutions are $\dfrac{-3 \pm \sqrt{29}}{2}$.

7. $3w^2 - 5w - 3 = 0$

$a = 3, b = -5, c = -3$

$$w = \frac{-(-5) \pm \sqrt{(-5)^2 - 4(3)(-3)}}{2(3)}$$

$$= \frac{5 \pm \sqrt{25 + 36}}{6}$$

$$= \frac{5 \pm \sqrt{61}}{6}$$

The solutions are $\dfrac{5 \pm \sqrt{61}}{6}$.

9. $5x^2 + 2x - 1 = 0$

$a = 5, b = 2, c = -1$

$$x = \frac{-2 \pm \sqrt{2^2 - 4(5)(-1)}}{2(5)}$$

$$= \frac{-2 \pm \sqrt{4 + 20}}{10} = \frac{-2 \pm \sqrt{24}}{10}$$

$$= \frac{-2 \pm 2\sqrt{6}}{10} = \frac{2\left(-1 \pm \sqrt{6}\right)}{2(5)}$$

$$= \frac{-1 \pm \sqrt{6}}{5}$$

The solutions are $\dfrac{-1 \pm \sqrt{6}}{5}$.

11. $-3m^2 + 6m - 2 = 0$

$a = -3, b = 6, c = -2$

$$m = \frac{-6 \pm \sqrt{6^2 - 4(-3)(-2)}}{2(-3)}$$

$$= \frac{-6 \pm \sqrt{36 - 24}}{-6} = \frac{-6 \pm \sqrt{12}}{-6}$$

$$= \frac{-6 \pm 2\sqrt{3}}{-6} = \frac{-2\left(3 \pm \sqrt{3}\right)}{-2(3)}$$

$$= \frac{3 \pm \sqrt{3}}{3}$$

The solutions are $\dfrac{3 \pm \sqrt{3}}{3}$.

13. $4x^2 + 2x + 3 = 0$

$a = 4, b = 2, c = 3$

$$x = \frac{-2 \pm \sqrt{2^2 - 4(4)(3)}}{2(4)}$$

$$= \frac{-2 \pm \sqrt{4 - 48}}{8}$$

$$= \frac{-2 \pm \sqrt{-44}}{8}$$

Since the radicand is negative, the equation has no real number solution.

15. $2x^2 + 5x = 4$

$2x^2 + 5x - 4 = 0$

$a = 2, b = 5, c = -4$

$$x = \frac{-5 \pm \sqrt{5^2 - 4(2)(-4)}}{2(2)}$$

$$= \frac{-5 \pm \sqrt{25 + 32}}{4}$$

$$= \frac{-5 \pm \sqrt{57}}{4}$$

The solutions are $\dfrac{-5 \pm \sqrt{57}}{4}$.

17. $-r^2 = r - 1$

$0 = r^2 + r - 1$

$a = 1, b = 1, c = -1$

$$r = \frac{-1 \pm \sqrt{1^2 - 4(1)(-1)}}{2(1)}$$

$$= \frac{-1 \pm \sqrt{1 + 4}}{2}$$

$$= \frac{-1 \pm \sqrt{5}}{2}$$

The solutions are $\dfrac{-1 \pm \sqrt{5}}{2}$.

19. $5x^2 + 3 = 2x$

$5x^2 - 2x + 3 = 0$

$a = 5, b = -2, c = 3$

$$x = \frac{-(-2) \pm \sqrt{(-2)^2 - 4(5)(3)}}{2(5)}$$

$$= \frac{2 \pm \sqrt{4 - 60}}{10}$$

$$= \frac{2 \pm \sqrt{-56}}{10}$$

Since the radicand is negative, the equation has no real number solution.

21.
$$x^2 = \frac{7}{3}x - 1$$

$$x^2 - \frac{7}{3}x + 1 = 0$$

$$3\left(x^2 - \frac{7}{3}x + 1\right) = 3(0)$$

$$3x^2 - 7x + 3 = 0$$
$$a = 3, b = -7, c = 3$$

$$x = \frac{-(-7) \pm \sqrt{(-7)^2 - 4(3)(3)}}{2(3)}$$

$$= \frac{7 \pm \sqrt{49 - 36}}{6}$$

$$= \frac{7 \pm \sqrt{13}}{6}$$

The solutions are $\frac{7 \pm \sqrt{13}}{6}$.

23. $\quad 3x(3x - 1) = 1$

$$9x^2 - 3x = 1$$

$$9x^2 - 3x - 1 = 0$$
$$a = 9, b = -3, c = -1$$

$$x = \frac{-(-3) \pm \sqrt{(-3)^2 - 4(9)(-1)}}{2(9)}$$

$$= \frac{3 \pm \sqrt{9 + 36}}{18} = \frac{3 \pm \sqrt{45}}{18}$$

$$= \frac{3 \pm 3\sqrt{5}}{18} = \frac{3(1 \pm \sqrt{5})}{3(6)}$$

$$= \frac{1 \pm \sqrt{5}}{6}$$

The solutions are $\frac{1 \pm \sqrt{5}}{6}$.

25. $\quad (t + 4)(t - 2) = 3$

$$t^2 + 2t - 8 = 3$$

$$t^2 + 2t - 11 = 0$$
$$a = 1, b = 2, c = -11$$

$$t = \frac{-2 \pm \sqrt{2^2 - 4(1)(-11)}}{2(1)}$$

$$= \frac{-2 \pm \sqrt{4 + 44}}{2} = \frac{-2 \pm \sqrt{48}}{2}$$

$$= \frac{-2 \pm 4\sqrt{3}}{2} = \frac{2(-1 \pm 2\sqrt{3})}{2(1)}$$

$$= -1 \pm 2\sqrt{3}$$

The solutions are $-1 \pm 2\sqrt{3}$.

27. $\quad 4x^2 - 3x - 9 = 0$
$$a = 4, b = -3, c = -9$$

$$x = \frac{-(-3) \pm \sqrt{(-3)^2 - 4(4)(-9)}}{2(4)}$$

$$= \frac{3 \pm \sqrt{9 + 144}}{8} = \frac{3 \pm \sqrt{153}}{8}$$

$$= \frac{3 \pm 3\sqrt{17}}{8}$$

The solutions are $\frac{3 - 3\sqrt{17}}{8} \approx -1.17$ and

$\frac{3 + 3\sqrt{17}}{8} \approx 1.92$.

29. $\quad 2.1r^2 + 6.8r - 17.1 = 0$
$$a = 2.1, b = 6.8, c = -17.1$$

$$r = \frac{-6.8 \pm \sqrt{6.8^2 - 4(2.1)(-17.1)}}{2(2.1)}$$

$$= \frac{-6.8 \pm \sqrt{46.24 + 143.64}}{4.2}$$

$$= \frac{-6.8 \pm \sqrt{189.88}}{4.2}$$

The solutions are $\frac{-6.8 - \sqrt{189.88}}{4.2} \approx -4.90$ and

$\frac{-6.8 + \sqrt{189.88}}{4.2} \approx 1.66$.

31. $-1.2x^2 = 2.8x - 12.9$

$$0 = 1.2x^2 + 2.8x - 12.9$$
$$a = 1.2, b = 2.8, c = -12.9$$

$$x = \frac{-2.8 \pm \sqrt{2.8^2 - 4(1.2)(-12.9)}}{2(1.2)}$$

$$= \frac{-2.8 \pm \sqrt{7.84 + 61.92}}{2.4}$$

$$= \frac{-2.8 \pm \sqrt{69.76}}{2.4}$$

The solutions are $\dfrac{-2.8 - \sqrt{69.76}}{2.4} \approx -4.65$ and

$\dfrac{-2.8 + \sqrt{69.76}}{2.4} \approx 2.31$.

33. $0.4x^2 - 3.4x + 17.4 = 54.9$

$$0.4x^2 - 3.4x - 37.5 = 0$$
$$a = 0.4, b = -3.4, c = -37.5$$

$$x = \frac{-(-3.4) \pm \sqrt{(-3.4)^2 - 4(0.4)(-37.5)}}{2(0.4)}$$

$$= \frac{3.4 \pm \sqrt{11.56 + 60}}{0.8}$$

$$= \frac{3.4 \pm \sqrt{71.56}}{0.8}$$

The solutions are $\dfrac{3.4 - \sqrt{71.56}}{0.8} \approx -6.32$ and

$\dfrac{3.4 + \sqrt{71.56}}{0.8} \approx 14.82$.

35. $y = 2x^2 - 5x + 1$

First we substitute 0 for y. Then we use the quadratic formula to solve for x.

$$0 = 2x^2 - 5x + 1$$
$$a = 2, b = -5, c = 1$$

$$x = \frac{-(-5) \pm \sqrt{(-5)^2 - 4(2)(1)}}{2(2)}$$

$$= \frac{5 \pm \sqrt{25 - 8}}{4}$$

$$= \frac{5 \pm \sqrt{17}}{4}$$

$x \approx 0.22$ or $x \approx 2.28$

So, the x-intercepts are about $(0.22, 0)$ and

$(2.28, 0)$.

37. $y = -5x^2 + 3x + 4$

First we substitute 0 for y. Then we use the quadratic formula to solve for x.

$$0 = -5x^2 + 3x + 4$$
$$a = -5, b = 3, c = 4$$

$$x = \frac{-3 \pm \sqrt{3^2 - 4(-5)(4)}}{2(-5)}$$

$$= \frac{-3 \pm \sqrt{9 + 80}}{-10} = \frac{-3 \pm \sqrt{89}}{-10}$$

$$= \frac{-1(3 \pm \sqrt{89})}{-1(10)} = \frac{3 \pm \sqrt{89}}{10}$$

$x \approx -0.64$ or $x \approx 1.24$

So, the x-intercepts are about $(-0.64, 0)$ and

$(1.24, 0)$.

39. $y = 3.7x^2 + 5.2x - 7.5$

First we substitute 0 for y. Then we use the quadratic formula to solve for x.

$$0 = 3.7x^2 + 5.2x - 7.5$$
$$a = 3.7, b = 5.2, c = -7.5$$

$$x = \frac{-5.2 \pm \sqrt{5.2^2 - 4(3.7)(-7.5)}}{2(3.7)}$$

$$= \frac{-5.2 \pm \sqrt{27.04 + 111}}{7.4} = \frac{-5.2 \pm \sqrt{138.04}}{7.4}$$

$x \approx -2.29$ or $x \approx 0.89$

So, the x-intercepts are about $(-2.29, 0)$ and

$(0.89, 0)$.

41. $y = -2.9x^2 - 1.9x + 8.4$

First we substitute 0 for y. Then we use the quadratic formula to solve for x.

$$0 = -2.9x^2 - 1.9x + 8.4$$
$$a = -2.9, b = -1.9, c = 8.4$$

$$x = \frac{-(-1.9) \pm \sqrt{(-1.9)^2 - 4(-2.9)(8.4)}}{2(-2.9)}$$

$$= \frac{1.9 \pm \sqrt{3.61 + 97.44}}{-5.8} = \frac{1.9 \pm \sqrt{101.05}}{-5.8}$$

$x \approx 1.41$ or $x \approx -2.06$

So, the x-intercepts are about $(-2.06, 0)$ and

$(1.41, 0)$.

43.
$$x^2 + 11x = -18$$
$$x^2 + 11x + 18 = 0$$
$$(x+9)(x+2) = 0$$
$$x+9 = 0 \quad \text{or} \quad x+2 = 0$$
$$x = -9 \qquad\qquad x = -2$$
The solutions are -9 and -2.

45.
$$(x+4)^2 = 13$$
$$x+4 = \pm\sqrt{13}$$
$$x = -4 \pm \sqrt{13}$$
The solutions are $-4 \pm \sqrt{13}$.

47.
$$3r^2 + 5r = 1$$
$$3r^2 + 5r - 1 = 0$$
$$a = 3, b = 5, c = -1$$
$$r = \frac{-5 \pm \sqrt{5^2 - 4(3)(-1)}}{2(3)}$$
$$= \frac{-5 \pm \sqrt{25+12}}{6} = \frac{-5 \pm \sqrt{37}}{6}$$
The solutions are $\dfrac{-5 \pm \sqrt{37}}{6}$.

49.
$$36x^2 - 49 = 0$$
$$(6x)^2 - 7^2 = 0$$
$$(6x-7)(6x+7) = 0$$
$$6x - 7 = 0 \quad \text{or} \quad 6x + 7 = 0$$
$$6x = 7 \qquad\qquad 6x = -7$$
$$x = \frac{7}{6} \qquad\qquad x = -\frac{7}{6}$$
The solutions are $-\dfrac{7}{6}$ and $\dfrac{7}{6}$.

51.
$$14x^2 = 21x$$
$$14x^2 - 21x = 0$$
$$7x(2x-3) = 0$$
$$7x = 0 \quad \text{or} \quad 2x - 3 = 0$$
$$x = 0 \qquad\qquad 2x = 3$$
$$x = \frac{3}{2}$$
The solutions are 0 and $\dfrac{3}{2}$.

53.
$$7x^2 - 3 = 2$$
$$7x^2 = 5$$
$$x^2 = \frac{5}{7}$$
$$x = \pm\sqrt{\frac{5}{7}} = \pm\frac{\sqrt{5}}{\sqrt{7}}$$
$$= \pm\frac{\sqrt{5}}{\sqrt{7}} \cdot \frac{\sqrt{7}}{\sqrt{7}}$$
$$= \pm\frac{\sqrt{35}}{7}$$
The solutions are $\pm\dfrac{\sqrt{35}}{7}$.

55.
$$6x^2 = 7x - 2$$
$$6x^2 - 7x + 2 = 0$$
$$(3x-2)(2x-1) = 0$$
$$3x - 2 = 0 \quad \text{or} \quad 2x - 1 = 0$$
$$3x = 2 \qquad\qquad 2x = 1$$
$$x = \frac{2}{3} \qquad\qquad x = \frac{1}{2}$$
The solutions are $\dfrac{1}{2}$ and $\dfrac{2}{3}$.

57.
$$6m^2 - 2m + 5 = 0$$
$$a = 6, b = -2, c = 5$$
$$m = \frac{-(-2) \pm \sqrt{(-2)^2 - 4(6)(5)}}{2(6)}$$
$$= \frac{2 \pm \sqrt{4-120}}{12} = \frac{2 \pm \sqrt{-116}}{12}$$
Since the radicand is negative, the equation has no real number solutions.

59.
$$\frac{7}{2}x^2 = x + \frac{3}{2}$$
$$2 \cdot \frac{7}{2}x^2 = 2\left(x + \frac{3}{2}\right)$$
$$7x^2 = 2x + 3$$
$$7x^2 - 2x - 3 = 0$$
$$a = 7, b = -2, c = -3$$

$$x = \frac{-(-2) \pm \sqrt{(-2)^2 - 4(7)(-3)}}{2(7)}$$

$$= \frac{2 \pm \sqrt{4 + 84}}{14} = \frac{2 \pm \sqrt{88}}{14}$$

$$= \frac{2 \pm 2\sqrt{22}}{14} = \frac{2(1 \pm \sqrt{22})}{2(7)}$$

$$= \frac{1 \pm \sqrt{22}}{7}$$

The solutions are $\dfrac{1 \pm \sqrt{22}}{7}$.

$$(w+3)^2 - 2w = 7$$
$$w^2 + 6w + 9 - 2w = 7$$

61. $\qquad w^2 + 4w + 2 = 0$

$$a = 1, b = 4, c = 2$$

$$w = \frac{-4 \pm \sqrt{4^2 - 4(1)(2)}}{2(1)}$$

$$= \frac{-4 \pm \sqrt{16 - 8}}{2} = \frac{-4 \pm \sqrt{8}}{2}$$

$$= \frac{-4 \pm 2\sqrt{2}}{2} = \frac{2(-2 \pm \sqrt{2})}{2(1)}$$

$$= -2 \pm \sqrt{2}$$

The solutions are $-2 \pm \sqrt{2}$.

63. $-x^2 + 2x + 3 = 0$

The graphs of $y = -x^2 + 2x + 3$ and $y = 0$ intersect only at $(-1, 0)$ and $(3, 0)$ whose x-coordinates are $x = -1$ and $x = 3$. Therefore, -1 and 3 are the solutions to the equation.

65. $-x^2 + 2x + 3 = 4$

The graphs of $y = -x^2 + 2x + 3$ and $y = 4$ intersect only at $(1, 4)$ whose x-coordinate is $x = 1$. Therefore, 1 is the solution to the equation.

67. $x^2 + 6x + 7 = 2$

The graphs of $y = x^2 + 6x + 7$ and $y = 2$ intersect only at $(-1, 2)$ and $(-5, 2)$ whose x-coordinates are $x = -1$ and $x = -5$. Therefore, -1 and -5 are the solutions to the equation.

69. $x^2 + 6x + 7 = -4$

The graphs of $y = x^2 + 6x + 7$ and $y = -4$ never intersect. Therefore, the equation has no real number solution.

71. $x^2 - 5 = 2x - 1$

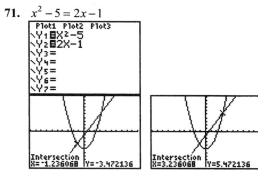

The intersection points are approximately $(-1.24, -3.47)$ and $(3.24, 5.47)$, whose x-coordinates are -1.24 and 3.24. Therefore, -1.24 and 3.24 are the approximate solutions to the equation.

73. $2x^2 - 5x + 1 = -x^2 + 3x + 3$

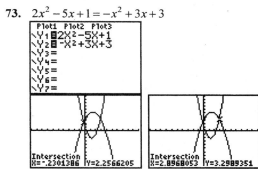

The intersection points are approximately $(-0.23, 2.26)$ and $(2.90, 3.30)$, whose x-coordinates are -0.23 and 2.90. Therefore, -0.23 and 2.90 are the approximate solutions to the equation.

75. The inputs $x = -2$ and $x = 4$ yield the output $y = -12$. Therefore, -2 and 4 are the solutions to the equation $-2x^2 + 4x + 4 = -12$.

77. The input $x = 1$ yields the output $y = 6$. Therefore, 1 is the solution to the equation $-2x^2 + 4x + 4 = 6$.

79. The student applied the quadratic formula before writing the quadratic equation in the general form $ax^2 + bx + c = 0$.

$$x^2 + 5x = 3$$

$$x^2 + 5x - 3 = 0$$

$$a = 1, b = 5, c = -3$$

$$x = \frac{-5 \pm \sqrt{5^2 - 4(1)(-3)}}{2(1)}$$

$$= \frac{-5 \pm \sqrt{25 + 12}}{2}$$

$$= \frac{-5 \pm \sqrt{37}}{2}$$

The solutions are $\dfrac{-5 \pm \sqrt{37}}{2}$.

81. a. $x^2 + 2x - 3 = 0$

$$(x + 3)(x - 1) = 0$$

$$x + 3 = 0 \quad \text{or} \quad x - 1 = 0$$

$$x = -3 \qquad\qquad x = 1$$

The solutions are -3 and 1.

b. $x^2 + 2x - 3 = 0$

$$a = 1, b = 2, c = -3$$

$$x = \frac{-2 \pm \sqrt{2^2 - 4(1)(-3)}}{2(1)}$$

$$= \frac{-2 \pm \sqrt{4 + 12}}{2} = \frac{-2 \pm \sqrt{16}}{2}$$

$$= \frac{-2 \pm 4}{2}$$

$$x = \frac{-2 - 4}{2} \quad \text{or} \quad x = \frac{-2 + 4}{2}$$

$$= \frac{-6}{2} \qquad\qquad = \frac{2}{2}$$

$$= -3 \qquad\qquad = 1$$

The solutions are -3 and 1.

c. $x^2 + 2x - 3 = 0$

$$x^2 + 2x = 3$$

Since $\left(\dfrac{2}{2}\right)^2 = 1^2 = 1$, we add 1 to both sides

of the equation to create a perfect square trinomial on the left side.

$$x^2 + 2x + 1 = 3 + 1$$

$$(x + 1)^2 = 4$$

$$x + 1 = \pm\sqrt{4}$$

$$x + 1 = \pm 2$$

$$x = -1 \pm 2$$

$$x = -1 - 2 \quad \text{or} \quad x = -1 + 2$$

$$= -3 \qquad\qquad = 1$$

The solutions are -3 and 1.

d. The results are the same.

e. Answers may vary.

83. Answers may vary.

85. $2x - 6 = 1$

$$2x = 7$$

$$x = \frac{7}{2}$$

The solution is $\dfrac{7}{2}$.

87. $2x^2 - 6 = 1$

$$2x^2 = 7$$

$$x^2 = \frac{7}{2}$$

$$x = \pm\sqrt{\frac{7}{2}} = \pm\frac{\sqrt{7}}{\sqrt{2}}$$

$$= \pm\frac{\sqrt{7}}{\sqrt{2}} \cdot \frac{\sqrt{2}}{\sqrt{2}} = \pm\frac{\sqrt{14}}{2}$$

The solutions are $\pm\dfrac{\sqrt{14}}{2}$.

89. $2x^2 - 6x = 1$

$$2x^2 - 6x - 1 = 0$$

$$a = 2, b = -6, c = -1$$

$$x = \frac{-(-6) \pm \sqrt{(-6)^2 - 4(2)(-1)}}{2(2)}$$

$$= \frac{6 \pm \sqrt{36 + 8}}{4} = \frac{6 \pm \sqrt{44}}{4}$$

$$= \frac{6 \pm 2\sqrt{11}}{4} = \frac{2(3 \pm \sqrt{11})}{2(2)}$$

$$= \frac{3 \pm \sqrt{11}}{2}$$

The solutions are $\dfrac{3 \pm \sqrt{11}}{2}$.

91. $y = 3x^2 - 2x - 4$

First we substitute 0 for y. Then we use the quadratic formula to solve for x.

$0 = 3x^2 - 2x - 4$

$a = 3, b = -2, c = -4$

$x = \dfrac{-(-2) \pm \sqrt{(-2)^2 - 4(3)(-4)}}{2(3)}$

$= \dfrac{2 \pm \sqrt{4 + 48}}{6} = \dfrac{2 \pm \sqrt{52}}{6}$

$= \dfrac{2 \pm 2\sqrt{13}}{6} = \dfrac{2\left(1 \pm \sqrt{13}\right)}{2(3)}$

$= \dfrac{1 \pm \sqrt{13}}{3}$

$x \approx -0.87$ or $x \approx 1.54$

So, the x-intercepts are about $(-0.87, 0)$ and $(1.54, 0)$.

93. $y = 3x - 2$

First we substitute 0 for y, then we solve for x.

$0 = 3x - 2$

$3x = 2$

$x = \dfrac{2}{3} \approx 0.67$

So, the x-intercept is about $(0.67, 0)$.

95.
$$x^2 = 5 - 3x$$

$x^2 + 3x - 5 = 0$

$a = 1, b = 3, c = -5$

$x = \dfrac{-3 \pm \sqrt{3^2 - 4(1)(-5)}}{2(1)}$

$= \dfrac{-3 \pm \sqrt{9 + 20}}{2}$

$= \dfrac{-3 \pm \sqrt{29}}{2}$

The solutions are $\dfrac{-3 \pm \sqrt{29}}{2}$.

Description: *quadratic equation in one variable*

97. $6x^2 - x - 1$

The leading coefficient is not 1, so we try factoring by grouping. $a \cdot c = 6(-1) = -6$, so we need two numbers whose product is -6 and

whose sum is -1. Since $-3(2) = -6$ and $-3 + 2 = -1$, we get

$6x^2 - x - 1 = 6x^2 - 3x + 2x - 1$

$= 3x(2x - 1) + 1(2x - 1)$

$= (3x + 1)(2x - 1)$

Description: *quadratic (or second degree) polynomial in one variable*

99. $3x - 4(7x - 5) + 3 = 3x - 28x + 20 + 3$

$= -25x + 23$

Description: *linear expression in one variable.*

Homework 9.6

1. a.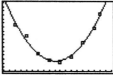

The graph appears to fit the data well.

b. Substitute 30 for r and solve for t.

$30 = 0.94t^2 - 28.5t + 228.8$

$0 = 0.94t^2 - 28.5t + 198.8$

$a = 0.94, b = -28.5, c = 198.8$

$t = \dfrac{-(-28.5) \pm \sqrt{(-28.5)^2 - 4(0.94)(198.8)}}{2(0.94)}$

$= \dfrac{28.5 \pm \sqrt{64.762}}{1.88}$

$t \approx 10.88$ or $t \approx 19.44$

The model predicts that the show's rating was 30 in 1991 and 1999.

c.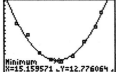

The vertex is approximately $(15.16, 12.78)$ so the model predicts that the show had its best ranking in 1995 with a rank of 13. Actually, the show had a better rank in 1996 when it had a rank of 11.

d. Substitute 24 for *t* and solve for *r*.

$$r = 0.94(24)^2 - 28.5(24) + 228.8$$

$$\approx 86$$

The model predicts that the show had a ranking of 86 in 2004. Barbara Walters likely left the show because of the low ranking and continued decline in the rankings.

3. a.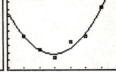

The model appears to fit the data well.

b.

The vertex is about $(9.90, 154.76)$, so the model predicts that the number of bank robberies in 2000 was 155, the lowest number of any year. (in reality the number was 136 in 2000)

c. Substitute 600 for *n* and solve for *t*.

$$600 = 25.9t^2 - 513t + 2695$$

$$0 = 25.9t^2 - 513t + 2095$$

$$a = 25.9, b = -513, c = 2095$$

$$t = \frac{-(-513) \pm \sqrt{(-513)^2 - 4(25.9)(2095)}}{2(25.9)}$$

$$= \frac{513 \pm \sqrt{46,127}}{51.8}$$

$$t \approx 5.76 \quad \text{or} \quad t \approx 14.05$$

The model predicts that the number of bank robberies in New York City was 600 in 1996 and 2004.

5. a.

The model appears to fit the data very well.

b. Substitute 23 for *d* and solve for *p*.

$$p = -0.00141(23)^2 + 0.34(23) + 69.5$$

$$\approx 76.57$$

The model predicts that about 76.6% of Internet users whose household income is $23,000 use email/instant messaging.

c. Substitute 80 for *p* and solve for *d*.

$$80 = -0.00141d^2 + 0.34d + 69.5$$

$$0 = -0.00141d^2 + 0.34d - 10.5$$

$$d = \frac{-0.34 \pm \sqrt{0.34^2 - 4(-0.00141)(-10.5)}}{2(-0.00141)}$$

$$= \frac{0.34 \pm \sqrt{0.05638}}{0.00282}$$

$$d \approx 36.4 \quad \text{or} \quad d \approx 204.8$$

The model predicts that **80%** of Internet users with household incomes of about $36,400 or $204,800 use email/instant messaging.

7. a.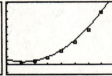

The model appears to fit the data very well.

b. For 2011, we have *t* = 21. Substitute 21 for *t* and solve for *v*.

$$v = 0.98(21)^2 - 16.1(21) + 68 \approx 162.1$$

The model predicts that the total value of goods in 2011 will be about $162.1 billion.

c. Subsitute 115 for *v* and solve for *t*.

$$115 = 0.98t^2 - 16.1t + 68$$

$$0 = 0.98t^2 - 16.1t - 47$$

$$a = 0.98, b = -16.1, c = -47$$

$$t = \frac{-(-16.1) \pm \sqrt{(-16.1)^2 - 4(0.98)(-47)}}{2(0.98)}$$

$$= \frac{16.1 \pm \sqrt{443.45}}{1.96}$$

$$t \approx -2.53 \quad \text{or} \quad t \approx 18.96$$

The model predicts that the total value of goods will be $115 billion in 2009. (we discard the other solution since it yields a year prior to 1996 when eBay began operating).

9. a.
```
WINDOW
 Xmin=-2
 Xmax=14
 Xscl=1
 Ymin=40
 Ymax=50
 Yscl=1
 Xres=1
```

The quadratic model appears to fit the data better.

b. For 1996 we have $t = 6$.

Linear:
$p = 0.45(6) + 41.54 = 44.24$

Quadratic:
$p = 0.044(6)^2 - 0.08(6) + 42.4$
$= 43.504$

Compared to the actual value for 1996, the quadratic model is better because the estimate is closer to the actual value of 43.5.

c. Linear:
$p = 0.45(0) + 41.54 = 41.54$
The p-intercept is $(0, 41.54)$.

Quadratic:
$p = 0.044(0)^2 - 0.08(0) + 42.4 = 42.4$
The p-intercept is $(0, 42.4)$

The p-intercept indicates the percentage in 1990. The quadratic model yields an estimate that is closer to the actual value of 42.2, so the quadratic model appears to be a better fit.

d. Subsitute 50 for p and solve for t using the quadratic model.
$50 = 0.044t^2 - 0.08t + 42.4$
$0 = 0.044t^2 - 0.08t - 7.6$
$a = 0.044, b = -0.08, c = -7.6$

$t = \dfrac{-(-0.08) \pm \sqrt{(-0.08)^2 - 4(0.044)(-7.6)}}{2(0.044)}$

$= \dfrac{0.08 \pm \sqrt{1.344}}{0.088}$

$t \approx -12.26$ or $t \approx 14.08$
The model predicts that women earned half of all law degrees in 2004 (model breakdown occurs for $t \approx -12.26$).

11. a.
```
WINDOW
 Xmin=-1
 Xmax=11
 Xscl=1
 Ymin=4
 Ymax=14
 Yscl=1
 Xres=1
```

The model appears to fit the data well.

b. For 1999, we have $t = 9$. Substitute 9 for t and solve for A.
$A = 0.073(9)^2 - 0.06(9) + 5.6$
≈ 10.97
The model predicts that corporations donated $10.97 billion in 1999.

$\dfrac{10.97}{190.2} \approx 0.0577$

Corporations accounted for about 5.77% of all donations in 1999.

c. Substitute 35 for A and solve for t.
$35 = 0.073t^2 - 0.06t + 5.6$
$0 = 0.073t^2 - 0.06t - 29.4$
$a = 0.073, b = -0.06, c = -29.4$

$t = \dfrac{-(-0.06) \pm \sqrt{(-0.06)^2 - 4(0.073)(-29.4)}}{2(0.073)}$

$= \dfrac{0.06 \pm \sqrt{8.5884}}{0.146}$

$t \approx -19.66$ or $t \approx 20.48$
The model predicts that corporations will donate $35 billion in 2010 (model breakdown occurs for $t \approx -19.66$).

13. a.
```
WINDOW
 Xmin=9
 Xmax=21
 Xscl=1
 Ymin=3
 Ymax=60
 Yscl=3
 Xres=1
```

The model appears to fit the data very well.

b. Substitute 45 for I and solve for t.
$45 = 0.175t^2 - 1.1t + 3.8$
$0 = 0.175t^2 - 1.1t - 41.2$
$a = 0.175, b = -1.1, c = -41.2$

$t = \dfrac{-(-1.1) \pm \sqrt{(-1.1)^2 - 4(0.175)(-41.2)}}{2(0.175)}$

$= \dfrac{1.1 \pm \sqrt{30.05}}{0.35}$

$t \approx -12.52$ or $t \approx 18.81$

The model predicts that people whose median annual income is \$45 thousand have about 19 years of full-time equivalent education (model breakdown occurs for $t \approx -12.52$ because the number of years of education cannot be negative).

c. i. With only a high school diploma, the median income is \$15.7 thousand.

$4(15.7) = 62.8$

So, over 4 years the student would earn an estimated \$62,800.

ii. Each year of college is estimated to cost \$4281.

$4(4281) = 17,124$

So, four years of college would cost the student \$17,124.

iii. Let x = the number of years the student would work after college. Then the high school graduate would work for $x + 4$ years.

We need to find when the total earnings for a high school graduate would equal the total earnings of a college graduate, less the cost of college.

$15.7(x + 4) = 31.0x - 17,124$

$15.7x + 62.8 = 31.0x - 17,124$

$79,924 = 15.3x$

$x = \dfrac{79,924}{15.3} \approx 5.2$

The student would need to work for about 5.2 years after college to be in the same financial position as if he had not gone to college.

iv. High School:
$37(15.7) = 580.9$

Bachelor's:
$33(31.0) - 17.124 = 1005.876$

$1005.876 - 580.9 = 424.976$
So, the student will earn \$424,976 (73%) more in his lifetime if he gets a Bachelor's degree than if he does not go to college.

v. Underestimate; our estimate is based on the assumption that the difference in income remains constant. Since those with larger incomes saw larger growth in income, the difference in income will actually increase.

15. Answers may vary.

17. a.

The data appear to be fairly linear.

Using the Regression feature of a graphing calculator, we find a reasonable model to be $n = 1.96t + 6.51$.

b. Substitute 40 for n and solve for t.

$40 = 1.96t + 6.51$

$33.49 = 1.96t$

$t = \dfrac{33.49}{1.96} \approx 17.09$

The model predicts that 40 thousand students were given extra time on the SAT in 2007.

c. For 2011, we have $t = 21$. Substitute 21 for t and solve for n.

$n = 1.96(21) + 6.51 = 41.16 + 6.51 = 47.67$

The model predicts that about 47.7 thousand students will be given extra time on the SAT in 2011.

19. a. Since percent of farms that owned or leased a computer has increased by 4 percentage points each year, we can model the situation by a linear equation. The slope is 4 percentage points per year. Since the percent that owned or leased computers in 2003 was 54%, the p-intercept is $(0, 54)$. So, a reasonable model is $p = 4t + 54$.

b. Substitute 7 for *t* (for the year 2010) and solve for *p*.

$$p = 4(7) + 54 = 28 + 54 = 82$$

The model predicts that 82% of farms in 2010 will own or lease computers.

c. Substitute 100 for *p* and solve for *t*.
$$100 = 4t + 54$$
$$46 = 4t$$
$$11.5 = t$$
The model predicts that all farms will own or lease computers in 2015.

21. Since the relationship is linear, we can compute the slope.
$$m = \frac{5.2 - 2.0}{2003 - 1986} = \frac{3.2}{17}$$
If we let *x* = the number of years since 1986, the *y*-intercept is $(0, 2.0)$. Therefore, the equation of the linear relationship is $y = \frac{3.2}{17}x + 2.0$.

For 2012, we have $x = 26$. Substitute 26 for *x* and solve for *y*.

$$y = \frac{3.2}{17}(26) + 2.0 \approx 6.9$$

The model predicts that 6.9% of radio listeners in 2012 will tune in to public radio.

23. – 31. Answers may vary.

Chapter 9 Review Exercises

1. $\sqrt{196} = 14$ because $14^2 = 196$

2. $\sqrt{64} = 8$ because $8^2 = 64$. Therefore,
$$-\sqrt{64} = -1\left(\sqrt{64}\right) = -1(8) = -8$$

3. $\sqrt{-25}$ is not a real number because the radicand, -25, is negative.

4. $-\sqrt{-81}$ is not a real number because the radicand, -81, is negative.

5. $\sqrt{95} \approx 9.75$

6. $-7.29\sqrt{38.36} \approx -45.15$

7. $\sqrt{84} = \sqrt{4 \cdot 21} = \sqrt{4}\sqrt{21} = 2\sqrt{21}$

8. $\sqrt{36x} = \sqrt{36 \cdot x} = \sqrt{36}\sqrt{x} = 6\sqrt{x}$

9. $\sqrt{18ab^2} = \sqrt{9b^2 \cdot 2a} = \sqrt{9b^2}\sqrt{2a} = 3b\sqrt{2a}$

10. $\sqrt{98m^2n^2} = \sqrt{49m^2n^2 \cdot 2}$
$$= \sqrt{49m^2n^2}\sqrt{2}$$
$$= 7mn\sqrt{2}$$

11. $36 < 39 < 49$ so
$$\sqrt{36} < \sqrt{39} < \sqrt{49}$$
$$6 < \sqrt{39} < 7$$
So, $\sqrt{39}$ is between 6 and 7.

12. $\sqrt{\frac{5}{8}} = \frac{\sqrt{5}}{\sqrt{8}} = \frac{\sqrt{5}}{2\sqrt{2}} = \frac{\sqrt{5}}{2\sqrt{2}} \cdot \frac{\sqrt{2}}{\sqrt{2}} = \frac{\sqrt{10}}{2 \cdot 2} = \frac{\sqrt{10}}{4}$

13. $\sqrt{\frac{50y}{x^2}} = \frac{\sqrt{50y}}{\sqrt{x^2}} = \frac{\sqrt{25 \cdot 2y}}{x} = \frac{\sqrt{25}\sqrt{2y}}{x} = \frac{5\sqrt{2y}}{x}$

14. $\frac{4}{\sqrt{x}} = \frac{4}{\sqrt{x}} \cdot \frac{\sqrt{x}}{\sqrt{x}} = \frac{4\sqrt{x}}{\sqrt{x^2}} = \frac{4\sqrt{x}}{x}$

15. $\sqrt{\frac{5b^2}{32}} = \frac{\sqrt{5b^2}}{\sqrt{32}} = \frac{\sqrt{b^2}\sqrt{5}}{\sqrt{16}\sqrt{2}} = \frac{b\sqrt{5}}{4\sqrt{2}}$
$$= \frac{b\sqrt{5}}{4\sqrt{2}} \cdot \frac{\sqrt{2}}{\sqrt{2}} = \frac{b\sqrt{10}}{4 \cdot 2}$$
$$= \frac{b\sqrt{10}}{8}$$

16. The student squared the numerator and denominator instead of multiplying the expression by $1 = \frac{\sqrt{7}}{\sqrt{7}}$ to rationalize the denominator.
$$\frac{3}{\sqrt{7}} = \frac{3}{\sqrt{7}} \cdot \frac{\sqrt{7}}{\sqrt{7}} = \frac{3\sqrt{7}}{\sqrt{49}} = \frac{3\sqrt{7}}{7}$$

17. $p^2 = 45$
$$p = \pm\sqrt{45}$$
$$p = \pm 3\sqrt{5}$$
The solutions are $\pm 3\sqrt{5}$.

18. $5t^2 = 7$

$$t^2 = \frac{7}{5}$$

$$t = \pm\sqrt{\frac{7}{5}} = \pm\frac{\sqrt{7}}{\sqrt{5}} = \pm\frac{\sqrt{7}}{\sqrt{5}} \cdot \frac{\sqrt{5}}{\sqrt{5}}$$

$$= \pm\frac{\sqrt{35}}{5}$$

The solutions are $\pm\dfrac{\sqrt{35}}{5}$.

19. $5x^2 + 4 = 12$

$$5x^2 = 8$$

$$x^2 = \frac{8}{5}$$

$$x = \pm\sqrt{\frac{8}{5}} = \pm\frac{\sqrt{8}}{\sqrt{5}} = \pm\frac{\sqrt{4}\sqrt{2}}{\sqrt{5}}$$

$$= \pm\frac{2\sqrt{2}}{\sqrt{5}} = \pm\frac{2\sqrt{2}}{\sqrt{5}} \cdot \frac{\sqrt{5}}{\sqrt{5}} =$$

$$= \pm\frac{2\sqrt{10}}{5}$$

The solutions are $\pm\dfrac{2\sqrt{10}}{5}$.

20. $(x+4)^2 = 27$

$$x+4 = \pm\sqrt{27}$$

$$x+4 = \pm3\sqrt{3}$$

$$x = -4 \pm 3\sqrt{3}$$

The solutions are $-4 \pm 3\sqrt{3}$.

21. $(w-6)^2 = -15$

Since the square of any real number is nonnegative, this equation has no real number solution.

22. $(m-8)^2 = 0$

$$m-8 = 0$$

$$m = 8$$

The solution is 8.

23. $a^2 + b^2 = c^2$

$$4^2 + 8^2 = c^2$$

$$16 + 64 = c^2$$

$$80 = c^2$$

$$c = \sqrt{80} = 4\sqrt{5}$$

24. $a^2 + b^2 = c^2$

$$a^2 + 3^2 = 6^2$$

$$a^2 + 9 = 36$$

$$a^2 = 27$$

$$a = \sqrt{27} = 3\sqrt{3}$$

25. $a^2 + b^2 = c^2$

$$3^2 + 7^2 = c^2$$

$$9 + 49 = c^2$$

$$58 = c^2$$

$$c = \sqrt{58}$$

26. $a^2 + b^2 = c^2$

$$a^2 + 5^2 = 9^2$$

$$a^2 + 25 = 81$$

$$a^2 = 56$$

$$a = \sqrt{56} = 2\sqrt{14}$$

27. First we define h to be the height (in feet) that the 12-foot ladder can reach. Then we draw a diagram that describes the situation.

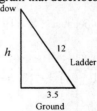

The triangle is a right triangle with one leg measuring 3.5 feet and a hypotenuse of 12 feet.

$$a^2 + b^2 = c^2$$

$$3.5^2 + b^2 = 12^2$$

$$12.25 + b^2 = 144$$

$$b^2 = 131.75$$

$$L = \sqrt{131.75} \approx 11.5$$

The 12-foot ladder is long enough to reach the bottom of the window.

28. $x^2 + 6x = 2$

Since $\left(\dfrac{6}{2}\right)^2 = 3^2 = 9$, we add 9 to both sides of

the equation so that the left side is a perfect square trinomial.
$$x^2 + 6x + 9 = 2 + 9$$
$$(x+3)^2 = 11$$
$$x + 3 = \pm\sqrt{11}$$
$$x = -3 \pm \sqrt{11}$$
The solutions are $-3 \pm \sqrt{11}$.

29. $x^2 + 2x = 17$

Since $\left(\dfrac{2}{2}\right)^2 = 1^2 = 1$, we add 1 to both sides of

the equation so that the left side is a perfect square trinomial.
$$x^2 + 2x + 1 = 17 + 1$$
$$(x+1)^2 = 18$$
$$x + 1 = \pm\sqrt{18}$$
$$x + 1 = \pm 3\sqrt{2}$$
$$x = -1 \pm 3\sqrt{2}$$
The solutions are $-1 \pm 3\sqrt{2}$.

30. $w^2 - 8w - 4 = 0$
$$w^2 - 8w = 4$$

Since $\left(\dfrac{-8}{2}\right)^2 = (-4)^2 = 16$, we add 16 to both

sides of the equation so that the left side is a perfect square trinomial.
$$w^2 - 8w + 16 = 4 + 16$$
$$(w-4)^2 = 20$$
$$w - 4 = \pm\sqrt{20}$$
$$w - 4 = \pm 2\sqrt{5}$$
$$w = 4 \pm 2\sqrt{5}$$
The solutions are $4 \pm 2\sqrt{5}$.

31. $t^2 - 12t + 40 = 0$
$$t^2 - 12t = -40$$

Since $\left(\dfrac{-12}{2}\right)^2 = (-6)^2 = 36$, we add 36 to both

sides of the equation so that the left side is a perfect square trinomial.
$$t^2 - 12t + 36 = -40 + 36$$
$$(t-6)^2 = -4$$

Since the square of any real number is nonnegative, this equation has no real number solution.

32. $2x^2 + 8x = 12$
$$x^2 + 4x = 6$$

Since $\left(\dfrac{4}{2}\right)^2 = 2^2 = 4$, we add 4 to both sides of

the equation so that the left side is a perfect square trinomial.
$$x^2 + 4x + 4 = 12 + 4$$
$$(x+2)^2 = 16$$
$$x + 2 = \pm\sqrt{16}$$
$$x + 2 = \pm 4$$
$$x = -2 \pm 4$$
$$x = -2 - 4 \quad \text{or} \quad x = -2 + 4$$
$$= -6 \qquad\qquad x = 2$$
The solutions are -6 and 2.

33. $3x^2 - 18x - 27 = 0$
$$x^2 - 6x - 9 = 0$$
$$x^2 - 6x = 9$$

Since $\left(\dfrac{-6}{2}\right)^2 = (-3)^2 = 9$, we add 9 to both

sides of the equation so that the left side is a perfect square trinomial.
$$x^2 - 6x + 9 = 9 + 9$$
$$(x-3)^2 = 18$$
$$x - 3 = \pm\sqrt{18}$$
$$x - 3 = \pm 3\sqrt{2}$$
$$x = 3 \pm 3\sqrt{2}$$
The solutions are $3 \pm 3\sqrt{2}$.

34. $3x^2 + 7x + 1 = 0$
$a = 3, b = 7, c = 1$

$x = \dfrac{-7 \pm \sqrt{7^2 - 4(3)(1)}}{2(3)}$

$= \dfrac{-7 \pm \sqrt{49 - 12}}{6}$

$= \dfrac{-7 \pm \sqrt{37}}{6}$

The solutions are $\dfrac{-7 \pm \sqrt{37}}{6}$.

35. $2x^2 - 5x - 4 = 0$
$a = 2, b = -5, c = -4$

$x = \dfrac{-(-5) \pm \sqrt{(-5)^2 - 4(2)(-4)}}{2(2)}$

$= \dfrac{5 \pm \sqrt{25 + 32}}{4}$

$= \dfrac{5 \pm \sqrt{57}}{4}$

The solutions are $\dfrac{5 \pm \sqrt{57}}{4}$.

36. $-5y^2 = 3 - 2y$
$5y^2 - 2y + 3 = 0$
$a = 5, b = -2, c = 3$

$y = \dfrac{-(-2) \pm \sqrt{(-2)^2 - 4(5)(3)}}{2(5)}$

$= \dfrac{2 \pm \sqrt{4 - 60}}{10}$

$= \dfrac{2 \pm \sqrt{-56}}{10}$

Since the radicand is negative, the equation has no real number solution.

37. $-p^2 + 3p = -5$
$p^2 - 3p - 5 = 0$
$a = 1, b = -3, c = -5$

$p = \dfrac{-(-3) \pm \sqrt{(-3)^2 - 4(1)(-5)}}{2(1)}$

$= \dfrac{3 \pm \sqrt{9 + 20}}{2} = \dfrac{3 \pm \sqrt{29}}{2}$

The solutions are $\dfrac{3 \pm \sqrt{29}}{2}$.

38. $2x^2 - x = \dfrac{3}{2}$

$2(2x^2 - x) = 2\left(\dfrac{3}{2}\right)$

$4x^2 - 2x = 3$
$4x^2 - 2x - 3 = 0$
$a = 4, b = -2, c = -3$

$x = \dfrac{-(-2) \pm \sqrt{(-2)^2 - 4(4)(-3)}}{2(4)}$

$= \dfrac{2 \pm \sqrt{4 + 48}}{8} = \dfrac{2 \pm \sqrt{52}}{8}$

$= \dfrac{2 \pm 2\sqrt{13}}{8} = \dfrac{2(1 \pm \sqrt{13})}{2(4)}$

$= \dfrac{1 \pm \sqrt{13}}{4}$

The solutions are $\dfrac{1 \pm \sqrt{13}}{4}$.

39. $2x(x - 1) = 5$
$2x^2 - 2x = 5$
$2x^2 - 2x - 5 = 0$
$a = 2, b = -2, c = -5$

$x = \dfrac{-(-2) \pm \sqrt{(-2)^2 - 4(2)(-5)}}{2(2)}$

$= \dfrac{2 \pm \sqrt{4 + 40}}{4} = \dfrac{2 \pm \sqrt{44}}{4}$

$= \dfrac{2 \pm 2\sqrt{11}}{4} = \dfrac{2(1 \pm \sqrt{11})}{2(2)}$

$= \dfrac{1 \pm \sqrt{11}}{2}$

The solutions are $\dfrac{1 \pm \sqrt{11}}{2}$.

40.
$$6w^2 = 8w + 5$$
$$6w^2 - 8w - 5 = 0$$
$$a = 6, b = -8, c = -5$$
$$w = \frac{-(-8) \pm \sqrt{(-8)^2 - 4(6)(-5)}}{2(6)}$$
$$= \frac{8 \pm \sqrt{64 + 120}}{12} = \frac{8 \pm \sqrt{184}}{12}$$
$$= \frac{8 \pm 2\sqrt{46}}{12} = \frac{2(4 \pm \sqrt{46})}{2(6)}$$
$$= \frac{4 \pm \sqrt{46}}{6}$$
$$w \approx -0.46 \quad \text{or} \quad w \approx 1.80$$
The solutions are approximately −0.46 and 1.80.

41.
$$-1.9t^2 - 5.4t + 27.9 = 14.1$$
$$-1.9t^2 - 5.4t + 13.8 = 0$$
$$1.9t^2 + 5.4t - 13.8 = 0$$
$$a = 1.9, b = 5.4, c = -13.8$$
$$t = \frac{-5.4 \pm \sqrt{5.4^2 - 4(1.9)(-13.8)}}{2(1.9)}$$
$$= \frac{-5.4 \pm \sqrt{29.16 + 104.88}}{3.8}$$
$$= \frac{-5.4 \pm \sqrt{134.04}}{3.8}$$
$$t \approx -4.47 \quad \text{or} \quad t \approx 1.63$$
The solutions are approximately −4.47 and 1.63.

42. In the second line, the student should have the entire right side divided by 2(2).
$$2x^2 + 3x - 7 = 0$$
$$a = 2, b = 3, c = -7$$
$$x = \frac{-3 \pm \sqrt{3^2 - 4(2)(-7)}}{2(2)}$$
$$= \frac{-3 \pm \sqrt{9 + 56}}{4} = \frac{-3 \pm \sqrt{65}}{4}$$
The solutions are $\frac{-3 \pm \sqrt{65}}{4}$.

43.
$$x^2 = 5x + 14$$
$$x^2 - 5x - 14 = 0$$
$$(x - 7)(x + 2) = 0$$
$$x - 7 = 0 \quad \text{or} \quad x + 2 = 0$$
$$x = 7 \qquad x = -2$$
The solutions are −2 and 7.

44.
$$3x^2 - 4x = 1$$
$$3x^2 - 4x - 1 = 0$$
$$a = 3, b = -4, c = -1$$
$$x = \frac{-(-4) \pm \sqrt{(-4)^2 - 4(3)(-1)}}{2(3)}$$
$$= \frac{4 \pm \sqrt{16 + 12}}{6} = \frac{4 \pm \sqrt{28}}{6}$$
$$= \frac{4 \pm 2\sqrt{7}}{6} = \frac{2(2 \pm \sqrt{7})}{2(3)}$$
$$= \frac{2 \pm \sqrt{7}}{3}$$
The solutions are $\frac{2 \pm \sqrt{7}}{3}$.

45.
$$r^2 + 13 = 0$$
$$r^2 = -13$$
Since the square of any real number is nonnegative, the equation has no real number solution.

46.
$$(t - 4)^2 = 24$$
$$t - 4 = \pm\sqrt{24}$$
$$t - 4 = \pm 2\sqrt{6}$$
$$t = 4 \pm 2\sqrt{6}$$
The solutions are $4 \pm 2\sqrt{6}$.

47.
$$x^2 = 2x + 35$$
$$x^2 - 2x - 35 = 0$$
$$(x - 7)(x + 5) = 0$$
$$x - 7 = 0 \quad \text{or} \quad x + 5 = 0$$
$$x = 7 \qquad x = -5$$
The solutions are −5 and 7.

48. $5x^2 + 8 = 20$

$$5x^2 = 12$$

$$x^2 = \frac{12}{5}$$

$$x = \pm\sqrt{\frac{12}{5}} = \pm\frac{\sqrt{12}}{\sqrt{5}} = \pm\frac{2\sqrt{3}}{\sqrt{5}}$$

$$= \pm\frac{2\sqrt{3}}{\sqrt{5}}\cdot\frac{\sqrt{5}}{\sqrt{5}} = \pm\frac{2\sqrt{15}}{5}$$

The solutions are $\pm\dfrac{2\sqrt{15}}{5}$.

49. $(x-2)^2 - 3x = 1$

$$x^2 - 4x + 4 - 3x = 1$$

$$x^2 - 7x + 3 = 0$$

$$a = 1, b = -7, c = 3$$

$$x = \frac{-(-7)\pm\sqrt{(-7)^2 - 4(1)(3)}}{2(1)}$$

$$= \frac{7\pm\sqrt{49-12}}{2}$$

$$= \frac{7\pm\sqrt{37}}{2}$$

The solutions are $\dfrac{7\pm\sqrt{37}}{2}$.

50. $3x^2 + 2x - 8 = 0$

$$(3x-4)(x+2) = 0$$

$$3x - 4 = 0 \quad\text{or}\quad x + 2 = 0$$

$$3x = 4 \qquad\qquad x = -2$$

$$x = \frac{4}{3}$$

The solutions are -2 and $\dfrac{4}{3}$.

51. $(w+7)^2 = 27$

$$w + 7 = \pm\sqrt{27}$$

$$w + 7 = \pm 3\sqrt{3}$$

$$w = -7 \pm 3\sqrt{3}$$

The solutions are $-7 \pm 3\sqrt{3}$.

52. $-2p^2 = 7p - 3$

$$2p^2 + 7p - 3 = 0$$

$$a = 2, b = 7, c = -3$$

$$p = \frac{-7\pm\sqrt{7^2 - 4(2)(-3)}}{2(2)}$$

$$= \frac{-7\pm\sqrt{49+24}}{4}$$

$$= \frac{-7\pm\sqrt{73}}{4}$$

The solutions are $\dfrac{-7\pm\sqrt{73}}{4}$.

53. $x^2 - 2x = \dfrac{5}{2}$

$$2\left(x^2 - 2x\right) = 2\left(\frac{5}{2}\right)$$

$$2x^2 - 4x = 5$$

$$2x^2 - 4x - 5 = 0$$

$$a = 2, b = -4, c = -5$$

$$x = \frac{-(-4)\pm\sqrt{(-4)^2 - 4(2)(-5)}}{2(2)}$$

$$= \frac{4\pm\sqrt{16+40}}{4} = \frac{4\pm\sqrt{56}}{4}$$

$$= \frac{4\pm 2\sqrt{14}}{4} = \frac{2\left(2\pm\sqrt{14}\right)}{2(2)}$$

$$= \frac{2\pm\sqrt{14}}{2}$$

The solutions are $\dfrac{2\pm\sqrt{14}}{2}$.

54. $4x(x-2) = -3$

$$4x^2 - 8x = -3$$

$$4x^2 - 8x + 3 = 0$$

$$(2x-1)(2x-3) = 0$$

$$2x - 1 = 0 \quad\text{or}\quad 2x - 3 = 0$$

$$2x = 1 \qquad\qquad 2x = 3$$

$$x = \frac{1}{2} \qquad\qquad x = \frac{3}{2}$$

The solutions are $\dfrac{1}{2}$ and $\dfrac{3}{2}$.

55. $\dfrac{1}{2}x^2 - 2x - 1 = -3$

The graphs of $y = \dfrac{1}{2}x^2 - 2x - 1$ and $y = -3$

intersect only at $(2, -3)$ whose x-coordinate is $x = 2$. Therefore, 2 is the solution to the equation.

56. $\dfrac{1}{2}x^2 - 2x - 1 = -4$

The graphs of $y = \dfrac{1}{2}x^2 - 2x - 1$ and $y = -4$ do not intersect. Therefore, the equation has no real solution.

57. $\dfrac{1}{2}x^2 - 2x - 1 = -1$

The graphs of $y = \dfrac{1}{2}x^2 - 2x - 1$ and $y = -1$

intersect only at $(0, -1)$ and $(4, -1)$ whose x-coordinates are $x = 0$ and $x = 4$. Therefore, 0 and 4 are the solutions to the equation.

58. $\dfrac{1}{2}x^2 - 2x - 1 = 5$

The graphs of $y = \dfrac{1}{2}x^2 - 2x - 1$ and $y = 5$

intersect only at $(-2, 5)$ and $(6, 5)$ whose x-coordinates are $x = -2$ and $x = 6$. Therefore, -2 and 6 are the solutions to the equation.

59. $7 - 2x^2 = x^2 - x - 8$

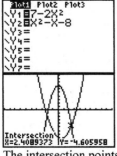

The intersection points are approximately $(-2.08, -1.62)$ and $(2.41, -4.61)$, whose x-coordinates are -2.08 and 2.41. Therefore, -2.08 and 2.41 are the approximate solutions to the equation.

60. $y = 3x^2 - 8x + 2$

First we substitute 0 for y. Then we use the quadratic formula to solve for x.

$0 = 3x^2 - 8x + 2$

$a = 3, b = -8, c = 2$

$x = \dfrac{-(-8) \pm \sqrt{(-8)^2 - 4(3)(2)}}{2(3)}$

$\quad = \dfrac{8 \pm \sqrt{64 - 24}}{6} = \dfrac{8 \pm \sqrt{40}}{6}$

$\quad = \dfrac{8 \pm 2\sqrt{10}}{6} = \dfrac{2\left(4 \pm \sqrt{10}\right)}{2(3)}$

$\quad = \dfrac{4 \pm \sqrt{10}}{3}$

$x \approx 0.28$ or $x \approx 2.39$

So, the x-intercepts are about $(0.28, 0)$ and $(2.39, 0)$.

61. $y = -3.9x^2 + 7.1x + 54.9$

$0 = -3.9x^2 + 7.1x + 54.9$

$a = -3.9, b = 7.1, c = 54.9$

$x = \dfrac{-7.1 \pm \sqrt{7.1^2 - 4(-3.9)(54.9)}}{2(-3.9)}$

$\quad = \dfrac{-7.1 \pm \sqrt{906.85}}{-7.8}$

$\quad = \dfrac{7.1 \pm \sqrt{906.85}}{7.8}$

$x \approx -2.95$ or $x \approx 4.77$

So, the x-intercepts are about $(-2.95, 0)$ and $(4.77, 0)$.

62. a. $\quad x^2 - 2x - 15 = 0$

$\quad (x + 3)(x - 5) = 0$

$\quad x + 3 = 0 \quad$ or $\quad x - 5 = 0$

$\quad\quad x = -3 \quad\quad\quad\quad x = 5$

The solutions are -3 and 5.

b. $x^2 - 2x - 15 = 0$

$a = 1, b = -2, c = -15$

$$x = \frac{-(-2) \pm \sqrt{(-2)^2 - 4(1)(-15)}}{2(1)}$$

$$= \frac{2 \pm \sqrt{4 + 60}}{2} = \frac{2 \pm \sqrt{64}}{2}$$

$$= \frac{2 \pm 8}{2}$$

$$x = \frac{2 - 8}{2} \quad \text{or} \quad x = \frac{2 + 8}{2}$$

$$= \frac{-6}{2} \qquad\qquad = \frac{10}{2}$$

$$= -3 \qquad\qquad\quad = 5$$

The solutions are -3 and 5.

c. $x^2 - 2x - 15 = 0$

$$x^2 - 2x = 15$$

Since $\left(\dfrac{-2}{2}\right)^2 = (-1)^2 = 1$, we add 1 to both

sides of the equation.

$$x^2 - 2x + 1 = 15 + 1$$

$$(x - 1)^2 = 16$$

$$x - 1 = \pm\sqrt{16}$$

$$x - 1 = \pm 4$$

$$x = 1 \pm 4$$

$$x = 1 - 4 = -3 \quad \text{or} \quad x = 1 + 4 = 5$$

The solutions are -3 and 5.

d. The answers are the same.

63. a.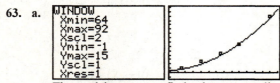

The model appears to fit the data very well.

b. Substitute 70 for a and solve for p.

$$p = 0.016(70)^2 - 2(70) + 63$$

$$= 1.4$$

The model predicts that 1.4% of Americans at age 70 have moderate or severe memory impairment.

c. $y = 0.016A^2 - 2A + 63$

Substitute 10 for p and solve for A.

$$10 = 0.016A^2 - 2A + 63$$

$$0 = 0.016A^2 - 2A + 53$$

$$a = 0.016, b = -2, c = 53$$

$$A = \frac{-(-2) \pm \sqrt{(-2)^2 - 4(0.016)(53)}}{2(0.016)}$$

$$= \frac{2 \pm \sqrt{0.608}}{0.032}$$

$A \approx 38.13$ or $A \approx 86.87$

The model predicts that 10% of Americans at age 87 will have moderate or sever memory impairment. (model breakdown occurs for $A \approx 38.13$ since this implies the percent with memory impairment is higher for young adults than for older adults)

64. a.

The model appears to fit the data very well.

b. For 2010, we have $t = 40$. Substitute 40 for t and solve for p.

$$p = -0.0176(40)^2 + 1.71(40) + 7.3 = 47.54$$

The model predicts that about 47.5% of medical degrees awarded in 2010 will be earned by women.

c. Substitute 48 for p and solve for t.

$$48 = -0.0176t^2 + 1.71t + 7.3$$

$$0.0176t^2 - 1.71t + 40.7 = 0$$

$$a = 0.0176, b = -1.71, c = 40.7$$

$$t = \frac{-(-1.71) \pm \sqrt{(-1.71)^2 - 4(0.0176)(40.7)}}{2(0.0176)}$$

$$= \frac{1.71 \pm \sqrt{0.05882}}{0.0352}$$

$t \approx 41.69$ or $t \approx 55.47$

The model predicts that women will be awarded 48% of medical degrees earned in 2012. (model breakdown occurs for $t = 55.47$ because, while the percent for women may peak, it is not likely that the percent will continually decrease once the peak is reached)

Chapter 9 Test

1. $\sqrt{121} = 11$ because $11^2 = 121$
   Therefore, $-\sqrt{121} = -1 \cdot \sqrt{121} = -1 \cdot 11 = -11$.

2. $-\sqrt{-9}$ is not a real number because the radicand is negative.

3. $\sqrt{48} = \sqrt{16 \cdot 3} = \sqrt{16}\sqrt{3} = 4\sqrt{3}$

4. $\sqrt{64x} = \sqrt{64 \cdot x} = \sqrt{64}\sqrt{x} = 8\sqrt{x}$

5. $\sqrt{45a^2b} = \sqrt{9a^2 \cdot 5b} = \sqrt{9a^2}\sqrt{5b} = 3a\sqrt{5b}$

6. $\dfrac{3}{\sqrt{7}} = \dfrac{3}{\sqrt{7}} \cdot \dfrac{\sqrt{7}}{\sqrt{7}} = \dfrac{3\sqrt{7}}{7}$

7. $\sqrt{\dfrac{2m^2}{n}} = \dfrac{\sqrt{2m^2}}{\sqrt{n}} = \dfrac{\sqrt{m^2}\sqrt{2}}{\sqrt{n}} = \dfrac{m\sqrt{2}}{\sqrt{n}}$
   $= \dfrac{m\sqrt{2}}{\sqrt{n}} \cdot \dfrac{\sqrt{n}}{\sqrt{n}} = \dfrac{m\sqrt{2n}}{n}$

8. $\sqrt{\dfrac{3}{20}} = \dfrac{\sqrt{3}}{\sqrt{20}} = \dfrac{\sqrt{3}}{2\sqrt{5}} = \dfrac{\sqrt{3}}{2\sqrt{5}} \cdot \dfrac{\sqrt{5}}{\sqrt{5}} = \dfrac{\sqrt{15}}{2 \cdot 5} = \dfrac{\sqrt{15}}{10}$

9. $a^2 + b^2 = c^2$
   $4^2 + 12^2 = c^2$
   $16 + 144 = c^2$
   $160 = c^2$
   $\sqrt{160} = c$
   $4\sqrt{10} = c$
   The hypotenuse has a length of $4\sqrt{10}$ units.

10. $a^2 + b^2 = c^2$
    $11^2 + b^2 = 17^2$
    $121 + b^2 = 289$
    $b^2 = 168$
    $b = \sqrt{168}$
    $b = 2\sqrt{42} \approx 12.96$
    The length of the picture frame is about 12.96 inches.

11. $3x^2 = 5$
    $x^2 = \dfrac{5}{3}$
    $x = \pm\sqrt{\dfrac{5}{3}} = \pm\dfrac{\sqrt{5}}{\sqrt{3}} = \pm\dfrac{\sqrt{5}}{\sqrt{3}} \cdot \dfrac{\sqrt{3}}{\sqrt{3}} = \pm\dfrac{\sqrt{15}}{3}$
    The solutions are $\pm\dfrac{\sqrt{15}}{3}$.

12. $-5x^2 + 4 = -28$
    $-5x^2 = -32$
    $x^2 = \dfrac{32}{5}$
    $x = \pm\sqrt{\dfrac{32}{5}} = \pm\dfrac{\sqrt{32}}{\sqrt{5}} = \pm\dfrac{4\sqrt{2}}{\sqrt{5}}$
    $= \pm\dfrac{4\sqrt{2}}{\sqrt{5}} \cdot \dfrac{\sqrt{5}}{\sqrt{5}} = \pm\dfrac{4\sqrt{10}}{5}$
    The solutions are $\pm\dfrac{4\sqrt{10}}{5}$.

13. $(t-5)^2 = 18$
    $t - 5 = \pm\sqrt{18}$
    $t - 5 = \pm 3\sqrt{2}$
    $t = 5 \pm 3\sqrt{2}$
    The solutions are $5 \pm 3\sqrt{2}$.

14. $2w^2 + 3w - 6 = 0$
    $a = 2, b = 3, c = -6$
    $w = \dfrac{-3 \pm \sqrt{3^2 - 4(2)(-6)}}{2(2)}$
    $= \dfrac{-3 \pm \sqrt{9 + 48}}{4}$
    $= \dfrac{-3 \pm \sqrt{57}}{4}$
    The solutions are $\dfrac{-3 \pm \sqrt{57}}{4}$.

15.
$$x(x-3)=40$$
$$x^2-3x=40$$
$$x^2-3x-40=0$$
$$(x-8)(x+5)=0$$
$$x-8=0 \quad \text{or} \quad x+5=0$$
$$x=8 \qquad\qquad x=-5$$
The solutions are -5 and 8.

16.
$$\frac{3}{2}x^2=x+2$$
$$2\left(\frac{3}{2}x^2\right)=2(x+2)$$
$$3x^2=2x+4$$
$$3x^2-2x-4=0$$
$$a=3,b=-2,c=-4$$
$$x=\frac{-(-2)\pm\sqrt{(-2)^2-4(3)(-4)}}{2(3)}$$
$$=\frac{2\pm\sqrt{4+48}}{6}=\frac{2\pm\sqrt{52}}{6}$$
$$=\frac{2\pm2\sqrt{13}}{6}=\frac{2\left(1\pm\sqrt{13}\right)}{2(3)}$$
$$=\frac{1\pm\sqrt{13}}{3}$$
The solutions are $\dfrac{1\pm\sqrt{13}}{3}$.

17. $5p^2-20p-35=0$
$$p^2-4p-7=0$$
$$p^2-4p=7$$
Since $\left(\dfrac{-4}{2}\right)^2=(-2)^2=4$, we add 4 to both sides

of the equation so that the left side is a perfect square trinomial.
$$p^2-4p+4=7+4$$
$$(p-2)^2=11$$
$$p-2=\pm\sqrt{11}$$
$$p=2\pm\sqrt{11}$$
The solutions are $2\pm\sqrt{11}$.

18. $1.4x^2-2.3x-38.5=7.4$
$$1.4x^2-2.3x-45.9=0$$
$$a=1.4,b=-2.3,c=-45.9$$
$$x=\frac{-(-2.3)\pm\sqrt{(-2.3)^2-4(1.4)(-45.9)}}{2(1.4)}$$
$$=\frac{2.3\pm\sqrt{5.29+257.04}}{2.8}$$
$$=\frac{2.3\pm\sqrt{262.33}}{2.8}$$
$$x\approx-4.96 \quad \text{or} \quad x\approx6.61$$
The approximate solutions are -4.96 and 6.61.

19. $y=-1.2x^2+37.9x-50.4$
Substitute 0 for y and solve for x.
$$0=-1.2x^2+37.9x-50.4$$
$$a=-1.2,b=37.9,c=-50.4$$
$$x=\frac{-37.9\pm\sqrt{37.9^2-4(-1.2)(-50.4)}}{2(-1.2)}$$
$$=\frac{-37.9\pm\sqrt{1194.49}}{-2.4}$$
$$x\approx30.19 \quad \text{or} \quad x\approx1.39$$
The approximate solutions are 30.19 and 1.39, so the approximate x-intercepts are $(1.39,0)$ and $(30.19,0)$.

20. The inputs $x=2$ and $x=4$ yield the output $y=13$. Therefore, 2 and 4 are the solutions to the equation $-x^2+6x+5=13$.

21. The inputs $x=1$ and $x=5$ yield the output $y=10$. Therefore, 1 and 5 are the solution to the equation $-x^2+6x+5=10$.

22. The input $x=3$ yields the output $y=14$. Therefore, 3 is the solution to the equation $-x^2+6x+5=14$.

23. The graph of $y=-x^2+6x+5$ reaches a maximum value of $y=14$ when $x=3$. There are no inputs that will yield $y=15$ so the equation $-x^2+6x+5=15$ has no real number solution.

24. In line 5, the student did not correctly cancel like factors. The student cancelled the 4 in the denominator with a factor of 4 in the first term of the numerator, but the factor does not occur in every term of the numerator as required.

$$2x^2 + 4x - 3 = 0$$

$$x = \frac{-4 \pm \sqrt{4^2 - 4(2)(-3)}}{2(2)}$$

$$x = \frac{-4 \pm \sqrt{16 + 24}}{4}$$

$$x = \frac{-4 \pm \sqrt{40}}{4}$$

$$x = \frac{-4 \pm 2\sqrt{10}}{4}$$

$$x = \frac{2(-2 \pm \sqrt{10})}{2(2)}$$

$$x = \frac{-2 \pm \sqrt{10}}{2}$$

The solutions are $\dfrac{-2 \pm \sqrt{10}}{2}$.

25. a.

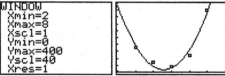

The model appears to fit the data very well.

b.

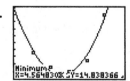

The vertex is approximately $(4.56, 14.84)$.

This means that the Walton's donations to education charities were approximately $14.8 million in 2000, the least of any year. This is an underestimate since the minimum amount was $35 million which occurred in 2000.

c. For 2003, we have $t = 8$. Substitute 8 for t and solve for d.

$$d = 56.3(8)^2 - 514(8) + 1188 = 679.2$$

The model predicts that total donations in 2003 by the Walton family to education charities will be approximately $679 million.

d. One billion is 1,000 million. Substitute 1000 for d and solve for t.

$$1000 = 56.3t^2 - 514t + 1188$$

$$0 = 56.3t^2 - 514t + 188$$

$$a = 56.3, b = -514, c = 188$$

$$t = \frac{-(-514) \pm \sqrt{(-514)^2 - 4(56.3)(188)}}{2(56.3)}$$

$$= \frac{514 \pm \sqrt{221858.4}}{112.6}$$

$$t \approx 0.38 \quad \text{or} \quad t \approx 8.75$$

The model predicts that the Walton family donated $1 billion to education charities in 2004.

Chapter 10
Rational Expressions and Equations

Homework 10.1

1. $\dfrac{7}{x}$

The number 0 is an excluded value because $\dfrac{7}{0}$ has division by 0. No other value of x leads to division by 0, so 0 is the only excluded value.

3. $\dfrac{x}{4}$

The denominator is a constant 4, so no substitution for x will lead to a division by 0. Thus, there are no excluded values.

5. $\dfrac{3}{x-4}$

The number 4 is an excluded value because $\dfrac{3}{4-4}$ has division by 0. No other value of x leads to division by 0, so 4 is the only excluded value.

7. $-\dfrac{x}{x+9}$

The number -9 is an excluded value because $-\dfrac{9}{-9+9}$ has division by 0. No other value of x leads to division by 0, so -9 is the only excluded value.

9. $\dfrac{2p}{3p-12}$

We set the denominator equal to 0 and solve for p.
$$3p-12=0$$
$$3p=12$$
$$p=4$$
So, the only excluded value is 4.

11. $\dfrac{x-4}{6x+8}$

We set the denominator equal to 0 and solve for x.
$$6x+8=0$$
$$6x=-8$$
$$x=\dfrac{-8}{6}=-\dfrac{4}{3}$$
So, the only excluded value is $-\dfrac{4}{3}$.

13. $\dfrac{2}{x^2+5x+6}$

We set the denominator equal to 0 and solve for x.
$$x^2+5x+6=0$$
$$(x+3)(x+2)=0$$
$$x+3=0 \quad \text{or} \quad x+2=0$$
$$x=-3 \quad \text{or} \quad x=-2$$
So, the excluded values are -3 and -2.

15. $\dfrac{r}{r^2-2r-35}$

We set the denominator equal to 0 and solve for r.
$$r^2-2r-35=0$$
$$(r+5)(r-7)=0$$
$$r+5=0 \quad \text{or} \quad r-7=0$$
$$r=-5 \quad \text{or} \quad r=7$$
So, the excluded values are -5 and 7.

17. $\dfrac{x-9}{x^2-10x+25}$

We set the denominator equal to 0 and solve for x.
$$x^2-10x+25=0$$
$$(x-5)(x-5)=0$$
$$x-5=0 \quad \text{or} \quad x-5=0$$
$$x=5 \quad \text{or} \quad x=5$$
So, the only excluded value is 5.

19. $\dfrac{3x-1}{x^2-16}$

We set the denominator equal to 0 and solve for x.
$$x^2-16=0$$
$$(x+4)(x-4)=0$$
$$x+4=0 \quad \text{or} \quad x-4=0$$
$$x=-4 \quad \text{or} \quad x=4$$
So, the excluded values are -4 and 4.

21. $\dfrac{c+4}{25c^2-49}$

We set the denominator equal to 0 and solve for c.

$$25c^2-49=0$$
$$(5c+7)(5c-7)=0$$
$$5c+7=0 \quad \text{or} \quad 5c-7=0$$
$$5c=-7 \quad \text{or} \quad 5c=7$$
$$c=-\frac{7}{5} \quad \text{or} \quad c=\frac{7}{5}$$

So, the excluded values are $-\dfrac{7}{5}$ and $\dfrac{7}{5}$.

23. $\dfrac{2x-5}{2x^2+13x+15}$

We set the denominator equal to 0 and solve for x.
$$2x^2+13x+15=0$$
$$(2x+3)(x+5)=0$$
$$2x+3=0 \quad \text{or} \quad x+5=0$$
$$2x=-3 \quad \text{or} \quad x=-5$$
$$x=-\frac{3}{2}$$

So, the excluded values are -5 and $-\dfrac{3}{2}$.

25. $\dfrac{3x}{6x^2-13x+6}$

We set the denominator equal to 0 and solve for x.
$$6x^2-13x+6=0$$
$$(3x-2)(2x-3)=0$$
$$3x-2=0 \quad \text{or} \quad 2x-3=0$$
$$3x=2 \quad \text{or} \quad 2x=3$$
$$x=\frac{2}{3} \quad \text{or} \quad x=\frac{3}{2}$$

So, the excluded values are $\dfrac{2}{3}$ and $\dfrac{3}{2}$.

27. $\dfrac{w-5}{w^3+2w^2-4w-8}$

We set the denominator equal to 0 and solve for w.
$$w^3+2w^2-4w-8=0$$
$$w^2(w+2)-4(w+2)=0$$
$$(w+2)(w^2-4)=0$$
$$(w+2)(w+2)(w-2)=0$$
$$w+2=0 \quad \text{or} \quad w+2=0 \quad \text{or} \quad w-2=0$$
$$w=-2 \quad \text{or} \quad w=-2 \quad \text{or} \quad w=2$$
So, the excluded values are -2 and 2.

29. $\dfrac{4x}{6}=\dfrac{2\cdot 2\cdot x}{2\cdot 3}=\dfrac{2x}{3}$

31. $\dfrac{12t^3}{15t}=\dfrac{2\cdot 2\cdot 3\cdot t\cdot t\cdot t}{3\cdot 5\cdot t}=\dfrac{2\cdot 2\cdot t\cdot t}{5}=\dfrac{4t^2}{5}$

33. $\dfrac{18x^3y}{27x^2y^4}=\dfrac{2\cdot 3\cdot 3\cdot x\cdot x\cdot x\cdot y}{3\cdot 3\cdot 3\cdot x\cdot x\cdot y\cdot y\cdot y\cdot y}$

$$=\dfrac{2\cdot x}{3\cdot y\cdot y\cdot y}$$

$$=\dfrac{2x}{3y^3}$$

35. $\dfrac{3x-6}{5x-10}=\dfrac{3(x-2)}{5(x-2)}=\dfrac{3}{5}$

37. $\dfrac{2x+12}{3x+18}=\dfrac{2(x+6)}{3(x+6)}=\dfrac{2}{3}$

39. $\dfrac{a^3+4a^2}{7a^2+28a}=\dfrac{a^2(a+4)}{7a(a+4)}=\dfrac{a\cdot a(a+4)}{7\cdot a(a+4)}=\dfrac{a}{7}$

41. $\dfrac{x^2-y^2}{3x+3y}=\dfrac{(x-y)(x+y)}{3(x+y)}=\dfrac{x-y}{3}$

43. $\dfrac{4x+8}{x^2+7x+10}=\dfrac{4(x+2)}{(x+5)(x+2)}=\dfrac{4}{x+5}$

45. $\dfrac{5x-35}{x^2-9x+14}=\dfrac{5(x-7)}{(x-2)(x-7)}=\dfrac{5}{x-2}$

47. $\dfrac{t^2+5t+4}{t^2+9t+20}=\dfrac{(t+1)(t+4)}{(t+5)(t+4)}=\dfrac{t+1}{t+5}$

49. $\dfrac{x^2-9x+14}{x^2-8x+7}=\dfrac{(x-7)(x-2)}{(x-7)(x-1)}=\dfrac{x-2}{x-1}$

51. $\dfrac{x^2-4}{x^2-3x-10}=\dfrac{(x-2)(x+2)}{(x-5)(x+2)}=\dfrac{x-2}{x-5}$

53. $\dfrac{x^3+8x^2+16x}{x^3-16x}=\dfrac{x(x^2+8x+16)}{x(x^2-16)}$

$$=\dfrac{x(x+4)(x+4)}{x(x+4)(x-4)}$$

$$=\dfrac{x+4}{x-4}$$

55. $\dfrac{6x-16}{9x^2-64} = \dfrac{2(3x-8)}{(3x+8)(3x-8)} = \dfrac{2}{3x+8}$

57. $\dfrac{-4w+8}{w^2+2w-8} = \dfrac{-4(w-2)}{(w+4)(w-2)} = \dfrac{-4}{w+4}$

59. $\dfrac{4x^2-25}{2x^2+x-15} = \dfrac{(2x-5)(2x+5)}{(2x-5)(x+3)} = \dfrac{2x+5}{x+3}$

61. $\dfrac{3x^2+9x+6}{6x^2+5x-1} = \dfrac{3(x^2+3x+2)}{(6x-1)(x+1)}$

$= \dfrac{3(x+2)(x+1)}{(6x-1)(x+1)}$

$= \dfrac{3(x+2)}{6x-1}$

63. $\dfrac{a^2+2ab+b^2}{a^2-b^2} = \dfrac{(a+b)(a+b)}{(a+b)(a-b)} = \dfrac{a+b}{a-b}$

65. $\dfrac{5x+10}{x^3+2x^2-3x-6} = \dfrac{5(x+2)}{x^2(x+2)-3(x+2)}$

$= \dfrac{5(x+2)}{(x+2)(x^2-3)}$

$= \dfrac{5}{x^2-3}$

67. $\dfrac{t^2+2t+1}{t^3+t^2-t-1} = \dfrac{(t+1)(t+1)}{t^2(t+1)-1(t+1)}$

$= \dfrac{(t+1)(t+1)}{(t+1)(t^2-1)}$

$= \dfrac{(t+1)(t+1)}{(t+1)(t+1)(t-1)}$

$= \dfrac{1}{t-1}$

69. $\dfrac{x^3+8}{x^2+7x+10} = \dfrac{x^3+2^3}{x^2+7x+10}$

$= \dfrac{(x+2)(x^2-x\cdot2+2^2)}{(x+2)(x+5)}$

$= \dfrac{(x+2)(x^2-2x+4)}{(x+2)(x+5)}$

$= \dfrac{x^2-2x+4}{x+5}$

71. $\dfrac{x^3-64}{x^2-16} = \dfrac{x^3-4^3}{x^2-16}$

$= \dfrac{(x-4)(x^2+x\cdot4+4^2)}{(x-4)(x+4)}$

$= \dfrac{(x-4)(x^2+4x+16)}{(x-4)(x+4)}$

$= \dfrac{x^2+4x+16}{x+4}$

73. a. To find the cost per student, *p*, we divide the total expense, $60, by the number of students, *n*.

Number of Students	Cost per Student (dollars per student)
n	*p*
10	$\dfrac{60}{10}=6$
20	$\dfrac{60}{20}=3$
30	$\dfrac{60}{30}=2$
40	$\dfrac{60}{40}=1.5$
n	$\dfrac{60}{n}$

Thus, $p=\dfrac{60}{n}$.

b. $\underbrace{p}_{\text{dollars per student}} = \dfrac{60}{n} \left.\begin{array}{l}\}\text{ dollars}\\ \}\text{ student}\end{array}\right.$

So, the units on both sides of the equation are dollars per student, suggesting the equation is correct.

c. Substitute 25 for *n*: $p=\dfrac{60}{25}=2.4$.

This means that the cost per student is $2.40 if 25 students go to the party.

d. As the value of *n* increases, the value of *p* decreases. This means that the greater the number of students who go to the party, the lower the cost per student will be.

75. a. The year 2000 is represented by $t = 20$, so we substitute 20 for t in the model:

$$p = \frac{-28(20)^2 + 2694(20) + 15{,}122}{50.8(20) + 275}$$

$$= \frac{57{,}802}{1291}$$

$$\approx 44.8$$

We estimate that 44.8% of prisoners were released in 2000.

b. From the table, in 2000, the number of prisoners released was 585 thousand and the total number of prisoners was 1331 thousand. The actual percentage of prisoners released was $\frac{585}{1331} \approx 0.440 = 44.0\%$.

c. The result from part (a) is an overestimate since it is larger than the actual percentage.

d. The year 2010 is represented by $t = 30$, so we substitute 30 for t in the model:

$$p = \frac{-28(30)^2 + 2694(30) + 15{,}122}{50.8(30) + 275}$$

$$= \frac{70{,}742}{1799}$$

$$\approx 39.3$$

We estimate that 39.3% of prisoners will be released in 2010.

77. No, the student is not correct. Explanations may vary.

79. Substitute 0 for x: $\frac{5}{0+2} = \frac{5}{2}$. Thus, 0 is not an excluded value.

81. Answers may vary. One possibility follows: The student incorrectly cancelled the x variables.

The expression $\frac{4x+3}{2x}$ is already simplified.

83 – 85. Examples may vary.

87. Answers may vary. One possibility follows: To find all excluded values of a rational expression, set the denominator of the expression equal to 0 and solve for the variable. The results are the excluded values of the expression.

89. $\dfrac{x^{-4} y^9}{5 w^{-2}} = \dfrac{y^9 w^2}{5 x^4}$

91. $\dfrac{4x - 12}{x^2 - 7x + 12} = \dfrac{4(x-3)}{(x-4)(x-3)} = \dfrac{4}{x-4}$

93. $\dfrac{6 - \sqrt{32}}{8} = \dfrac{6 - 4\sqrt{2}}{8} = \dfrac{2\left(3 - 2\sqrt{2}\right)}{2 \cdot 4} = \dfrac{3 - 2\sqrt{2}}{4}$

95. $2p^3 - 3p^2 - 18p + 27 = p^2(2p - 3) - 9(2p - 3)$
$$= (p^2 - 9)(2p - 3)$$
$$= (p - 3)(p + 3)(2p - 3)$$

This is a cubic (or third-degree) polynomial in one variable.

97. $y = x^2 - 2x$

First, we list some solutions in the table. Then we sketch a curve that contains the points corresponding to the solutions.

x	y
-3	$(-3)^2 - 2(-3) = 15$
-2	$(-2)^2 - 2(-2) = 8$
-1	$(-1)^2 - 2(-1) = 3$
0	$(0)^2 - 2(0) = 0$
1	$(1)^2 - 2(1) = -1$
2	$(2)^2 - 2(2) = 0$
3	$(3)^2 - 2(3) = 3$

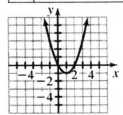

This is a quadratic (or second-degree) equation in two variables.

99. $x(3x - 2) = 4$
$$3x^2 - 2x = 4$$
$$3x^2 - 2x - 4 = 0$$

We substitute 3 for a, -2 for b, and -4 for c in the quadratic formula:

$$x = \frac{-b \pm \sqrt{b^2 - 4ac}}{2a}$$

$$x = \frac{-(-2) \pm \sqrt{(-2)^2 - 4(3)(-4)}}{2(3)}$$

$$x = \frac{2 \pm \sqrt{4 + 48}}{6}$$

$$x = \frac{2 \pm \sqrt{52}}{6}$$

$$x = \frac{2 \pm 2\sqrt{13}}{6}$$

$$x = \frac{2\left(1 \pm \sqrt{13}\right)}{2 \cdot 3}$$

$$x = \frac{1 \pm \sqrt{13}}{3}$$

This is a quadratic (or second-degree) equation in one variable.

Homework 10.2

1. $\dfrac{3}{x} \cdot \dfrac{5}{x} = \dfrac{3 \cdot 5}{x \cdot x} = \dfrac{15}{x^2}$

3. $\dfrac{x}{6} \div \dfrac{3}{2x} = \dfrac{x}{6} \cdot \dfrac{2x}{3} = \dfrac{x \cdot 2 \cdot x}{2 \cdot 3 \cdot 3} = \dfrac{x \cdot x}{3 \cdot 3} = \dfrac{x^2}{9}$

5. $\dfrac{6a^2}{7} \cdot \dfrac{21}{5a^8} = \dfrac{2 \cdot 3 \cdot a^2 \cdot 3 \cdot 7}{7 \cdot 5 \cdot a^8} = \dfrac{2 \cdot 3 \cdot 3}{5 \cdot a^{8-2}} = \dfrac{18}{5a^6}$

7. $\dfrac{2}{x-3} \cdot \dfrac{x-4}{x+5} = \dfrac{2(x-4)}{(x-3)(x+5)}$

9. $\dfrac{k-2}{k-6} \div \dfrac{k+6}{k+4} = \dfrac{k-2}{k-6} \cdot \dfrac{k+4}{k+6} = \dfrac{(k-2)(k+4)}{(k-6)(k+6)}$

11. $\dfrac{6}{7x-14} \cdot \dfrac{5x-10}{9} = \dfrac{2 \cdot 3}{7(x-2)} \cdot \dfrac{5(x-2)}{3 \cdot 3}$
$= \dfrac{2 \cdot 3 \cdot 5(x-2)}{7(x-2) \cdot 3 \cdot 3}$
$= \dfrac{2 \cdot 5}{7 \cdot 3}$
$= \dfrac{10}{21}$

13. $\dfrac{3x+18}{x-6} \div \dfrac{x+6}{2x-12} = \dfrac{3x+18}{x-6} \cdot \dfrac{2x-12}{x+6}$
$= \dfrac{3(x+6)}{x-6} \cdot \dfrac{2(x-6)}{x+6}$
$= \dfrac{3(x+6) \cdot 2(x-6)}{(x-6) \cdot (x+6)}$
$= \dfrac{3 \cdot 2}{1}$
$= 6$

15. $\dfrac{4w^6}{w+3} \cdot \dfrac{w+5}{2w^2} = \dfrac{2 \cdot 2 \cdot w^6 \cdot (w+5)}{(w+3) \cdot 2 \cdot w^2}$
$= \dfrac{2 \cdot w^{6-2} \cdot (w+5)}{(w+3)}$
$= \dfrac{2w^4(w+5)}{w+3}$

17. $\dfrac{(x-4)(x+1)}{(x-7)(x+2)} \cdot \dfrac{5(x+2)}{3(x-4)} = \dfrac{(x-4)(x+1) \cdot 5(x+2)}{(x-7)(x+2) \cdot 3(x-4)}$
$= \dfrac{(x+1) \cdot 5}{(x-7) \cdot 3}$
$= \dfrac{5(x+1)}{3(x-7)}$

19. $\dfrac{4(x-4)^2}{(x+5)^2} \div \dfrac{14(x-4)}{15(x+5)} = \dfrac{4(x-4)^2}{(x+5)^2} \cdot \dfrac{15(x+5)}{14(x-4)}$
$= \dfrac{2 \cdot 2 \cdot (x-4)^2 \cdot 3 \cdot 5 \cdot (x+5)}{(x+5)^2 \cdot 2 \cdot 7 \cdot (x-4)}$
$= \dfrac{2 \cdot (x-4) \cdot 3 \cdot 5}{(x+5) \cdot 7}$
$= \dfrac{30(x-4)}{7(x+5)}$

21. $\dfrac{4t^7}{3t-9} \cdot \dfrac{5t-15}{8t^3} = \dfrac{4t^7}{3(t-3)} \cdot \dfrac{5(t-3)}{8t^3}$
$= \dfrac{2 \cdot 2 \cdot t^7 \cdot 5(t-3)}{3(t-3) \cdot 2 \cdot 2 \cdot 2 \cdot t^3}$
$= \dfrac{t^{7-3} \cdot 5}{3 \cdot 2}$
$= \dfrac{5t^4}{6}$

23. $\dfrac{8x^2}{x^2-49} \div \dfrac{4x^5}{3x+21} = \dfrac{8x^2}{x^2-49} \cdot \dfrac{3x+21}{4x^5}$
$= \dfrac{8x^2}{(x-7)(x+7)} \cdot \dfrac{3(x+7)}{4x^5}$
$= \dfrac{2 \cdot 2 \cdot 2 \cdot x^2 \cdot 3(x+7)}{(x-7)(x+7) \cdot 2 \cdot 2 \cdot x^5}$
$= \dfrac{2 \cdot 3}{(x-7)x^{5-2}}$
$= \dfrac{6}{x^3(x-7)}$

25.
$$\frac{15a^4b}{8ab^5} \div \frac{25ab^3}{4a} = \frac{15a^4b}{8ab^5} \cdot \frac{4a}{25ab^3}$$
$$= \frac{3\cdot5\cdot a^4\cdot b\cdot2\cdot2\cdot a}{2\cdot2\cdot2\cdot a\cdot b^5\cdot5\cdot5\cdot a\cdot b^3}$$
$$= \frac{3\cdot a^{4+1-1-1}}{2\cdot5\cdot b^{5+3-1}}$$
$$= \frac{3a^3}{10b^7}$$

27.
$$\frac{x^2-3x-28}{x^2+4x-45} \cdot \frac{x+9}{x-7} = \frac{(x-7)(x+4)}{(x+9)(x-5)} \cdot \frac{x+9}{x-7}$$
$$= \frac{(x-7)(x+4)(x+9)}{(x+9)(x-5)(x-7)}$$
$$= \frac{x+4}{x-5}$$

29.
$$\frac{t^2-36}{t^2-81} \div \frac{t+6}{t-9} = \frac{t^2-36}{t^2-81} \cdot \frac{t-9}{t+6}$$
$$= \frac{(t-6)(t+6)}{(t-9)(t+9)} \cdot \frac{t-9}{t+6}$$
$$= \frac{(t-6)(t+6)(t-9)}{(t-9)(t+9)(t+6)}$$
$$= \frac{t-6}{t+9}$$

31.
$$\frac{x^2+6x+8}{x^2-5x} \cdot \frac{4x-20}{3x+6} = \frac{(x+2)(x+4)}{x(x-5)} \cdot \frac{4(x-5)}{3(x+2)}$$
$$= \frac{(x+2)(x+4)\cdot4(x-5)}{x(x-5)\cdot3(x+2)}$$
$$= \frac{4(x+4)}{3x}$$

33.
$$\frac{x^2-4}{x^2-7x+12} \div \frac{x^2-8x+12}{x^2-2x-3}$$
$$= \frac{x^2-4}{x^2-7x+12} \cdot \frac{x^2-2x-3}{x^2-8x+12}$$
$$= \frac{(x-2)(x+2)}{(x-4)(x-3)} \cdot \frac{(x-3)(x+1)}{(x-6)(x-2)}$$
$$= \frac{(x-2)(x+2)(x-3)(x+1)}{(x-4)(x-3)(x-6)(x-2)}$$
$$= \frac{(x+2)(x+1)}{(x-4)(x-6)}$$

35.
$$\frac{a^2+6a+9}{a^2+6a} \cdot \frac{a^2+11a+30}{4a^2-36}$$
$$= \frac{(a+3)(a+3)}{a(a+6)} \cdot \frac{(a+6)(a+5)}{4(a^2-9)}$$
$$= \frac{(a+3)(a+3)}{a(a+6)} \cdot \frac{(a+6)(a+5)}{4(a-3)(a+3)}$$
$$= \frac{(a+3)(a+3)(a+6)(a+5)}{a(a+6)\cdot4(a-3)(a+3)}$$
$$= \frac{(a+3)(a+5)}{4a(a-3)}$$

37.
$$\frac{x^2+x-6}{x^2+9x+8} \cdot \frac{x^2+5x-24}{x^2-7x+10}$$
$$= \frac{(x+3)(x-2)}{(x+8)(x+1)} \cdot \frac{(x+8)(x-3)}{(x-5)(x-2)}$$
$$= \frac{(x+3)(x-2)(x+8)(x-3)}{(x+8)(x+1)(x-5)(x-2)}$$
$$= \frac{(x+3)(x-3)}{(x+1)(x-5)}$$

39.
$$\frac{x^2+16}{x^2+8x+16} \div \frac{4x}{x^2-16}$$
$$= \frac{x^2+16}{x^2+8x+16} \cdot \frac{x^2-16}{4x}$$
$$= \frac{x^2+16}{(x+4)(x+4)} \cdot \frac{(x-4)(x+4)}{4x}$$
$$= \frac{(x^2+16)(x-4)(x+4)}{(x+4)(x+4)\cdot4x}$$
$$= \frac{(x^2+16)(x-4)}{4x(x+4)}$$

41.
$$\frac{9x^2-25}{x^2-8x+16} \cdot \frac{x^2-4x}{3x^2-2x-5}$$
$$= \frac{(3x-5)(3x+5)}{(x-4)(x-4)} \cdot \frac{x(x-4)}{(3x-5)(x+1)}$$
$$= \frac{(3x-5)(3x+5)\cdot x(x-4)}{(x-4)(x-4)(3x-5)(x+1)}$$
$$= \frac{x(3x+5)}{(x-4)(x+1)}$$

43. $\dfrac{3p^2+15p+12}{4p^2} \div \dfrac{9p+36}{6p^4}$

$= \dfrac{3p^2+15p+12}{4p^2} \cdot \dfrac{6p^4}{9p+36}$

$= \dfrac{3(p^2+5p+4)}{2\cdot2\cdot p^2} \cdot \dfrac{2\cdot3\cdot p^4}{9(p+4)}$

$= \dfrac{3(p+4)(p+1)}{2\cdot2\cdot p^2} \cdot \dfrac{2\cdot3\cdot p^4}{3\cdot3\cdot(p+4)}$

$= \dfrac{3(p+4)(p+1)\cdot2\cdot3\cdot p^4}{2\cdot2\cdot p^2\cdot3\cdot3\cdot(p+4)}$

$= \dfrac{(p+1)\cdot p^{4-2}}{2}$

$= \dfrac{p^2(p+1)}{2}$

45. $\dfrac{x^2-25}{x^3-3x^2-4x} \div \dfrac{x^2-10x+25}{x^2-4x}$

$= \dfrac{x^2-25}{x^3-3x^2-4x} \cdot \dfrac{x^2-4x}{x^2-10x+25}$

$= \dfrac{(x-5)(x+5)}{x(x^2-3x-4)} \cdot \dfrac{x(x-4)}{(x-5)(x-5)}$

$= \dfrac{(x-5)(x+5)}{x(x-4)(x+1)} \cdot \dfrac{x(x-4)}{(x-5)(x-5)}$

$= \dfrac{(x-5)(x+5)\cdot x(x-4)}{x(x-4)(x+1)(x-5)(x-5)}$

$= \dfrac{x+5}{(x+1)(x-5)}$

47. $\dfrac{-6x+12}{x^2+10x+21} \cdot \dfrac{x^2-9}{-3x+6}$

$\dfrac{-6(x-2)}{(x+7)(x+3)} \cdot \dfrac{(x-3)(x+3)}{-3(x-2)}$

$= \dfrac{-3\cdot2\cdot(x-2)(x-3)(x+3)}{(x+7)(x+3)\cdot(-3)(x-2)}$

$= \dfrac{2(x-3)}{x+7}$

49. $(b^2-25)\cdot\dfrac{2b}{b^2-6b+5}$

$= (b-5)(b+5)\cdot\dfrac{2b}{(b-5)(b-1)}$

$= \dfrac{(b-5)(b+5)\cdot2b}{(b-5)(b-1)}$

$= \dfrac{2b(b+5)}{b-1}$

51. $\dfrac{x^2+8x}{x-4} \div (x^2+16x+64)$

$= \dfrac{x^2+8x}{x-4} \cdot \dfrac{1}{x^2+16x+64}$

$= \dfrac{x(x+8)}{x-4} \cdot \dfrac{1}{(x+8)(x+8)}$

$= \dfrac{x(x+8)}{(x-4)(x+8)(x+8)}$

$= \dfrac{x}{(x-4)(x+8)}$

53. $\dfrac{a^2-b^2}{4a^2-9b^2} \cdot \dfrac{2a-3b}{a+b}$

$= \dfrac{(a-b)(a+b)}{(2a-3b)(2a+3b)} \cdot \dfrac{2a-3b}{a+b}$

$= \dfrac{(a-b)(a+b)(2a-3b)}{(2a-3b)(2a+3b)(a+b)}$

$= \dfrac{a-b}{2a+3b}$

55. $\dfrac{x^2-y^2}{3x+6y} \div \dfrac{x^2+2xy+y^2}{x^2+2xy}$

$= \dfrac{x^2-y^2}{3x+6y} \cdot \dfrac{x^2+2xy}{x^2+2xy+y^2}$

$= \dfrac{(x-y)(x+y)}{3(x+2y)} \cdot \dfrac{x(x+2y)}{(x+y)(x+y)}$

$= \dfrac{(x-y)(x+y)\cdot x(x+2y)}{3(x+2y)(x+y)(x+y)}$

$= \dfrac{x(x-y)}{3(x+y)}$

57. $\dfrac{x^3-2x^2+3x-6}{x^2-3x} \cdot \dfrac{3x-9}{x^2-4}$

$= \dfrac{x^2(x-2)+3(x-2)}{x(x-3)} \cdot \dfrac{3(x-3)}{(x-2)(x+2)}$

$= \dfrac{(x^2+3)(x-2)}{x(x-3)} \cdot \dfrac{3(x-3)}{(x-2)(x+2)}$

$= \dfrac{(x^2+3)(x-2)\cdot3(x-3)}{x(x-3)(x-2)(x+2)}$

$= \dfrac{3(x^2+3)}{x(x+2)}$

59. $\dfrac{x^3+27}{6x-3} \div \dfrac{2x^2-6x+18}{2x^2-x}$

$= \dfrac{x^3+27}{6x-3} \cdot \dfrac{2x^2-x}{2x^2-6x+18}$

$= \dfrac{(x+3)(x^2-3x+9)}{3(2x-1)} \cdot \dfrac{x(2x-1)}{2(x^2-3x+9)}$

$= \dfrac{(x+3)(x^2-3x+9)\cdot x(2x-1)}{3(2x-1)\cdot 2(x^2-3x+9)}$

$= \dfrac{x(x+3)}{6}$

61. a. $\dfrac{x^2-1}{x^2-4} \cdot \dfrac{x^2+3x+2}{x^2-3x+2}$

$= \dfrac{(x-1)(x+1)}{(x-2)(x+2)} \cdot \dfrac{(x+2)(x+1)}{(x-2)(x-1)}$

$= \dfrac{(x-1)(x+1)(x+2)(x+1)}{(x-2)(x+2)(x-2)(x-1)}$

$= \dfrac{(x+1)^2}{(x-2)^2}$

b. $\dfrac{x^2-1}{x^2-4} \div \dfrac{x^2+3x+2}{x^2-3x+2}$

$= \dfrac{x^2-1}{x^2-4} \cdot \dfrac{x^2-3x+2}{x^2+3x+2}$

$= \dfrac{(x-1)(x+1)}{(x-2)(x+2)} \cdot \dfrac{(x-2)(x-1)}{(x+2)(x+1)}$

$= \dfrac{(x-1)(x+1)(x-2)(x-1)}{(x-2)(x+2)(x+2)(x+1)}$

$= \dfrac{(x-1)^2}{(x+2)^2}$

63. $\dfrac{3\text{ feet}}{1} \cdot \dfrac{12\text{ inches}}{1\text{ foot}} = 36$ inches

The height of the net is 36 inches at the center.

65. $\dfrac{10\text{ kilometers}}{1} \cdot \dfrac{1\text{ mile}}{1.61\text{ kilometers}} \approx 6.21$ miles

The race is 6.21 miles long.

67. $\dfrac{16.3\text{ pounds}}{1\text{ year}} \cdot \dfrac{16\text{ ounces}}{1\text{ pound}} \cdot \dfrac{1\text{ year}}{365\text{ days}} \approx 0.71 \dfrac{\text{ounces}}{\text{day}}$

Americans consumed an average of 0.71 ounces of fish and shellfish per day.

69. $\dfrac{6.29\text{ km}}{1\text{ liter}} \cdot \dfrac{1\text{ mile}}{1.61\text{ km}} \cdot \dfrac{0.946\text{ liter}}{1\text{ quart}} \cdot \dfrac{4\text{ quarts}}{1\text{ gallon}}$

$\approx 14.78 \dfrac{\text{miles}}{\text{gallon}}$

The car's gas mileage is 14.78 miles per gallon.

71. $\dfrac{71.2\text{ mg}}{12\text{ ounces}} \cdot \dfrac{1\text{ gram}}{1000\text{ mg}} \cdot \dfrac{8\text{ ounces}}{1\text{ cup}} \cdot \dfrac{4\text{ cups}}{1\text{ quart}} \cdot \dfrac{4\text{ quarts}}{1\text{ gallon}}$

$\approx 0.76 \dfrac{\text{grams}}{\text{gallon}}$

There is 0.76 gram of caffeine in 1 gallon of Jolt®.

73. $\dfrac{x^3}{12} \cdot \dfrac{3}{x} = \dfrac{x^3\cdot 3}{2\cdot 2\cdot 3\cdot x} = \dfrac{x^{3-1}}{2\cdot 2} = \dfrac{x^2}{4}$

75. $\dfrac{x-2}{x^2-3x-18} \div \dfrac{x+4}{x^2+4x+3}$

$= \dfrac{x-2}{x^2-3x-18} \cdot \dfrac{x^2+4x+3}{x+4}$

$= \dfrac{x-2}{(x-6)(x+3)} \cdot \dfrac{(x+3)(x+1)}{x+4}$

$= \dfrac{(x-2)(x+3)(x+1)}{(x-6)(x+3)(x+4)}$

$= \dfrac{(x-2)(x+1)}{(x-6)(x+4)}$

77. Answers may vary. One possibility follows: The student failed to simplify. The correct answer is:

$\dfrac{x+2}{x+4} \cdot \dfrac{x+4}{x+6} = \dfrac{(x+2)(x+4)}{(x+4)(x+6)} = \dfrac{x+2}{x+6}$

79. Answers may vary. One possibility follows: Substitute 0 for x:

$\dfrac{x}{3} \cdot \dfrac{x+4}{x-7} = \dfrac{x^2+4}{3x-7}$

$\dfrac{0}{3} \cdot \dfrac{0+4}{0-7} \overset{?}{=} \dfrac{0^2+4}{3(0)-7}$

$0 \overset{?}{=} -\dfrac{4}{7}$ False

Since the result is false, the result is incorrect. The correct multiplication is:

$\dfrac{x}{3} \cdot \dfrac{x+4}{x-7} = \dfrac{x(x+4)}{3(x-7)}$ or $\dfrac{x^2+4x}{3x-21}$

81. Answers may vary.

83. To multiply two rational expressions, begin by factoring the numerators and denominators. Next, multiply by using the property $\dfrac{A}{B} \cdot \dfrac{C}{D} = \dfrac{AC}{BD}$, where B and D are nonzero. Finally, simplify the result. Examples may vary.

85. $(2x-6)(x^2-7x+12)$

$= 2x(x^2) + 2x(-7x) + 2x(12) + (-6)(x^2) +$
$(-6)(-7x) + (-6)(12)$

$= 2x^3 - 14x^2 + 24x - 6x^2 + 42x - 72$

$= 2x^3 - 14x^2 - 6x^2 + 24x + 42x - 72$

$= 2x^3 - 20x^2 + 66x - 72$

87. $(2x-6) \div (x^2 - 7x + 12) = \dfrac{2x-6}{x^2-7x+12}$

$= \dfrac{2(x-3)}{(x-4)(x-3)}$

$= \dfrac{2}{x-4}$

89. $-3(x-5)^2 = -3(x^2 - 2 \cdot x \cdot 5 + 5^2)$

$= -3(x^2 - 10x + 25)$

$= -3x^2 + 30x - 75$

This is a quadratic (or second-degree) polynomial in one variable.

91. $-3(x-5)^2 = -24$

$\dfrac{-3(x-5)^2}{-3} = \dfrac{-24}{-3}$

$(x-5)^2 = 8$

$x - 5 = \pm\sqrt{8}$

$x - 5 = \pm 2\sqrt{2}$

$x = 5 \pm 2\sqrt{2}$

This is a quadratic (or second-degree) equation in one variable.

93. Substitute 2 for x:

$-3(x-5)^2 = -3(2-5)^2$

$= -3(-3)^2$

$= -3(9)$

$= -27$

This is a quadratic (or second-degree) polynomial in one variable.

Homework 10.3

1. $\dfrac{7}{x} + \dfrac{2}{x} = \dfrac{7+2}{x} = \dfrac{9}{x}$

3. $\dfrac{2x}{x-1} + \dfrac{6x}{x-1} = \dfrac{2x+6x}{x-1} = \dfrac{8x}{x-1}$

5. $\dfrac{3x-2}{x+3} + \dfrac{5x+4}{x+3} = \dfrac{3x-2+5x+4}{x+3} = \dfrac{8x+2}{x+3}$

7. $\dfrac{t^2}{t+5} + \dfrac{7t+10}{t+5} = \dfrac{t^2+7t+10}{t+5} = \dfrac{(t+5)(t+2)}{t+5} = t+2$

9. $\dfrac{x}{x^2-4} + \dfrac{2}{x^2-4} = \dfrac{x+2}{x^2-4} = \dfrac{x+2}{(x-2)(x+2)} = \dfrac{1}{x-2}$

11. $\dfrac{x^2-5x}{x^2+5x+6} + \dfrac{4x-12}{x^2+5x+6} = \dfrac{x^2-5x+4x-12}{x^2+5x+6}$

$= \dfrac{x^2-x-12}{x^2+5x+6}$

$= \dfrac{(x-4)(x+3)}{(x+3)(x+2)}$

$= \dfrac{x-4}{x+2}$

13. $\dfrac{3r^2-5r}{r^2+6r+9} + \dfrac{-2r^2+r-21}{r^2+6r+9} = \dfrac{3r^2-5r-2r^2+r-21}{r^2+6r+9}$

$= \dfrac{r^2-4r-21}{r^2+6r+9}$

$= \dfrac{(r-7)(r+3)}{(r+3)(r+3)}$

$= \dfrac{r-7}{r+3}$

15. $\dfrac{3}{x} + \dfrac{5}{2x} = \dfrac{3}{x} \cdot \dfrac{2}{2} + \dfrac{5}{2x} = \dfrac{6}{2x} + \dfrac{5}{2x} = \dfrac{6+5}{2x} = \dfrac{11}{2x}$

17. $\dfrac{3}{2w} + \dfrac{5}{6} = \dfrac{3}{2 \cdot w} + \dfrac{5}{2 \cdot 3}$

$= \dfrac{3}{2 \cdot w} \cdot \dfrac{3}{3} + \dfrac{5}{2 \cdot 3} \cdot \dfrac{w}{w}$

$= \dfrac{9}{6w} + \dfrac{5w}{6w}$

$= \dfrac{9+5w}{6w}$

$= \dfrac{5w+9}{6w}$

19. $\dfrac{5x}{6} + \dfrac{3}{4x} = \dfrac{5x}{2 \cdot 3} + \dfrac{3}{2 \cdot 2 \cdot x}$

$\qquad = \dfrac{5x}{2 \cdot 3} \cdot \dfrac{2x}{2x} + \dfrac{3}{2 \cdot 2 \cdot x} \cdot \dfrac{3}{3}$

$\qquad = \dfrac{10x^2}{12x} + \dfrac{9}{12x}$

$\qquad = \dfrac{10x^2 + 9}{12x}$

21. $\dfrac{5}{8x^3} + \dfrac{3}{10x} = \dfrac{5}{2 \cdot 2 \cdot 2 \cdot x \cdot x \cdot x} + \dfrac{3}{2 \cdot 5 \cdot x}$

$\qquad = \dfrac{5}{2 \cdot 2 \cdot 2 \cdot x \cdot x \cdot x} \cdot \dfrac{5}{5} + \dfrac{3}{2 \cdot 5 \cdot x} \cdot \dfrac{2 \cdot 2 \cdot x \cdot x}{2 \cdot 2 \cdot x \cdot x}$

$\qquad = \dfrac{25}{40x^3} + \dfrac{12x^2}{40x^3}$

$\qquad = \dfrac{25 + 12x^2}{40x^3}$

$\qquad = \dfrac{12x^2 + 25}{40x^3}$

23 $\dfrac{a}{b} + \dfrac{b}{a} = \dfrac{a}{b} \cdot \dfrac{a}{a} + \dfrac{b}{a} \cdot \dfrac{b}{b} = \dfrac{a^2}{ab} + \dfrac{b^2}{ab} = \dfrac{a^2 + b^2}{ab}$

25. $\dfrac{5}{4x} + \dfrac{2}{x+3} = \dfrac{5}{4x} \cdot \dfrac{x+3}{x+3} + \dfrac{2}{x+3} \cdot \dfrac{4x}{4x}$

$\qquad = \dfrac{5(x+3)}{4x(x+3)} + \dfrac{8x}{4x(x+3)}$

$\qquad = \dfrac{5(x+3) + 8x}{4x(x+3)}$

$\qquad = \dfrac{5x + 15 + 8x}{4x(x+3)}$

$\qquad = \dfrac{13x + 15}{4x(x+3)}$

27. $\dfrac{3}{r+2} + \dfrac{4}{r-5} = \dfrac{3}{r+2} \cdot \dfrac{r-5}{r-5} + \dfrac{4}{r-5} \cdot \dfrac{r+2}{r+2}$

$\qquad = \dfrac{3(r-5)}{(r+2)(r-5)} + \dfrac{4(r+2)}{(r-5)(r+2)}$

$\qquad = \dfrac{3(r-5) + 4(r+2)}{(r-5)(r+2)}$

$\qquad = \dfrac{3r - 15 + 4r + 8}{(r-5)(r+2)}$

$\qquad = \dfrac{7r - 7}{(r-5)(r+2)}$

$\qquad = \dfrac{7(r-1)}{(r-5)(r+2)}$

29. $\dfrac{6}{x-1} + \dfrac{3}{5x} = \dfrac{6}{x-1} \cdot \dfrac{5x}{5x} + \dfrac{3}{5x} \cdot \dfrac{x-1}{x-1}$

$\qquad = \dfrac{30x}{5x(x-1)} + \dfrac{3(x-1)}{5x(x-1)}$

$\qquad = \dfrac{30x + 3(x-1)}{5x(x-1)}$

$\qquad = \dfrac{30x + 3x - 3}{5x(x-1)}$

$\qquad = \dfrac{33x - 3}{5x(x-1)}$

$\qquad = \dfrac{3(11x - 1)}{5x(x-1)}$

31. $\dfrac{5x}{x-4} + \dfrac{2}{x+2} = \dfrac{5x}{x-4} \cdot \dfrac{x+2}{x+2} + \dfrac{2}{x+2} \cdot \dfrac{x-4}{x-4}$

$\qquad = \dfrac{5x(x+2)}{(x-4)(x+2)} + \dfrac{2(x-4)}{(x+2)(x-4)}$

$\qquad = \dfrac{5x(x+2) + 2(x-4)}{(x-4)(x+2)}$

$\qquad = \dfrac{5x^2 + 10x + 2x - 8}{(x-4)(x+2)}$

$\qquad = \dfrac{5x^2 + 12x - 8}{(x-4)(x+2)}$

33. $\dfrac{1}{a+b} + \dfrac{1}{a-b} = \dfrac{1}{a+b} \cdot \dfrac{a-b}{a-b} + \dfrac{1}{a-b} \cdot \dfrac{a+b}{a+b}$

$\qquad = \dfrac{a-b}{(a+b)(a-b)} + \dfrac{a+b}{(a-b)(a+b)}$

$\qquad = \dfrac{a-b+a+b}{(a+b)(a-b)}$

$\qquad = \dfrac{2a}{(a+b)(a-b)}$

35. $\dfrac{x}{x+4} + \dfrac{2}{5x+20} = \dfrac{x}{x+4} + \dfrac{2}{5(x+4)}$

$\qquad = \dfrac{x}{x+4} \cdot \dfrac{5}{5} + \dfrac{2}{5(x+4)}$

$\qquad = \dfrac{5x}{5(x+4)} + \dfrac{2}{5(x+4)}$

$\qquad = \dfrac{5x + 2}{5(x+4)}$

37.
$$\frac{p}{3p-9}+\frac{-1}{p-3}=\frac{p}{3(p-3)}+\frac{-1}{p-3}$$

$$=\frac{p}{3(p-3)}+\frac{-1}{p-3}\cdot\frac{3}{3}$$

$$=\frac{p}{3(p-3)}+\frac{-3}{3(p-3)}$$

$$=\frac{p-3}{3(p-3)}$$

$$=\frac{1}{3}$$

39.
$$\frac{2x}{5x-25}+\frac{4}{3x-15}=\frac{2x}{5(x-5)}+\frac{4}{3(x-5)}$$

$$=\frac{2x}{5(x-5)}\cdot\frac{3}{3}+\frac{4}{3(x-5)}\cdot\frac{5}{5}$$

$$=\frac{6x}{15(x-5)}+\frac{20}{15(x-5)}$$

$$=\frac{6x+20}{15(x-5)}$$

$$=\frac{2(3x+10)}{15(x-5)}$$

41.
$$\frac{6}{x^2-1}+\frac{3}{x+1}=\frac{6}{(x-1)(x+1)}+\frac{3}{x+1}$$

$$=\frac{6}{(x-1)(x+1)}+\frac{3}{x+1}\cdot\frac{x-1}{x-1}$$

$$=\frac{6}{(x-1)(x+1)}+\frac{3(x-1)}{(x+1)(x-1)}$$

$$=\frac{6+3(x-1)}{(x-1)(x+1)}$$

$$=\frac{6+3x-3}{(x+1)(x-1)}$$

$$=\frac{3x+3}{(x+1)(x-1)}$$

$$=\frac{3(x+1)}{(x+1)(x-1)}$$

$$=\frac{3}{x-1}$$

43.
$$\frac{t^2+2t}{t^2+11t+18}+\frac{4}{t+9}=\frac{t^2+2t}{(t+9)(t+2)}+\frac{4}{t+9}$$

$$=\frac{t^2+2t}{(t+9)(t+2)}+\frac{4}{t+9}\cdot\frac{t+2}{t+2}$$

$$=\frac{t^2+2t}{(t+9)(t+2)}+\frac{4(t+2)}{(t+9)(t+2)}$$

$$=\frac{t^2+2t+4(t+2)}{(t+9)(t+2)}$$

$$=\frac{t^2+2t+4t+8}{(t+9)(t+2)}$$

$$=\frac{t^2+6t+8}{(t+9)(t+2)}$$

$$=\frac{(t+4)(t+2)}{(t+9)(t+2)}$$

$$=\frac{t+4}{t+9}$$

45.
$$\frac{x}{2x-6}+\frac{3}{x^2-9}=\frac{x}{2(x-3)}+\frac{3}{(x-3)(x+3)}$$

$$=\frac{x}{2(x-3)}\cdot\frac{x+3}{x+3}+\frac{3}{(x-3)(x+3)}\cdot\frac{2}{2}$$

$$=\frac{x(x+3)}{2(x-3)(x+3)}+\frac{6}{2(x-3)(x+3)}$$

$$=\frac{x(x+3)+6}{2(x-3)(x+3)}$$

$$=\frac{x^2+3x+6}{2(x-3)(x+3)}$$

47.
$$\frac{x}{x^2-4}+\frac{1}{x^2+2x}=\frac{x}{(x-2)(x+2)}+\frac{1}{x(x+2)}$$

$$=\frac{x}{(x-2)(x+2)}\cdot\frac{x}{x}+\frac{1}{x(x+2)}\cdot\frac{x-2}{x-2}$$

$$=\frac{x^2}{x(x-2)(x+2)}+\frac{x-2}{x(x+2)(x-2)}$$

$$=\frac{x^2+x-2}{x(x-2)(x+2)}$$

$$=\frac{(x+2)(x-1)}{x(x-2)(x+2)}$$

$$=\frac{x-1}{x(x-2)}$$

49. $\dfrac{4}{(a-3)(a+1)}+\dfrac{2}{(a+1)(a+4)}$

$=\dfrac{4}{(a-3)(a+1)}\cdot\dfrac{a+4}{a+4}+\dfrac{2}{(a+1)(a+4)}\cdot\dfrac{a-3}{a-3}$

$=\dfrac{4(a+4)}{(a-3)(a+1)(a+4)}+\dfrac{2(a-3)}{(a+1)(a+4)(a-3)}$

$=\dfrac{4(a+4)+2(a-3)}{(a-3)(a+1)(a+4)}$

$=\dfrac{4a+16+2a-6}{(a-3)(a+1)(a+4)}$

$=\dfrac{6a+10}{(a-3)(a+1)(a+4)}=\dfrac{2(3a+5)}{(a-3)(a+1)(a+4)}$

51. $\dfrac{3}{x^2-16}+\dfrac{5}{x^2+5x+4}$

$=\dfrac{3}{(x-4)(x+4)}+\dfrac{5}{(x+4)(x+1)}$

$=\dfrac{3}{(x-4)(x+4)}\cdot\dfrac{x+1}{x+1}+\dfrac{5}{(x+4)(x+1)}\cdot\dfrac{x-4}{x-4}$

$=\dfrac{3(x+1)}{(x-4)(x+4)(x+1)}+\dfrac{5(x-4)}{(x+4)(x+1)(x-4)}$

$=\dfrac{3(x+1)+5(x-4)}{(x-4)(x+1)(x+4)}$

$=\dfrac{3x+3+5x-20}{(x-4)(x+1)(x+4)}=\dfrac{8x-17}{(x-4)(x+1)(x+4)}$

53. $\dfrac{4x}{x^2+3x-18}+\dfrac{2}{x^2+10x+24}$

$=\dfrac{4x}{(x+6)(x-3)}+\dfrac{2}{(x+6)(x+4)}$

$=\dfrac{4x}{(x+6)(x-3)}\cdot\dfrac{x+4}{x+4}+\dfrac{2}{(x+6)(x+4)}\cdot\dfrac{x-3}{x-3}$

$=\dfrac{4x(x+4)}{(x+6)(x-3)(x+4)}+\dfrac{2(x-3)}{(x+6)(x+4)(x-3)}$

$=\dfrac{4x(x+4)+2(x-3)}{(x-3)(x+4)(x+6)}$

$=\dfrac{4x^2+16x+2x-6}{(x-3)(x+4)(x+6)}$

$=\dfrac{4x^2+18x-6}{(x-3)(x+4)(x+6)}=\dfrac{2(2x^2+9x-3)}{(x-3)(x+4)(x+6)}$

55. $\dfrac{x}{x^2+9x+20}+\dfrac{-4}{x^2+8x+15}$

$=\dfrac{x}{(x+5)(x+4)}+\dfrac{-4}{(x+5)(x+3)}$

$=\dfrac{x}{(x+5)(x+4)}\cdot\dfrac{x+3}{x+3}+\dfrac{-4}{(x+5)(x+3)}\cdot\dfrac{x+4}{x+4}$

$=\dfrac{x(x+3)}{(x+5)(x+4)(x+3)}+\dfrac{-4(x+4)}{(x+5)(x+3)(x+4)}$

$=\dfrac{x(x+3)-4(x+4)}{(x+3)(x+4)(x+5)}$

$=\dfrac{x^2+3x-4x-16}{(x+3)(x+4)(x+5)}$

$=\dfrac{x^2-x-16}{(x+3)(x+4)(x+5)}$

57. $3+\dfrac{w-2}{w+5}=3\cdot\dfrac{w+5}{w+5}+\dfrac{w-2}{w+5}$

$=\dfrac{3(w+5)}{w+5}+\dfrac{w-2}{w+5}$

$=\dfrac{3(w+5)+w-2}{w+5}$

$=\dfrac{3w+15+w-2}{w+5}$

$=\dfrac{4w+13}{w+5}$

59. $\dfrac{x-3}{x+5}+\dfrac{x+2}{x-4}=\dfrac{x-3}{x+5}\cdot\dfrac{x-4}{x-4}+\dfrac{x+2}{x-4}\cdot\dfrac{x+5}{x+5}$

$=\dfrac{(x-3)(x-4)}{(x+5)(x-4)}+\dfrac{(x+2)(x+5)}{(x-4)(x+5)}$

$=\dfrac{(x-3)(x-4)+(x+2)(x+5)}{(x-4)(x+5)}$

$=\dfrac{x^2-4x-3x+12+x^2+5x+2x+10}{(x-4)(x+5)}$

$=\dfrac{2x^2+22}{(x-4)(x+5)}$

$=\dfrac{2(x^2+11)}{(x-4)(x+5)}$

61. $\dfrac{y-2}{y-3}+\dfrac{y+3}{y+2}=\dfrac{y-2}{y-3}\cdot\dfrac{y+2}{y+2}+\dfrac{y+3}{y+2}\cdot\dfrac{y-3}{y-3}$

$\qquad=\dfrac{(y-2)(y+2)}{(y-3)(y+2)}+\dfrac{(y+3)(y-3)}{(y+2)(y-3)}$

$\qquad=\dfrac{(y-2)(y+2)+(y+3)(y-3)}{(y-3)(y+2)}$

$\qquad=\dfrac{y^2+2y-2y-4+y^2-3y+3y-9}{(y-3)(y+2)}$

$\qquad=\dfrac{2y^2-13}{(y-3)(y+2)}$

63. $\dfrac{x-6}{x-5}+\dfrac{2x}{x^2-2x-15}=\dfrac{x-6}{x-5}+\dfrac{2x}{(x-5)(x+3)}$

$\qquad=\dfrac{x-6}{x-5}\cdot\dfrac{x+3}{x+3}+\dfrac{2x}{(x-5)(x+3)}$

$\qquad=\dfrac{(x-6)(x+3)}{(x-5)(x+3)}+\dfrac{2x}{(x-5)(x+3)}$

$\qquad=\dfrac{(x-6)(x+3)+2x}{(x-5)(x+3)}$

$\qquad=\dfrac{x^2+3x-6x-18+2x}{(x-5)(x+3)}$

$\qquad=\dfrac{x^2-x-18}{(x-5)(x+3)}$

65. $\dfrac{5}{b^2-7b+12}+\dfrac{b+2}{b-4}=\dfrac{5}{(b-4)(b-3)}+\dfrac{b+2}{b-4}$

$\qquad=\dfrac{5}{(b-4)(b-3)}+\dfrac{b+2}{b-4}\cdot\dfrac{b-3}{b-3}$

$\qquad=\dfrac{5}{(b-4)(b-3)}+\dfrac{(b+2)(b-3)}{(b-4)(b-3)}$

$\qquad=\dfrac{5+(b+2)(b-3)}{(b-4)(b-3)}$

$\qquad=\dfrac{5+b^2-3b+2b-6}{(b-4)(b-3)}$

$\qquad=\dfrac{b^2-b-1}{(b-4)(b-3)}$

67. $\dfrac{x-2}{4x+12}+\dfrac{5}{x^2-9}=\dfrac{x-2}{4(x+3)}+\dfrac{5}{(x-3)(x+3)}$

$\qquad=\dfrac{x-2}{4(x+3)}\cdot\dfrac{x-3}{x-3}+\dfrac{5}{(x-3)(x+3)}\cdot\dfrac{4}{4}$

$\qquad=\dfrac{(x-2)(x-3)}{4(x+3)(x-3)}+\dfrac{20}{4(x-3)(x+3)}$

$\qquad=\dfrac{(x-2)(x-3)+20}{4(x-3)(x+3)}$

$\qquad=\dfrac{x^2-3x-2x+6+20}{4(x-3)(x+3)}$

$\qquad=\dfrac{x^2-5x+26}{4(x-3)(x+3)}$

69. $\dfrac{2x}{x-y}+\dfrac{2xy}{x^2-2xy+y^2}=\dfrac{2x}{x-y}+\dfrac{2xy}{(x-y)^2}$

$\qquad=\dfrac{2x}{x-y}\cdot\dfrac{x-y}{x-y}+\dfrac{2xy}{(x-y)^2}$

$\qquad=\dfrac{2x(x-y)}{(x-y)^2}+\dfrac{2xy}{(x-y)^2}$

$\qquad=\dfrac{2x(x-y)+2xy}{(x-y)^2}$

$\qquad=\dfrac{2x^2-2xy+2xy}{(x-y)^2}$

$\qquad=\dfrac{2x^2}{(x-y)^2}$

71. $\dfrac{2x}{x^3-4x^2+2x-8}+\dfrac{5}{3x^2+6}$

$\qquad=\dfrac{2x}{x^2(x-4)+2(x-4)}+\dfrac{5}{3(x^2+2)}$

$\qquad=\dfrac{2x}{(x-4)(x^2+2)}+\dfrac{5}{3(x^2+2)}$

$\qquad=\dfrac{2x}{(x-4)(x^2+2)}\cdot\dfrac{3}{3}+\dfrac{5}{3(x^2+2)}\cdot\dfrac{x-4}{x-4}$

$\qquad=\dfrac{6x}{3(x-4)(x^2+2)}+\dfrac{5(x-4)}{3(x^2+2)(x-4)}$

$\qquad=\dfrac{6x+5(x-4)}{3(x-4)(x^2+2)}$

$\qquad=\dfrac{6x+5x-20}{3(x-4)(x^2+2)}$

$\qquad=\dfrac{11x-20}{3(x-4)(x^2+2)}$

73. $\dfrac{5x}{x^3-8}+\dfrac{4}{x^2-4}$

$=\dfrac{5x}{(x-2)(x^2+2x-4)}+\dfrac{4}{(x-2)(x+2)}$

$=\dfrac{5x}{(x-2)(x^2+2x-4)}\cdot\dfrac{x+2}{x+2}+\dfrac{4}{(x-2)(x+2)}\cdot\dfrac{x^2+2x-4}{x^2+2x-4}$

$=\dfrac{5x(x+2)}{(x-2)(x^2+2x-4)(x+2)}+\dfrac{4(x^2+2x-4)}{(x-2)(x+2)(x^2+2x-4)}$

$=\dfrac{5x(x+2)+4(x^2+2x-4)}{(x-2)(x+2)(x^2+2x-4)}$

$=\dfrac{5x^2+10x+4x^2+8x-16}{(x-2)(x+2)(x^2+2x-4)}$

$=\dfrac{9x^2+18x-16}{(x-2)(x+2)(x^2+2x-4)}$

75. a. The total illumination is represented by:

$\dfrac{18}{d^2}+\dfrac{18}{(2d)^2}$

b. $\dfrac{18}{d^2}+\dfrac{18}{(2d)^2}=\dfrac{18}{d^2}+\dfrac{18}{4d^2}$

$=\dfrac{18}{d^2}\cdot\dfrac{4}{4}+\dfrac{18}{4d^2}$

$=\dfrac{72}{4d^2}+\dfrac{18}{4d^2}$

$=\dfrac{90}{4d^2}$

$=\dfrac{45}{2d^2}$

c. Substitute 1.2 for d:

$\dfrac{45}{2(1.2)^2}=\dfrac{45}{2(1.44)}=\dfrac{45}{2.88}\approx 15.63$

This means that the total illumination is 15.62 W/m^2 when the person is 1.2 meters away from the closer light and $2(1.2)=2.4$ meters away from the other light.

77. Answers may vary. One possibility follows: When forming a common denominator, the student should multiply by $\dfrac{x+3}{x+3}$ and $\dfrac{x+2}{x+2}$, not $\dfrac{1}{x+3}$ and $\dfrac{1}{x+2}$. The correct result is:

$\dfrac{3}{x+2}+\dfrac{5}{x+3}=\dfrac{3}{x+2}\cdot\dfrac{x+3}{x+3}+\dfrac{5}{x+3}\cdot\dfrac{x+2}{x+2}$

$=\dfrac{3(x+3)}{(x+2)(x+3)}+\dfrac{5(x+2)}{(x+3)(x+2)}$

$=\dfrac{3(x+3)+5(x+2)}{(x+2)(x+3)}$

$=\dfrac{3x+9+5x+10}{(x+2)(x+3)}$

$=\dfrac{8x+19}{(x+2)(x+3)}$

79. The work is correct. However, there is a better way. Explanations may vary. One follows: The student could factor the denominator of the first rational expression and find a more simple common denominator as follows:

$\dfrac{2}{x^2+2x}+\dfrac{3}{x+2}=\dfrac{2}{x(x+2)}+\dfrac{3}{x+2}$

$=\dfrac{2}{x(x+2)}+\dfrac{3}{x+2}\cdot\dfrac{x}{x}$

$=\dfrac{2}{x(x+2)}+\dfrac{3x}{x(x+2)}$

$=\dfrac{2+3x}{x(x+2)}$

$=\dfrac{3x+2}{x(x+2)}$

81. To add two rational expressions with different denominators, begin by factoring the denominators of the expressions if possible and determine which factors are missing. Next, use the property $\dfrac{A}{A}=1$, where A is nonzero, to introduce missing factors. Then, add the expressions by using the property $\dfrac{A}{B}+\dfrac{C}{B}=\dfrac{A+C}{B}$, where B is nonzero. Finally, simplify the result. Examples may vary.

83. $(4x^2-7x+2)+(-3x^2+2x+4)$

$=4x^2-7x+2-3x^2+2x+4$

$=4x^2-3x^2-7x+2x+2+4$

$=x^2-5x+6$

85. $\dfrac{4x^2-7x+2}{x-3}+\dfrac{-3x^2+2x+4}{x-3}$

$=\dfrac{4x^2-7x+2-3x^2+2x+4}{x-3}$

$=\dfrac{x^2-5x+6}{x-3}$

$=\dfrac{(x-3)(x-2)}{x-3}$

$=x-2$

87. $y=3x-2$

$y=5x+4$

Substitute $5x+4$ for y in the first equation and solve for x.

$5x+4=3x-2$

$2x+4=-2$

$2x=-6$

$x=-3$

Substitute -3 for x in the equation $y=5x+4$ and solve for y.

$y=5(-3)+4=-15+4=-11$

The solution is $(-3,-11)$.

This is a system of two linear equations in two variables.

89. The equation $y=3x-2$ is in slope-intercept

form, so the slope is $m=3=\dfrac{3}{1}$ and the y-

intercept is $(0,-2)$. We first plot $(0,-2)$. From this point, we move 1 unit to the right and 3 units up, where we plot the point $(1,1)$. We then sketch the line that contains these two points.

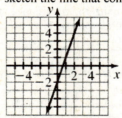

This is a linear equation in two variables.

91. $3x-2=5x+4$

$-2x-2=4$

$-2x=6$

$x=-3$

This is a linear equation in one variable.

Homework 10.4

1. $\dfrac{6}{x}-\dfrac{4}{x}=\dfrac{6-4}{x}=\dfrac{2}{x}$

3. $\dfrac{9x}{x-2}-\dfrac{2x}{x-2}=\dfrac{9x-2x}{x-2}=\dfrac{7x}{x-2}$

5. $\dfrac{x}{x^2-9}-\dfrac{3}{x^2-9}=\dfrac{x-3}{x^2-9}=\dfrac{x-3}{(x-3)(x+3)}=\dfrac{1}{x+3}$

7. $\dfrac{3r}{r+6}-\dfrac{7r-4}{r+6}=\dfrac{3r-(7r-4)}{r+6}$

$=\dfrac{3r-7r+4}{r+6}$

$=\dfrac{-4r+4}{r+6}$

$=\dfrac{-4(r-1)}{r+6}$

$=-\dfrac{4(r-1)}{r+6}$

9. $\dfrac{x^2}{x+1}-\dfrac{2x+3}{x+1}=\dfrac{x^2-(2x+3)}{x+1}$

$=\dfrac{x^2-2x-3}{x+1}$

$=\dfrac{(x-3)(x+1)}{x+1}$

$=x-3$

11. $\dfrac{x^2+7x}{x^2-2x-8}-\dfrac{3x+32}{x^2-2x-8}=\dfrac{x^2+7x-(3x+32)}{x^2-2x-8}$

$=\dfrac{x^2+7x-3x-32}{x^2-2x-8}$

$=\dfrac{x^2+4x-32}{x^2-2x-8}$

$=\dfrac{(x+8)(x-4)}{(x-4)(x+2)}$

$=\dfrac{x+8}{x+2}$

13. $\dfrac{4a^2-5a-12}{a^2+8a+16} - \dfrac{3a^2-6a}{a^2+8a+16}$

$= \dfrac{4a^2-5a-12-(3a^2-6a)}{a^2+8a+16}$

$= \dfrac{4a^2-5a-12-3a^2+6a}{a^2+8a+16}$

$= \dfrac{a^2+a-12}{a^2+8a+16}$

$= \dfrac{(a+4)(a-3)}{(a+4)^2}$

$= \dfrac{a-3}{a+4}$

15. $\dfrac{3}{4x} - \dfrac{2}{x} = \dfrac{3}{2\cdot2\cdot x} - \dfrac{2}{x}$

$= \dfrac{3}{2\cdot2\cdot x} - \dfrac{2}{x}\cdot\dfrac{2\cdot2}{2\cdot2}$

$= \dfrac{3}{4x} - \dfrac{8}{4x}$

$= \dfrac{3-8}{4x}$

$= \dfrac{-5}{4x} = -\dfrac{5}{4x}$

17. $\dfrac{5}{2b} - \dfrac{3}{8} = \dfrac{5}{2b} - \dfrac{3}{2\cdot2\cdot2}$

$= \dfrac{5}{2b}\cdot\dfrac{4}{4} - \dfrac{3}{2\cdot2\cdot2}\cdot\dfrac{b}{b}$

$= \dfrac{20}{8b} - \dfrac{3b}{8b}$

$= \dfrac{20-3b}{8b}$

$= \dfrac{-3b+20}{8b}$

19. $\dfrac{5x}{8} - \dfrac{1}{6x} = \dfrac{5x}{2\cdot2\cdot2} - \dfrac{1}{2\cdot3\cdot x}$

$= \dfrac{5x}{2\cdot2\cdot2}\cdot\dfrac{3x}{3x} - \dfrac{1}{2\cdot3\cdot x}\cdot\dfrac{2\cdot2}{2\cdot2}$

$= \dfrac{15x^2}{24x} - \dfrac{4}{24x}$

$= \dfrac{15x^2-4}{24x}$

21. $\dfrac{5}{10x^4} - \dfrac{3}{15x^2} = \dfrac{5}{2\cdot5\cdot x\cdot x\cdot x\cdot x} - \dfrac{3}{3\cdot5\cdot x\cdot x}$

$= \dfrac{1}{2\cdot x\cdot x\cdot x\cdot x} - \dfrac{1}{5\cdot x\cdot x}$

$= \dfrac{1}{2\cdot x\cdot x\cdot x\cdot x}\cdot\dfrac{5}{5} - \dfrac{1}{5\cdot x\cdot x}\cdot\dfrac{2\cdot x\cdot x}{2\cdot x\cdot x}$

$= \dfrac{5}{10x^4} - \dfrac{2x^3}{10x^4}$

$= \dfrac{5-2x^2}{10x^4} = \dfrac{-2x^2+5}{10x^4}$

23 $\dfrac{a}{2b} - \dfrac{b}{3a} = \dfrac{a}{2b}\cdot\dfrac{3a}{3a} - \dfrac{b}{3a}\cdot\dfrac{2b}{2b}$

$= \dfrac{3a^2}{6ab} - \dfrac{2b^2}{6ab}$

$= \dfrac{3a^2-2b^2}{6ab}$

25. $\dfrac{3}{p-2} - \dfrac{5}{4p} = \dfrac{3}{p-2}\cdot\dfrac{4p}{4p} - \dfrac{5}{4p}\cdot\dfrac{p-2}{p-2}$

$= \dfrac{12p}{4p(p-2)} - \dfrac{5(p-2)}{4p(p-2)}$

$= \dfrac{12p-5(p-2)}{4p(p-2)}$

$= \dfrac{12p-5p+10}{4p(p-2)}$

$= \dfrac{7p+10}{4p(p-2)}$

27. $\dfrac{7}{x-1} - \dfrac{3}{x+4} = \dfrac{7}{x-1}\cdot\dfrac{x+4}{x+4} - \dfrac{3}{x+4}\cdot\dfrac{x-1}{x-1}$

$= \dfrac{7(x+4)}{(x-1)(x+4)} - \dfrac{3(x-1)}{(x+4)(x-1)}$

$= \dfrac{7(x+4)-3(x-1)}{(x-1)(x+4)}$

$= \dfrac{7x+28-3x+3}{(x-1)(x+4)}$

$= \dfrac{4x+31}{(x-1)(x+4)}$

29.
$$\frac{3x}{x-2} - \frac{5}{x+3} = \frac{3x}{x-2} \cdot \frac{x+3}{x+3} - \frac{5}{x+3} \cdot \frac{x-2}{x-2}$$
$$= \frac{3x(x+3)}{(x-2)(x+3)} - \frac{5(x-2)}{(x+3)(x-2)}$$
$$= \frac{3x(x+3)-5(x-2)}{(x-2)(x+3)}$$
$$= \frac{3x^2+9x-5x+10}{(x-2)(x+3)}$$
$$= \frac{3x^2+4x+10}{(x-2)(x+3)}$$

31.
$$\frac{1}{a+b} - \frac{1}{a-b} = \frac{1}{a+b} \cdot \frac{a-b}{a-b} - \frac{1}{a-b} \cdot \frac{a+b}{a+b}$$
$$= \frac{a-b}{(a+b)(a-b)} - \frac{a+b}{(a-b)(a+b)}$$
$$= \frac{a-b-(a+b)}{(a+b)(a-b)}$$
$$= \frac{a-b-a-b}{(a+b)(a-b)}$$
$$= \frac{-2b}{(a+b)(a-b)}$$

33.
$$\frac{c}{2c-8} - \frac{3}{c-4} = \frac{c}{2(c-4)} - \frac{3}{c-4}$$
$$= \frac{c}{2(c-4)} - \frac{3}{c-4} \cdot \frac{2}{2}$$
$$= \frac{c}{2(c-4)} - \frac{6}{2(c-4)}$$
$$= \frac{c-6}{2(c-4)}$$

35.
$$\frac{4x}{6x-24} - \frac{7}{4x-16} = \frac{4x}{6(x-4)} - \frac{7}{4(x-4)}$$
$$= \frac{4x}{2\cdot3(x-4)} - \frac{7}{2\cdot2(x-4)}$$
$$= \frac{4x}{2\cdot3(x-4)} \cdot \frac{2}{2} - \frac{7}{2\cdot2(x-4)} \cdot \frac{3}{3}$$
$$= \frac{8x}{12(x-4)} - \frac{21}{12(x-4)}$$
$$= \frac{8x-21}{12(x-4)}$$

37.
$$\frac{x}{x-1} - \frac{2}{x^2-1} = \frac{x}{x-1} - \frac{2}{(x-1)(x+1)}$$
$$= \frac{x}{x-1} \cdot \frac{x+1}{x+1} - \frac{2}{(x-1)(x+1)}$$
$$= \frac{x(x+1)}{(x-1)(x+1)} - \frac{2}{(x-1)(x+1)}$$
$$= \frac{x(x+1)-2}{(x-1)(x+1)}$$
$$= \frac{x^2+x-2}{(x-1)(x+1)}$$
$$= \frac{(x+2)(x-1)}{(x-1)(x+1)}$$
$$= \frac{x+2}{x+1}$$

39.
$$\frac{3x-1}{x^2+2x-15} - \frac{2}{x+5} = \frac{3x-1}{(x+5)(x-3)} - \frac{2}{x+5}$$
$$= \frac{3x-1}{(x+5)(x-3)} - \frac{2}{x+5} \cdot \frac{x-3}{x-3}$$
$$= \frac{3x-1}{(x+5)(x-3)} - \frac{2(x-3)}{(x+5)(x-3)}$$
$$= \frac{3x-1-2(x-3)}{(x+5)(x-3)}$$
$$= \frac{3x-1-2x+6}{(x+5)(x-3)}$$
$$= \frac{x+5}{(x+5)(x-3)}$$
$$= \frac{1}{x-3}$$

41.
$$\frac{t}{3t-21} - \frac{4}{t^2-49} = \frac{t}{3(t-7)} - \frac{4}{(t-7)(t+7)}$$
$$= \frac{t}{3(t-7)} \cdot \frac{t+7}{t+7} - \frac{4}{(t-7)(t+7)} \cdot \frac{3}{3}$$
$$= \frac{t(t+7)}{3(t-7)(t+7)} - \frac{12}{3(t-7)(t+7)}$$
$$= \frac{t(t+7)-12}{3(t-7)(t+7)}$$
$$= \frac{t^2+7t-12}{3(t-7)(t+7)}$$

43. $\dfrac{4x}{x^2-25} - \dfrac{2}{x^2+5x}$

$= \dfrac{4x}{(x-5)(x+5)} - \dfrac{2}{x(x+5)}$

$= \dfrac{4x}{(x-5)(x+5)} \cdot \dfrac{x}{x} - \dfrac{2}{x(x+5)} \cdot \dfrac{x-5}{x-5}$

$= \dfrac{4x^2}{x(x-5)(x+5)} - \dfrac{2(x-5)}{x(x+5)(x-5)}$

$= \dfrac{4x^2 - 2(x-5)}{x(x-5)(x+5)}$

$= \dfrac{4x^2 - 2x + 10}{x(x-5)(x+5)}$

$= \dfrac{2(2x^2 - x + 5)}{x(x-5)(x+5)}$

45. $\dfrac{3}{(x-5)(x+2)} - \dfrac{4}{(x+2)(x+4)}$

$= \dfrac{3}{(x-5)(x+2)} \cdot \dfrac{x+4}{x+4} - \dfrac{4}{(x+2)(x+4)} \cdot \dfrac{x-5}{x-5}$

$= \dfrac{3(x+4)}{(x-5)(x+2)(x+4)} - \dfrac{4(x-5)}{(x+2)(x+4)(x-5)}$

$= \dfrac{3(x+4) - 4(x-5)}{(x-5)(x+2)(x+4)}$

$= \dfrac{3x+12 - 4x + 20}{(x-5)(x+2)(x+4)}$

$= \dfrac{-x+32}{(x-5)(x+2)(x+4)}$

47. $\dfrac{7}{x^2-5x+6} - \dfrac{2}{x^2-3x}$

$= \dfrac{7}{(x-3)(x-2)} - \dfrac{2}{x(x-3)}$

$= \dfrac{7}{(x-3)(x-2)} \cdot \dfrac{x}{x} - \dfrac{2}{x(x-3)} \cdot \dfrac{x-2}{x-2}$

$= \dfrac{7x}{x(x-3)(x-2)} - \dfrac{2(x-2)}{x(x-3)(x-2)}$

$= \dfrac{7x - 2(x-2)}{x(x-3)(x-2)}$

$= \dfrac{7x - 2x + 4}{x(x-3)(x-2)}$

$= \dfrac{5x+4}{x(x-3)(x-2)}$

49. $\dfrac{5b}{b^2+3b-10} - \dfrac{3}{b^2+4b-12}$

$= \dfrac{5b}{(b+5)(b-2)} - \dfrac{3}{(b+6)(b-2)}$

$= \dfrac{5b}{(b+5)(b-2)} \cdot \dfrac{b+6}{b+6} - \dfrac{3}{(b+6)(b-2)} \cdot \dfrac{b+5}{b+5}$

$= \dfrac{5b(b+6)}{(b+5)(b-2)(b+6)} - \dfrac{3(b+5)}{(b+6)(b-2)(b+5)}$

$= \dfrac{5b(b+6) - 3(b+5)}{(b-2)(b+5)(b+6)}$

$= \dfrac{5b^2 + 30b - 3b - 15}{(b-2)(b+5)(b+6)}$

$= \dfrac{5b^2 + 27b - 15}{(b-2)(b+5)(b+6)}$

51. $\dfrac{2x}{x^2+11x+18} - \dfrac{5}{x^2-5x-14}$

$= \dfrac{2x}{(x+9)(x+2)} - \dfrac{5}{(x-7)(x+2)}$

$= \dfrac{2x}{(x+9)(x+2)} \cdot \dfrac{x-7}{x-7} - \dfrac{5}{(x-7)(x+2)} \cdot \dfrac{x+9}{x+9}$

$= \dfrac{2x(x-7)}{(x+9)(x+2)(x-7)} - \dfrac{5(x+9)}{(x-7)(x+2)(x+9)}$

$= \dfrac{2x(x-7) - 5(x+9)}{(x-7)(x+2)(x+9)}$

$= \dfrac{2x^2 - 14x - 5x - 45}{(x-7)(x+2)(x+9)}$

$= \dfrac{2x^2 - 19x - 45}{(x-7)(x+2)(x+9)}$

53. $\dfrac{x+3}{x-6} - 4 = \dfrac{x+3}{x-6} - 4 \cdot \dfrac{x-6}{x-6}$

$= \dfrac{x+3}{x-6} - \dfrac{4(x-6)}{x-6}$

$= \dfrac{x+3 - 4(x-6)}{x-6}$

$= \dfrac{x+3 - 4x + 24}{x-6}$

$= \dfrac{-3x+27}{x-6}$

$= \dfrac{-3(x-9)}{x-6}$

$= -\dfrac{3(x-9)}{x-6}$

55. $\dfrac{x+2}{x-4} - \dfrac{x-3}{x+1} = \dfrac{x+2}{x-4} \cdot \dfrac{x+1}{x+1} - \dfrac{x-3}{x+1} \cdot \dfrac{x-4}{x-4}$

$= \dfrac{(x+2)(x+1)}{(x-4)(x+1)} - \dfrac{(x-3)(x-4)}{(x+1)(x-4)}$

$= \dfrac{(x+2)(x+1) - (x-3)(x-4)}{(x-4)(x+1)}$

$= \dfrac{x^2 + x + 2x + 2 - (x^2 - 4x - 3x + 12)}{(x-4)(x+1)}$

$= \dfrac{x^2 + x + 2x + 2 - x^2 + 4x + 3x - 12)}{(x-4)(x+1)}$

$= \dfrac{10x - 10}{(x-4)(x+1)} = \dfrac{10(x-1)}{(x-4)(x+1)}$

57. $\dfrac{x+2}{x-4} - \dfrac{4}{x^2 - 9x + 20} = \dfrac{x+2}{x-4} - \dfrac{4}{(x-5)(x-4)}$

$= \dfrac{x+2}{x-4} \cdot \dfrac{x-5}{x-5} - \dfrac{4}{(x-5)(x-4)}$

$= \dfrac{(x+2)(x-5)}{(x-5)(x-4)} - \dfrac{4}{(x-5)(x-4)}$

$= \dfrac{(x+2)(x-5) - 4}{(x-5)(x-4)}$

$= \dfrac{x^2 - 5x + 2x - 10 - 4}{(x-5)(x-4)}$

$= \dfrac{x^2 - 3x - 14}{(x-5)(x-4)}$

59. $\dfrac{5t}{t^2 - 10t + 21} - \dfrac{t+4}{t-7} = \dfrac{5t}{(t-7)(t-3)} - \dfrac{t+4}{t-7}$

$= \dfrac{5t}{(t-7)(t-3)} - \dfrac{t+4}{t-7} \cdot \dfrac{t-3}{t-3}$

$= \dfrac{5t}{(t-7)(t-3)} - \dfrac{(t+4)(t-3)}{(t-7)(t-3)}$

$= \dfrac{5t - (t+4)(t-3)}{(t-7)(t-3)}$

$= \dfrac{5t - (t^2 - 3t + 4t - 12)}{(t-7)(t-3)}$

$= \dfrac{5t - t^2 + 3t - 4t + 12}{(t-7)(t-3)}$

$= \dfrac{-t^2 + 4t + 12}{(t-7)(t-3)}$

$= \dfrac{-(t^2 - 4t - 12)}{(t-7)(t-3)}$

$= -\dfrac{(t-6)(t+2)}{(t-7)(t-3)}$

61. $\dfrac{x-4}{3x+3} - \dfrac{6}{x^2 - 1} = \dfrac{x-4}{3(x+1)} - \dfrac{6}{(x-1)(x+1)}$

$= \dfrac{x-4}{3(x+1)} \cdot \dfrac{x-1}{x-1} - \dfrac{6}{(x-1)(x+1)} \cdot \dfrac{3}{3}$

$= \dfrac{(x-4)(x-1)}{3(x+1)(x-1)} - \dfrac{18}{3(x-1)(x+1)}$

$= \dfrac{(x-4)(x-1) - 18}{3(x-1)(x+1)}$

$= \dfrac{x^2 - x - 4x + 4 - 18}{3(x-1)(x+1)}$

$= \dfrac{x^2 - 5x - 14}{3(x-1)(x+1)}$

$= \dfrac{(x-7)(x+2)}{3(x-1)(x+1)}$

63. $\dfrac{3x}{x+y} - \dfrac{3xy}{2x^2 + 3xy + y^2}$

$= \dfrac{3x}{x+y} - \dfrac{3xy}{(2x+y)(x+y)}$

$= \dfrac{3x}{x+y} \cdot \dfrac{2x+y}{2x+y} - \dfrac{3xy}{(2x+y)(x+y)}$

$= \dfrac{3x(2x+y)}{(x+y)(2x+y)} - \dfrac{3xy}{(2x+y)(x+y)}$

$= \dfrac{3x(2x+y) - 3xy}{(2x+y)(x+y)}$

$= \dfrac{6x^2 + 3xy - 3xy}{(2x+y)(x+y)}$

$= \dfrac{6x^2}{(2x+y)(x+y)}$

65. $\dfrac{2}{x^3 - 6x^2 - 3x + 18} - \dfrac{3x}{5x^2 - 15}$

$= \dfrac{2}{x^2(x-6) - 3(x-6)} - \dfrac{3x}{5(x^2 - 3)}$

$= \dfrac{2}{(x^2 - 3)(x-6)} - \dfrac{3x}{5(x^2 - 3)}$

$= \dfrac{2}{(x^2 - 3)(x-6)} \cdot \dfrac{5}{5} - \dfrac{3x}{5(x^2 - 3)} \cdot \dfrac{x-6}{x-6}$

$= \dfrac{10}{5(x^2 - 3)(x-6)} - \dfrac{3x(x-6)}{5(x^2 - 3)(x-6)}$

$= \dfrac{10 - 3x(x-6)}{5(x^2 - 3)(x-6)}$

$= \dfrac{10 - 3x^2 + 18x}{5(x^2 - 3)(x-6)} = \dfrac{-3x^2 + 18x + 10}{5(x^2 - 3)(x-6)}$

67. $\dfrac{3x}{x^3+1}-\dfrac{2}{x^2+2x+1}$

$=\dfrac{3x}{(x+1)(x^2-x+1)}-\dfrac{2}{(x+1)^2}$

$=\dfrac{3x}{(x+1)(x^2-x+1)}\cdot\dfrac{x+1}{x+1}-\dfrac{2}{(x+1)^2}\cdot\dfrac{x^2-x+1}{x^2-x+1}$

$=\dfrac{3x(x+1)}{(x+1)^2(x^2-x+1)}-\dfrac{2(x^2-x+1)}{(x+1)^2(x^2-x+1)}$

$=\dfrac{3x(x+1)-2(x^2-x+1)}{(x+1)^2(x^2-x+1)}$

$=\dfrac{3x^2+3x-2x^2+2x-2}{(x+1)^2(x^2-x+1)}$

$=\dfrac{x^2+5x-2}{(x+1)^2(x^2-x+1)}$

69. Answers may vary. One possibility follows: The student failed to distribute the subtraction properly through the numerator of the second rational expression. The correct difference is:

$\dfrac{3x}{x+2}-\dfrac{5x+7}{x+2}=\dfrac{3x-(5x+7)}{x+2}$

$=\dfrac{3x-5x-7}{x+2}$

$=\dfrac{-2x-7}{x+2}$

71. Answers may vary. One possibility follows: To subtract two rational expressions with a common denominator, subtract the numerators and keep the common denominator.

73. $\dfrac{x+4}{x+3}\div\dfrac{x-7}{x+3}=\dfrac{x+4}{x+3}\cdot\dfrac{x+3}{x+7}$

$=\dfrac{(x+4)(x+3)}{(x+3)(x+7)}$

$=\dfrac{x+4}{x+7}$

75. $\dfrac{x+4}{x+3}-\dfrac{x-7}{x+3}=\dfrac{x+4-(x-7)}{x+3}$

$=\dfrac{x+4-x+7}{x+3}$

$=\dfrac{11}{x+3}$

77. $\dfrac{x^2-9}{x^2+10x+25}\cdot\dfrac{x^2+5x}{x^2-4x-21}$

$=\dfrac{(x-3)(x+3)}{(x+5)(x+5)}\cdot\dfrac{x(x+5)}{(x-7)(x+3)}$

$=\dfrac{(x-3)(x+3)\cdot x(x+5)}{(x+5)(x+5)\cdot(x-7)(x+3)}$

$=\dfrac{x(x-3)}{(x+5)(x-7)}$

79. $\dfrac{2x}{x^2-4x+4}+\dfrac{4}{x^2-9x+14}$

$=\dfrac{2x}{(x-2)^2}+\dfrac{4}{(x-7)(x-2)}$

$=\dfrac{2x}{(x-2)(x-2)}\cdot\dfrac{x-7}{x-7}+\dfrac{4}{(x-7)(x-2)}\cdot\dfrac{x-2}{x-2}$

$=\dfrac{2x(x-7)}{(x-2)^2(x-7)}+\dfrac{4(x-2)}{(x-7)(x-2)^2}$

$=\dfrac{2x(x-7)+4(x-2)}{(x-7)(x-2)^2}$

$=\dfrac{2x^2-14x+4x-8}{(x-7)(x-2)^2}$

$=\dfrac{2x^2-10x-8}{(x-7)(x-2)^2}$

$=\dfrac{2(x^2-5x-4)}{(x-7)(x-2)^2}$

81. $\dfrac{x^2-10x+16}{5x^2-3x}\div\dfrac{x^2-3x-40}{25x^2-9}$

$=\dfrac{x^2-10x+16}{5x^2-3x}\cdot\dfrac{25x^2-9}{x^2-3x-40}$

$=\dfrac{(x-8)(x-2)}{x(5x-3)}\cdot\dfrac{(5x-3)(5x+3)}{(x-8)(x+5)}$

$=\dfrac{(x-8)(x-2)\cdot(5x-3)(5x+3)}{x(5x-3)\cdot(x-8)(x+5)}$

$=\dfrac{(x-2)(5x+3)}{x(x+5)}$

83. $\dfrac{3}{x^2+4x-21}-\dfrac{5x}{2x-6}$

$$=\dfrac{3}{(x+7)(x-3)}-\dfrac{5x}{2(x-3)}$$

$$=\dfrac{3}{(x+7)(x-3)}\cdot\dfrac{2}{2}-\dfrac{5x}{2(x-3)}\cdot\dfrac{x+7}{x+7}$$

$$=\dfrac{6}{2(x+7)(x-3)}-\dfrac{5x(x+7)}{2(x-3)(x+7)}$$

$$=\dfrac{6-5x(x+7)}{2(x-3)(x+7)}$$

$$=\dfrac{6-5x^2-35x}{2(x-3)(x+7)}$$

$$=\dfrac{-5x^2-35x+6}{2(x-3)(x+7)}$$

85. a. $\dfrac{x}{2}+\dfrac{4}{x}=\dfrac{x}{2}\cdot\dfrac{x}{x}+\dfrac{4}{x}\cdot\dfrac{2}{2}=\dfrac{x^2}{2x}+\dfrac{8}{2x}=\dfrac{x^2+8}{2x}$

b. $\dfrac{x}{2}-\dfrac{4}{x}=\dfrac{x}{2}\cdot\dfrac{x}{x}-\dfrac{4}{x}\cdot\dfrac{2}{2}=\dfrac{x^2}{2x}-\dfrac{8}{2x}=\dfrac{x^2-8}{2x}$

c. $\dfrac{x}{2}\cdot\dfrac{4}{x}=\dfrac{x\cdot4}{2\cdot x}=\dfrac{x\cdot2\cdot2}{2\cdot x}=2$

d. $\dfrac{x}{2}\div\dfrac{4}{x}=\dfrac{x}{2}\cdot\dfrac{x}{4}=\dfrac{x\cdot x}{2\cdot4}=\dfrac{x^2}{8}$

87. $\quad 3x^2+2x=8$

$\quad 3x^2+2x-8=0$

$\quad (3x-4)(x+2)=0$

$\quad 3x-4=0\quad$ or $\quad x+2=0$

$\qquad 3x=4\quad$ or $\qquad x=-2$

$\qquad x=\dfrac{4}{3}$

This is a quadratic (or second-degree) equation in one variable.

89. $3x^2+2x-8=(3x-4)(x+2)$

This is a quadratic (or second-degree) polynomial in one variable.

91. $\dfrac{5}{3x^2+2x-8}-\dfrac{3x}{x^2-4}$

$$=\dfrac{5}{(3x-4)(x+2)}-\dfrac{3x}{(x-2)(x+2)}$$

$$=\dfrac{5}{(3x-4)(x+2)}\cdot\dfrac{x-2}{x-2}-\dfrac{3x}{(x-2)(x+2)}\cdot\dfrac{3x-4}{3x-4}$$

$$=\dfrac{5(x-2)}{(3x-4)(x+2)(x-2)}-\dfrac{3x(3x-4)}{(x-2)(x+2)(3x-4)}$$

$$=\dfrac{5(x-2)-3x(3x-4)}{(x-2)(x+2)(3x-4)}$$

$$=\dfrac{5x-10-9x^2+12x}{(x-2)(x+2)(3x-4)}$$

$$=\dfrac{-9x^2+17x-10}{(x-2)(x+2)(3x-4)}$$

This is a rational expression in one variable.

Homework 10.5

1. $\dfrac{3}{x}-2=\dfrac{7}{x}$

We note that 0 is an excluded value.

$$x\cdot\left(\dfrac{3}{x}-2\right)=x\cdot\left(\dfrac{7}{x}\right)$$

$$x\cdot\dfrac{3}{x}-x\cdot2=x\cdot\dfrac{7}{x}$$

$$3-2x=7$$

$$-2x=4$$

$$x=-2$$

Since -2 is not an excluded value, we conclude that -2 is the solution of the equation.

3. $5-\dfrac{4}{x}=3+\dfrac{2}{x}$

We note that 0 is an excluded value.

$$x\cdot\left(5-\dfrac{4}{x}\right)=x\cdot\left(3+\dfrac{2}{x}\right)$$

$$x\cdot5-x\cdot\dfrac{4}{x}=x\cdot3+x\cdot\dfrac{2}{x}$$

$$5x-4=3x+2$$

$$2x=6$$

$$x=3$$

Since 3 is not an excluded value, we conclude that 3 is the solution of the equation.

5. $\dfrac{5}{p-1} = \dfrac{2p+1}{p-1}$

We note that 1 is an excluded value.

$$(p-1)\cdot\left(\frac{5}{p-1}\right) = (p-1)\cdot\left(\frac{2p+1}{p-1}\right)$$
$$5 = 2p+1$$
$$4 = 2p$$
$$2 = p$$

Since 2 is not an excluded value, we conclude that 2 is the solution of the equation.

7. $\dfrac{8x+4}{x+2} = \dfrac{5x-2}{x+2}$

We note that -2 is an excluded value.

$$(x+2)\cdot\left(\frac{8x+4}{x+2}\right) = (x+2)\cdot\left(\frac{5x-2}{x+2}\right)$$
$$8x+4 = 5x-2$$
$$3x = -6$$
$$x = -2$$

Our result -2 is not a solution because it is an excluded value. Since this is the only possible solution, we conclude that no number is a solution. That is, the solution is the empty set.

9. $\dfrac{w+2}{w-4} + 3 = \dfrac{2}{w-4}$

We note that 4 is an excluded value.

$$(w-4)\cdot\left(\frac{w+2}{w-4}+3\right) = (w-4)\cdot\left(\frac{2}{w-4}\right)$$
$$(w-4)\cdot\frac{w+2}{w-4} + (w-4)\cdot 3 = 2$$
$$w+2+3w-12 = 2$$
$$4w-10 = 2$$
$$4w = 12$$
$$w = 3$$

Since 3 is not an excluded value, we conclude that 3 is the solution of the equation.

11. $\dfrac{2}{x} + \dfrac{5}{4} = \dfrac{3}{x}$

We note that 0 is an excluded value.

$$4x\cdot\left(\frac{2}{x}+\frac{5}{4}\right) = 4x\cdot\left(\frac{3}{x}\right)$$
$$4x\cdot\frac{2}{x} + 4x\cdot\frac{5}{4} = 4x\cdot\frac{3}{x}$$
$$8+5x = 12$$
$$5x = 4$$
$$x = \frac{4}{5}$$

Since $\dfrac{4}{5}$ is not an excluded value, we conclude

that $\dfrac{4}{5}$ is the solution of the equation.

13. $\dfrac{5}{6x} - \dfrac{1}{2} = \dfrac{3}{4x}$

We note that 0 is an excluded value.

$$12x\cdot\left(\frac{5}{6x}-\frac{1}{2}\right) = 12x\cdot\left(\frac{3}{4x}\right)$$
$$12x\cdot\frac{5}{6x} - 12x\cdot\frac{1}{2} = 12x\cdot\frac{3}{4x}$$
$$2\cdot 5 - 6x\cdot 1 = 3\cdot 3$$
$$10-6x = 9$$
$$-6x = -1$$
$$x = \frac{-1}{-6}$$
$$x = \frac{1}{6}$$

Since $\dfrac{1}{6}$ is not an excluded value, we conclude

that $\dfrac{1}{6}$ is the solution of the equation.

15. $\dfrac{4}{x-2} = \dfrac{2}{x+3}$

We note that -3 and 2 are excluded values.

$$(x-2)(x+3)\cdot\left(\frac{4}{x-2}\right) = (x-2)(x+3)\cdot\left(\frac{2}{x+3}\right)$$
$$(x+3)\cdot 4 = (x-2)\cdot 2$$
$$4x+12 = 2x-4$$
$$2x = -16$$
$$x = -8$$

Since -8 is not an excluded value, we conclude that -8 is the solution of the equation.

17. $\dfrac{2r+7}{4r} = \dfrac{5}{3}$

We note that 0 is an excluded value.

$$12r \cdot \left(\dfrac{2r+7}{4r}\right) = 12r \cdot \left(\dfrac{5}{3}\right)$$

$$3 \cdot (2r+7) = 4r \cdot 5$$

$$6r + 21 = 20r$$

$$21 = 14r$$

$$r = \dfrac{21}{14}$$

$$r = \dfrac{3}{2}$$

Since $\dfrac{3}{2}$ is not an excluded value, we conclude

that $\dfrac{3}{2}$ is the solution of the equation.

19. $\dfrac{5}{x+3} + \dfrac{3}{4} = 2$

We note that -3 is an excluded value.

$$4(x+3) \cdot \left(\dfrac{5}{x+3} + \dfrac{3}{4}\right) = 4(x+3) \cdot (2)$$

$$4(x+3) \cdot \dfrac{5}{x+3} + 4(x+3) \cdot \dfrac{3}{4} = 8(x+3)$$

$$4 \cdot 5 + (x+3) \cdot 3 = 8x + 24$$

$$20 + 3x + 9 = 8x + 24$$

$$3x + 29 = 8x + 24$$

$$-5x = -5$$

$$x = 1$$

Since 1 is not an excluded value, we conclude that 1 is the solution of the equation.

---

21. $\dfrac{2}{x-3} + \dfrac{1}{x+3} = \dfrac{5}{x^2-9}$

$$\dfrac{2}{x-3} + \dfrac{1}{x+3} = \dfrac{5}{(x-3)(x+3)}$$

We note that -3 and 3 are excluded values.

$$(x-3)(x+3) \cdot \left(\dfrac{2}{x-3} + \dfrac{1}{x+3}\right) = (x-3)(x+3) \cdot \left(\dfrac{5}{(x-3)(x+3)}\right)$$

$$(x-3)(x+3) \cdot \dfrac{2}{x-3} + (x-3)(x+3) \cdot \dfrac{1}{x+3} = 5$$

$$(x+3) \cdot 2 + (x-3) \cdot 1 = 5$$

$$2x + 6 + x - 3 = 5$$

$$3x + 3 = 5$$

$$3x = 2$$

$$x = \dfrac{2}{3}$$

Since 1 is not an excluded value, we conclude that 1 is the solution of the equation.

23. $\dfrac{4}{x+2} + \dfrac{3}{x+1} = \dfrac{3}{x^2+3x+2}$

$$\dfrac{4}{x+2} + \dfrac{3}{x+1} = \dfrac{3}{(x+2)(x+1)}$$

We note that -2 and -1 are excluded values.

$$(x+2)(x+1) \cdot \left(\dfrac{4}{x+2} + \dfrac{3}{x+1}\right) = (x+2)(x+1) \cdot \left(\dfrac{3}{(x+2)(x+1)}\right)$$

$$(x+2)(x+1) \cdot \dfrac{4}{x+2} + (x+2)(x+1) \cdot \dfrac{3}{x+1} = 3$$

$$(x+1)\cdot 4 + (x+2)\cdot 3 = 3$$
$$4x+4+3x+6 = 3$$
$$7x+10 = 3$$
$$7x = -7$$
$$x = -1$$

Our result -1 is not a solution because it is an excluded value. Since this is the only possible solution, we conclude that no number is a solution. That is, the solution is the empty set.

25.
$$\frac{5}{x^2-4} + \frac{2}{x+2} = \frac{4}{x-2}$$

$$\frac{5}{(x-2)(x+2)} + \frac{2}{x+2} = \frac{4}{x-2}$$

We note that -2 and 2 are excluded values.

$$(x-2)(x+2)\cdot\left(\frac{5}{(x-2)(x+2)} + \frac{2}{x+2}\right) = (x-2)(x+2)\cdot\left(\frac{4}{x-2}\right)$$

$$(x-2)(x+2)\cdot\frac{5}{(x-2)(x+2)} + (x-2)(x+2)\cdot\frac{2}{x+2} = (x+2)\cdot 4$$

$$5+(x-2)\cdot 2 = 4x+8$$
$$5+2x-4 = 4x+8$$
$$2x+1 = 4x+8$$
$$-2x = 7$$
$$x = -\frac{7}{2}$$

Since $-\frac{7}{2}$ is not an excluded value, we conclude that $-\frac{7}{2}$ is the solution of the equation.

27.
$$\frac{3}{y-4} - \frac{4}{y-3} = \frac{3}{y^2-7y+12}$$

$$\frac{3}{y-4} - \frac{4}{y-3} = \frac{3}{(y-4)(y-3)}$$

We note that 3 and 4 are excluded values.

$$(y-4)(y-3)\cdot\left(\frac{3}{y-4} - \frac{4}{y-3}\right) = (y-4)(y-3)\cdot\left(\frac{3}{(y-4)(y-3)}\right)$$

$$(y-4)(y-3)\cdot\frac{3}{y-4} - (y-4)(y-3)\cdot\frac{4}{y-3} = 3$$

$$(y-3)\cdot 3 - (y-4)\cdot 4 = 3$$
$$3y-9-4y+16 = 3$$
$$-y+7 = 3$$
$$-y = -4$$
$$y = 4$$

Our result 4 is not a solution because it is an excluded value. Since this is the only possible solution, we conclude that no number is a solution. That is, the solution is the empty set.

29.

$$\frac{4}{x^2 - 3x} - \frac{5}{x} = \frac{7}{x - 3}$$

$$\frac{4}{x(x-3)} - \frac{5}{x} = \frac{7}{x - 3}$$

We note that 0 and 3 are excluded values.

$$x(x-3) \cdot \left(\frac{4}{x(x-3)} - \frac{5}{x} \right) = x(x-3) \cdot \left(\frac{7}{x-3} \right)$$

$$x(x-3) \cdot \frac{4}{x(x-3)} - x(x-3) \cdot \frac{5}{x} = x \cdot 7$$

$$4 - (x-3) \cdot 5 = 7x$$

$$4 - 5x + 15 = 7x$$

$$-5x + 19 = 7x$$

$$19 = 12x$$

$$x = \frac{19}{12}$$

Since $\frac{19}{12}$ is not an excluded value, we conclude that $\frac{19}{12}$ is the solution of the equation.

31.

$$\frac{3}{2x - 6} + \frac{5x}{6x - 18} = \frac{2}{4x - 12}$$

$$\frac{3}{2(x-3)} + \frac{5x}{6(x-3)} = \frac{2}{4(x-3)}$$

We note that 3 is an excluded value.

$$12(x-3) \cdot \left(\frac{3}{2(x-3)} + \frac{5x}{6(x-3)} \right) = 12(x-3) \cdot \left(\frac{2}{4(x-3)} \right)$$

$$12(x-3) \cdot \frac{3}{2(x-3)} + 12(x-3) \cdot \frac{5x}{6(x-3)} = 3 \cdot 2$$

$$6 \cdot 3 + 2 \cdot 5x = 6$$

$$18 + 10x = 6$$

$$10x = -12$$

$$x = \frac{-12}{10} = -\frac{6}{5}$$

Since $-\frac{6}{5}$ is not an excluded value, we conclude that $-\frac{6}{5}$ is the solution of the equation.

33.

$$\frac{2}{x^2 - x - 6} - \frac{4}{x + 2} = \frac{3}{2x - 6}$$

$$\frac{2}{(x-3)(x+2)} - \frac{4}{x+2} = \frac{3}{2(x-3)}$$

We note that -2 and 3 are excluded values.

$$2(x-3)(x+2) \cdot \left(\frac{2}{(x-3)(x+2)} - \frac{4}{x+2} \right) = 2(x-3)(x+2) \cdot \left(\frac{3}{2(x-3)} \right)$$

$$2(x-3)(x+2) \cdot \frac{2}{(x-3)(x+2)} - 2(x-3)(x+2) \cdot \frac{4}{x+2} = (x+2) \cdot 3$$

$$2 \cdot 2 - 2(x-3) \cdot 4 = 3x + 6$$
$$4 - 8(x-3) = 3x + 6$$
$$4 - 8x + 24 = 3x + 6$$
$$-8x + 28 = 3x + 6$$
$$-11x = -22$$
$$x = 2$$

Since 2 is not an excluded value, we conclude that 2 is the solution of the equation.

---

35. $\dfrac{3}{x} = \dfrac{4}{x^2} - 1$

We note that 0 is an excluded value.
$$x^2 \cdot \left(\dfrac{3}{x}\right) = x^2 \cdot \left(\dfrac{4}{x^2} - 1\right)$$
$$x \cdot 3 = x^2 \cdot \left(\dfrac{4}{x^2}\right) - x^2 \cdot 1$$
$$3x = 4 - x^2$$
$$x^2 + 3x - 4 = 0$$
$$(x+4)(x-1) = 0$$
$$x + 4 = 0 \quad \text{or} \quad x - 1 = 0$$
$$x = -4 \quad \text{or} \quad x = 1$$

Since −4 and 1 are neither excluded values, we conclude that −4 and 1 are the solutions of the equation.

37. $1 = \dfrac{15}{t^2} - \dfrac{2}{t}$

We note that 0 is an excluded value.
$$t^2 \cdot 1 = t^2 \cdot \left(\dfrac{15}{t^2} - \dfrac{2}{t}\right)$$
$$t^2 = t^2 \cdot \left(\dfrac{15}{t^2}\right) - t^2 \cdot \left(\dfrac{2}{t}\right)$$
$$t^2 = 15 - t \cdot 2$$
$$t^2 = 15 - 2t$$
$$t^2 + 2t - 15 = 0$$
$$(t+5)(t-3) = 0$$
$$t + 5 = 0 \quad \text{or} \quad t - 3 = 0$$
$$t = -5 \quad \text{or} \quad t = 3$$

Since −5 and 3 are neither excluded values, we conclude that −5 and 3 are the solutions of the equation.

39. $\dfrac{2}{x^2} + \dfrac{3}{x} = 5$

We note that 0 is an excluded value.
$$x^2 \cdot \left(\dfrac{2}{x^2} + \dfrac{3}{x}\right) = x^2 \cdot 5$$
$$x^2 \cdot \left(\dfrac{2}{x^2}\right) + x^2 \cdot \left(\dfrac{3}{x}\right) = 5x^2$$
$$2 + x \cdot 3 = 5x^2$$
$$2 + 3x = 5x^2$$
$$0 = 5x^2 - 3x - 2$$
$$0 = (5x+2)(x-1)$$
$$5x + 2 = 0 \quad \text{or} \quad x - 1 = 0$$
$$5x = -2 \quad \text{or} \quad x = 1$$
$$x = -\dfrac{2}{5}$$

Since $-\dfrac{2}{5}$ and 1 are neither excluded values, we conclude that $-\dfrac{2}{5}$ and 1 are the solutions of the equation.

41. $\dfrac{x-3}{x+2} = \dfrac{x+5}{x-1}$

We note that −2 and 1 are excluded values.
$$(x+2)(x-1) \cdot \left(\dfrac{x-3}{x+2}\right) = (x+2)(x-1) \cdot \left(\dfrac{x+5}{x-1}\right)$$
$$(x-1)(x-3) = (x+2)(x+5)$$
$$x^2 - 3x - x + 3 = x^2 + 5x + 2x + 10$$
$$x^2 - 4x + 3 = x^2 + 7x + 10$$
$$-4x + 3 = 7x + 10$$
$$-11x + 3 = 10$$
$$-11x = 7$$
$$x = -\dfrac{7}{11}$$

Since $-\dfrac{7}{11}$ is not an excluded value, we conclude that $-\dfrac{7}{11}$ is the solution of the equation.

43. $\dfrac{x}{x+1} = \dfrac{3}{x-1} + \dfrac{2}{x^2-1}$

$\dfrac{x}{x+1} = \dfrac{3}{x-1} + \dfrac{2}{(x-1)(x+1)}$

We note that -1 and 1 are excluded values.

$(x-1)(x+1)\cdot\left(\dfrac{x}{x+1}\right) = (x-1)(x+1)\cdot\left(\dfrac{3}{x-1} + \dfrac{2}{(x-1)(x+1)}\right)$

$(x-1)\cdot x = (x-1)(x+1)\cdot\left(\dfrac{3}{x-1}\right) + (x-1)(x+1)\cdot\left(\dfrac{2}{(x-1)(x+1)}\right)$

$x^2 - x = (x+1)\cdot 3 + 2$

$x^2 - x = 3x + 3 + 2$

$x^2 - x = 3x + 5$

$x^2 - 4x - 5 = 0$

$(x-5)(x+1) = 0$

$x - 5 = 0$ or $x + 1 = 0$

$x = 5$ or $x = -1$

Since -1 is an excluded value, it is not a solution. We conclude that 5 is the only solution of the equation.

45. $\dfrac{r}{r-3} - \dfrac{2}{r+5} = \dfrac{10}{r^2+2r-15}$

$\dfrac{r}{r-3} - \dfrac{2}{r+5} = \dfrac{10}{(r+5)(r-3)}$

We note that -5 and 3 are excluded values.

$(r+5)(r-3)\cdot\left(\dfrac{r}{r-3} - \dfrac{2}{r+5}\right) = (r+5)(r-3)\cdot\left(\dfrac{10}{(r+5)(r-3)}\right)$

$(r+5)(r-3)\cdot\left(\dfrac{r}{r-3}\right) - (r+5)(r-3)\cdot\left(\dfrac{2}{r+5}\right) = 10$

$(r+5)\cdot r - (r-3)\cdot 2 = 10$

$r^2 + 5r - 2r + 6 = 10$

$r^2 + 3r - 4 = 0$

$(r+4)(r-1) = 0$

$r + 4 = 0$ or $r - 1 = 0$

$r = -4$ or $r = 1$

Since -4 and 1 are neither excluded values, we conclude that -4 and 1 are the solutions of the equation.

47.
$$\frac{1}{x-2} = \frac{2x}{x+1} - \frac{6}{x^2-x-2}$$

$$\frac{1}{x-2} = \frac{2x}{x+1} - \frac{6}{(x-2)(x+1)}$$

We note that -1 and 2 are excluded values.

$$(x-2)(x+1) \cdot \left(\frac{1}{x-2}\right) = (x-2)(x+1) \cdot \left(\frac{2x}{x+1} - \frac{6}{(x-2)(x+1)}\right)$$

$$(x+1) \cdot 1 = (x-2)(x+1) \cdot \left(\frac{2x}{x+1}\right) - (x-2)(x+1) \cdot \left(\frac{6}{(x-2)(x+1)}\right)$$

$$x+1 = (x-2) \cdot 2x - 6$$

$$x+1 = 2x^2 - 4x - 6$$

$$0 = 2x^2 - 5x - 7$$

$$0 = (2x-7)(x+1)$$

$$2x-7 = 0 \quad \text{or} \quad x+1 = 0$$

$$2x = 7 \quad \text{or} \qquad x = -1$$

$$x = \frac{7}{2}$$

Since -1 is an excluded value, it is not a solution. We conclude that $\frac{7}{2}$ is the only solution of the equation.

49.
$$\frac{2}{x^2-9} = \frac{x}{x-3} - \frac{x-5}{x+3}$$

$$\frac{2}{(x-3)(x+3)} = \frac{x}{x-3} - \frac{x-5}{x+3}$$

We note that -3 and 3 are excluded values.

$$(x-3)(x+3) \cdot \left(\frac{2}{(x-3)(x+3)}\right) = (x-3)(x+3) \cdot \left(\frac{x}{x-3} - \frac{x-5}{x+3}\right)$$

$$2 = (x-3)(x+3) \cdot \left(\frac{x}{x-3}\right) - (x-3)(x+3) \cdot \left(\frac{x-5}{x+3}\right)$$

$$2 = (x+3) \cdot x - (x-3) \cdot (x-5)$$

$$2 = x^2 + 3x - (x^2 - 5x - 3x + 15)$$

$$2 = x^2 + 3x - x^2 + 5x + 3x - 15$$

$$2 = 11x - 15$$

$$17 = 11x$$

$$x = \frac{17}{11}$$

Since $\frac{17}{11}$ is not an excluded value, we conclude that $\frac{17}{11}$ is the solution of the equation.

51. $\dfrac{3}{x-4}+\dfrac{7}{x-3}=\dfrac{x+4}{x-3}$

We note that 3 and 4 are excluded values.

$$(x-4)(x-3)\cdot\left(\dfrac{3}{x-4}+\dfrac{7}{x-3}\right)=(x-4)(x-3)\cdot\left(\dfrac{x+4}{x-3}\right)$$

$$(x-4)(x-3)\cdot\left(\dfrac{3}{x-4}\right)+(x-4)(x-3)\cdot\left(\dfrac{7}{x-3}\right)=(x-4)\cdot(x+4)$$

$$(x-3)\cdot 3+(x-4)\cdot 7=x^2+4x-4x-16$$

$$3x-9+7x-28=x^2-16$$

$$10x-37=x^2-16$$

$$0=x^2-10x+21$$

$$0=(x-3)(x-7)$$

$$x-3=0 \quad\text{or}\quad x-7=0$$

$$x=3 \quad\text{or}\quad x=7$$

Since 3 is an excluded value, it is not a solution. We conclude that 7 is the only solution of the equation.

53. $\dfrac{2}{m-3}+\dfrac{5m}{m^2-9}=\dfrac{4}{m+3}-\dfrac{3}{m^2-9}$

$$\dfrac{2}{m-3}+\dfrac{5m}{(m-3)(m+3)}=\dfrac{4}{m+3}-\dfrac{3}{(m-3)(m+3)}$$

We note that -3 and 3 are excluded values.

$$(m-3)(m+3)\cdot\left(\dfrac{2}{m-3}+\dfrac{5m}{(m-3)(m+3)}\right)=(m-3)(m+3)\cdot\left(\dfrac{4}{m+3}-\dfrac{3}{(m-3)(m+3)}\right)$$

$$(m-3)(m+3)\left(\dfrac{2}{m-3}\right)+(m-3)(m+3)\left(\dfrac{5m}{(m-3)(m+3)}\right)=(m-3)(m+3)\left(\dfrac{4}{m+3}\right)-(m-3)(m+3)\left(\dfrac{3}{(m-3)(m+3)}\right)$$

$$(m+3)\cdot 2+5m=(m-3)\cdot 4-3$$

$$2m+6+5m=4m-12-3$$

$$7m+6=4m-15$$

$$3m+6=-15$$

$$3m=-21$$

$$m=-7$$

Since -7 is not an excluded value, we conclude that -7 is the solution of the equation.

55. a. The year 2003 is represented by $t=13$. We substitute 13 for t in the model:

$$D=\dfrac{120(13)+460}{2.06(13)+84.82}=\dfrac{1560+460}{26.78+84.82}=\dfrac{2020}{111.21}\approx 18.1$$

Using the model, we predict the average household debt in 2003 was $18.1 thousand.

To find the actual average household debt in 2003, we divide the total consumer debt by the number of households:

$$\dfrac{\$2.0\text{ trillion}}{111.6\text{ million households}}=\dfrac{\$2,000,000,000,000}{111,600,000\text{ households}}\approx\$17,900/\text{household}$$

The actual average household debt in 2003 was approximately $17.9 thousand.

Our result from the model was an overestimate.

b. The year 2010 is represented by $t = 20$. We substitute 20 for t in the model:

$$D = \frac{120(20) + 460}{2.06(20) + 84.82} = \frac{2400 + 460}{41.2 + 84.82} = \frac{2860}{126.02} \approx 22.7$$

We predict the average household debt in 2010 will be \$22.7 thousand.

c. We substitute 24 for D and solve for t:

$$24 = \frac{120t + 460}{2.06t + 84.82}$$

$$(2.06t + 84.82) \cdot 24 = (2.06t + 84.82)\left(\frac{120t + 460}{2.06t + 84.82}\right)$$

$$49.44t + 2035.68 = 120t + 460$$

$$-70.56t = -1575.68$$

$$t \approx 22.3$$

Now, $1990 + 22.3 = 2012.3$.

We predict that average household debt will be \$24 thousand in about the year 2012.

57. We substitute 38 for p and solve for t:

$$38 = \frac{-28t^2 + 2694t + 15,122}{50.8t + 275}$$

$$(50.8t + 275) \cdot 38 = (50.8t + 275)\left(\frac{-28t^2 + 2694t + 15,122}{50.8t + 275}\right)$$

$$1930.4t + 10,450 = -28t^2 + 2694t + 15,122$$

$$28t^2 - 763.6t - 4672 = 0$$

We use the quadratic formula to solve the equation:

$$t = \frac{-(-763.6) \pm \sqrt{(-763.6)^2 - 4(28)(-4672)}}{2(28)}$$

$$t = \frac{763.6 \pm \sqrt{583,084.96 + 523,264}}{56}$$

$$t = \frac{763.6 \pm \sqrt{1,106,348.96}}{56}$$

$$t \approx \frac{763.6 \pm 1051.8312}{56}$$

$$t \approx \frac{763.6 - 1051.8312}{56} \quad \text{or} \quad t \approx \frac{763.6 + 1051.8312}{56}$$

$$t \approx \frac{-288.2312}{56} \quad \text{or} \quad t \approx \frac{1815.4312}{56}$$

$$t \approx -5.15 \quad \text{or} \quad t \approx 32.42$$

We disregard -5.15. Now, $1980 + 32.42 = 2012.42$.

We predict that 38% of prisoners will be released in the year 2012.

59. Answers may vary. One possibility follows: The student has confused adding rational expression with solving rational equations. The student should not multiply by the common denominator. The correct solution is:

$$\frac{5}{x+2} + \frac{3}{x} = \frac{5}{x+2} \cdot \frac{x}{x} + \frac{3}{x} \cdot \frac{x+2}{x+2} = \frac{5x}{(x+2)x} + \frac{3(x+2)}{x(x+2)} = \frac{5x + 3(x+2)}{x(x+2)} = \frac{5x + 3x + 6}{x(x+2)} = \frac{8x + 6}{x(x+2)} = \frac{2(4x+3)}{x(x+2)}$$

61. Answers may vary. One possibility follows: The student has simplified the right side of the equation but has not solved the equation. The correct solution is:

$$\frac{7}{x^2 + x - 20} = \frac{4}{x+5} - \frac{2}{x-4}$$

$$\frac{7}{(x+5)(x-4)} = \frac{4}{x+5} - \frac{2}{x-4}$$

We note that -5 and 4 are excluded values.

$$(x+5)(x-4) \cdot \left(\frac{7}{(x+5)(x-4)} \right) = (x+5)(x-4) \cdot \left(\frac{4}{x+5} - \frac{2}{x-4} \right)$$

$$7 = (x+5)(x-4) \cdot \left(\frac{4}{x+5} \right) - (x+5)(x-4) \cdot \left(\frac{2}{x-4} \right)$$

$$7 = (x-4) \cdot 4 - (x+5) \cdot 2$$

$$7 = 4x - 16 - 2x - 10$$

$$7 = 2x - 26$$

$$33 = 2x$$

$$x = \frac{33}{2}$$

Since $\frac{33}{2}$ is not an excluded value, we conclude that $\frac{33}{2}$ is the solution of the equation.

63. Answers may vary. One possibility follows:

When simplifying a rational expression, we cannot multiply it by the LCD because doing so will cause a non-equivalent expression. We can multiply the numerator and denominator of the rational expression by the same expression, but generally will not be the LCD.

When solving a rational equation, we can multiply both sides of the equation by the LCD. Doing so will gives a simpler equation to solve, but we have to be sure to discard any solutions that are excluded values.

65. $\dfrac{7}{x} + \dfrac{3}{x} = \dfrac{7+3}{x} = \dfrac{10}{x}$

67. $\dfrac{7}{x} + \dfrac{3}{x} = 1$

We note that 0 is an excluded value.

$$x \cdot \left(\frac{7}{x} + \frac{3}{x} \right) = x \cdot 1$$

$$x \cdot \left(\frac{7}{x} \right) + x \cdot \left(\frac{3}{x} \right) = x$$

$$7 + 3 = x$$

$$10 = x$$

Since 10 is not an excluded value, we conclude that 10 is the solution of the equation.

69. $\dfrac{2}{x-3} + \dfrac{3x}{x+2} = \dfrac{2}{x^2 - x - 6}$

$$\frac{2}{x-3} + \frac{3x}{x+2} = \frac{2}{(x-3)(x+2)}$$

We note that -2 and 3 are excluded values.

$$(x-3)(x+2)\cdot\left(\frac{2}{x-3}+\frac{3x}{x+2}\right)=(x-3)(x+2)\cdot\left(\frac{2}{(x-3)(x+2)}\right)$$

$$(x-3)(x+2)\cdot\left(\frac{2}{x-3}\right)+(x-3)(x+2)\cdot\left(\frac{3x}{x+2}\right)=2$$

$$(x+2)\cdot 2+(x-3)\cdot 3x=2$$

$$2x+4+3x^2-9x=2$$

$$3x^2-7x+2=0$$

$$(3x-1)(x-2)=0$$

$$3x-1=0 \quad \text{or} \quad x-2=0$$

$$3x=1 \quad \text{or} \quad x=2$$

$$x=\frac{1}{3}$$

Since $\frac{1}{3}$ and 2 are neither excluded values, we conclude that $\frac{1}{3}$ and 2 are the solutions of the equation.

71.
$$\frac{2}{x-3}+\frac{3x}{x+2}=\frac{2}{x-3}\cdot\frac{x+2}{x+2}+\frac{3x}{x+2}\cdot\frac{x-3}{x-3}$$

$$=\frac{2(x+2)}{(x-3)(x+2)}+\frac{3x(x-3)}{(x+2)(x-3)}$$

$$=\frac{2(x+2)+3x(x-3)}{(x-3)(x+2)}$$

$$=\frac{2x+4+3x^2-9x}{(x-3)(x+2)}$$

$$=\frac{3x^2-7x+4}{(x-3)(x+2)}$$

$$=\frac{(3x-4)(x-1)}{(x-3)(x+2)}$$

73.
$$3x^2-2x=4$$

$$3x^2-2x-4=0$$

We use the quadratic formula to solve the equation, with $a=3$, $b=-2$, and $c=-4$:

$$x=\frac{-(-2)\pm\sqrt{(-2)^2-4(3)(-4)}}{2(3)}$$

$$=\frac{2\pm\sqrt{4+48}}{6}=\frac{2\pm\sqrt{52}}{6}$$

$$=\frac{2\pm 2\sqrt{13}}{6}=\frac{2(1\pm\sqrt{13})}{2(3)}$$

$$=\frac{1\pm\sqrt{13}}{3}$$

The solutions are $\frac{1\pm\sqrt{13}}{3}$.

75.
$$3x^3+8=2x^2+12x$$

$$3x^3-2x^2-12x+8=0$$

$$x^2(3x-2)-4(3x-2)=0$$

$$(x^2-4)(3x-2)=0$$

$$(x-2)(x+2)(3x-2)=0$$

$$(x-2)(x+2)(3x-2)=0$$

$$x-2=0 \quad \text{or} \quad x+2=0 \quad \text{or} \quad 3x-2=0$$

$$x=2 \quad \text{or} \quad x=-2 \quad \text{or} \quad 3x=2$$

$$x=\frac{2}{3}$$

The solutions are -2, $\frac{2}{3}$, and 2.

77.
$$2x^2+1=8$$

$$2x^2=7$$

$$x^2=\frac{7}{2}$$

$$x=\pm\sqrt{\frac{7}{2}}=\pm\frac{\sqrt{7}}{\sqrt{2}}\cdot\frac{\sqrt{2}}{\sqrt{2}}=\pm\frac{\sqrt{14}}{2}$$

79.
$$3(5m-2n)-4(m+3n)+n$$

$$=15m-6n-4m-12n+n$$

$$=15m-4m-6n-12n+n$$

$$=11m-17n$$

This is a linear (or first-degree) polynomial in two variables.

81. $2x^3 - 18x^2 + 28x = 2x(x^2 - 9x + 14)$
$$= 2x(x-7)(x-2)$$

This is a cubic (or third-degree) polynomial in one variable.

83. $\dfrac{2}{x-4} - \dfrac{3}{x+4} = \dfrac{5}{x^2 - 16}$

$$\dfrac{2}{x-4} - \dfrac{3}{x+4} = \dfrac{5}{(x-4)(x+4)}$$

We note that -4 and 4 are excluded values.

$$(x-4)(x+4) \cdot \left(\dfrac{2}{x-4} - \dfrac{3}{x+4} \right) = (x-4)(x+4) \cdot \left(\dfrac{5}{(x-4)(x+4)} \right)$$

$$(x-4)(x+4) \cdot \left(\dfrac{2}{x-4} \right) - (x-4)(x+4) \cdot \left(\dfrac{3}{x+4} \right) = 5$$

$$(x+4) \cdot 2 - (x-4) \cdot 3 = 5$$

$$2x + 8 - 3x + 12 = 5$$

$$-x + 20 = 5$$

$$-x = -15$$

$$x = 15$$

Since 15 is not an excluded value, we conclude that 15 is the solution of the equation. This is a rational equation in one variable.

Homework 10.6

1. Let x be the cost for 6 months' use of AOL.

$$\dfrac{95.60}{4} = \dfrac{x}{6}$$

$$12\left(\dfrac{95.60}{4} \right) = 12\left(\dfrac{x}{6} \right)$$

$$286.80 = 2x$$

$$143.40 = x$$

The person will pay $143.40 for 6 months' use of AOL.

3. Let x be the number of cups that contain 7 grams of sugar.

$$\dfrac{0.75}{4} = \dfrac{x}{7}$$

$$28\left(\dfrac{0.75}{4} \right) = 28\left(\dfrac{x}{7} \right)$$

$$5.25 = 4x$$

$$1.31 \approx x$$

Approximately 1.31 cubs of Post Grape-Nuts Flakes® cereal contains 7 grams of sugar.

5. Let x be the number of ounces of acid in 6 ounces of the solution.

$$\dfrac{10}{4} = \dfrac{6}{x}$$

$$4x\left(\dfrac{10}{4} \right) = 4x\left(\dfrac{6}{x} \right)$$

$$10x = 24$$

$$x = 2.4$$

There are 2.4 ounces of acid in 6 ounces of the solution.

7. Let x be the enrollment at the college.

$$\dfrac{3}{5} = \dfrac{21,720}{x}$$

$$5x\left(\dfrac{3}{5} \right) = 5x\left(\dfrac{21,720}{x} \right)$$

$$3x = 108,600$$

$$x = 36,200$$

The enrollment at the college is 36,200 students.

9. Let x be the number of inches on the map that represent 270 miles.

$$\frac{1.25}{50} = \frac{x}{270}$$

$$1350\left(\frac{1.25}{50}\right) = 1350\left(\frac{x}{270}\right)$$

$$33.75 = 5x$$

$$6.75 = x$$

On the map, 6.75 inches represent 270 miles.

11. Let x be the number of U.S. dollars that can be exchanged for 260 euros.

$$\frac{4.05}{5} = \frac{260}{x}$$

$$5x\left(\frac{4.05}{5}\right) = 5x\left(\frac{260}{x}\right)$$

$$4.05x = 1300$$

$$x \approx 320.99$$

Approximately 320.99 U.S. dollars can be exchanged for 260 euros.

13. Let x be the weight of the person on Earth who weights 28.5 pounds on the Moon.

$$\frac{29.8}{180} = \frac{28.5}{x}$$

$$180x\left(\frac{29.8}{180}\right) = 180x\left(\frac{28.5}{x}\right)$$

$$29.8x = 5130$$

$$x \approx 172$$

The person's weighs approximately 172 pounds on Earth if she weights 28.5 pounds on the Moon.

15. a. Let x be the amount the person will pay for 7 months of phone service.

$$\frac{175.50}{3} = \frac{x}{7}$$

$$21\left(\frac{175.50}{3}\right) = 21\left(\frac{x}{7}\right)$$

$$1228.50 = 3x$$

$$409.50 = x$$

The person will pay \$409.50 for 7 months of phone service.

b. Answers may vary. One possibility follows: We assumed that the person's monthly bill for phone service is constant. If the person begins to make a lot of long-distance calls during the 7-month period, then our estimate will be an underestimate.

17. a. Let x be the number of adults in Deerfield, Illinois who have online access at home.

$$\frac{x}{13,590} = \frac{1335}{2022}$$

$$13,590\left(\frac{x}{13,590}\right) = 13,590\left(\frac{1335}{2022}\right)$$

$$x \approx 8973$$

Approximately 8973 adults in Deerfield, Illinois have online access at home.

b. Answers may vary. One possibility follows: We assumed that the ratio of adults in Deerfield, Illinois who have internet access at home is the same as that for the entire United States. Since the median household income for Deerfield, Illinois is a lot higher than the median income for the United States, the estimate from part (a) is likely an underestimate, since adults from Deerfield would be more likely to afford it.

19. $\dfrac{26 \text{ tons}}{40 \text{ mph}} = 0.65$ tons per mph

$\dfrac{55 \text{ tons}}{60 \text{ mph}} = 0.92$ tons per mph

No, the speed and the force of impact are not proportional for a 4000-pound car, since the ratios are not the same.

21. a. $\dfrac{27,764}{1294} \approx 21.5$

The ratio of the FTE enrollment to the number of FTE faculty at LSU in fall 2003 was 21.5:1. This ratio is greater than 17:1.

b. Let x by the number of FTE faculty who would need to be hired for the 17:1 ratio.

$$\frac{27,764}{1294+x} = \frac{17}{1}$$

$$(1294+x)\left(\frac{27,764}{1294+x}\right) = (1294+x)\left(\frac{17}{1}\right)$$

$$27,764 = 21,998 + 17x$$

$$5766 = 17x$$

$$339.2 \approx x$$

LSU would need to hire approximately 339.2 FTE faculty to achieve the 17:1 ratio.

$$339.2 \text{ FTE} \cdot \left(\frac{\$73,125}{\text{FTE}}\right) = \$24,804,000$$

The cost would be about \$24.8 million in order to achieve the 17:1 ratio.

c. Let x be the reduction in FTE enrollment for the 17:1 ratio.

$$\frac{27,764 - x}{1294} = \frac{17}{1}$$

$$1294\left(\frac{27,764 - x}{1294}\right) = 1294\left(\frac{17}{1}\right)$$

$$27,764 - x = 21,998$$

$$-x = -5766$$

$$x = 5766$$

FTE enrollment would need to be reduced by 5766 students in order to achieve the 17:1 ratio.

$$5766 \text{ FTE} \cdot \left(\frac{\$3345}{\text{FTE}}\right) = \$19,287,270$$

The cost would be about $19.3 million in order to achieve the 17:1 ratio.

d. The cheaper way to reduce the ratio would be to reduce the FTE enrollment.

23.

$$\frac{3}{x} = \frac{8}{13}$$

$$\frac{3}{x} \cdot 13x = \frac{8}{13} \cdot 13x$$

$$39 = 8x$$

$$\frac{39}{8} = x$$

$$4.88 \approx x$$

The length is approximately 4.88 inches.

25.

$$\frac{x}{8} = \frac{13}{7}$$

$$\frac{x}{8} \cdot 8 = \frac{13}{7} \cdot 8$$

$$x = \frac{104}{7}$$

$$x \approx 14.86$$

The length is approximately 14.86 meters.

27.

$$\frac{9}{x} = \frac{8}{5}$$

$$\frac{9}{x} \cdot 5x = \frac{8}{5} \cdot 5x$$

$$45 = 8x$$

$$\frac{45}{8} = x$$

$$5.63 \approx x$$

The length is approximately 5.63 feet.

29. a.

$$\frac{y}{x} = k$$

$$\frac{y}{x} \cdot x = k \cdot x$$

$$y = kx$$

The y-intercept is $(0, 0)$.

b. $C = 15t$

The variable t and C are proportional. Explanations may vary. One possibility follows: Since $\frac{C}{t} = 15$, a constant, the variables t and C are proportional.

c. $C = 10t + 25$

The variable t and C are not proportional. Explanations may vary. One possibility follows: Since $\frac{C}{t}$ is not a constant, the variables t and C are not proportional.

d. Substitute 2 for t: $C = 10(2) + 25 = 45$

It costs $45 to rent a bike for 2 hours.

Substitute 3 for t: $C = 10(3) + 25 = 55$

It costs $55 to rent a bike for 3 hours.

For 2 hours, the ratio is $\frac{\$45}{2 \text{ hours}} = \$22.50$ per hour. For 3 hours, the ratio is $\frac{\$45}{3 \text{ hours}} = \$15.00$ per hour. Since the ratios are not the same, the variable t and C are not proportional. This conclusion agrees with our conclusion in part (c).

31. a. The slope is $\frac{n}{m}$.

b. The slope is $\frac{q}{p}$.

c. Answers may vary. One possibility follows: Since the triangles are similar, we have:

$$\frac{n}{q} = \frac{m}{p}$$

$$\frac{n}{q} \cdot pq = \frac{m}{p} \cdot pq$$

$$np = mq$$

$$\frac{np}{mp} = \frac{mq}{mp}$$

$$\frac{n}{m} = \frac{q}{p}$$

d. Answers may vary. One possibility follows: The results from parts (a), (b), and (c) show that for any two distinct points on a line will have slopes that are proportional. Thus, the slope will simplify to the same result no matter which two points are used.

33. $2x^2 + 7x - 15 = 0$

$(2x - 3)(x + 5) = 0$

$2x - 3 = 0$ or $x + 5 = 0$

$2x = 3$ or $x = -5$

$x = \dfrac{3}{2}$

The solutions are -5 and $\dfrac{3}{2}$. This is a quadratic (or second-degree) equation in one variable.

35. $2x^2 + 7x - 15 = (2x - 3)(x + 5)$

This is a quadratic (or second-degree) polynomial in one variable.

37. Substitute -2 for x and simplify:

$2x^2 + 7x - 15 = 2(-2)^2 + 7(-2) - 15$

$\qquad\qquad\qquad = 2(4) + 7(-2) - 15$

$\qquad\qquad\qquad = 8 - 14 - 15$

$\qquad\qquad\qquad = -21$

This is a quadratic (or second-degree) polynomial in one variable.

Homework 10.7

1. $w = kt$

3. $F = \dfrac{k}{r}$

5. $c = ks^2$

7. $B = \dfrac{k}{t^3}$

9. H varies directly as u.

11. P varies inversely as V.

13. E varies directly as c^2.

15. F varies inversely as r^2.

17. Since y varies directly as x, we have $y = kx$.
We find k by substituting 4 for x and 8 for y:

$y = kx$

$8 = k(4)$

$2 = k$

The equation is $y = 2x$.

19. Since F varies inversely as r, we have $F = \dfrac{k}{r}$.
We find k by substituting 6 for r and 2 for F:

$F = \dfrac{k}{r}$

$2 = \dfrac{k}{6}$

$12 = k$

The equation is $F = \dfrac{12}{r}$.

21. Since H varies directly as $\sqrt{u}$, we have $H = k\sqrt{u}$.
We find k by substituting 4 for u and 6 for H

$6 = k\sqrt{4}$

$6 = 2k$

$3 = k$

The equation is $H = 3\sqrt{u}$.

23. Since C varies inversely as x^2, we have $C = \dfrac{k}{x^2}$.
We find k by substituting 3 for x and 2 for C:

$C = \dfrac{k}{x^2}$

$2 = \dfrac{k}{3^2}$

$2 = \dfrac{k}{9}$

$18 = k$

The equation is $C = \dfrac{18}{x^2}$.

25. Since y varies directly as x, we have $y = kx$.
We find k by substituting 5 for x and 20 for y:

$y = kx$

$20 = k(5)$

$4 = k$

The equation is $y = 4x$. We find the value of y when $x = 3$ by substituting 3 for x:

$y = 4(3) = 12$

So, $y = 12$ when $x = 3$.

27. Since P varies inversely as V, we have $P = \dfrac{k}{V}$.

We find k by substituting 6 for P and 3 for V:

$$P = \frac{k}{V}$$
$$6 = \frac{k}{3}$$
$$18 = k$$

The equation is $P = \dfrac{18}{V}$. We find the value of P

when $V = 2$ by substituting 2 for V:

$$P = \frac{18}{2} = 9$$

So, $P = 9$ when $V = 2$.

29. Since d varies directly as t^2, we have $d = kt^2$.
We find k by substituting 2 for t and 12 for d:

$$d = kt^2$$
$$12 = k(2)^2$$
$$12 = 4k$$
$$3 = k$$

The equation is $d = 3t^2$. We find the value of d
when $t = 4$ by substituting 4 for t:

$$d = 3(4)^2 = 3(16) = 48$$

So, $d = 48$ when $t = 4$.

31. Since F varies inversely as r^2, we have $F = \dfrac{k}{r^2}$.

We find k by substituting 4 for r and 1 for F:

$$F = \frac{k}{r^2}$$
$$1 = \frac{k}{(4)^2}$$
$$1 = \frac{k}{16}$$
$$16 = k$$

The equation is $F = \dfrac{16}{r^2}$. We find the value of r

when $F = 4$ by substituting 4 for F and solving
for r:

$$4 = \frac{16}{r^2}$$
$$4r^2 = 16$$
$$r^2 = 4$$
$$r = \pm\sqrt{4} = \pm 2$$

So, $r = \pm 2$ when $F = 4$.

33. Since w varies directly as t with a positive
variation constant k, the value of w increases as
the value of t increases.

35. Since z varies inversely as u with a positive
variation constant k and the values of u are
positive, the value of z decreases as the value of
u increases.

37. Since the conduction of electricity varies directly
with the humidity with a positive variation
constant k, the conduction of electricity increases
if the humidity increases.

39. Since the demand for refinancing home loans
varies inversely with the rate of interest with a
positive variation constant k (and the rate of
interest is positive), the demand for refinancing
home loans decreases if the interest rate increases.

41. Answers may vary. One possibility follows:
The spin off means that as the outside temperature
increases, the number of doors left open decreases.
Explanations may vary.

43. We let C be the cost of tuition for x credit hours.
Our model has the form $C = kx$. We find k by
substituting 12 for x and 1053 for C:

$$C = kx$$
$$1053 = k(12)$$
$$87.75 = k$$

The model is $C = 87.75x$. We find the total cost
of 15 credit hours by substituting 15 for x:
$$C = 87.75(15) = 1316.25$$

So, the total cost of 15 credit hours is \$1316.25.

45. We let V be the volume of gasoline pumped in t
seconds. Our model has the form $V = kt$. We
find k by substituting 81 for t and 10 for V:

$$V = kt$$
$$10 = k(81)$$
$$\frac{10}{81} = k$$

The model is $V = \dfrac{10}{81}t$. We find the volume of

gasoline that can be pumped in 104 seconds by
substituting 104 for t:

$$V = \frac{10}{81}(104) \approx 12.8$$

So, 12.8 gallons of gasoline can be pumped in
104 seconds.

47. We let w be the weight of a pepperoni, mushroom, and garlic pizza with radius r inches. Our model has the form $w = kr^2$. We find k by substituting $\dfrac{15}{2} = 7.5$ for r and 32 for w:

$$w = kr^2$$
$$32 = k(7.5)^2$$
$$32 = 56.25k$$
$$0.5689 \approx k$$

The model is $w = 0.5689r^2$. We find the weight of a 13-diameter pizza by substituting $\dfrac{13}{2} = 6.5$ for r: $w = 0.5689(6.5)^2 \approx 24$

So, the weight of a 13-inch-diameter pepperoni, mushroom, and garlic pizza is 24 ounces.

49. a. More force must be exerted on a wrench with a shorter handle. Explanations may vary. One possibility follows:
Since the force varies inversely as the length of the handle, this means the force decreases as the length increases, and the force increases as the length decreases.

b. We let F be the force needed when the handle is x inches long. Our model has the form $F = \dfrac{k}{x}$. We find k by substituting 6 for x and 40 for F:

$$F = \frac{k}{x}$$
$$40 = \frac{k}{6}$$
$$240 = k$$

The model is $F = \dfrac{240}{x}$. We find the force needed when the handle is 8 inches by substituting 8 for x:

$$F = \frac{240}{8} = 30$$

So, 30 pounds of force are needed when the handle is 8 inches long.

51. We let B be brightness of light at a distance of d. Our model has the form $B = \dfrac{k}{d^2}$. We find k by substituting 90 for d and 0.55 for B:

$$B = \frac{k}{d^2}$$
$$0.55 = \frac{k}{(90)^2}$$
$$0.55 = \frac{k}{8100}$$
$$4455 = k$$

The model is $B = \dfrac{4455}{d^2}$. We find the distance when the brightness of light is 0.26 mW/cm^2 by substituting 0.26 for B and solving for d:

$$B = \frac{4455}{d^2}$$
$$0.26 = \frac{4455}{d^2}$$
$$0.26d^2 = 4455$$
$$d^2 = \frac{4455}{0.26}$$
$$d^2 \approx 17{,}134.6154$$
$$d = \pm\sqrt{17{,}134.6154} \approx \pm130.9$$

Since distance cannot be negative, we discard -130.9. So, the brightness is 0.26 mW/cm^2 at a distance of 130.9 centimeters.

53. a.

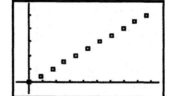

b. A direct variation model is of the form $F = kS$. We need only one point to find k. Since all of the points in the scattergram fit the pattern, any of the points will likely generate a reasonable model. We choose the point (0.173, 5.0):

$$F = ks$$
$$5.0 = 0.173k$$
$$28.90 \approx k$$

Thus, the model is $F = 28.90S$.

c. The force varies directly as the stretch.

d. We substitute 4.2 for F and solve for S:
$$4.2 = 28.90S$$
$$0.146 \approx S$$
So, the force will be 4.2 newtons for a stretch of 0.145 meter.

55. a.

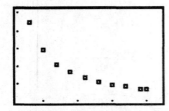

b. An inverse variation model is of the form $P = \dfrac{k}{V}$. We need only one point to find k.

Since all of the points in the scattergram fit the pattern, any of the points will likely generate a reasonable model. We choose the point (15, 0.48):

$$P = \frac{k}{V}$$

$$15 = \frac{k}{0.48}$$

$$7.2 = k$$

Thus, the model is $P = \dfrac{7.2}{V}$.

c. The pressure varies inversely as the volume.

d. We substitute 0.9 for P and solve for V:

$$0.9 = \frac{7.2}{V}$$

$$0.9V = 7.2$$

$$V = 8.0$$

So, the pressure will be 0.9 atm when the volume is 8.0 cm^3.

57. a. Answers may vary. One possible equation follows: $y = \dfrac{30}{x}$.

i. Substitute 3 for x: $y = \dfrac{30}{3} = 10$.

There is one output for this input.

ii. Substitute 5 for x: $y = \dfrac{30}{5} = 6$.

There is one output for this input.

iii. Substitute −2 for x: $y = \dfrac{30}{-2} = -15$.

There is one output for this input.

b. For our equation, $y = \dfrac{30}{x}$, one output originates from a single input. Explanations may vary.

c. For any equation in which y varies inversely as x, one output originates from a single input. Explanations may vary.

59. a. Yes, if y varies directly as x, then x and y are linearly related. Explanations may vary. One possibility follows: If y varies directly as x, then $y = kx$. This equation is of the form $y = mx + b$, with $m = k$ and $b = 0$. That is the slope is the constant of proportionality and the y-intercept is (0, 0).

b. No, if w and t are linearly related, w does not necessarily vary directly as t. Explanations may vary. One possibility follows: If w and t are linearly related, then $w = mt + b$, but this is only of the form $w = kt$ if $b = 0$.

61. a. Let c be the cost of 6 pillows:

$$\frac{4}{50} = \frac{6}{c}$$

$$\frac{4}{50} \cdot 50c = \frac{6}{c} \cdot 50c$$

$$4c = 300$$

$$c = 75$$

The cost of 6 pillows is $75.

b. Our model has the form $c = kn$. We find k by substituting 4 for n and 50 for c:

$$c = kn$$

$$50 = k(4)$$

$$12.5 = k$$

The model is $c = 12.5n$.

c. We find the cost of 6 pillows by substituting 6 for n in our model from part (b):

$$c = 12.5(6) = 75$$

So, the cost of 6 pillows is $75. This is the same result as that from part (a).

63.

$$2w^2 - 3w - 5 = 0$$

$$(2w - 5)(w + 1) = 0$$

$$2w - 5 = 0 \quad \text{or} \quad w + 1 = 0$$

$$2w = 5 \quad \text{or} \quad w = -1$$

$$w = \frac{5}{2}$$

The solutions are −1 and $\dfrac{5}{2}$. This is a quadratic (or second-degree) equation in one variable.

65. Substitute -1 for w and simplify:

$$2w^2 - 3w - 5 = 2(-1)^2 - 3(-1) - 5$$
$$= 2(1) - 3(-1) - 5$$
$$= 2 + 3 - 5$$
$$= 0$$

This is a quadratic (or second-degree) polynomial in one variable.

67. $2w^2 - 3w - 5 = (2w - 5)(w + 1)$

This is a quadratic (or second-degree) polynomial in one variable.

Homework 10.8

1. $\dfrac{\dfrac{4}{5}}{\dfrac{8}{3}} = \dfrac{4}{5} \div \dfrac{8}{3} = \dfrac{4}{5} \cdot \dfrac{3}{8} = \dfrac{2 \cdot 2}{5} \cdot \dfrac{3}{2 \cdot 2 \cdot 2}$

$$= \dfrac{2 \cdot 2 \cdot 3}{5 \cdot 2 \cdot 2 \cdot 2} = \dfrac{3}{5 \cdot 2} = \dfrac{3}{10}$$

3. $\dfrac{\dfrac{x}{4}}{\dfrac{x}{7}} = \dfrac{x}{4} \div \dfrac{x}{7} = \dfrac{x}{4} \cdot \dfrac{7}{x} = \dfrac{x \cdot 7}{4 \cdot x} = \dfrac{7}{4}$

5. $\dfrac{\dfrac{6}{5x}}{\dfrac{3}{7x}} = \dfrac{6}{5x} \div \dfrac{3}{7x} = \dfrac{6}{5x} \cdot \dfrac{7x}{3} = \dfrac{2 \cdot 3}{5x} \cdot \dfrac{7x}{3}$

$$= \dfrac{2 \cdot 3 \cdot 7 \cdot x}{5 \cdot x \cdot 3} = \dfrac{2 \cdot 7}{5} = \dfrac{14}{5}$$

7. $\dfrac{\dfrac{w}{6}}{\dfrac{w^2}{9}} = \dfrac{w}{6} \div \dfrac{w^2}{9} = \dfrac{w}{6} \cdot \dfrac{9}{w^2} = \dfrac{w}{2 \cdot 3} \cdot \dfrac{3 \cdot 3}{w^2}$

$$= \dfrac{w \cdot 3 \cdot 3}{2 \cdot 3 \cdot w^2} = \dfrac{3}{2 \cdot w^{2-1}} = \dfrac{3}{2w}$$

9. $\dfrac{\dfrac{15x}{8}}{\dfrac{25x^3}{12}} = \dfrac{15x}{8} \div \dfrac{25x^3}{12} = \dfrac{15x}{8} \cdot \dfrac{12}{25x^3}$

$$= \dfrac{3 \cdot 5 \cdot x}{2 \cdot 2 \cdot 2} \cdot \dfrac{2 \cdot 2 \cdot 3}{5 \cdot 5 \cdot x^3} = \dfrac{3 \cdot 5 \cdot x \cdot 2 \cdot 2 \cdot 3}{2 \cdot 2 \cdot 2 \cdot 5 \cdot 5 \cdot x^3}$$

$$= \dfrac{3 \cdot 3}{2 \cdot 5 \cdot x^{3-1}} = \dfrac{9}{10x^2}$$

11. $\dfrac{\dfrac{14}{x^2 - 9}}{\dfrac{21}{x - 3}} = \dfrac{14}{x^2 - 9} \div \dfrac{21}{x - 3} = \dfrac{14}{x^2 - 9} \cdot \dfrac{x - 3}{21}$

$$= \dfrac{2 \cdot 7}{(x - 3)(x + 3)} \cdot \dfrac{x - 3}{3 \cdot 7}$$

$$= \dfrac{2 \cdot 7 \cdot (x - 3)}{(x - 3)(x + 3) \cdot 3 \cdot 7}$$

$$= \dfrac{2}{(x + 3) \cdot 3} = \dfrac{2}{3(x + 3)}$$

13. $\dfrac{\dfrac{5a - 10}{4}}{\dfrac{3a - 6}{2}} = \dfrac{5a - 10}{4} \div \dfrac{3a - 6}{2} = \dfrac{5a - 10}{4} \cdot \dfrac{2}{3a - 6}$

$$= \dfrac{5(a - 2)}{2 \cdot 2} \cdot \dfrac{2}{3(a - 2)} = \dfrac{5 \cdot (a - 2) \cdot 2}{2 \cdot 2 \cdot 3 \cdot (a - 2)}$$

$$= \dfrac{5}{2 \cdot 3} = \dfrac{5}{6}$$

15. $\dfrac{\dfrac{x^2 - 2x - 15}{6x}}{\dfrac{x - 5}{10}} = \dfrac{x^2 - 2x - 15}{6x} \div \dfrac{x - 5}{10}$

$$= \dfrac{x^2 - 2x - 15}{6x} \cdot \dfrac{10}{x - 5}$$

$$= \dfrac{(x - 5)(x + 3)}{2 \cdot 3 \cdot x} \cdot \dfrac{2 \cdot 5}{x - 5}$$

$$= \dfrac{(x - 5)(x + 3) \cdot 2 \cdot 5}{2 \cdot 3 \cdot x \cdot (x - 5)}$$

$$= \dfrac{(x + 3) \cdot 5}{3 \cdot x} = \dfrac{5(x + 3)}{3x}$$

17. $\dfrac{\dfrac{4}{x} + \dfrac{2}{x}}{\dfrac{9}{x} - \dfrac{7}{x}} = \dfrac{\dfrac{4}{x} + \dfrac{2}{x}}{\dfrac{9}{x} - \dfrac{7}{x}} \cdot \dfrac{x}{x} = \dfrac{\dfrac{4}{x} \cdot x + \dfrac{2}{x} \cdot x}{\dfrac{9}{x} \cdot x - \dfrac{7}{x} \cdot x} = \dfrac{4 + 2}{9 - 7} = \dfrac{6}{2} = 3$

19. $\dfrac{\dfrac{3}{8} + \dfrac{1}{4}}{\dfrac{1}{2} + \dfrac{5}{8}} = \dfrac{\dfrac{3}{8} + \dfrac{1}{4}}{\dfrac{1}{2} + \dfrac{5}{8}} \cdot \dfrac{8}{8} = \dfrac{\dfrac{3}{8} \cdot 8 + \dfrac{1}{4} \cdot 8}{\dfrac{1}{2} \cdot 8 + \dfrac{5}{8} \cdot 8} = \dfrac{3 + 2}{4 + 5} = \dfrac{5}{9}$

21. $\dfrac{\dfrac{7}{4x}+\dfrac{1}{x}}{\dfrac{3}{2x}+\dfrac{5}{x}}=\dfrac{\dfrac{7}{4x}+\dfrac{1}{x}}{\dfrac{3}{2x}+\dfrac{5}{x}}\cdot\dfrac{4x}{4x}=\dfrac{\dfrac{7}{4x}\cdot 4x+\dfrac{1}{x}\cdot 4x}{\dfrac{3}{2x}\cdot 4x+\dfrac{5}{x}\cdot 4x}$

$\qquad =\dfrac{7+4}{6+20}=\dfrac{11}{26}$

23. $\dfrac{\dfrac{5}{3r}-\dfrac{3}{2r}}{\dfrac{1}{2r}-\dfrac{4}{3r}}=\dfrac{\dfrac{5}{3r}-\dfrac{3}{2r}}{\dfrac{1}{2r}-\dfrac{4}{3r}}\cdot\dfrac{6r}{6r}=\dfrac{\dfrac{5}{3r}\cdot 6r-\dfrac{3}{2r}\cdot 6r}{\dfrac{1}{2r}\cdot 6r-\dfrac{4}{3r}\cdot 6r}$

$\qquad =\dfrac{10-9}{3-8}=\dfrac{1}{-5}=-\dfrac{1}{5}$

25. $\dfrac{\dfrac{2}{3}-\dfrac{4}{3x}}{\dfrac{5}{6x}}=\dfrac{\dfrac{2}{3}-\dfrac{4}{3x}}{\dfrac{5}{6x}}\cdot\dfrac{6x}{6x}=\dfrac{\dfrac{2}{3}\cdot 6x-\dfrac{4}{3x}\cdot 6x}{\dfrac{5}{6x}\cdot 6x}$

$\qquad =\dfrac{4x-8}{5}=\dfrac{4(x-2)}{5}$

27. $\dfrac{2+\dfrac{5}{x}}{4-\dfrac{1}{x}}=\dfrac{2+\dfrac{5}{x}}{4-\dfrac{1}{x}}\cdot\dfrac{x}{x}=\dfrac{2\cdot x+\dfrac{5}{x}\cdot x}{4\cdot x-\dfrac{1}{x}\cdot x}=\dfrac{2x+5}{4x-1}$

29. $\dfrac{\dfrac{2}{b}-\dfrac{3}{b^2}}{\dfrac{4}{b}+\dfrac{5}{b^2}}=\dfrac{\dfrac{2}{b}-\dfrac{3}{b^2}}{\dfrac{4}{b}+\dfrac{5}{b^2}}\cdot\dfrac{b^2}{b^2}=\dfrac{\dfrac{2}{b}\cdot b^2-\dfrac{3}{b^2}\cdot b^2}{\dfrac{4}{b}\cdot b^2+\dfrac{5}{b^2}\cdot b^2}=\dfrac{2b-3}{4b+5}$

31. $\dfrac{\dfrac{1}{x}-\dfrac{4}{x^3}}{\dfrac{1}{x}-\dfrac{2}{x^2}}=\dfrac{\dfrac{1}{x}-\dfrac{4}{x^3}}{\dfrac{1}{x}-\dfrac{2}{x^2}}\cdot\dfrac{x^3}{x^3}=\dfrac{\dfrac{1}{x}\cdot x^3-\dfrac{4}{x^3}\cdot x^3}{\dfrac{1}{x}\cdot x^3-\dfrac{2}{x^2}\cdot x^3}$

$\qquad =\dfrac{x^2-4}{x^2-2x}=\dfrac{(x-2)(x+2)}{x(x-2)}=\dfrac{x+2}{x}$

33. $\dfrac{\dfrac{3}{4x}+\dfrac{1}{x^2}}{\dfrac{5}{2x}+\dfrac{1}{x^2}}=\dfrac{\dfrac{3}{4x}+\dfrac{1}{x^2}}{\dfrac{5}{2x}+\dfrac{1}{x^2}}\cdot\dfrac{4x^2}{4x^2}=\dfrac{\dfrac{3}{4x}\cdot 4x^2+\dfrac{1}{x^2}\cdot 4x^2}{\dfrac{5}{2x}\cdot 4x^2+\dfrac{1}{x^2}\cdot 4x^2}$

$\qquad =\dfrac{3x+4}{10x+4}=\dfrac{3x+4}{2(5x+2)}$

35. $\dfrac{\dfrac{r}{3}-\dfrac{3}{r}}{\dfrac{r}{2}-\dfrac{2}{r}}=\dfrac{\dfrac{r}{3}-\dfrac{3}{r}}{\dfrac{r}{2}-\dfrac{2}{r}}\cdot\dfrac{6r}{6r}=\dfrac{\dfrac{r}{3}\cdot 6r-\dfrac{3}{r}\cdot 6r}{\dfrac{r}{2}\cdot 6r-\dfrac{2}{r}\cdot 6r}=\dfrac{2r^2-18}{3r^2-12}$

$\qquad =\dfrac{2(r^2-9)}{3(r^2-4)}=\dfrac{2(r-3)(r+3)}{3(r-2)(r+2)}$

37. $\dfrac{2-\dfrac{8}{x^2}}{1-\dfrac{2}{x}}=\dfrac{2-\dfrac{8}{x^2}}{1-\dfrac{2}{x}}\cdot\dfrac{x^2}{x^2}=\dfrac{2\cdot x^2-\dfrac{8}{x^2}\cdot x^2}{1\cdot x^2-\dfrac{2}{x}\cdot x^2}=\dfrac{2x^2-8}{x^2-2x}$

$\qquad =\dfrac{2(x^2-4)}{x(x-2)}=\dfrac{2(x-2)(x+2)}{x(x-2)}=\dfrac{2(x+2)}{x}$

39. Answers may vary. One possibility follows: The error is in the statement that follows:

$$x\div\left(\dfrac{1}{x}+\dfrac{1}{3}\right)=x\cdot\left(\dfrac{x}{1}+\dfrac{3}{1}\right)$$

Instead, the rational expression within the parentheses should be combined into a single expression before multiplying by the reciprocal. The correct simplification follows:

$$\dfrac{x}{\dfrac{1}{x}+\dfrac{1}{3}}=x\div\left(\dfrac{1}{x}+\dfrac{1}{3}\right)=x\div\left(\dfrac{1}{x}\cdot\dfrac{3}{3}+\dfrac{1}{3}\cdot\dfrac{x}{x}\right)$$

$$=x\div\left(\dfrac{3}{3x}+\dfrac{x}{3x}\right)=x\div\left(\dfrac{3+x}{3x}\right)=x\cdot\dfrac{3x}{3+x}$$

$$=\dfrac{x\cdot 3x}{3+x}=\dfrac{3x^2}{x+3}$$

41. Answers may vary. One possibility follows: A complex rational expression is a rational expression whose numerator or denominator (or both) contains a rational expression. Examples may vary.

43. $\dfrac{x^2-9}{3x+15}\div\dfrac{x-3}{2x+10}=\dfrac{x^2-9}{3x+15}\cdot\dfrac{2x+10}{x-3}$

$\qquad =\dfrac{(x-3)(x+3)}{3(x+5)}\cdot\dfrac{2(x+5)}{x-3}$

$\qquad =\dfrac{(x-3)(x+3)\cdot 2(x+5)}{3(x+5)\cdot(x-3)}$

$\qquad =\dfrac{2(x+3)}{3}$

45. $\dfrac{\dfrac{x^2-9}{3x+15}}{\dfrac{x-3}{2x+10}} = \dfrac{x^2-9}{3x+15} \div \dfrac{x-3}{2x+10} = \dfrac{x^2-9}{3x+15} \cdot \dfrac{2x+10}{x-3}$

$\qquad = \dfrac{(x-3)(x+3)}{3(x+5)} \cdot \dfrac{2(x+5)}{x-3}$

$\qquad = \dfrac{(x-3)(x+3) \cdot 2(x+5)}{3(x+5) \cdot (x-3)}$

$\qquad = \dfrac{2(x+3)}{3}$

47. $\dfrac{2x^{-1}}{5 \cdot 3^{-1}} = \dfrac{2 \cdot 3^1}{5 \cdot x^1} = \dfrac{2 \cdot 3}{5 \cdot x} = \dfrac{6}{5x}$

49. $\dfrac{2+x^{-1}}{5+3^{-1}} = \dfrac{2+\dfrac{1}{x}}{5+\dfrac{1}{3}} = \dfrac{2+\dfrac{1}{x}}{5+\dfrac{1}{3}} \cdot \dfrac{3x}{3x} = \dfrac{2 \cdot 3x + \dfrac{1}{x} \cdot 3x}{5 \cdot 3x + \dfrac{1}{3} \cdot 3x}$

$\qquad = \dfrac{6x+3}{15x+x} = \dfrac{6x+3}{16x} = \dfrac{3(2x+1)}{16x}$

51 – 59. Answers may vary.

Chapter 10 Review Exercises

1. $\dfrac{4}{x}$

The number 0 is an excluded value because $\dfrac{4}{0}$ has division by 0. No other value of x leads to division by 0, so 0 is the only excluded value.

2. $\dfrac{x}{7}$

The denominator is a constant 7, so no substitution for x will lead to a division by 0. Thus, there are no excluded values.

3. $\dfrac{9}{3x-5}$

We set the denominator equal to 0 and solve for x.

$3x - 5 = 0$

$\qquad 3x = 5$

$\qquad x = \dfrac{5}{3}$

So, the only excluded value is $\dfrac{5}{3}$.

4. $\dfrac{x-5}{x^2-6x+8}$

We set the denominator equal to 0 and solve for x.

$x^2 - 6x + 8 = 0$

$(x-4)(x-2) = 0$

$x - 4 = 0 \quad$ or $\quad x - 2 = 0$

$\qquad x = 4 \quad$ or $\qquad x = 2$

So, the excluded values are 2 and 4.

5. $\dfrac{5t-6}{t^2+6t+9}$

We set the denominator equal to 0 and solve for t.

$t^2 + 6t + 9 = 0$

$(t+3)(t+3) = 0$

$t + 3 = 0 \quad$ or $\quad t + 3 = 0$

$\qquad t = -3 \quad$ or $\qquad t = -3$

So, the only excluded value is -3.

6. $\dfrac{7w}{4w^2-9}$

We set the denominator equal to 0 and solve for w.

$4w^2 - 9 = 0$

$(2w+3)(2w-3) = 0$

$2w + 3 = 0 \qquad$ or $\qquad 2w - 3 = 0$

$\quad 2w = -3 \qquad$ or $\qquad 2w = 3$

$\quad w = -\dfrac{3}{2} \qquad$ or $\qquad w = \dfrac{3}{2}$

So, the excluded values are $-\dfrac{3}{2}$ and $\dfrac{3}{2}$.

7. $\dfrac{2x+3}{5x^2+18x-8}$

We set the denominator equal to 0 and solve for x.

$5x^2 + 18x - 8 = 0$

$(5x-2)(x+4) = 0$

$5x - 2 = 0 \quad$ or $\quad x + 4 = 0$

$\quad 5x = 2 \quad$ or $\qquad x = -4$

$\quad x = \dfrac{2}{5}$

So, the excluded values are -4 and $\dfrac{2}{5}$.

8. $\dfrac{x+9}{2x^3+5x^2-18x-45}$

We set the denominator equal to 0 and solve for x.

$$2x^3+5x^2-18x-45=0$$

$$x^2(2x+5)-9(2x+5)=0$$

$$(2x+5)(x^2-9)=0$$

$$(2x+5)(x+3)(x-3)=0$$

$$2x+5=0 \quad\text{or}\quad x+3=0 \quad\text{or}\quad x-3=0$$

$$2x=-5 \quad\text{or}\qquad x=-3 \quad\text{or}\qquad x=3$$

$$x=-\frac{5}{2}$$

So, the excluded values are $-\dfrac{5}{2}$, -3, and 3.

9. $\dfrac{28x^3y^5}{35x^7y^2} = \dfrac{2\cdot 2\cdot 7\cdot x\cdot x\cdot x\cdot y\cdot y\cdot y\cdot y\cdot y}{5\cdot 7\cdot x\cdot x\cdot x\cdot x\cdot x\cdot x\cdot x\cdot y\cdot y}$

$\qquad = \dfrac{2\cdot 2\cdot y\cdot y\cdot y}{5\cdot x\cdot x\cdot x\cdot x}$

$\qquad = \dfrac{4y^3}{5x^4}$

10. $\dfrac{7x+21}{x^2+x-6} = \dfrac{7(x+3)}{(x+3)(x-2)} = \dfrac{7}{x-2}$

11. $\dfrac{w^2+7w+12}{w^2-16} = \dfrac{(w+4)(w+3)}{(w+4)(w-4)} = \dfrac{w+3}{w-4}$

12. $\dfrac{2a^2-2ab}{a^2-b^2} = \dfrac{2a(a-b)}{(a-b)(a+b)} = \dfrac{2a}{a+b}$

13. $\dfrac{x^2-8x+12}{3x^2-16x-12} = \dfrac{(x-6)(x-2)}{(3x+2)(x-6)} = \dfrac{x-2}{3x+2}$

14. $\dfrac{x^2+10x+25}{2x^3-3x^2-50x+75} = \dfrac{(x+5)(x+5)}{x^2(2x-3)-25(2x-3)}$

$\qquad = \dfrac{(x+5)(x+5)}{(x^2-25)(2x-3)}$

$\qquad = \dfrac{(x+5)(x+5)}{(x-5)(x+5)(2x-3)}$

$\qquad = \dfrac{x+5}{(x-5)(2x-3)}$

15. $\dfrac{m-4}{m+3}\cdot\dfrac{m+2}{m-3} = \dfrac{(m-4)(m+2)}{(m+3)(m-3)}$

16. $\dfrac{25b^3}{b^2-b}\cdot\dfrac{b^2-1}{35b} = \dfrac{5\cdot 5\cdot b\cdot b\cdot b}{b(b-1)}\cdot\dfrac{(b-1)(b+1)}{5\cdot 7\cdot b}$

$\qquad = \dfrac{5\cdot 5\cdot b\cdot b\cdot b\cdot(b-1)(b+1)}{b\cdot(b-1)\cdot 5\cdot 7\cdot b}$

$\qquad = \dfrac{5\cdot b\cdot(b+1)}{7}$

$\qquad = \dfrac{5b(b+1)}{7}$

17. $\dfrac{x^3+8x^2+16x}{x^2-10x+21}\cdot\dfrac{4x-12}{3x+12}$

$\qquad = \dfrac{x(x^2+8x+16)}{(x-7)(x-3)}\cdot\dfrac{4(x-3)}{3(x+4)}$

$\qquad = \dfrac{x(x+4)(x+4)}{(x-7)(x-3)}\cdot\dfrac{4(x-3)}{3(x+4)}$

$\qquad = \dfrac{x(x+4)(x+4)\cdot 4(x-3)}{(x-7)(x-3)\cdot 3(x+4)}$

$\qquad = \dfrac{x(x+4)\cdot 4}{(x-7)\cdot 3}$

$\qquad = \dfrac{4x(x+4)}{3(x-7)}$

18. $\dfrac{-3x+12}{x^2-1}\cdot\dfrac{-2x-2}{x^2-16} = \dfrac{-3(x-4)}{(x-1)(x+1)}\cdot\dfrac{-2(x+1)}{(x-4)(x+4)}$

$\qquad = \dfrac{-3(x-4)\cdot(-2)(x+1)}{(x-1)(x+1)\cdot(x-4)(x+4)}$

$\qquad = \dfrac{-3\cdot(-2)}{(x-1)\cdot(x+4)}$

$\qquad = \dfrac{6}{(x-1)(x+4)}$

19. $\dfrac{5x-15}{6x^5}\div\dfrac{2x-6}{4x^2} = \dfrac{5x-15}{6x^5}\cdot\dfrac{4x^2}{2x-6}$

$\qquad = \dfrac{5(x-3)}{2\cdot 3\cdot x^5}\cdot\dfrac{2\cdot 2\cdot x^2}{2(x-3)}$

$\qquad = \dfrac{5(x-3)\cdot 2\cdot 2\cdot x^2}{2\cdot 3\cdot x^5\cdot 2(x-3)}$

$\qquad = \dfrac{5}{3\cdot x^{5-2}}$

$\qquad = \dfrac{5}{3x^3}$

20. $\dfrac{9x^2-36}{3x+3} \div \dfrac{2x^2-6x-20}{2x^2+7x+5}$

$= \dfrac{9x^2-36}{3x+3} \cdot \dfrac{2x^2+7x+5}{2x^2-6x-20}$

$= \dfrac{9(x^2-4)}{3(x+1)} \cdot \dfrac{(2x+5)(x+1)}{2(x^2-3x-10)}$

$= \dfrac{3\cdot3\cdot(x-2)(x+2)}{3\cdot(x+1)} \cdot \dfrac{(2x+5)(x+1)}{2(x-5)(x+2)}$

$= \dfrac{3\cdot3(x-2)(x+2)\cdot(2x+5)(x+1)}{3(x+1)\cdot2(x-5)(x+2)}$

$= \dfrac{3(x-2)(2x+5)}{2(x-5)}$

21. $\dfrac{7t+14}{t-7} \div (3t^2+2t-8) = \dfrac{7t+14}{t-7} \cdot \dfrac{1}{3t^2+2t-8}$

$= \dfrac{7(t+2)}{t-7} \cdot \dfrac{1}{(3t-4)(t+2)}$

$= \dfrac{7(t+2)}{(t-7)\cdot(3t-4)(t+2)}$

$= \dfrac{7}{(t-7)(3t-4)}$

22. $\dfrac{a^2-9b^2}{4a^2-8ab} \div \dfrac{a^2-4ab-21b^2}{a^2-4ab+4b^2}$

$= \dfrac{a^2-9b^2}{4a^2-8ab} \cdot \dfrac{a^2-4ab+4b^2}{a^2-4ab-21b^2}$

$= \dfrac{(a-3b)(a+3b)}{4a(a-2b)} \cdot \dfrac{(a-2b)(a-2b)}{(a-7b)(a+3b)}$

$= \dfrac{(a-3b)(a+3b)\cdot(a-2b)(a-2b)}{4a(a-2b)\cdot(a-7b)(a+3b)}$

$= \dfrac{(a-3b)(a-2b)}{4a(a-7b)}$

23. $\dfrac{x^2}{x+7}+\dfrac{5x-14}{x+7} = \dfrac{x^2+5x-14}{x+7}$

$= \dfrac{(x+7)(x-2)}{x+7}$

$= x-2$

24. $\dfrac{3}{4x}+\dfrac{5}{6x^3} = \dfrac{3}{2\cdot2\cdot x}+\dfrac{5}{2\cdot3\cdot x\cdot x\cdot x}$

$= \dfrac{3}{2\cdot2\cdot x}\cdot\dfrac{3\cdot x\cdot x}{3\cdot x\cdot x}+\dfrac{5}{2\cdot3\cdot x\cdot x\cdot x}\cdot\dfrac{2}{2}$

$= \dfrac{9x^2}{12x^3}+\dfrac{10}{12x^3}$

$= \dfrac{9x^2+10}{12x^3}$

25. $\dfrac{5}{x^2+7x+6}+\dfrac{2x}{x^2-3x-4}$

$= \dfrac{5}{(x+6)(x+1)}+\dfrac{2x}{(x-4)(x+1)}$

$= \dfrac{5}{(x+6)(x+1)}\cdot\dfrac{x-4}{x-4}+\dfrac{2x}{(x-4)(x+1)}\cdot\dfrac{x+6}{x+6}$

$= \dfrac{5(x-4)}{(x+6)(x+1)(x-4)}+\dfrac{2x(x+6)}{(x-4)(x+1)(x+6)}$

$= \dfrac{5(x-4)+2x(x+6)}{(x-4)(x+1)(x+6)}$

$= \dfrac{5x-20+2x^2+12x}{(x-4)(x+1)(x+6)}$

$= \dfrac{2x^2+17x-20}{(x-4)(x+1)(x+6)}$

26. $\dfrac{x+4}{x-2}+\dfrac{2}{x^2+4x-12}$

$= \dfrac{x+4}{x-2}+\dfrac{2}{(x+6)(x-2)}$

$= \dfrac{x+4}{x-2}\cdot\dfrac{x+6}{x+6}+\dfrac{2}{(x+6)(x-2)}$

$= \dfrac{(x+4)(x+6)}{(x-2)(x+6)}+\dfrac{2}{(x+6)(x-2)}$

$= \dfrac{(x+4)(x+6)+2}{(x-2)(x+6)}$

$= \dfrac{x^2+6x+4x+24+2}{(x-2)(x+6)}$

$= \dfrac{x^2+10x+26}{(x-2)(x+6)}$

27. $\dfrac{p^2}{p^2-4} - \dfrac{4p+12}{p^2-4} = \dfrac{p^2-(4p+12)}{p^2-4}$

$\qquad\qquad = \dfrac{p^2-4p-12}{(p-2)(p+2)}$

$\qquad\qquad = \dfrac{(p-6)(p+2)}{(p-2)(p+2)}$

$\qquad\qquad = \dfrac{p-6}{p-2}$

28. $\dfrac{x-4}{x^2+2x-3} - \dfrac{x+2}{x^2-6x+5}$

$\quad = \dfrac{x-4}{(x+3)(x-1)} - \dfrac{x+2}{(x-5)(x-1)}$

$\quad = \dfrac{x-4}{(x+3)(x-1)}\cdot\dfrac{x-5}{x-5} - \dfrac{x+2}{(x-5)(x-1)}\cdot\dfrac{x+3}{x+3}$

$\quad = \dfrac{(x-4)(x-5)}{(x+3)(x-1)(x-5)} - \dfrac{(x+2)(x+3)}{(x-5)(x-1)(x+3)}$

$\quad = \dfrac{(x-4)(x-5)-(x+2)(x+3)}{(x+3)(x-1)(x-5)}$

$\quad = \dfrac{x^2-5x-4x+20-(x^2+3x+2x+6)}{(x-5)(x-1)(x+3)}$

$\quad = \dfrac{x^2-5x-4x+20-x^2-3x-2x-6}{(x-5)(x-1)(x+3)}$

$\quad = \dfrac{-14x+14}{(x-5)(x-1)(x+3)}$

$\quad = \dfrac{-14(x-1)}{(x-5)(x-1)(x+3)}$

$\quad = \dfrac{-14}{(x-5)(x+3)} = -\dfrac{14}{(x-5)(x+3)}$

29. $\dfrac{2xy}{x^2-10xy+25y^2} - \dfrac{2y}{x-5y}$

$\quad = \dfrac{2xy}{(x-5y)^2} - \dfrac{2y}{x-5y}$

$\quad = \dfrac{2xy}{(x-5y)^2} - \dfrac{2y}{x-5y}\cdot\dfrac{x-5y}{x-5y}$

$\quad = \dfrac{2xy}{(x-5y)^2} - \dfrac{2y(x-5y)}{(x-5y)^2}$

$\quad = \dfrac{2xy-2y(x-5y)}{(x-5y)^2}$

$\quad = \dfrac{2xy-2xy+10y^2}{(x-5y)^2}$

$\quad = \dfrac{10y^2}{(x-5y)^2}$

30. $\dfrac{y}{y^3+64} - \dfrac{3}{y^2+4y}$

$\quad = \dfrac{y}{(y+4)(y^2-4y+16)} - \dfrac{3}{y(y+4)}$

$\quad = \dfrac{y}{(y+4)(y^2-4y+16)}\cdot\dfrac{y}{y} - \dfrac{3}{y(y+4)}\cdot\dfrac{y^2-4y+16}{y^2-4y+16}$

$\quad = \dfrac{y^2}{y(y+4)(y^2-4y+16)} - \dfrac{3(y^2-4y+16)}{y(y+4)(y^2-4y+16)}$

$\quad = \dfrac{y^2-3(y^2-4y+16)}{y(y+4)(y^2-4y+16)}$

$\quad = \dfrac{y^2-3y^2+12y-48}{y(y+4)(y^2-4y+16)}$

$\quad = \dfrac{-2y^2+12y-48}{y(y+4)(y^2-4y+16)}$

$\quad = \dfrac{-2(y^2-6y+24)}{y(y+4)(y^2-4y+16)}$

$\quad = -\dfrac{2(y^2-6y+24)}{y(y+4)(y^2-4y+16)}$

31. $\dfrac{8x^2+4x}{9x^4}\cdot\dfrac{12x^9}{10x+5} = \dfrac{4x(2x+1)}{3\cdot3\cdot x^4}\cdot\dfrac{4\cdot3\cdot x^9}{5(2x+1)}$

$\qquad\qquad = \dfrac{4x(2x+1)\cdot4\cdot3\cdot x^9}{3\cdot3\cdot x^4\cdot5(2x+1)}$

$\qquad\qquad = \dfrac{4\cdot4\cdot x^{1+9-4}}{3\cdot5}$

$\qquad\qquad = \dfrac{16x^6}{15}$

32. $\dfrac{x^2-36}{x^2+13x+36} \div \dfrac{x^2+12x+36}{9x+36}$

$\quad = \dfrac{x^2-36}{x^2+13x+36}\cdot\dfrac{9x+36}{x^2+12x+36}$

$\quad = \dfrac{(x-6)(x+6)}{(x+9)(x+4)}\cdot\dfrac{9(x+4)}{(x+6)(x+6)}$

$\quad = \dfrac{(x-6)(x+6)\cdot9(x+4)}{(x+9)(x+4)\cdot(x+6)(x+6)}$

$\quad = \dfrac{9(x-6)}{(x+9)(x+6)}$

33. $\dfrac{5m}{m^2+4m-45}+\dfrac{3}{m^2+6m-27}$

$=\dfrac{5m}{(m+9)(m-5)}+\dfrac{3}{(m+9)(m-3)}$

$=\dfrac{5m}{(m+9)(m-5)}\cdot\dfrac{m-3}{m-3}+\dfrac{3}{(m+9)(m-3)}\cdot\dfrac{m-5}{m-5}$

$=\dfrac{5m(m-3)}{(m+9)(m-5)(m-3)}+\dfrac{3(m-5)}{(m+9)(m-3)(m-5)}$

$=\dfrac{5m(m-3)+3(m-5)}{(m+9)(m-5)(m-3)}$

$=\dfrac{5m^2-15m+3m-15}{(m+9)(m-5)(m-3)}=\dfrac{5m^2-12m-15}{(m+9)(m-5)(m-3)}$

34. $\dfrac{c-5}{c+2}-\dfrac{c+3}{c-1}=\dfrac{c-5}{c+2}\cdot\dfrac{c-1}{c-1}-\dfrac{c+3}{c-1}\cdot\dfrac{c+2}{c+2}$

$=\dfrac{(c-5)(c-1)}{(c+2)(c-1)}-\dfrac{(c+3)(c+2)}{(c-1)(c+2)}$

$=\dfrac{(c-5)(c-1)-(c+3)(c+2)}{(c+2)(c-1)}$

$=\dfrac{c^2-c-5c+5-(c^2+2c+3c+6)}{(c-1)(c+2)}$

$=\dfrac{c^2-c-5c+5-c^2-2c-3c-6}{(c-1)(c+2)}$

$=\dfrac{-11c-1}{(c-1)(c+2)}$

35. Answers may vary. One possibility follows: The student failed to distribute the subtraction properly through the numerator of the second rational expression. The correct difference is:

$\dfrac{2x}{x+4}-\dfrac{7x-3}{x+4}=\dfrac{2x-(7x-3)}{x+4}$

$=\dfrac{2x-7x+3}{x+4}=\dfrac{-5x+3}{x+4}$

36. $\dfrac{0.62\text{ pound}}{1}\cdot\dfrac{16\text{ ounces}}{1\text{ pound}}=9.92$ ounces

A TI-84 graphing calculator weighs 9.92 ounces.

37. $\dfrac{121\text{ gallons}}{1\text{ year}}\cdot\dfrac{4\text{ quarts}}{1\text{ gallon}}\cdot\dfrac{4\text{ cups}}{1\text{ quart}}\cdot\dfrac{1\text{ year}}{365\text{ days}}$

$\approx 5.30\ \dfrac{\text{cups}}{\text{day}}$

Americans consume an average 5.30 cups of water per day.

38. $\dfrac{3x-2}{x+4}=\dfrac{9}{x+4}$

We note that -4 is an excluded value.

$(x+4)\cdot\left(\dfrac{3x-2}{x+4}\right)=(x+4)\cdot\left(\dfrac{9}{x+4}\right)$

$3x-2=9$

$3x=11$

$x=\dfrac{11}{3}$

Since $\dfrac{11}{3}$ is not an excluded value, we conclude that $\dfrac{11}{3}$ is the solution of the equation.

39. $\dfrac{5}{2x}-\dfrac{4}{x}=\dfrac{3}{2}$

We note that 0 is an excluded value.

$2x\cdot\left(\dfrac{5}{2x}-\dfrac{4}{x}\right)=2x\cdot\left(\dfrac{3}{2}\right)$

$2x\cdot\dfrac{5}{2x}-2x\cdot\dfrac{4}{x}=3x$

$5-8=3x$

$-3=3x$

$-1=x$

Since -1 is not an excluded value, we conclude that -1 is the solution of the equation.

40. $\dfrac{3}{r-1}+\dfrac{2}{4r-4}=\dfrac{7}{4}$

$\dfrac{3}{r-1}+\dfrac{2}{4(r-1)}=\dfrac{7}{4}$

We note that 1 is an excluded value.

$4(r-1)\cdot\left(\dfrac{3}{r-1}+\dfrac{2}{4(r-1)}\right)=4(r-1)\cdot\left(\dfrac{7}{4}\right)$

$4(r-1)\cdot\left(\dfrac{3}{r-1}\right)+4(r-1)\cdot\left(\dfrac{2}{4(r-1)}\right)=(r-1)\cdot7$

$4\cdot3+2=7r-7$

$12+2=7r-7$

$14=7r-7$

$21=7r$

$3=r$

Since 3 is not an excluded value, we conclude that 3 is the solution of the equation.

41.

$$\frac{3}{t-2} + \frac{2}{t+2} = \frac{4}{t^2-4}$$

$$\frac{3}{t-2} + \frac{2}{t+2} = \frac{4}{(t-2)(t+2)}$$

We note that -2 and 2 are excluded values.

$$(t-2)(t+2) \cdot \left(\frac{3}{t-2} + \frac{2}{t+2}\right) = (t-2)(t+2) \cdot \left(\frac{4}{(t-2)(t+2)}\right)$$

$$(t-2)(t+2) \cdot \left(\frac{3}{t-2}\right) + (t-2)(t+2) \cdot \left(\frac{2}{t+2}\right) = 4$$

$$(t+2) \cdot 3 + (t-2) \cdot 2 = 4$$

$$3t + 6 + 2t - 4 = 4$$

$$5t + 2 = 4$$

$$5t = 2$$

$$t = \frac{2}{5}$$

Since $\frac{2}{5}$ is not an excluded value, we conclude that $\frac{2}{5}$ is the solution of the equation.

42.

$$\frac{5}{x-4} - \frac{2}{x-7} = \frac{4}{x^2-11x+28}$$

$$\frac{5}{x-4} - \frac{2}{x-7} = \frac{4}{(x-7)(x-4)}$$

We note that 4 and 7 are excluded values.

$$(x-7)(x-4) \cdot \left(\frac{5}{x-4} - \frac{2}{x-7}\right) = (x-7)(x-4) \cdot \left(\frac{4}{(x-7)(x-4)}\right)$$

$$(x-7)(x-4) \cdot \left(\frac{5}{x-4}\right) - (x-7)(x-4) \cdot \left(\frac{2}{x-7}\right) = 4$$

$$(x-7) \cdot 5 - (x-4) \cdot 2 = 4$$

$$5x - 35 - 2x + 8 = 4$$

$$3x - 27 = 4$$

$$3x = 31$$

$$x = \frac{31}{3}$$

Since $\frac{31}{3}$ is not an excluded value, we conclude that $\frac{31}{3}$ is the solution of the equation.

43. $\dfrac{3}{x+3} - \dfrac{2}{x-3} = \dfrac{-12}{x^2-9}$

$$\dfrac{3}{x+3} - \dfrac{2}{x-3} = \dfrac{-12}{(x-3)(x+3)}$$

We note that -3 and 3 are excluded values.

$$(x-3)(x+3)\cdot\left(\dfrac{3}{x+3} - \dfrac{2}{x-3}\right) = (x-3)(x+3)\cdot\left(\dfrac{-12}{(x-3)(x+3)}\right)$$

$$(x-3)(x+3)\cdot\left(\dfrac{3}{x+3}\right) - (x-3)(x+3)\cdot\left(\dfrac{2}{x-3}\right) = -12$$

$$(x-3)\cdot 3 - (x+3)\cdot 2 = -12$$

$$3x - 9 - 2x - 6 = -12$$

$$x - 15 = -12$$

$$x = 3$$

Our result 3 is not a solution because it is an excluded value. Since this is the only possible solution, we conclude that no number is a solution. That is, the solution is the empty set.

44. $\dfrac{4}{x} = \dfrac{3}{x^2} - 2$

We note that 0 is an excluded value.

$$x^2 \cdot\left(\dfrac{4}{x}\right) = x^2\cdot\left(\dfrac{3}{x^2} - 2\right)$$

$$4x = x^2\cdot\left(\dfrac{3}{x^2}\right) - x^2\cdot 2$$

$$4x = 3 - 2x^2$$

$$2x^2 + 4x - 3 = 0$$

We substitute 2 for a, 4 for b, and -3 for c in the quadratic formula:

$$x = \dfrac{-b \pm \sqrt{b^2 - 4ac}}{2a}$$

$$x = \dfrac{-(4) \pm \sqrt{(4)^2 - 4(2)(-3)}}{2(2)} = \dfrac{-4 \pm \sqrt{16+24}}{4} = \dfrac{-4 \pm \sqrt{40}}{4} = \dfrac{-4 \pm 2\sqrt{10}}{4} = \dfrac{2\left(-2 \pm \sqrt{10}\right)}{4} = \dfrac{-2 \pm \sqrt{10}}{2}$$

Neither of these are excluded values, so we conclude that $\dfrac{-2 \pm \sqrt{10}}{2}$ are the solutions of the equation.

45. $\dfrac{-3}{x+6} + \dfrac{2}{x+1} = \dfrac{x-2}{x+6}$

We note that -6 and -1 are excluded values.

$$(x+6)(x+1)\cdot\left(\dfrac{-3}{x+6} + \dfrac{2}{x+1}\right) = (x+6)(x+1)\cdot\left(\dfrac{x-2}{x+6}\right)$$

$$(x+6)(x+1)\cdot\left(\dfrac{-3}{x+6}\right) + (x+6)(x+1)\cdot\left(\dfrac{2}{x+1}\right) = (x+1)(x-2)$$

$$(x+1)\cdot(-3) + (x+6)\cdot 2 = x^2 - 2x + x - 2$$

$$-3x - 3 + 2x + 12 = x^2 - x - 2$$

$$11 = x^2$$

$$\pm\sqrt{11} = x$$

Neither of these are excluded values, so we conclude that $\pm\sqrt{11}$ are the solutions of the equation.

46. a. The year 2003 is represented by $t = 23$. We substitute 23 for t in the model:

$$M = \frac{288.6(23) - 1493}{0.22(23) + 11.4} = \frac{5144.8}{16.46} \approx 313$$

Using the model, we predict the average amount of money spent on textbooks in 2003 was $313 per student.

To find the actual average amount of money spent on textbooks per student in 2003, we divide the total amount spent on textbooks debt by the number of students:

$$\frac{\$5086 \text{ million}}{16.6 \text{ million students}} \approx \$306 / \text{student}$$

The actual average amount of money spent on textbooks in 2003 was approximately $306 per student.

Our result from the model ias an overestimate.

b. The year 2011 is represented by $t = 31$. We substitute 20 for t in the model:

$$M = \frac{288.6(31) - 1493}{0.22(31) + 11.4} = \frac{7453.6}{18.22} \approx 409$$

We predict the average amount of money spent on textbooks in 2011 will be $409 per student.

c. We substitute 450 for M and solve for t:

$$450 = \frac{288.6t - 1493}{0.22t + 11.4}$$

$$(0.22t + 11.4) \cdot 450 = (0.22t + 11.4)\left(\frac{288.6t - 1493}{0.22t + 11.4}\right)$$

$$99t + 5130 = 288.6t - 1493$$

$$-189.6t = -6623$$

$$t \approx 35$$

Now, $1980 + 35 = 2015$.

We predict that average amount of money spent on textbooks will be $450 per student in approximately the year 2015.

47. Let x be the number of ounces of diced tomatoes used to make 7 servings of chicken cacciatore.

$$\frac{4}{14} = \frac{7}{x}$$

$$14x\left(\frac{4}{14}\right) = 14x\left(\frac{7}{x}\right)$$

$$4x = 98$$

$$x = 24.5$$

So, 24.5 ounces of diced tomatoes should be used to make 7 servings of chicken cacciatore.

48.
$$\frac{x}{15} = \frac{11}{6}$$

$$\frac{x}{15} \cdot 15 = \frac{11}{6} \cdot 15$$

$$x = 27.5$$

The length is 27.5 yards.

49. $A = kw$

50. $F = \dfrac{k}{r^2}$

51. V varies directly as r^3.

52. I varies inversely as d^2.

53. Since w varies directly as t^2, we have $w = kt^2$. We find k by substituting 3 for t and 18 for w:

$$w = kt^2$$

$$18 = k(3)^2$$

$$18 = 9k$$

$$2 = k$$

The equation is $w = 2t^2$. We find the value of t when $w = 50$ by substituting 50 for w and solving for t:

$$50 = 2t^2$$

$$25 = t^2$$

$$t = \pm\sqrt{25} = \pm 5$$

So, $t = \pm 5$ when $w = 50$.

54. Since H varies inversely as d, we have $H = \dfrac{k}{d}$.

We find k by substituting 3 for d and 4 for H:

$$H = \frac{k}{d}$$

$$4 = \frac{k}{3}$$

$$12 = k$$

The equation is $H = \dfrac{12}{d}$. We find the value of H when $d = 6$ by substituting 6 for d:

$$H = \frac{12}{6} = 2$$

So, $H = 2$ when $d = 6$.

55. Since w varies directly as p with a positive variation constant k, the value of w decreases as the value of p decreases.

56. Since C varies inversely as t with a positive variation constant k and the values of t are positive, the value of C decreases as the value of t increases.

57. We let R be the total royalties from selling x books. Our model has the form $R = kx$. We find k by substituting 5000 for x and 47,500 for R:
$$R = kx$$
$$47,500 = k(5000)$$
$$9.5 = k$$
The model is $R = 9.5x$. We find the total royalties from selling 7500 books by substituting 7500 for x:
$$R = 9.5(7500) = 71,250$$
So, the total royalties for selling 7500 books would be $71,250.

58. We let t be time it takes the commuter to drive to work at average driving speed s. Our model has the form $t = \dfrac{k}{s}$. We find k by substituting 45 for s and 40 for t:
$$t = \frac{k}{s}$$
$$40 = \frac{k}{45}$$
$$1800 = k$$
The model is $t = \dfrac{1800}{s}$. We find the time when average speed is 50 mph by substituting 50 for s:
$$t = \frac{1800}{50} = 36$$
So, it will take the commuter 36 minutes to drive to work at an average driving speed of 50 mph.

59. a.

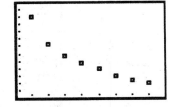

b. An inverse variation model is of the form $A = \dfrac{k}{D}$. We need only one point to find k. Since all of the points in the scattergram fit the pattern fairly, any of the points will likely generate a reasonable model. We arbitrarily choose the point $(9, 6.0)$:
$$A = \frac{k}{D}$$
$$6.0 = \frac{k}{9}$$
$$54 = k$$
Thus, the model is $A = \dfrac{54}{D}$.

c. The apparent height varies inversely as the distance.

d. We substitute 12 for D:
$$A = \frac{54}{12} = 4.5$$
So, the apparent height of the painting will be 4.5 inches when the person is standing 12 feet away from it.

e. Let x be the actual height of the painting. The ratio of the actual height, x, to the distance at which the person stands from the painting, 12 feet, is equal to the ratio of the apparent height, 4.5 inches, to the person's arm length, 2 feet. We set up and solve the proportion accordingly:
$$\frac{x}{12} = \frac{4.5}{2}$$
$$x = \frac{4.5}{2} \cdot 12 = 27$$
The actual height of the painting is approximately 27 inches.

60.
$$\frac{\dfrac{12}{x^2}}{\dfrac{9}{x^3}} = \frac{12}{x^2} \div \frac{9}{x^3} = \frac{12}{x^2} \cdot \frac{x^3}{9} = \frac{2 \cdot 2 \cdot 3}{x^2} \cdot \frac{x^3}{3 \cdot 3}$$
$$= \frac{2 \cdot 2 \cdot 3 \cdot x^3}{x^2 \cdot 3 \cdot 3} = \frac{2 \cdot 2 \cdot x^{3-2}}{3} = \frac{4x}{3}$$

61.
$$\frac{\dfrac{x^2 - 6x - 16}{25x}}{\dfrac{x - 8}{35}} = \frac{x^2 - 6x - 16}{25x} \div \frac{x - 8}{35}$$
$$= \frac{x^2 - 6x - 16}{25x} \cdot \frac{35}{x - 8}$$
$$= \frac{(x - 8)(x + 2)}{5 \cdot 5 \cdot x} \cdot \frac{5 \cdot 7}{x - 8}$$
$$= \frac{(x - 8)(x + 2) \cdot 5 \cdot 7}{5 \cdot 5 \cdot x \cdot (x - 8)} = \frac{7(x + 2)}{5x}$$

62. $\dfrac{5-\dfrac{2}{w}}{1-\dfrac{3}{w}} = \dfrac{5-\dfrac{2}{w}}{1-\dfrac{3}{w}} \cdot \dfrac{w}{w} = \dfrac{5 \cdot w - \dfrac{2}{w} \cdot w}{1 \cdot w - \dfrac{3}{w} \cdot w} = \dfrac{5w-2}{w-3}$

63. $\dfrac{\dfrac{3}{2b}+\dfrac{1}{b^2}}{\dfrac{1}{3b}-\dfrac{2}{b^2}} = \dfrac{\dfrac{3}{2b}+\dfrac{1}{b^2}}{\dfrac{1}{3b}-\dfrac{2}{b^2}} \cdot \dfrac{6b^2}{6b^2}$

$= \dfrac{\dfrac{3}{2b} \cdot 6b^2 + \dfrac{1}{b^2} \cdot 6b^2}{\dfrac{1}{3b} \cdot 6b^2 - \dfrac{2}{b^2} \cdot 6b^2}$

$= \dfrac{9b+6}{2b-12} = \dfrac{3(3b+2)}{2(b-6)}$

Chapter 10 Test

1. $\dfrac{x}{2}$

The denominator is a constant 2, so no substitution for x will lead to a division by 0. Thus, there are no excluded values.

2. $\dfrac{w-2}{w+9}$

We set the denominator equal to 0 and solve for w.
$w+9=0$
$\quad w=-9$
So, the only excluded value is -9.

3. $\dfrac{x-5}{x^2+3x-54}$

We set the denominator equal to 0 and solve for x.
$x^2+3x-54=0$
$(x+9)(x-6)=0$
$x+9=0 \quad \text{or} \quad x-6=0$
$\quad x=-9 \quad \text{or} \quad\quad x=6$
So, the excluded values are -9 and 6.

4. Answers may vary.

5. $\dfrac{p^2-16}{p^2-9p+20} = \dfrac{(p-4)(p+4)}{(p-4)(p-5)} = \dfrac{p+4}{p-5}$

6. $\dfrac{5m^2+10m}{3m^3-2m^2-12m-54} = \dfrac{5m(m+2)}{m^2(3m-2)-4(3m-2)}$

$= \dfrac{5m(m+2)}{(m^2-4)(3m-2)}$

$= \dfrac{5m(m+2)}{(m-2)(m+2)(3m-2)}$

$= \dfrac{5m}{(m-2)(3m-2)}$

7. $\dfrac{x^2+5x+6}{x^2-10x+25} \cdot \dfrac{x^2-25}{x^2-2x-8}$

$= \dfrac{(x+3)(x+2)}{(x-5)(x-5)} \cdot \dfrac{(x-5)(x+5)}{(x-4)(x+2)}$

$= \dfrac{(x+3)(x+2) \cdot (x-5)(x+5)}{(x-5)(x-5) \cdot (x-4)(x+2)}$

$= \dfrac{(x+3)(x+5)}{(x-5)(x-4)}$

8. $\dfrac{x^2-2xy}{20x^6y} \div \dfrac{xy-2y^2}{45x^4y^2} = \dfrac{x^2-2xy}{20x^6y} \cdot \dfrac{45x^4y^2}{xy-2y^2}$

$= \dfrac{x(x-2y)}{2 \cdot 2 \cdot 5 \cdot x^6y} \cdot \dfrac{3 \cdot 3 \cdot 5 \cdot x^4y^2}{y(x-2y)}$

$= \dfrac{x(x-2y) \cdot 3 \cdot 3 \cdot 5 \cdot x^4y^2}{2 \cdot 2 \cdot 5 \cdot x^6y \cdot y(x-2y)}$

$= \dfrac{3 \cdot 3}{2 \cdot 2 \cdot x^{6-1-5}}$

$= \dfrac{9}{4x}$

9. $\dfrac{5a}{a^2-a-20} + \dfrac{2}{a^2-3a-10}$

$= \dfrac{5a}{(a-5)(a+4)} + \dfrac{2}{(a-5)(a+2)}$

$= \dfrac{5a}{(a-5)(a+4)} \cdot \dfrac{a+2}{a+2} + \dfrac{2}{(a-5)(a+2)} \cdot \dfrac{a+4}{a+4}$

$= \dfrac{5a(a+2)}{(a-5)(a+4)(a+2)} + \dfrac{2(a+4)}{(a-5)(a+2)(a+4)}$

$= \dfrac{5a(a+2)+2(a+4)}{(a-5)(a+2)(a+4)}$

$= \dfrac{5a^2+10a+2a+8}{(a-5)(a+2)(a+4)}$

$= \dfrac{5a^2+12a+8}{(a-5)(a+2)(a+4)}$

10.
$$\frac{t}{t^2-49} - \frac{5}{2t-14} = \frac{t}{(t-7)(t+7)} - \frac{5}{2(t-7)}$$

$$= \frac{t}{(t-7)(t+7)} \cdot \frac{2}{2} - \frac{5}{2(t-7)} \cdot \frac{t+7}{t+7}$$

$$= \frac{2t}{2(t-7)(t+7)} - \frac{5(t+7)}{2(t-7)(t+7)}$$

$$= \frac{2t-5(t+7)}{2(t-7)(t+7)}$$

$$= \frac{2t-5t-35}{2(t-7)(t+7)}$$

$$= \frac{-3t-35}{2(t-7)(t+7)}$$

11.
$$\frac{9x^2-4}{-2x-8} \div \frac{6x^2-x-2}{-3x-12}$$

$$= \frac{9x^2-4}{-2x-8} \cdot \frac{-3x-12}{6x^2-x-2}$$

$$= \frac{(3x-2)(3x+2)}{-2(x+4)} \cdot \frac{-3(x+4)}{(3x-2)(2x+1)}$$

$$= \frac{(3x-2)(3x+2)\cdot(-3)(x+4)}{-2(x+4)\cdot(3x-2)(2x+1)}$$

$$= \frac{(3x+2)\cdot(-3)}{-2(2x+1)}$$

$$= \frac{-3(3x+2)}{-2(2x+1)} = \frac{3(3x+2)}{2(2x+1)}$$

12.
$$\frac{x-1}{x+4} - \frac{x+2}{x-7} = \frac{x-1}{x+4} \cdot \frac{x-7}{x-7} - \frac{x+2}{x-7} \cdot \frac{x+4}{x+4}$$

$$= \frac{(x-1)(x-7)}{(x+4)(x-7)} - \frac{(x+2)(x+4)}{(x-7)(x+4)}$$

$$= \frac{(x-1)(x-7)-(x+2)(x+4)}{(x-7)(x+4)}$$

$$= \frac{x^2-7x-x+7-(x^2+4x+2x+8)}{(x-7)(x+4)}$$

$$= \frac{x^2-7x-x+7-x^2-4x-2x-8}{(x-7)(x+4)}$$

$$= \frac{-14x-1}{(x-7)(x+4)}$$

13.
$$\frac{4}{p-2} + \frac{3}{5} = 1$$

We note that 2 is an excluded value.

$$5(p-2)\cdot\left(\frac{4}{p-2}+\frac{3}{5}\right) = 5(p-2)\cdot 1$$

$$5(p-2)\cdot\left(\frac{4}{p-2}\right) + 5(p-2)\cdot\left(\frac{3}{5}\right) = 5(p-2)$$

$$5\cdot 4 + (p-2)\cdot 3 = 5p-10$$

$$20 + 3p - 6 = 5p - 10$$

$$3p + 14 = 5p - 10$$

$$-2p = -24$$

$$p = 12$$

Since 12 is not an excluded value, we conclude that 12 is the solution of the equation.

14.
$$\frac{3}{w^2-4w-21} - \frac{5}{w-7} = \frac{2}{3w+9}$$

$$\frac{3}{(w-7)(w+3)} - \frac{5}{w-7} = \frac{2}{3(w+3)}$$

We note that −3 and 7 are excluded values.

$$3(w-7)(w+3)\left(\frac{3}{(w-7)(w+3)} - \frac{5}{w-7}\right) = 3(w-7)(w+3)\left(\frac{2}{3(w+3)}\right)$$

$$3(w-7)(w+3)\left(\frac{3}{(w-7)(w+3)}\right) - 3(w-7)(w+3)\left(\frac{5}{w-7}\right) = 3(w-7)(w+3)\left(\frac{2}{3(w+3)}\right)$$

$$3\cdot 3 - 3(w+3)\cdot 5 = (w-7)\cdot 2$$

$$9 - 15w - 45 = 2w - 14$$

$$-15w - 36 = 2w - 14$$

$$-17w = 22$$

$$w = -\frac{22}{17}$$

Since $-\frac{22}{17}$ is not an excluded value, we conclude that $-\frac{22}{17}$ is the solution of the equation.

15. $\dfrac{130 \text{ km}}{1 \text{ hour}} \cdot \dfrac{1 \text{ mile}}{1.61 \text{ km}} \approx 80.75 \dfrac{\text{miles}}{\text{hour}}$

So, the speed limit is 80.75 miles per hour.

16. a. The year 2000 is represented by $t = 20$. We substitute 20 for t in the model:

$$p = \frac{-22(20) + 804}{1.28(20) + 80.45} = \frac{364}{106.05} \approx 3.4$$

Using the model, we predict that about 3.4% of households were burglarized in 2000.

To find the actual percentage of households burglarized in 2000, we divide the number of households burglarized by the total number of households:

$$\frac{3.33 \text{ million}}{104.7 \text{ million}} \approx 0.032 \text{ or } 3.2\%$$

In 2000, approximately 3.2% of households were actually burglarized.

Our result from the model is an overestimate.

b. The year 2010 is represented by $t = 30$. We substitute 30 for t in the model:

$$p = \frac{-22(30) + 804}{1.28(30) + 80.45} = \frac{144}{118.85} \approx 1.2$$

We predict that about 1.2% of households will be burglarized in 2010.

c. We substitute 1 for p and solve for t:

$$1 = \frac{-22t + 804}{1.28t + 80.45}$$

$$(1.28t + 80.45) \cdot 1 = (1.28t + 80.45)\left(\frac{-22t + 804}{1.28t + 80.45}\right)$$

$$1.28t + 80.45 = -22t + 804$$

$$23.28t = 723.55$$

$$t \approx 31$$

Now, $1980 + 31 = 2011$.

We predict that 1% of households will be burglarized in the year 2011.

17. Let x be the amount of gasoline required to travel 400 miles on highways.

$$\frac{153}{3} = \frac{400}{x}$$

$$3x\left(\frac{153}{3}\right) = 3x\left(\frac{400}{x}\right)$$

$$153x = 1200$$

$$x \approx 7.84$$

Approximately 7.84 gallons of gasoline are required to travel 400 miles on highways.

18. $\dfrac{x}{9} = \dfrac{14}{8}$

$$\frac{x}{9} \cdot 9 = \frac{14}{8} \cdot 9$$

$$x = 15.75$$

The length is 15.75 meters.

19. Since H varies directly as t, we have $H = kt$. We find k by substituting 3 for t and 12 for H:

$$H = kt$$

$$12 = k(3)$$

$$4 = k$$

The equation is $H = 4t$. We find the value of H when $t = 6$ by substituting 6 for t:

$$H = 4(6) = 24$$

So, $H = 24$ when $t = 6$.

20. Since p varies inversely as w^2, we have $p = \dfrac{k}{w^2}$.

We find k by substituting 2 for w and 9 for p:

$$p = \frac{k}{w^2}$$

$$9 = \frac{k}{(2)^2}$$

$$9 = \frac{k}{4}$$

$$36 = k$$

The equation is $p = \dfrac{36}{w^2}$. We find the value of w when $p = 4$ by substituting 4 for p and solving for w:

$$4 = \frac{36}{w^2}$$

$$4 \cdot w^2 = \frac{36}{w^2} \cdot w^2$$

$$4w^2 = 36$$

$$w^2 = 9$$

$$w = \pm\sqrt{9} = \pm 3$$

So, $w = \pm 3$ when $p = 4$.

21. Since c varies inversely as u with a positive variation constant k and the values of u are positive, the value of c decreases as the value of u increases.

22. We let x be the number of credit hours and T be the total in-district tuition. Our model has the form $T = kx$. We find k by substituting 12 for x and 732 for T:

$$T = kx$$

$$732 = k(12)$$

$$61 = k$$

The model is $T = 61x$. We find the number of credit hours in which a student can enroll for $960 by substituting 960 for T and solving for x:

$$960 = 61x$$

$$15.73 \approx x$$

So, the student can enroll in at most 15 credit hours for $960. (Note that enrolling in 16 credit hours would cost $976, which is more than $960.)

23.
$$\frac{\dfrac{4}{3x} - \dfrac{1}{x^2}}{\dfrac{1}{2x} - \dfrac{5}{x^2}} = \frac{\dfrac{4}{3x} - \dfrac{1}{x^2}}{\dfrac{1}{2x} - \dfrac{5}{x^2}} \cdot \frac{6x^2}{6x^2}$$

$$= \frac{\dfrac{4}{3x} \cdot 6x^2 - \dfrac{1}{x^2} \cdot 6x^2}{\dfrac{1}{2x} \cdot 6x^2 - \dfrac{5}{x^2} \cdot 6x^2}$$

$$= \frac{8x - 6}{3x - 30}$$

$$= \frac{2(4x - 3)}{3(x - 10)}$$

Chapter 11
More Radical Expressions and Equations

Homework 11.1

1. $4\sqrt{3} + 5\sqrt{3} = (4+5)\sqrt{3} = 9\sqrt{3}$

3. $-3\sqrt{5} + 9\sqrt{5} = (-3+9)\sqrt{5} = 6\sqrt{5}$

5. $4\sqrt{x} + \sqrt{x} = 4\sqrt{x} + 1\sqrt{x} = (4+1)\sqrt{x} = 5\sqrt{x}$

7. $\sqrt{7} + \sqrt{7} = 1\sqrt{7} + 1\sqrt{7} = (1+1)\sqrt{7} = 2\sqrt{7}$

9. $2\sqrt{6} - 9\sqrt{6} = (2-9)\sqrt{6} = -7\sqrt{6}$

11. $-4\sqrt{5t} - 7\sqrt{5t} = (-4-7)\sqrt{5t} = -11\sqrt{5t}$

13. $7\sqrt{5} - 3\sqrt{7}$

Since the radicals $7\sqrt{5}$ and $3\sqrt{7}$ have different radicands, they are not like radicals and we cannot use the distributive law. The expression is already in simplified form.

15. $5\sqrt{2} - \sqrt{2} + 2\sqrt{2} = (5-1+2)\sqrt{2} = 6\sqrt{2}$

17. $-4\sqrt{7} + 8\sqrt{7} - 3\sqrt{5} = (-4+8)\sqrt{7} - 3\sqrt{5}$
$= 4\sqrt{7} - 3\sqrt{5}$

19. $7\sqrt{p} - 4\sqrt{p} - 3\sqrt{p} = (7-4-3)\sqrt{p} = 0 \cdot \sqrt{p} = 0$

21. $\sqrt{18} + 4\sqrt{2} = \sqrt{9 \cdot 2} + 4\sqrt{2}$
$= \sqrt{9}\sqrt{2} + 4\sqrt{2}$
$= 3\sqrt{2} + 4\sqrt{2}$
$= 7\sqrt{2}$

23. $5\sqrt{6} - \sqrt{24} = 5\sqrt{6} - \sqrt{4 \cdot 6}$
$= 5\sqrt{6} - \sqrt{4}\sqrt{6}$
$= 5\sqrt{6} - 2\sqrt{6}$
$= 3\sqrt{6}$

25. $-4\sqrt{5} - \sqrt{20} = -4\sqrt{5} - \sqrt{4 \cdot 5}$
$= -4\sqrt{5} - \sqrt{4}\sqrt{5}$
$= -4\sqrt{5} - 2\sqrt{5}$
$= -6\sqrt{5}$

27. $\sqrt{28} + \sqrt{63} = \sqrt{4 \cdot 7} + \sqrt{9 \cdot 7}$
$= \sqrt{4}\sqrt{7} + \sqrt{9}\sqrt{7}$
$= 2\sqrt{7} + 3\sqrt{7}$
$= 5\sqrt{7}$

29. $2\sqrt{27} + 5\sqrt{12} = 2\sqrt{9 \cdot 3} + 5\sqrt{4 \cdot 3}$
$= 2\sqrt{9}\sqrt{3} + 5\sqrt{4}\sqrt{3}$
$= 2 \cdot 3\sqrt{3} + 5 \cdot 2\sqrt{3}$
$= 6\sqrt{3} + 10\sqrt{3}$
$= 16\sqrt{3}$

31. $5\sqrt{8} - 5\sqrt{18} = 5\sqrt{4 \cdot 2} - 5\sqrt{9 \cdot 2}$
$= 5\sqrt{4}\sqrt{2} - 5\sqrt{9}\sqrt{2}$
$= 5 \cdot 2\sqrt{2} - 5 \cdot 3\sqrt{2}$
$= 10\sqrt{2} - 15\sqrt{2}$
$= -5\sqrt{2}$

33. $\sqrt{20} - 4\sqrt{5} + \sqrt{45} = \sqrt{4 \cdot 5} - 4\sqrt{5} + \sqrt{9 \cdot 5}$
$= \sqrt{4}\sqrt{5} - 4\sqrt{5} + \sqrt{9}\sqrt{5}$
$= 2\sqrt{5} - 4\sqrt{5} + 3\sqrt{5}$
$= 1\sqrt{5}$
$= \sqrt{5}$

35. $5\sqrt{12} + 4\sqrt{75} - 2\sqrt{3} = 5\sqrt{4 \cdot 3} + 4\sqrt{25 \cdot 3} - 2\sqrt{3}$
$= 5\sqrt{4}\sqrt{3} + 4\sqrt{25}\sqrt{3} - 2\sqrt{3}$
$= 5 \cdot 2\sqrt{3} + 4 \cdot 5\sqrt{3} - 2\sqrt{3}$
$= 10\sqrt{3} + 20\sqrt{3} - 2\sqrt{3}$
$= 28\sqrt{3}$

37. $\sqrt{4x} + \sqrt{81x} = \sqrt{4 \cdot x} + \sqrt{81 \cdot x}$
$= \sqrt{4}\sqrt{x} + \sqrt{81}\sqrt{x}$
$= 2\sqrt{x} + 9\sqrt{x}$
$= 11\sqrt{x}$

39. $2\sqrt{9x} - 3\sqrt{4x} = 2\sqrt{9 \cdot x} - 3\sqrt{4 \cdot x}$
$= 2\sqrt{9}\sqrt{x} - 3\sqrt{4}\sqrt{x}$
$= 2 \cdot 3\sqrt{x} - 3 \cdot 2\sqrt{x}$
$= 6\sqrt{x} - 6\sqrt{x}$
$= 0$

41. $\sqrt{50x} - \sqrt{20} - \sqrt{32x}$
$= \sqrt{25 \cdot 2x} - \sqrt{4 \cdot 5} - \sqrt{16 \cdot 2x}$
$= \sqrt{25}\sqrt{2x} - \sqrt{4}\sqrt{5} - \sqrt{16}\sqrt{2x}$
$= 5\sqrt{2x} - 2\sqrt{5} - 4\sqrt{2x}$
$= 5\sqrt{2x} - 4\sqrt{2x} - 2\sqrt{5}$
$= \sqrt{2x} - 2\sqrt{5}$

43. $3\sqrt{12t} - 2\sqrt{44} + 5\sqrt{27t}$
$= 3\sqrt{4 \cdot 3t} - 2\sqrt{4 \cdot 11} + 5\sqrt{9 \cdot 3t}$
$= 3\sqrt{4}\sqrt{3t} - 2\sqrt{4}\sqrt{11} + 5\sqrt{9}\sqrt{3t}$
$= 3 \cdot 2\sqrt{3t} - 2 \cdot 2\sqrt{11} + 5 \cdot 3\sqrt{3t}$
$= 6\sqrt{3t} - 4\sqrt{11} + 15\sqrt{3t}$
$= 6\sqrt{3t} + 15\sqrt{3t} - 4\sqrt{11}$
$= 21\sqrt{3t} - 4\sqrt{11}$

45. $x\sqrt{20} + x\sqrt{45} = x\sqrt{4 \cdot 5} + x\sqrt{9 \cdot 5}$
$= x\sqrt{4}\sqrt{5} + x\sqrt{9}\sqrt{5}$
$= x \cdot 2\sqrt{5} + x \cdot 3\sqrt{5}$
$= 2x\sqrt{5} + 3x\sqrt{5}$
$= 5x\sqrt{5}$

47. $x\sqrt{40} - 5\sqrt{10x^2} = x\sqrt{4 \cdot 10} - 5\sqrt{x^2 \cdot 10}$
$= x\sqrt{4}\sqrt{10} - 5\sqrt{x^2}\sqrt{10}$
$= x \cdot 2\sqrt{10} - 5 \cdot x\sqrt{10}$
$= 2x\sqrt{10} - 5x\sqrt{10}$
$= -3x\sqrt{10}$

49. $\sqrt{5w^2} - \sqrt{45w^2} = \sqrt{w^2 \cdot 5} - \sqrt{9w^2 \cdot 5}$
$= \sqrt{w^2}\sqrt{5} - \sqrt{9w^2}\sqrt{5}$
$= w\sqrt{5} - 3w\sqrt{5}$
$= -2w\sqrt{5}$

51. $2\sqrt{75x^2} + 3\sqrt{12x^2} = 2\sqrt{25x^2 \cdot 3} + 3\sqrt{4x^2 \cdot 3}$
$= 2\sqrt{25x^2}\sqrt{3} + 3\sqrt{4x^2}\sqrt{3}$
$= 2 \cdot 5x\sqrt{3} + 3 \cdot 2x\sqrt{3}$
$= 10x\sqrt{3} + 6x\sqrt{3}$
$= 16x\sqrt{3}$

53. $\sqrt{3x^2} - 2x\sqrt{3} + 5\sqrt{4x}$
$= \sqrt{x^2 \cdot 3} - 2x\sqrt{3} + 5\sqrt{4 \cdot x}$
$= \sqrt{x^2}\sqrt{3} - 2x\sqrt{3} + 5\sqrt{4}\sqrt{x}$
$= x\sqrt{3} - 2x\sqrt{3} + 5 \cdot 2\sqrt{x}$
$= -x\sqrt{3} + 10\sqrt{x}$

55. $5\sqrt{48y^2} - 3\sqrt{20y} - 2y\sqrt{12}$
$= 5\sqrt{16y^2 \cdot 3} - 3\sqrt{4 \cdot 5y} - 2y\sqrt{4 \cdot 3}$
$= 5\sqrt{16y^2}\sqrt{3} - 3\sqrt{4}\sqrt{5y} - 2y\sqrt{4}\sqrt{3}$
$= 5 \cdot 4y\sqrt{3} - 3 \cdot 2\sqrt{5y} - 2y \cdot 2\sqrt{3}$
$= 20y\sqrt{3} - 6\sqrt{5y} - 4y\sqrt{3}$
$= 20y\sqrt{3} - 4y\sqrt{3} - 6\sqrt{5y}$
$= 16y\sqrt{3} - 6\sqrt{5y}$

57. a. Substitute 1427.0 for d in the equation
$T = 0.0005443d\sqrt{d}$ and solve for T.
$T = 0.0005443(1427.0)\sqrt{1427.0}$
≈ 29.34
The period of Saturn is about 29.3 years.

b. $\dfrac{59}{29.34} \approx 2.01$

A person who is 59 years old in "Earth years" would be about 2 years old in "Saturn years".

59. a. $n = 308\sqrt{t} + 1050$

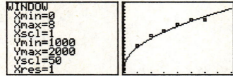

The model appears to fit the data pretty well.

b. For 2010, we have $t = 2010 - 1998 = 12$. Substitute 12 for t in the equation and solve for n.

$n = 308\sqrt{12} + 1050 \approx 2117$

The model predicts that there will be about 2117 agents in Tuscon in 2010.

c.

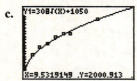

The model predicts that there will be 2000 agents in Tuscon in 2008 (about 10 years after 1998).

61. The student incorrectly added the radicands.

$2\sqrt{5} + 4\sqrt{5} = (2+4)\sqrt{5} = 6\sqrt{5}$

63. $4\sqrt{7} + 5\sqrt{7} = (4+5)\sqrt{7} = 9\sqrt{7}$

65. $5x - 7x = (5-7)x = -2x$

67. $5\sqrt{x} - 7\sqrt{x} = (5-7)\sqrt{x} = -2\sqrt{x}$

69. $\sqrt{12x^2} + \sqrt{3x^2} = \sqrt{4x^2 \cdot 3} + \sqrt{x^2 \cdot 3}$
$= \sqrt{4x^2}\sqrt{3} + \sqrt{x^2}\sqrt{3}$
$= 2x\sqrt{3} + x\sqrt{3}$
$= 3x\sqrt{3}$

71. $12x^2 + 3x^2 = (12+3)x^2 = 15x^2$

73. $t(t-2) = 7$
$t^2 - 2t = 7$
$t^2 - 2t + 1 = 7 + 1$
$(t-1)^2 = 8$
$t - 1 = \pm\sqrt{8}$
$t - 1 = \pm 2\sqrt{2}$
$t = 1 \pm 2\sqrt{2}$

The solutions are $1 - 2\sqrt{2}$ and $1 + 2\sqrt{2}$.
Description: *quadratic equation in one variable*

75. $\dfrac{3w}{w^2 - 9} - \dfrac{1}{4w + 12} = \dfrac{3w}{(w-3)(w+3)} - \dfrac{1}{4(w+3)}$

$= \dfrac{12w}{4(w-3)(w+3)} - \dfrac{(w-3)}{4(w-3)(w+3)}$

$= \dfrac{12w - (w-3)}{4(w-3)(w+3)}$

$= \dfrac{12w - w + 3}{4(w-3)(w+3)}$

$= \dfrac{11w + 3}{4(w-3)(w+3)}$

Description: *rational expression in one variable*

77. $x^3 - 10x^2 + 25x = x(x^2 - 10x + 25)$
$= x(x-5)(x-5)$
$= x(x-5)^2$

Description: *cubic (or third degree) polynomial in one variable*

Homework 11.2

1. $\sqrt{2} \cdot \sqrt{5} = \sqrt{2 \cdot 5} = \sqrt{10}$

3. $-\sqrt{6} \cdot \sqrt{2} = -\sqrt{6 \cdot 2} = -\sqrt{12}$
$= -\sqrt{4 \cdot 3} = -\sqrt{4}\sqrt{3}$
$= -2\sqrt{3}$

5. $\sqrt{17} \cdot \sqrt{17} = \sqrt{17 \cdot 17} = \sqrt{17^2} = 17$

7. $2\sqrt{5}\cdot\sqrt{10} = 2\sqrt{5\cdot 10} = 2\sqrt{50}$

$\qquad = 2\sqrt{25\cdot 2} = 2\sqrt{25}\sqrt{2}$

$\qquad = 2\cdot 5\sqrt{2}$

$\qquad = 10\sqrt{2}$

9. $7\sqrt{2}\left(-3\sqrt{14}\right) = 7\cdot(-3)\cdot\sqrt{2}\cdot\sqrt{14}$

$\qquad = -21\sqrt{2\cdot 14}$

$\qquad = -21\sqrt{28}$

$\qquad = -21\sqrt{4\cdot 7}$

$\qquad = -21\sqrt{4}\sqrt{7}$

$\qquad = -21\cdot 2\sqrt{7}$

$\qquad = -42\sqrt{7}$

11. $5\sqrt{t}\cdot 7\sqrt{t} = 5\cdot 7\cdot\sqrt{t}\cdot\sqrt{t}$

$\qquad = 35\sqrt{t\cdot t}$

$\qquad = 35\sqrt{t^2}$

$\qquad = 35t$

13. $\sqrt{7t}\sqrt{3t} = \sqrt{7t\cdot 3t}$

$\qquad = \sqrt{21t^2} = \sqrt{t^2}\sqrt{21}$

$\qquad = t\sqrt{21}$

15. $\left(8\sqrt{x}\right)^2 = 8^2\cdot\left(\sqrt{x}\right)^2$

$\qquad = 64x$

17. $\sqrt{ab}\sqrt{bc} = \sqrt{ab\cdot bc}$

$\qquad = \sqrt{ab^2 c}$

$\qquad = \sqrt{b^2\cdot ac}$

$\qquad = \sqrt{b^2}\sqrt{ac}$

$\qquad = b\sqrt{ac}$

19. $\sqrt{5}\left(1+\sqrt{7}\right) = \sqrt{5}\cdot 1 + \sqrt{5}\cdot\sqrt{7}$

$\qquad = \sqrt{5}+\sqrt{35}$

21. $-\sqrt{7}\left(\sqrt{2}+\sqrt{5}\right) = -\sqrt{7}\cdot\sqrt{2}-\sqrt{7}\cdot\sqrt{5}$

$\qquad = -\sqrt{14}-\sqrt{35}$

23. $5\sqrt{c}\left(9-\sqrt{c}\right) = 5\sqrt{c}\cdot 9 - 5\sqrt{c}\cdot\sqrt{c}$

$\qquad = 45\sqrt{c}-5c$

25. $-4\sqrt{5}\left(3\sqrt{2}-\sqrt{5}\right) = -4\sqrt{5}\cdot 3\sqrt{2} - \left(-4\sqrt{5}\right)\cdot\sqrt{5}$

$\qquad = -12\sqrt{5}\sqrt{2}+4\sqrt{5}\sqrt{5}$

$\qquad = -12\sqrt{10}+4\cdot 5$

$\qquad = -12\sqrt{10}+20$

$\qquad = 20-12\sqrt{10}$

27. $\left(4+\sqrt{5}\right)\left(2-\sqrt{5}\right) = 4\cdot 2 - 4\sqrt{5}+2\sqrt{5}-\sqrt{5}\sqrt{5}$

$\qquad = 8-4\sqrt{5}+2\sqrt{5}-5$

$\qquad = 8-2\sqrt{5}-5$

$\qquad = 3-2\sqrt{5}$

29. $\left(\sqrt{3}-\sqrt{5}\right)\left(\sqrt{3}+\sqrt{2}\right)$

$\qquad = \sqrt{3}\sqrt{3}+\sqrt{3}\sqrt{2}-\sqrt{5}\sqrt{3}-\sqrt{5}\sqrt{2}$

$\qquad = 3+\sqrt{6}-\sqrt{15}-\sqrt{10}$

31. $\left(\sqrt{x}-\sqrt{7}\right)\left(\sqrt{x}-\sqrt{2}\right)$

$\qquad = \sqrt{x}\sqrt{x}-\sqrt{x}\sqrt{2}-\sqrt{7}\sqrt{x}+\sqrt{7}\sqrt{2}$

$\qquad = x-\sqrt{2x}-\sqrt{7x}+\sqrt{14}$

33. $\left(y+\sqrt{2}\right)\left(y-\sqrt{11}\right)$

$\qquad = y\cdot y - y\sqrt{11}+\sqrt{2}\cdot y - \sqrt{2}\sqrt{11}$

$\qquad = y^2 - y\sqrt{11}+y\sqrt{2}-\sqrt{22}$

35. $\left(4-2\sqrt{5}\right)\left(2-3\sqrt{5}\right)$

$\qquad = 4\cdot 2 - 4\left(3\sqrt{5}\right)-\left(2\sqrt{5}\right)2+\left(2\sqrt{5}\right)\left(3\sqrt{5}\right)$

$\qquad = 8-4\cdot 3\sqrt{5}-2\cdot 2\sqrt{5}+2\cdot 3\sqrt{5}\sqrt{5}$

$\qquad = 8-12\sqrt{5}-4\sqrt{5}+6\cdot 5$

$\qquad = 8-16\sqrt{5}+30$

$\qquad = 38-16\sqrt{5}$

37. $\left(\sqrt{3}+\sqrt{7}\right)\left(\sqrt{3}-\sqrt{7}\right) = \left(\sqrt{3}\right)^2-\left(\sqrt{7}\right)^2$

$\qquad = 3-7$

$\qquad = -4$

39. $\left(r+\sqrt{3}\right)\left(r-\sqrt{3}\right) = r^2-\left(\sqrt{3}\right)^2$

$\qquad = r^2-3$

41. $\left(5\sqrt{2}-2\sqrt{3}\right)\left(5\sqrt{2}+2\sqrt{3}\right)=\left(5\sqrt{2}\right)^2-\left(2\sqrt{3}\right)^2$

$$=5^2\left(\sqrt{2}\right)^2-2^2\left(\sqrt{3}\right)^2$$
$$=25\cdot 2-4\cdot 3$$
$$=50-12$$
$$=38$$

43. $\left(3\sqrt{a}-5\sqrt{b}\right)\left(3\sqrt{a}+5\sqrt{b}\right)=\left(3\sqrt{a}\right)^2-\left(5\sqrt{b}\right)^2$

$$=3^2\left(\sqrt{a}\right)^2-5^2\left(\sqrt{b}\right)^2$$
$$=9a-25b$$

45. $\left(4+\sqrt{7}\right)^2=4^2+2\cdot 4\cdot\sqrt{7}+\left(\sqrt{7}\right)^2$

$$=16+8\sqrt{7}+7$$
$$=23+8\sqrt{7}$$

47. $\left(\sqrt{3}-\sqrt{5}\right)^2=\left(\sqrt{3}\right)^2-2\cdot\sqrt{3}\cdot\sqrt{5}+\left(\sqrt{5}\right)^2$

$$=3-2\sqrt{15}+5$$
$$=8-2\sqrt{15}$$

49. $\left(b+\sqrt{2}\right)^2=b^2+2\cdot b\cdot\sqrt{2}+\left(\sqrt{2}\right)^2$

$$=b^2+2b\sqrt{2}+2$$

51. $\left(\sqrt{x}-\sqrt{6}\right)^2=\left(\sqrt{x}\right)^2-2\cdot\sqrt{x}\cdot\sqrt{6}+\left(\sqrt{6}\right)^2$

$$=x-2\sqrt{6x}+6$$

53. $\left(2\sqrt{5}-3\sqrt{2}\right)^2$

$$=\left(2\sqrt{5}\right)^2-2\left(2\sqrt{5}\right)\left(3\sqrt{2}\right)+\left(3\sqrt{2}\right)^2$$
$$=2^2\left(\sqrt{5}\right)^2-2\cdot 2\cdot 3\sqrt{5}\sqrt{2}+3^2\left(\sqrt{2}\right)^2$$
$$=4\cdot 5-12\sqrt{10}+9\cdot 2$$
$$=20-12\sqrt{10}+18$$
$$=38-12\sqrt{10}$$

55. The student incorrectly distributed the square across a sum.

$$\left(x+\sqrt{3}\right)^2=x^2+2\cdot x\cdot\sqrt{3}+\left(\sqrt{3}\right)^2$$
$$=x^2+2x\sqrt{3}+3$$

57. The student incorrectly distributed the 3 into the radicand.

$$3\left(2\sqrt{5}\right)=3\cdot 2\sqrt{5}=6\sqrt{5}$$

59. a.
 i. True; $\sqrt{ab}=\sqrt{a}\sqrt{b}$ if $a,b\geq 0$

 ii. False; $\sqrt{a+b}$ cannot be simplified further.

 iii. True; $\left(ab\right)^2=a^2b^2$

 iv. False; $\left(a+b\right)^2=a^2+2ab+b^2$

 v. True; $\dfrac{1}{ab}=\left(ab\right)^{-1}=a^{-1}b^{-1}$

 vi. False; $\dfrac{1}{a+b}=\left(a+b\right)^{-1}$ cannot be simplified further.

b. Answers may vary; exponents can distriube across a product, but not a sum.

61. $3\sqrt{x}+2\sqrt{x}=\left(3+2\right)\sqrt{x}=5\sqrt{x}$

63. $\left(3\sqrt{x}\right)\left(2\sqrt{x}\right)=3\cdot 2\cdot\sqrt{x}\sqrt{x}$

$$=6x$$

65. $\left(3+\sqrt{x}\right)\left(2+\sqrt{x}\right)=3\cdot 2+3\cdot\sqrt{x}+\sqrt{x}\cdot 2+\sqrt{x}\sqrt{x}$

$$=6+3\sqrt{x}+2\sqrt{x}+x$$
$$=6+5\sqrt{x}+x$$
$$=x+5\sqrt{x}+6$$

67. $\left(3+x\right)\left(2+x\right)=3\cdot 2+3\cdot x+x\cdot 2+x^2$

$$=6+3x+2x+x^2$$
$$=x^2+5x+6$$

69. $\left(3\sqrt{x}\right)^2=3^2\left(\sqrt{x}\right)^2=9x$

71. $\left(3+\sqrt{x}\right)^2=3^2+2\cdot 3\cdot\sqrt{x}+\left(\sqrt{x}\right)^2$

$$=9+6\sqrt{x}+x$$
$$=x+6\sqrt{x}+9$$

73. $(3+x)^2 = 3^2 + 2 \cdot 3 \cdot x + x^2$

$\qquad = 9 + 6x + x^2$

$\qquad = x^2 + 6x + 9$

75. $y = -\dfrac{2}{5}x - 1$

This is a linear equation of the form $y = mx + b$.

The slope is $m = -\dfrac{2}{5}$ and the y-intercept is

$(0, -1)$. Using the slope, a second point on the graph would be $(0 + 5, -1 - 2) = (5, -3)$. Plot the points $(0, -1)$ and $(5, -3)$, then connect with a straight line.

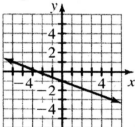

Description: *linear equation in two variables*

77. $\qquad \dfrac{p}{p+4} - \dfrac{4}{p-4} = \dfrac{p^2 + 16}{p^2 - 16}$

$(p^2 - 16)\left(\dfrac{p}{p+4} - \dfrac{4}{p-4}\right) = (p^2 - 16)\left(\dfrac{p^2 + 16}{p^2 - 16}\right)$

$\qquad p(p-4) - 4(p+4) = p^2 + 16$

$\qquad p^2 - 4p - 4p - 16 = p^2 + 16$

$\qquad p^2 - 8p - 16 = p^2 + 16$

$\qquad -8p = 32$

$\qquad p = -4$

Since -4 makes a denominator equal 0 in the original equation, it must be excluded from the solution set. Therefore, the equation has no real number solution.

Description: *rational equation in one variable*

79. $\dfrac{x^2 - 1}{x^2 + 5x + 6} \div \dfrac{x^2 - 3x - 4}{x^2 + 3x}$

$= \dfrac{(x-1)(x+1)}{(x+3)(x+2)} \cdot \dfrac{x(x+3)}{(x-4)(x+1)}$

$= \dfrac{x(x-1)}{(x+2)(x-4)}$

Description: *rational expression in one variable*

Homework 11.3

1. $\qquad \sqrt{x} = 6$

$\qquad \left(\sqrt{x}\right)^2 = 6^2$

$\qquad x = 36$

Check: $\sqrt{36} \overset{?}{=} 6$

$\qquad\qquad 6 \overset{?}{=} 6$ true

The solution is 36.

3. $\qquad \sqrt{5x+1} = 4$

$\qquad \left(\sqrt{5x+1}\right)^2 = 4^2$

$\qquad 5x + 1 = 16$

$\qquad 5x = 15$

$\qquad x = 3$

Check: $\sqrt{5(3)+1} \overset{?}{=} 4$

$\qquad\qquad \sqrt{16} \overset{?}{=} 4$

$\qquad\qquad 4 \overset{?}{=} 4$ true

The solution is 3.

5. $\sqrt{x} = -7$

Since the principle square root must be nonnegative, the equation has no solution.

Note: The same result would be obtained if we used the usual method of squaring both sides, provided we check our potential solutions.

$\qquad \sqrt{x} = -7$

$\qquad \left(\sqrt{x}\right)^2 = (-7)^2$

$\qquad x = 49$

Check: $\sqrt{49} \overset{?}{=} -7$

$\qquad\qquad 7 \overset{?}{=} -7$ false

The equation has no solution.

7.
$$\sqrt{3t-2} = 2$$
$$\left(\sqrt{3t-2}\right)^2 = 2^2$$
$$3t-2 = 4$$
$$3t = 6$$
$$t = 2$$

Check: $\sqrt{3(2)-2} \overset{?}{=} 2$

$$\sqrt{4} \overset{?}{=} 2$$
$$2 \overset{?}{=} 2 \quad \text{true}$$

The solution is 2.

9. $\sqrt{x+4} = -2$

Since the principle square root must be nonnegative, the equation has no solution.

Note: The same result would be obtained if we used the usual method of squaring both sides, provided we check our potential solutions.

$$\sqrt{x+4} = -2$$
$$\left(\sqrt{x+4}\right)^2 = (-2)^2$$
$$x+4 = 4$$
$$x = 0$$

Check: $\sqrt{0+4} \overset{?}{=} -2$

$$\sqrt{4} \overset{?}{=} -2$$
$$2 \overset{?}{=} -2 \quad \text{false}$$

The equation has no solution.

11.
$$\sqrt{5x-2} = \sqrt{3x+8}$$
$$\left(\sqrt{5x-2}\right)^2 = \left(\sqrt{3x+8}\right)^2$$
$$5x-2 = 3x+8$$
$$5x = 3x+10$$
$$2x = 10$$
$$x = 5$$

Check: $\sqrt{5(5)-2} \overset{?}{=} \sqrt{3(5)+8}$

$$\sqrt{23} \overset{?}{=} \sqrt{23} \quad \text{true}$$

The solution is 5.

13.
$$\sqrt{25p^2+4p-8} = 5p$$
$$\left(\sqrt{25p^2+4p-8}\right)^2 = (5p)^2$$
$$25p^2+4p-8 = 25p^2$$
$$4p-8 = 0$$
$$4p = 8$$
$$p = 2$$

Check: $\sqrt{25(2)^2+4(2)-8} \overset{?}{=} 5(2)$

$$\sqrt{100} \overset{?}{=} 10$$
$$10 \overset{?}{=} 10 \quad \text{true}$$

The solution is 2.

15.
$$\sqrt{x^2-6x} = 4$$
$$\left(\sqrt{x^2-6x}\right)^2 = 4^2$$
$$x^2-6x = 16$$
$$x^2-6x-16 = 0$$
$$(x-8)(x+2) = 0$$
$$x-8 = 0 \quad \text{or} \quad x+2 = 0$$
$$x = 8 \qquad\qquad x = -2$$

Check: $\sqrt{8^2-6(8)} \overset{?}{=} 4 \qquad \sqrt{(-2)^2-6(-2)} \overset{?}{=} 4$

$$\sqrt{16} \overset{?}{=} 4 \qquad\qquad \sqrt{16} \overset{?}{=} 4$$
$$4 \overset{?}{=} 4 \quad \text{true} \qquad\quad 4 \overset{?}{=} 4 \quad \text{true}$$

The solutions are -2 and 8.

17. $\sqrt{x}+3 = 7$

$$\sqrt{x} = 4$$
$$\left(\sqrt{x}\right)^2 = 4^2$$
$$x = 16$$

Check: $\sqrt{16}+3 \overset{?}{=} 7$

$$4+3 \overset{?}{=} 7$$
$$7 \overset{?}{=} 7 \quad \text{true}$$

The solution is 16.

19. $3\sqrt{x} = 6$

$\sqrt{x} = 2$

$\left(\sqrt{x}\right)^2 = 2^2$

$x = 4$

Check: $3\sqrt{4} \overset{?}{=} 6$

$3(2) \overset{?}{=} 6$

$6 \overset{?}{=} 6$ true

The solution is 4.

21. $5\sqrt{w} + 3 = 23$

$5\sqrt{w} = 20$

$\sqrt{w} = 4$

$\left(\sqrt{w}\right)^2 = 4^2$

$w = 16$

Check: $5\sqrt{16} + 3 \overset{?}{=} 23$

$5(4) + 3 \overset{?}{=} 23$

$23 \overset{?}{=} 23$ true

The solution is 16.

23. $-3\sqrt{x} + 7 = 2$

$-3\sqrt{x} = -5$

$\sqrt{x} = \dfrac{5}{3}$

$\left(\sqrt{x}\right)^2 = \left(\dfrac{5}{3}\right)^2$

$x = \dfrac{25}{9}$

Check: $-3\sqrt{\tfrac{25}{9}} + 7 \overset{?}{=} 2$

$-3\left(\dfrac{5}{3}\right) + 7 \overset{?}{=} 2$

$2 \overset{?}{=} 2$ true

The solution is $\dfrac{25}{9}$.

25. $4\sqrt{x} - 3 = -5$

$4\sqrt{x} = -2$

$\sqrt{x} = -\dfrac{1}{2}$

Since the principle square root must be nonnegative, the equation has no solution.

27. $5 + \sqrt{a-3} = 8$

$\sqrt{a-3} = 3$

$\left(\sqrt{a-3}\right)^2 = 3^2$

$a - 3 = 9$

$a = 12$

Check: $5 + \sqrt{12-3} \overset{?}{=} 8$

$5 + \sqrt{9} \overset{?}{=} 8$

$5 + 3 \overset{?}{=} 8$

$8 \overset{?}{=} 8$ true

The solution is 12.

29. $\sqrt{x-2} = x - 2$

$\left(\sqrt{x-2}\right)^2 = \left(x-2\right)^2$

$x - 2 = x^2 - 4x + 4$

$0 = x^2 - 5x + 6$

$0 = (x-3)(x-2)$

$x - 3 = 0$ or $x - 2 = 0$

$x = 3$ $x = 2$

Check: $\sqrt{2-2} \overset{?}{=} 2-2$ $\sqrt{3-2} \overset{?}{=} 3-2$

$\sqrt{0} \overset{?}{=} 0$ $\sqrt{1} \overset{?}{=} 1$

$0 \overset{?}{=} 0$ true $1 \overset{?}{=} 1$ true

The solutions are 2 and 3.

31.

$$\sqrt{x+3} = x+1$$

$$\left(\sqrt{x+3}\right)^2 = (x+1)^2$$

$$x+3 = x^2+2x+1$$

$$0 = x^2+x-2$$

$$0 = (x+2)(x-1)$$

$$x+2=0 \quad \text{or} \quad x-1=0$$

$$x=-2 \qquad\quad x=1$$

Check: $\sqrt{1+3} \overset{?}{=} 1+1 \qquad \sqrt{-2+3} \overset{?}{=} -2+1$

$\qquad\qquad\; \sqrt{4} \overset{?}{=} 2 \qquad\qquad\; \sqrt{1} \overset{?}{=} -1$

$\qquad\qquad\; 2 \overset{?}{=} 2$ true $\qquad\quad 1 \overset{?}{=} -1$ false

The only solution is 1.

33.

$$\sqrt{4r-3}-r = -2$$

$$\sqrt{4r-3} = r-2$$

$$\left(\sqrt{4r-3}\right)^2 = (r-2)^2$$

$$4r-3 = r^2-4r+4$$

$$0 = r^2-8r+7$$

$$0 = (r-7)(r-1)$$

$$r-7=0 \quad \text{or} \quad r-1=0$$

$$r=7 \qquad\quad r=1$$

Check: $\sqrt{4(7)-3}-7 \overset{?}{=} -2 \quad \sqrt{4(1)-3}-1 \overset{?}{=} -2$

$\qquad\qquad\;\; \sqrt{25}-7 \overset{?}{=} -2 \qquad\quad \sqrt{1}-1 \overset{?}{=} -2$

$\qquad\qquad\;\; 5-7 \overset{?}{=} -2 \qquad\qquad 1-1 \overset{?}{=} -2$

$\qquad\qquad\;\; -2 \overset{?}{=} -2$ true $\qquad\quad 0 \overset{?}{=} -2$ false

The only solution is 7.

35.

$$\sqrt{(x+5)(x+2)} = x+3$$

$$\left(\sqrt{(x+5)(x+2)}\right)^2 = (x+3)^2$$

$$(x+5)(x+2) = (x+3)^2$$

$$x^2+7x+10 = x^2+6x+9$$

$$7x+10 = 6x+9$$

$$x+10 = 9$$

$$x = -1$$

Check: $\sqrt{(-1+5)(-1+2)} \overset{?}{=} -1+3$

$\qquad\qquad\qquad \sqrt{4} \overset{?}{=} 2$

$\qquad\qquad\qquad 2 \overset{?}{=} 2$ true

The solution is -1.

37.

$$\sqrt{x+5} = \sqrt{x}+1$$

$$\left(\sqrt{x+5}\right)^2 = \left(\sqrt{x}+1\right)^2$$

$$x+5 = \left(\sqrt{x}\right)^2+2\sqrt{x}+1$$

$$x+5 = x+2\sqrt{x}+1$$

$$4 = 2\sqrt{x}$$

$$2 = \sqrt{x}$$

$$2^2 = \left(\sqrt{x}\right)^2$$

$$4 = x$$

Check: $\sqrt{4+5} \overset{?}{=} \sqrt{4}+1$

$\qquad\qquad \sqrt{9} \overset{?}{=} 2+1$

$\qquad\qquad 3 \overset{?}{=} 3$ true

The solution is 4.

39.

$$\sqrt{m-5}+2 = \sqrt{m+3}$$

$$\left(\sqrt{m-5}+2\right)^2 = \left(\sqrt{m+3}\right)^2$$

$$\left(\sqrt{m-5}\right)^2+4\sqrt{m-5}+4 = m+3$$

$$m-5+4\sqrt{m-5} = m-1$$

$$4\sqrt{m-5} = 4$$

$$\sqrt{m-5} = 1$$

$$\left(\sqrt{m-5}\right)^2 = 1^2$$

$$m-5 = 1$$

$$m = 6$$

Check: $\sqrt{6-5}+2 \overset{?}{=} \sqrt{6+3}$

$\qquad\qquad \sqrt{1}+2 \overset{?}{=} \sqrt{9}$

$\qquad\qquad\;\; 1+2 \overset{?}{=} 3$

$\qquad\qquad\qquad 3 \overset{?}{=} 3$ true

The solution is 6.

41. $y = 1 + \sqrt{x}$

We list some solutions in the table below. We choose perfect squares as values of x, because we can find their principle square roots mentally. Since the radicand of $\sqrt{x}$ must be nonnegative, we cannot choose any negative values for x.

x	y
0	$1 + \sqrt{0} = 1$
1	$1 + \sqrt{1} = 2$
4	$1 + \sqrt{4} = 3$
9	$1 + \sqrt{9} = 4$
16	$1 + \sqrt{16} = 5$

Plot the points that correspond to the solutions we found, then connect the points with a smooth curve.

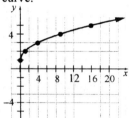

43. The graph of $y = \sqrt{x+4} - 1$ intersects the graph of $y = -1$ only at the point $(-4, -1)$, whose x-coordinate is -4. Therefore, the only solution to the equation is -4.

45. The graph of $y = \sqrt{x+4} - 1$ does not intersect the graph of $y = -2$. Therefore the equation has no solution.

47. $\sqrt{x} - 4 = 3 - x$

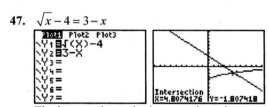

The intersection point is approximately $(4.81, -1.81)$ whose x-coordinate is 4.81.

Therefore, the solution to the equation is approximately 4.81.

49. $\sqrt{x} + 3 = x^2 - 5$

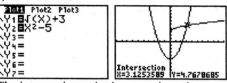

The intersection point is approximately $(3.13, 4.77)$, whose x-coordinate is 3.13.

Therefore, the solution to the equation is approximately 3.13.

51. The student incorrectly squared the right side of the equation.

$$\sqrt{x^2 + x + 6} = x + 4$$
$$\left(\sqrt{x^2 + x + 6}\right)^2 = (x+4)^2$$
$$x^2 + x + 6 = x^2 + 8x + 16$$
$$0 = 7x + 10$$
$$7x = -10$$
$$x = -\frac{10}{7}$$

Check: $\sqrt{\left(-\frac{10}{7}\right)^2 + \left(-\frac{10}{7}\right) + 6} \overset{?}{=} \left(-\frac{10}{7}\right) + 4$

$$\sqrt{\frac{324}{49}} \overset{?}{=} \frac{18}{7}$$

$$\frac{18}{7} \overset{?}{=} \frac{18}{7} \quad \text{true}$$

The solution is $-\dfrac{10}{7}$.

53. a.

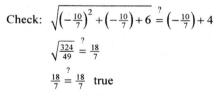

The model appears to fit the data reasonably well.

b. Let $t = 0$ and solve for c.
$$c = 50\sqrt{0} + 28 = 28$$
The c-intercept is $(0, 28)$. This indicates that there were 28 million acres of genetically modified crops in 1997.

c. For 2010, we have $t = 13$.

$c = 50\sqrt{13} + 28 \approx 208.3$

The model predicts that in 2010 there will be about 208.3 million acres of genetically modified crops. Yes, this is larger than the land area of Texas.

d. Substitute 225 for c and solve for t.

$225 = 50\sqrt{t} + 28$

$197 = 50\sqrt{t}$

$\dfrac{197}{50} = \sqrt{t}$

$\left(\dfrac{197}{50}\right)^2 = t$ or $t \approx 15.52$

The model predicts that there will be 225 million acres of genetically modified crops in 2013 (16 years after 1997).

55. a.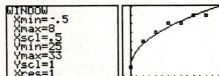

The model appears to fit the data reasonably well.

b. For 2011, we have $t = 12$.

$n = 2.65\sqrt{12} + 26 \approx 35.2$

The model predicts that in 2011 there will be about 35.2 thousand McDonald's restaurants.

c. $\dfrac{35200}{192} \approx 183.3$

According to the model, there will be about 183.3 McDonald's restaurants per country in 2011.

d. Substitute 36 for n and solve for t.

$36 = 2.65\sqrt{n} + 26$

$10 = 2.65\sqrt{n}$

$\dfrac{10}{2.65} = \sqrt{n}$

$\left(\dfrac{10}{2.65}\right)^2 = n$ or $n \approx 14.2$

The model predicts that there will be 36 thousand McDonalds' restaurants in 2013.

57. a.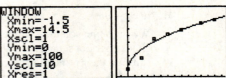

The model appears to fit the data reasonably well.

b. Substitute 100 for p and solve for t.

$100 = 20\sqrt{t} + 11$

$89 = 20\sqrt{t}$

$\dfrac{89}{20} = \sqrt{t}$

$\left(\dfrac{89}{20}\right)^2 = t$ or $t \approx 19.8$

The model predicts that all tires will be reused or recycled in 2010.

c. Since the plant would not be operational until 2008 and the model predicts that all tires will be reused or recycled beginning in 2010, it does not make sense to build the plant.

59. $2x^2 + 4x = 7$

$2x^2 + 4x - 7 = 0$

$a = 2, b = 4, c = -7$

$x = \dfrac{-4 \pm \sqrt{4^2 - 4(2)(-7)}}{2(2)}$

$= \dfrac{-4 \pm \sqrt{72}}{4}$

$= \dfrac{-4 \pm 6\sqrt{2}}{4}$

$= \dfrac{-2 \pm 3\sqrt{2}}{2}$

The solutions are $\dfrac{-2 \pm 3\sqrt{2}}{2}$.

61. $\sqrt{x+2} = x-4$

$$\left(\sqrt{x+2}\right)^2 = (x-4)^2$$

$$x+2 = x^2 - 8x + 16$$

$$0 = x^2 - 9x + 14$$

$$0 = (x-7)(x-2)$$

$$x-7=0 \quad \text{or} \quad x-2=0$$

$$x=7 \qquad\qquad x=2$$

Check: $\sqrt{7+2} \overset{?}{=} 7-4 \qquad \sqrt{2+2} \overset{?}{=} 2-4$

$\qquad\qquad \sqrt{9} \overset{?}{=} 3 \qquad\qquad \sqrt{4} \overset{?}{=} -2$

$\qquad\qquad 3 = 3 \ \text{ true} \qquad\quad 2 = -2 \ \text{ false}$

The only solution is 7.

63.
$$\frac{3}{x-2} - \frac{5x}{x^2-4} = \frac{7}{x+2}$$

$$\left(x^2-4\right)\left(\frac{3}{x-2} - \frac{5x}{x^2-4}\right) = \left(x^2-4\right)\left(\frac{7}{x+2}\right)$$

$$3(x+2) - 5x = 7(x-2)$$

$$3x + 6 - 5x = 7x - 14$$

$$-2x + 6 = 7x - 14$$

$$-9x = -20$$

$$x = \frac{20}{9}$$

The solution is $\dfrac{20}{9}$.

65. – 73. Answers may vary.

Chapter 11 Review Exercises

1. $6\sqrt{3} + 9\sqrt{3} = (6+9)\sqrt{3}$

$\qquad\qquad\quad = 15\sqrt{3}$

2. $-4\sqrt{2x} + 3\sqrt{2x} = (-4+3)\sqrt{2x}$

$\qquad\qquad\qquad = -1\sqrt{2x}$

$\qquad\qquad\qquad = -\sqrt{2x}$

3. $7\sqrt{5x} - 9\sqrt{5x} = (7-9)\sqrt{5x}$

$\qquad\qquad\qquad = -2\sqrt{5x}$

4. $8\sqrt{3} + 4\sqrt{6}$

Since the radicals $8\sqrt{3}$ and $4\sqrt{6}$ have different radicands, they are not like radicals and we cannot use the distributive law. The expression is already in simplified form.

5. $2\sqrt{5} - 3\sqrt{7} - 8\sqrt{5} = 2\sqrt{5} - 8\sqrt{5} - 3\sqrt{7}$

$\qquad\qquad\qquad\qquad = (2-8)\sqrt{5} - 3\sqrt{7}$

$\qquad\qquad\qquad\qquad = -6\sqrt{5} - 3\sqrt{7}$

6. $\sqrt{27} + 8\sqrt{3} = \sqrt{9 \cdot 3} + 8\sqrt{3}$

$\qquad\qquad\qquad = \sqrt{9}\sqrt{3} + 8\sqrt{3}$

$\qquad\qquad\qquad = 3\sqrt{3} + 8\sqrt{3}$

$\qquad\qquad\qquad = 11\sqrt{3}$

7. $-3\sqrt{6} - \sqrt{24} = -3\sqrt{6} - \sqrt{4 \cdot 6}$

$\qquad\qquad\qquad = -3\sqrt{6} - \sqrt{4}\sqrt{6}$

$\qquad\qquad\qquad = -3\sqrt{6} - 2\sqrt{6}$

$\qquad\qquad\qquad = -5\sqrt{6}$

8. $\sqrt{32} + \sqrt{18} = \sqrt{16 \cdot 2} + \sqrt{9 \cdot 2}$

$\qquad\qquad\qquad = \sqrt{16}\sqrt{2} + \sqrt{9}\sqrt{2}$

$\qquad\qquad\qquad = 4\sqrt{2} + 3\sqrt{2}$

$\qquad\qquad\qquad = 7\sqrt{2}$

9. $\sqrt{45} - \sqrt{20} = \sqrt{9 \cdot 5} - \sqrt{4 \cdot 5}$

$\qquad\qquad\qquad = \sqrt{9}\sqrt{5} - \sqrt{4}\sqrt{5}$

$\qquad\qquad\qquad = 3\sqrt{5} - 2\sqrt{5}$

$\qquad\qquad\qquad = 1\sqrt{5}$

$\qquad\qquad\qquad = \sqrt{5}$

10. $5\sqrt{8} + 3\sqrt{50} = 5\sqrt{4 \cdot 2} + 3\sqrt{25 \cdot 2}$

$\qquad\qquad\qquad = 5\sqrt{4}\sqrt{2} + 3\sqrt{25}\sqrt{2}$

$\qquad\qquad\qquad = 5 \cdot 2\sqrt{2} + 3 \cdot 5\sqrt{2}$

$\qquad\qquad\qquad = 10\sqrt{2} + 15\sqrt{2}$

$\qquad\qquad\qquad = 25\sqrt{2}$

11. $2\sqrt{40} - 5\sqrt{90} = 2\sqrt{4 \cdot 10} - 5\sqrt{9 \cdot 10}$
$$= 2\sqrt{4}\sqrt{10} - 5\sqrt{9}\sqrt{10}$$
$$= 2 \cdot 2\sqrt{10} - 5 \cdot 3\sqrt{10}$$
$$= 4\sqrt{10} - 15\sqrt{10}$$
$$= -11\sqrt{10}$$

12. $\sqrt{32} - 5\sqrt{3} - \sqrt{72} = \sqrt{16 \cdot 2} - 5\sqrt{3} - \sqrt{36 \cdot 2}$
$$= \sqrt{16}\sqrt{2} - 5\sqrt{3} - \sqrt{36}\sqrt{2}$$
$$= 4\sqrt{2} - 5\sqrt{3} - 6\sqrt{2}$$
$$= 4\sqrt{2} - 6\sqrt{2} - 5\sqrt{3}$$
$$= -2\sqrt{2} - 5\sqrt{3}$$

13. $7\sqrt{5} + 4\sqrt{20} - 8\sqrt{24} = 7\sqrt{5} + 4\sqrt{4 \cdot 5} - 8\sqrt{4 \cdot 6}$
$$= 7\sqrt{5} + 4\sqrt{4}\sqrt{5} - 8\sqrt{4}\sqrt{6}$$
$$= 7\sqrt{5} + 4 \cdot 2\sqrt{5} - 8 \cdot 2\sqrt{6}$$
$$= 7\sqrt{5} + 8\sqrt{5} - 16\sqrt{6}$$
$$= 15\sqrt{5} - 16\sqrt{6}$$

14. $\sqrt{64x} + \sqrt{81x} = \sqrt{64 \cdot x} + \sqrt{81 \cdot x}$
$$= \sqrt{64}\sqrt{x} + \sqrt{81}\sqrt{x}$$
$$= 8\sqrt{x} + 9\sqrt{x}$$
$$= 17\sqrt{x}$$

15. $2\sqrt{25x} - 3\sqrt{49x} = 2\sqrt{25}\sqrt{x} - 3\sqrt{49}\sqrt{x}$
$$= 2 \cdot 5\sqrt{x} - 3 \cdot 7\sqrt{x}$$
$$= 10\sqrt{x} - 21\sqrt{x}$$
$$= -11\sqrt{x}$$

16. $\sqrt{12x} + \sqrt{8} - \sqrt{75x} = \sqrt{4 \cdot 3x} + \sqrt{4 \cdot 2} - \sqrt{25 \cdot 3x}$
$$= \sqrt{4}\sqrt{3x} + \sqrt{4}\sqrt{2} - \sqrt{25}\sqrt{3x}$$
$$= 2\sqrt{3x} + 2\sqrt{2} - 5\sqrt{3x}$$
$$= 2\sqrt{3x} - 5\sqrt{3x} + 2\sqrt{2}$$
$$= -3\sqrt{3x} + 2\sqrt{2}$$

17. $w\sqrt{28} - w\sqrt{63} = w\sqrt{4 \cdot 7} - w\sqrt{9 \cdot 7}$
$$= w\sqrt{4}\sqrt{7} - w\sqrt{9}\sqrt{7}$$
$$= w \cdot 2\sqrt{7} - w \cdot 3\sqrt{7}$$
$$= 2w\sqrt{7} - 3w\sqrt{7}$$
$$= -w\sqrt{7}$$

18. $\sqrt{7a^2} + \sqrt{28a^2} = \sqrt{a^2 \cdot 7} + \sqrt{4a^2 \cdot 7}$
$$= \sqrt{a^2}\sqrt{7} + \sqrt{4a^2}\sqrt{7}$$
$$= a\sqrt{7} + 2a\sqrt{7}$$
$$= 3a\sqrt{7}$$

19. $5\sqrt{3x^2} - 3x\sqrt{48} = 5\sqrt{x^2 \cdot 3} - 3x\sqrt{16 \cdot 3}$
$$= 5\sqrt{x^2}\sqrt{3} - 3x\sqrt{16}\sqrt{3}$$
$$= 5x\sqrt{3} - 3x \cdot 4\sqrt{3}$$
$$= 5x\sqrt{3} - 12x\sqrt{3}$$
$$= -7x\sqrt{3}$$

20. $2\sqrt{8x} - 4\sqrt{45x^2} + x\sqrt{20}$
$$= 2\sqrt{4 \cdot 2x} - 4\sqrt{9x^2 \cdot 5} + x\sqrt{4 \cdot 5}$$
$$= 2\sqrt{4}\sqrt{2x} - 4\sqrt{9x^2}\sqrt{5} + x\sqrt{4}\sqrt{5}$$
$$= 2 \cdot 2\sqrt{2x} - 4 \cdot 3x\sqrt{5} + x \cdot 2\sqrt{5}$$
$$= 4\sqrt{2x} - 12x\sqrt{5} + 2x\sqrt{5}$$
$$= 4\sqrt{2x} - 10x\sqrt{5}$$

21. $\sqrt{7a}\sqrt{5b} = \sqrt{7a \cdot 5b}$
$$= \sqrt{35ab}$$

22. $4\sqrt{3}\left(-2\sqrt{6}\right) = -8\sqrt{3 \cdot 6}$
$$= -8\sqrt{18}$$
$$= -8\sqrt{9 \cdot 2}$$
$$= -8\sqrt{9}\sqrt{2}$$
$$= -8 \cdot 3\sqrt{2}$$
$$= -24\sqrt{2}$$

23. $\left(5\sqrt{x}\right)^2 = 5^2\left(\sqrt{x}\right)^2 = 25x$

24. $\sqrt{2}\left(\sqrt{3} + \sqrt{7}\right) = \sqrt{2}\sqrt{3} + \sqrt{2}\sqrt{7}$
$$= \sqrt{2 \cdot 3} + \sqrt{2 \cdot 7}$$
$$= \sqrt{6} + \sqrt{14}$$

25. $-\sqrt{3}\left(\sqrt{7} - \sqrt{5}\right) = -\sqrt{3}\sqrt{7} + \sqrt{3}\sqrt{5}$
$$= -\sqrt{3 \cdot 7} + \sqrt{3 \cdot 5}$$
$$= -\sqrt{21} + \sqrt{15}$$

26. $\sqrt{x}\left(\sqrt{x}-\sqrt{2}\right) = \sqrt{x}\sqrt{x}-\sqrt{x}\sqrt{2}$

$\qquad = \sqrt{x\cdot x}-\sqrt{x\cdot 2}$

$\qquad = \sqrt{x^2}-\sqrt{2x}$

$\qquad = x-\sqrt{2x}$

27. $2\sqrt{7}\left(5\sqrt{3}+\sqrt{7}\right) = 2\sqrt{7}\left(5\sqrt{3}\right)+2\sqrt{7}\left(\sqrt{7}\right)$

$\qquad = 10\sqrt{7\cdot 3}+2\sqrt{7\cdot 7}$

$\qquad = 10\sqrt{21}+2\cdot 7$

$\qquad = 10\sqrt{21}+14$

28. $\left(\sqrt{5}-3\right)\left(\sqrt{5}+6\right)$

$\qquad = \sqrt{5}\sqrt{5}+\sqrt{5}\cdot 6-3\sqrt{5}-3\cdot 6$

$\qquad = 5+6\sqrt{5}-3\sqrt{5}-18$

$\qquad = 3\sqrt{5}-13$

29. $\left(\sqrt{2}-9\right)\left(\sqrt{2}-8\right)$

$\qquad = \sqrt{2}\sqrt{2}-\sqrt{2}\cdot 8-9\sqrt{2}+9\cdot 8$

$\qquad = 2-8\sqrt{2}-9\sqrt{2}+72$

$\qquad = 74-17\sqrt{2}$

30. $\left(\sqrt{2}-\sqrt{7}\right)\left(\sqrt{3}-\sqrt{5}\right)$

$\qquad = \sqrt{2}\sqrt{3}-\sqrt{2}\sqrt{5}-\sqrt{7}\sqrt{3}+\sqrt{7}\sqrt{5}$

$\qquad = \sqrt{6}-\sqrt{10}-\sqrt{21}+\sqrt{35}$

31. $\left(\sqrt{b}+8\right)\left(\sqrt{b}-1\right)$

$\qquad = \sqrt{b}\sqrt{b}-\sqrt{b}\cdot 1+8\sqrt{b}-8\cdot 1$

$\qquad = b-\sqrt{b}+8\sqrt{b}-8$

$\qquad = b+7\sqrt{b}-8$

32. $\left(t+\sqrt{3}\right)\left(t+\sqrt{5}\right)$

$\qquad = t^2+t\sqrt{5}+\sqrt{3}\cdot t+\sqrt{3}\sqrt{5}$

$\qquad = t^2+t\sqrt{5}+t\sqrt{3}+\sqrt{15}$

33. $\left(3\sqrt{7}-1\right)\left(2\sqrt{7}-2\right)$

$\qquad = 3\sqrt{7}\cdot 2\sqrt{7}-3\sqrt{7}\cdot 2-1\cdot 2\sqrt{7}+1\cdot 2$

$\qquad = 3\cdot 2\sqrt{7^2}-3\cdot 2\sqrt{7}-1\cdot 2\sqrt{7}+1\cdot 2$

$\qquad = 6\cdot 7-6\sqrt{7}-2\sqrt{7}+2$

$\qquad = 42-8\sqrt{7}+2$

$\qquad = 44-8\sqrt{7}$

34. $\left(\sqrt{5}-\sqrt{7}\right)\left(\sqrt{5}+\sqrt{7}\right) = \left(\sqrt{5}\right)^2-\left(\sqrt{7}\right)^2$

$\qquad = 5-7$

$\qquad = -2$

35. $\left(4\sqrt{a}+3\sqrt{b}\right)\left(4\sqrt{a}-3\sqrt{b}\right)$

$\qquad = \left(4\sqrt{a}\right)^2-\left(3\sqrt{b}\right)^2$

$\qquad = 4^2\left(\sqrt{a}\right)^2-3^2\left(\sqrt{b}\right)^2$

$\qquad = 16a-9b$

36. $\left(b-\sqrt{3}\right)\left(b+\sqrt{3}\right) = b^2-\left(\sqrt{3}\right)^2$

$\qquad = b^2-3$

37. $\left(4-\sqrt{5}\right)^2 = 4^2-2(4)\left(\sqrt{5}\right)+\left(\sqrt{5}\right)^2$

$\qquad = 16-8\sqrt{5}+5$

$\qquad = 21-8\sqrt{5}$

38. $\left(\sqrt{x}+6\right)^2 = \left(\sqrt{x}\right)^2+2\left(\sqrt{x}\right)(6)+6^2$

$\qquad = x+12\sqrt{x}+36$

39. $\left(x+\sqrt{3}\right)^2 = x^2+2x\sqrt{3}+\left(\sqrt{3}\right)^2$

$\qquad = x^2+2x\sqrt{3}+3$

40. $\left(3\sqrt{5}-4\sqrt{2}\right)^2 = \left(3\sqrt{5}\right)^2-2\left(3\sqrt{5}\right)\left(4\sqrt{2}\right)+\left(4\sqrt{2}\right)^2$

$\qquad = 3^2\left(\sqrt{5}\right)^2-2\cdot 3\cdot 4\sqrt{5\cdot 2}+4^2\left(\sqrt{2}\right)^2$

$\qquad = 9\cdot 5-24\sqrt{10}+16\cdot 2$

$\qquad = 45-24\sqrt{10}+32$

$\qquad = 77-24\sqrt{10}$

41. $\sqrt{9x^2 - 5x + 20} = 3x$

$\left(\sqrt{9x^2 - 5x + 20}\right)^2 = (3x)^2$

$9x^2 - 5x + 20 = 9x^2$

$-5x + 20 = 0$

$-5x = -20$

$x = 4$

Check: $\sqrt{9(4)^2 - 5(4) + 20} \stackrel{?}{=} 3(4)$

$\sqrt{144} \stackrel{?}{=} 12$

$12 \stackrel{?}{=} 12$ true

The solution is 4.

42. $\sqrt{2x - 8} = 6$

$\left(\sqrt{2x - 8}\right)^2 = 6^2$

$2x - 8 = 36$

$2x = 44$

$x = 22$

Check: $\sqrt{2(22) - 8} \stackrel{?}{=} 6$

$\sqrt{36} \stackrel{?}{=} 6$

$6 \stackrel{?}{=} 6$ true

The solution is 22.

43. $\sqrt{3r + 8} = -7$

Since the principle square root must be nonnegative, the equation has no solution.

44. $\sqrt{4p + 3} = \sqrt{6p - 2}$

$\left(\sqrt{4p + 3}\right)^2 = \left(\sqrt{6p - 2}\right)^2$

$4p + 3 = 6p - 2$

$-2p + 3 = -2$

$-2p = -5$

$p = \frac{5}{2}$

Check: $\sqrt{4\left(\frac{5}{2}\right) + 3} \stackrel{?}{=} \sqrt{6\left(\frac{5}{2}\right) - 2}$

$\sqrt{13} \stackrel{?}{=} \sqrt{13}$ true

The solution is $\frac{5}{2}$.

45. $\sqrt{x^2 - 3x} = 2$

$\left(\sqrt{x^2 - 3x}\right)^2 = 2^2$

$x^2 - 3x = 4$

$x^2 - 3x - 4 = 0$

$(x - 4)(x + 1) = 0$

$x - 4 = 0$ or $x + 1 = 0$

$x = 4 \qquad\qquad x = -1$

Check: $\sqrt{4^2 - 3(4)} \stackrel{?}{=} 2 \qquad \sqrt{(-1)^2 - 3(-1)} \stackrel{?}{=} 2$

$\sqrt{4} \stackrel{?}{=} 2 \qquad\qquad \sqrt{4} \stackrel{?}{=} 2$

$2 \stackrel{?}{=} 2$ true $\qquad\quad 2 \stackrel{?}{=} 2$ true

The solutions are -1 and 4.

46. $\sqrt{x} + 2 = 9$

$\sqrt{x} = 7$

$\left(\sqrt{x}\right)^2 = 7^2$

$x = 49$

Check: $\sqrt{49} + 2 \stackrel{?}{=} 9$

$7 + 2 \stackrel{?}{=} 9$

$9 \stackrel{?}{=} 9$ true

The solution is 49.

47. $2\sqrt{x} - 5 = 11$

$2\sqrt{x} = 16$

$\sqrt{x} = 8$

$\left(\sqrt{x}\right)^2 = 8^2$

$x = 64$

Check: $2\sqrt{64} - 5 \stackrel{?}{=} 11$

$2(8) - 5 \stackrel{?}{=} 11$

$11 \stackrel{?}{=} 11$ true

The solution is 64.

48. $-3\sqrt{x} + 4 = 16$

$\qquad -3\sqrt{x} = 12$

$\qquad \sqrt{x} = -4$

Since the principle square root must be nonnegative, the equation has no solution.

49. $3 + \sqrt{y-2} = 5$

$\qquad \sqrt{y-2} = 2$

$\qquad \left(\sqrt{y-2}\right)^2 = 2^2$

$\qquad y - 2 = 4$

$\qquad y = 6$

Check: $3 + \sqrt{6-2} \stackrel{?}{=} 5$

$\qquad 3 + \sqrt{4} \stackrel{?}{=} 5$

$\qquad 3 + 2 \stackrel{?}{=} 5$

$\qquad 5 \stackrel{?}{=} 5$ true

The solution is 6.

50. $\sqrt{p+5} = p + 3$

$\qquad \left(\sqrt{p+5}\right)^2 = \left(p+3\right)^2$

$\qquad p + 5 = p^2 + 6p + 9$

$\qquad 0 = p^2 + 5p + 4$

$\qquad 0 = \left(p+1\right)\left(p+4\right)$

$\quad p + 1 = 0 \quad$ or $\quad p + 4 = 0$

$\qquad p = -1 \qquad\qquad p = -4$

Check: $\sqrt{-1+5} \stackrel{?}{=} -1+3$ or $\sqrt{-4+5} \stackrel{?}{=} -4+3$

$\qquad\quad \sqrt{4} \stackrel{?}{=} 2 \qquad\qquad \sqrt{1} \stackrel{?}{=} -1$

$\qquad\quad 2 \stackrel{?}{=} 2$ true $\qquad\quad 1 \stackrel{?}{=} -1$ false

The only solution is $p = -1$.

51. $\sqrt{(x+3)(x-4)} = x - 2$

$\qquad \left(\sqrt{(x+3)(x-4)}\right)^2 = \left(x-2\right)^2$

$\qquad (x+3)(x-4) = \left(x-2\right)^2$

$\qquad x^2 - x - 12 = x^2 - 4x + 4$

$\qquad -x - 12 = -4x + 4$

$\qquad 3x - 12 = 4$

$\qquad 3x = 16$

$\qquad x = \dfrac{16}{3}$

Check: $\sqrt{\left(\frac{16}{3}+3\right)\left(\frac{16}{3}-4\right)} \stackrel{?}{=} \frac{16}{3} - 2$

$\qquad \sqrt{\frac{25}{3} \cdot \frac{4}{3}} \stackrel{?}{=} \frac{10}{3}$

$\qquad \sqrt{\frac{100}{9}} \stackrel{?}{=} \frac{10}{3}$

$\qquad \frac{10}{3} \stackrel{?}{=} \frac{10}{3}$ true

The solution is $\dfrac{16}{3}$.

52. $1 + \sqrt{2x-3} = x$

$\qquad \sqrt{2x-3} = x - 1$

$\qquad \left(\sqrt{2x-3}\right)^2 = \left(x-1\right)^2$

$\qquad 2x - 3 = x^2 - 2x + 1$

$\qquad 0 = x^2 - 4x + 4$

$\qquad 0 = \left(x-2\right)^2$

$\qquad x = 2$

Check: $1 + \sqrt{2(2)-3} \stackrel{?}{=} 2$

$\qquad 1 + \sqrt{1} \stackrel{?}{=} 2$

$\qquad 1 + 1 \stackrel{?}{=} 2$

$\qquad 2 \stackrel{?}{=} 2$ true

The solution is 2.

53. $\sqrt{5w+1} - w = -1$

$\sqrt{5w+1} = w - 1$

$\left(\sqrt{5w+1}\right)^2 = (w-1)^2$

$5w + 1 = w^2 - 2w + 1$

$0 = w^2 - 7w$

$0 = w(w-7)$

$w = 0 \quad$ or $\quad w = 7$

Check: $\sqrt{5(0)+1} - 0 \overset{?}{=} -1 \qquad \sqrt{5(7)+1} - 7 \overset{?}{=} -1$

$\sqrt{1} - 0 \overset{?}{=} -1 \qquad \sqrt{36} - 7 \overset{?}{=} -1$

$1 \overset{?}{=} -1 \text{ false} \qquad 6 - 7 \overset{?}{=} -1$

$-1 \overset{?}{=} -1 \text{ true}$

The only solution is $w = 7$.

54. $\sqrt{t+5} = \sqrt{t} + 1$

$\left(\sqrt{t+5}\right)^2 = \left(\sqrt{t}+1\right)^2$

$t + 5 = t + 2\sqrt{t} + 1$

$4 = 2\sqrt{t}$

$2 = \sqrt{t}$

$2^2 = \left(\sqrt{t}\right)^2$

$4 = t$

Check: $\sqrt{4+5} \overset{?}{=} \sqrt{4} + 1$

$\sqrt{9} \overset{?}{=} 2 + 1$

$3 \overset{?}{=} 3 \text{ true}$

The solution is 4.

55. $-2\sqrt{x} + 3 = \frac{1}{2}x^2 - 6$

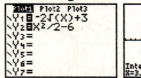
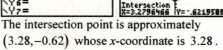

The intersection point is approximately $(3.28, -0.62)$ whose x-coordinate is 3.28.

Therefore, the solution to the equation is approximately 3.28.

56. a.

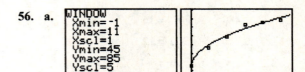

The model appears to fit the data reasonably well.

b. For 2011, we have $t = 18$.

$p = 9\sqrt{18} + 52.3 \approx 90.5$

The model predicts that in 2011 about 90.5% of the FDA budget will go towards new drug approval.

c. Substitute 100 for p and solve for t.

$100 = 9\sqrt{t} + 52.3$

$47.7 = 9\sqrt{t}$

$\sqrt{t} = 5.3$

$t = (5.3)^2$

$t = 28.09$

The model predicts that all of the FDA budget will go to new drug approval in 2021. Model breakdown has occurred since it is unlikely that the entire budget will be allocated to new drug approval.

Chapter 11 Test

1. $5\sqrt{3} - 3\sqrt{2} - 7\sqrt{3} = 5\sqrt{3} - 7\sqrt{3} - 3\sqrt{2}$

$= -2\sqrt{3} - 3\sqrt{2}$

2. $-3\sqrt{20x} - 2\sqrt{45x} = -3\sqrt{4\cdot5x} - 2\sqrt{9\cdot5x}$

$= -3\sqrt{4}\sqrt{5x} - 2\sqrt{9}\sqrt{5x}$

$= -3\cdot2\sqrt{5x} - 2\cdot3\sqrt{5x}$

$= -6\sqrt{5x} - 6\sqrt{5x}$

$= -12\sqrt{5x}$

3. $5\sqrt{45} - 5\sqrt{18x} - 2\sqrt{32x}$

$= 5\sqrt{9\cdot5} - 5\sqrt{9\cdot2x} - 2\sqrt{16\cdot2x}$

$= 5\sqrt{9}\sqrt{5} - 5\sqrt{9}\sqrt{2x} - 2\sqrt{16}\sqrt{2x}$

$= 5\cdot3\sqrt{5} - 5\cdot3\sqrt{2x} - 2\cdot4\sqrt{2x}$

$= 15\sqrt{5} - 15\sqrt{2x} - 8\sqrt{2x}$

$= 15\sqrt{5} - 23\sqrt{2x}$

4.
$$3\sqrt{24b^2} - 7\sqrt{6b^2} = 3\sqrt{4b^2 \cdot 6} - 7\sqrt{b^2 \cdot 6}$$
$$= 3\sqrt{4b^2}\sqrt{6} - 7\sqrt{b^2}\sqrt{6}$$
$$= 3 \cdot 2b\sqrt{6} - 7b\sqrt{6}$$
$$= 6b\sqrt{6} - 7b\sqrt{6}$$
$$= -b\sqrt{6}$$

5.
$$-8\sqrt{14} \cdot 5\sqrt{2} = -8 \cdot 5\sqrt{14 \cdot 2}$$
$$= -40\sqrt{28}$$
$$= -40\sqrt{4 \cdot 7}$$
$$= -40\sqrt{4}\sqrt{7}$$
$$= -40 \cdot 2\sqrt{7}$$
$$= -80\sqrt{7}$$

6.
$$\sqrt{x}\left(\sqrt{x} - 3\right) = \sqrt{x}\sqrt{x} - \sqrt{x} \cdot 3$$
$$= x - 3\sqrt{x}$$

7.
$$\left(\sqrt{5} - 2\right)\left(\sqrt{5} + 4\right) = \sqrt{5}\sqrt{5} + \sqrt{5} \cdot 4 - 2\sqrt{5} - 2 \cdot 4$$
$$= 5 + 4\sqrt{5} - 2\sqrt{5} - 8$$
$$= 2\sqrt{5} - 3$$

8.
$$\left(\sqrt{2} - \sqrt{5}\right)\left(\sqrt{2} + \sqrt{7}\right)$$
$$= \sqrt{2}\sqrt{2} + \sqrt{2}\sqrt{7} - \sqrt{5}\sqrt{2} - \sqrt{5}\sqrt{7}$$
$$= 2 + \sqrt{14} - \sqrt{10} - \sqrt{35}$$

9.
$$\left(5\sqrt{a} - 2\sqrt{b}\right)\left(5\sqrt{a} + 2\sqrt{b}\right) = \left(5\sqrt{a}\right)^2 - \left(2\sqrt{b}\right)^2$$
$$= 5^2\left(\sqrt{a}\right)^2 - 2^2\left(\sqrt{b}\right)^2$$
$$= 25a - 4b$$

10.
$$\left(4 + \sqrt{3}\right)^2 = 4^2 + 2(4)\left(\sqrt{3}\right) + \left(\sqrt{3}\right)^2$$
$$= 16 + 8\sqrt{3} + 3$$
$$= 19 + 8\sqrt{3}$$

11.
$$\left(\sqrt{x} - \sqrt{5}\right)^2 = \left(\sqrt{x}\right)^2 - 2\left(\sqrt{x}\right)\left(\sqrt{5}\right) + \left(\sqrt{5}\right)^2$$
$$= x - 2\sqrt{5x} + 5$$

12.
$$\left(3\sqrt{2} + 2\sqrt{3}\right)^2$$
$$= \left(3\sqrt{2}\right)^2 + 2\left(3\sqrt{2}\right)\left(2\sqrt{3}\right) + \left(2\sqrt{3}\right)^2$$
$$= 3^2\left(\sqrt{2}\right)^2 + 2 \cdot 3 \cdot 2\sqrt{2 \cdot 3} + 2^2\left(\sqrt{3}\right)^2$$
$$= 9 \cdot 2 + 12\sqrt{6} + 4 \cdot 3$$
$$= 18 + 12\sqrt{6} + 12$$
$$= 30 + 12\sqrt{6}$$

13.
$$\sqrt{3x - 5} = 4$$
$$\left(\sqrt{3x - 5}\right)^2 = 4^2$$
$$3x - 5 = 16$$
$$3x = 21$$
$$x = 7$$
Check: $\sqrt{3(7) - 5} \overset{?}{=} 4$
$$\sqrt{16} \overset{?}{=} 4$$
$$4 \overset{?}{=} 4 \text{ true}$$
The solution is 7.

14.
$$\sqrt{2x + 7} = \sqrt{5x - 8}$$
$$\left(\sqrt{2x + 7}\right)^2 = \left(\sqrt{5x - 8}\right)^2$$
$$2x + 7 = 5x - 8$$
$$-3x + 7 = -8$$
$$-3x = -15$$
$$x = 5$$
Check: $\sqrt{2(5) + 7} \overset{?}{=} \sqrt{5(5) - 8}$
$$\sqrt{17} \overset{?}{=} \sqrt{17} \text{ true}$$
The solution is 5.

15.
$$4\sqrt{t} - 3 = 5$$
$$4\sqrt{t} = 8$$
$$\sqrt{t} = 2$$
$$\left(\sqrt{t}\right)^2 = 2^2$$
$$t = 4$$
Check: $4\sqrt{4} - 3 \overset{?}{=} 5$
$$4 \cdot 2 - 3 \overset{?}{=} 5$$
$$5 \overset{?}{=} 5 \text{ true}$$
The solution is 4.

16. $3 + \sqrt{w+5} = 9$

$\sqrt{w+5} = 6$

$\left(\sqrt{w+5}\right)^2 = 6^2$

$w + 5 = 36$

$w = 31$

Check: $3 + \sqrt{31+5} \overset{?}{=} 9$

$3 + \sqrt{36} \overset{?}{=} 9$

$3 + 6 \overset{?}{=} 9$

$9 = 9$ true

The solution is 31.

17. $\sqrt{x-3} = x - 5$

$\left(\sqrt{x-3}\right)^2 = (x-5)^2$

$x - 3 = x^2 - 10x + 25$

$0 = x^2 - 11x + 28$

$0 = (x-7)(x-4)$

$x - 7 = 0$ or $x - 4 = 0$

$x = 7$ $\qquad x = 4$

Check: $\sqrt{7-3} \overset{?}{=} 7-5 \qquad \sqrt{4-3} \overset{?}{=} 4-5$

$\sqrt{4} \overset{?}{=} 2 \qquad\qquad \sqrt{1} \overset{?}{=} -1$

$2 = 2$ true $\qquad 1 = -1$ false

The only solution is $x = 7$.

18. $\sqrt{x+8} = \sqrt{x} + 2$

$\left(\sqrt{x+8}\right)^2 = \left(\sqrt{x}+2\right)^2$

$x + 8 = x + 4\sqrt{x} + 4$

$4 = 4\sqrt{x}$

$1 = \sqrt{x}$

$1^2 = \left(\sqrt{x}\right)^2$

$1 = x$

Check: $\sqrt{1+8} \overset{?}{=} \sqrt{1} + 2$

$\sqrt{9} \overset{?}{=} 1 + 2$

$3 = 3$ true

The solution is 1.

19. The graph of $y = \sqrt{x+5} - 3$ intersects the graph of $y = 0$ only at the point $(4, 0)$, whose x-coordinate is 4. Therefore, the only solution to the equation is 4.

20. The graph of $y = \sqrt{x+5} - 3$ intersects the graph of $y = -1$ only at the point $(-1, -1)$, whose x-coordinate is -1. Therefore, the only solution to the equation is -1.

21. The graph of $y = \sqrt{x+5} - 3$ intersects the graph of $y = -3$ only at the point $(-5, -3)$, whose x-coordinate is -5. Therefore, the only solution to the equation is -5.

22. The graph of $y = \sqrt{x+5} - 3$ does not intersect the graph of $y = -4$. Therefore the equation has no solution.

23. a.

```
WINDOW
Xmin=-2
Xmax=20
Xscl=2
Ymin=15
Ymax=50
Yscl=5
Xres=1
```

The model appears to fit the data reasonably well.

b. For 2010, we have $t = 28$.

$p = 5.5\sqrt{28} + 21 \approx 50.1$

The model predicts that in 2010 about 50.1% of high school students will take foreign-language instruction.

c. Substitute 50 for p and solve for t.

$50 = 5.5\sqrt{t} + 21$

$29 = 5.5\sqrt{t}$

$\sqrt{t} = \dfrac{29}{5.5}$

$\left(\sqrt{t}\right)^2 = \left(\dfrac{29}{5.5}\right)^2$

$t \approx 27.8$

The model predicts that half of high school students will take foreign-language instruction in the year 2010.

Cumulative Review of Chapters 1 – 11

1. $-3ab^2(5a-2b) = -3ab^2(5a) + 3ab^2(2b)$
$= -15a^2b^2 + 6ab^3$
This is a fourth-degree polynomial.

2. $7 - 2(4p+3) - 5p = 7 - 8p - 6 - 5p$
$= -8p - 5p + 7 - 6$
$= -13p + 1$
This is a linear (or first-degree) polynomial.

3. $(3x^2 - 5x + 1) + (6x^2 - 2x - 4)$
$= 3x^2 + 6x^2 - 5x - 2x + 1 - 4$
$= 9x^2 - 7x - 3$
This is a quadratic (or second-degree) polynomial.

4. $(4x^3 - x^2 + x) - (5x^3 - 2x + 1)$
$= 4x^3 - x^2 + x - 5x^3 + 2x - 1$
$= 4x^3 - 5x^3 - x^2 + x + 2x - 1$
$= -x^3 - x^2 + 3x - 1$
This is a cubic (or third-degree) polynomial.

5. $(2m - 5n)(6m + 7n)$
$= 2m \cdot 6m + 2m \cdot 7n - 5n \cdot 6m - 5n \cdot 7n$
$= 12m^2 + 14mn - 30mn - 35n^2$
$= 12m^2 - 16mn - 35n^2$
This is a quadratic (or second-degree) polynomial.

6. $(3x + 2)(x^2 - 4x + 3)$
$= 3x \cdot x^2 - 3x \cdot 4x + 3x \cdot 3 + 2 \cdot x^2 - 2 \cdot 4x + 2 \cdot 3$
$= 3x^3 - 12x^2 + 9x + 2x^2 - 8x + 6$
$= 3x^3 - 12x^2 + 2x^2 + 9x - 8x + 6$
$= 3x^3 - 10x^2 + x + 6$
This is a cubic (or third-degree) polynomial.

7. $(2p + 3q)^2 = (2p)^2 + 2(2p)(3q) + (3q)^2$
$= 4p^2 + 12pq + 9q^2$
This is a quadratic (or second-degree) polynomial.

8. $\dfrac{5m-10}{m^3 - 2m^2 - 15m} \cdot \dfrac{3m^2 + 9m}{m^2 - 4}$
$= \dfrac{5(m-2)}{m(m^2 - 2m - 15)} \cdot \dfrac{3m(m+3)}{(m-2)(m+2)}$
$= \dfrac{5(m-2)}{m(m-5)(m+3)} \cdot \dfrac{3m(m+3)}{(m-2)(m+2)}$
$= \dfrac{5(m-2) \cdot 3m(m+3)}{m(m-5)(m+3) \cdot (m-2)(m+2)}$
$= \dfrac{5 \cdot 3}{(m-5)(m+2)}$
$= \dfrac{15}{(m-5)(m+2)}$
This is a rational expression.

9. $\dfrac{2w^2 + 5w - 3}{w^2 - 10w + 25} \div \dfrac{4w^2 - 1}{2w - 10}$
$= \dfrac{2w^2 + 5w - 3}{w^2 - 10w + 25} \cdot \dfrac{2w - 10}{4w^2 - 1}$
$= \dfrac{(2w-1)(w+3)}{(w-5)(w-5)} \cdot \dfrac{2(w-5)}{(2w-1)(2w+1)}$
$= \dfrac{(2w-1)(w+3) \cdot 2(w-5)}{(w-5)(w-5) \cdot (2w-1)(2w+1)}$
$= \dfrac{(w+3) \cdot 2}{(w-5) \cdot (2w+1)}$
$= \dfrac{2(w+3)}{(w-5)(2w+1)}$
This is a rational expression.

10. $\dfrac{2x}{x^2 - 36} + \dfrac{4}{5x - 30}$
$= \dfrac{2x}{(x-6)(x+6)} + \dfrac{4}{5(x-6)}$
$= \dfrac{2x}{(x-6)(x+6)} \cdot \dfrac{5}{5} + \dfrac{4}{5(x-6)} \cdot \dfrac{x+6}{x+6}$
$= \dfrac{2x \cdot 5}{5(x-6)(x+6)} + \dfrac{4(x+6)}{5(x-6)(x+6)}$
$= \dfrac{2x \cdot 5 + 4(x+6)}{5(x-6)(x+6)}$
$= \dfrac{10x + 4x + 24}{5(x-6)(x+6)}$
$= \dfrac{14x + 24}{5(x-6)(x+6)} = \dfrac{2(7x+12)}{5(x-6)(x+6)}$
This is a rational expression.

11. $\dfrac{3}{x^2-x-6}-\dfrac{5x}{x^2+6x+8}$

$=\dfrac{3}{(x-3)(x+2)}-\dfrac{5x}{(x+4)(x+2)}$

$=\dfrac{3}{(x-3)(x+2)}\cdot\dfrac{x+4}{x+4}-\dfrac{5x}{(x+4)(x+2)}\cdot\dfrac{x-3}{x-3}$

$=\dfrac{3(x+4)}{(x-3)(x+2)(x+4)}-\dfrac{5x(x-3)}{(x+4)(x+2)(x-3)}$

$=\dfrac{3(x+4)-5x(x-3)}{(x-3)(x+2)(x+4)}$

$=\dfrac{3x+12-5x^2+15x}{(x-3)(x+2)(x+4)}$

$=\dfrac{-5x^2+18x+12}{(x-3)(x+2)(x+4)}$

This is a rational expression.

12. $\sqrt{\dfrac{3}{7}}=\dfrac{\sqrt{3}}{\sqrt{7}}=\dfrac{\sqrt{3}}{\sqrt{7}}\cdot\dfrac{\sqrt{7}}{\sqrt{7}}=\dfrac{\sqrt{21}}{\sqrt{49}}=\dfrac{\sqrt{21}}{7}$

This is a radical expression.

13. $2\sqrt{27}-5\sqrt{12}=2\sqrt{9\cdot3}-5\sqrt{4\cdot3}$

$=2\sqrt{9}\sqrt{3}-5\sqrt{4}\sqrt{3}$

$=2\cdot3\cdot\sqrt{3}-5\cdot2\cdot\sqrt{3}$

$=6\sqrt{3}-10\sqrt{3}$

$=-4\sqrt{3}$

This is a radical expression.

14. $\left(\sqrt{x}-\sqrt{2}\right)\left(\sqrt{x}+\sqrt{5}\right)$

$=\sqrt{x}\sqrt{x}+\sqrt{x}\sqrt{5}-\sqrt{2}\sqrt{x}-\sqrt{2}\sqrt{5}$

$=x+\sqrt{5x}-\sqrt{2x}-\sqrt{10}$

This is a radical expression.

15. $\left(2\sqrt{b}-5\sqrt{c}\right)\left(2\sqrt{b}+5\sqrt{c}\right)=\left(2\sqrt{b}\right)^2-\left(5\sqrt{c}\right)^2$

$=4b-25c$

This is a radical expression.

16. $\left(\sqrt{3}-\sqrt{7}\right)^2=\left(\sqrt{3}\right)^2-2\sqrt{3}\sqrt{7}+\left(\sqrt{7}\right)^2$

$=3-2\sqrt{21}+7$

$=10-2\sqrt{21}$

This is a radical expression.

17. $y=-3(x+2)^2+1$

$y=-3(x^2+2\cdot x\cdot2+2^2)+1$

$y=-3(x^2+4x+4)+1$

$y=-3x^2-12x-12+1$

$y=-3x^2-12x-11$

18. $\left(3x^4y^8\right)^3\left(2xy^4\right)=3^3x^{4\cdot3}y^{8\cdot3}\cdot2xy^4$

$=27x^{12}y^{24}\cdot2xy^4$

$=27\cdot2x^{12+1}y^{24+4}$

$=54x^{13}y^{28}$

19. $\sqrt{50x^2y}=\sqrt{25x^2\cdot2y}=\sqrt{25x^2}\sqrt{2y}=5x\sqrt{2y}$

20. $\sqrt{\dfrac{t}{7}}=\dfrac{\sqrt{t}}{\sqrt{7}}=\dfrac{\sqrt{t}}{\sqrt{7}}\cdot\dfrac{\sqrt{7}}{\sqrt{7}}=\dfrac{\sqrt{7t}}{\sqrt{49}}=\dfrac{\sqrt{7t}}{7}$

21. We substitute -2 for a, 8 for b, and -4 for c.

$ac^2-\dfrac{b}{c}=(-2)(-4)^2-\dfrac{8}{(-4)}$

$=(-2)(16)-\dfrac{8}{(-4)}$

$=-32-(-2)$

$=-32+2$

$=-30$

22. We substitute -2 for a, 8 for b, and -4 for c.

$\dfrac{a+bc}{ab-c}=\dfrac{-2+8(-4)}{-2(8)-(-4)}=\dfrac{-2+(-32)}{-16+4}=\dfrac{-34}{-12}=\dfrac{17}{6}$

23. $16p^2-81q^2=(4p)^2-(9q)^2$

$=(4p-9q)(4p+9q)$

24. $a^2-6a-27$

The leading coefficient is 1, so we need to find two numbers whose product is -27 and whose sum is -6. Since $(-9)(3)=-27$ and $-9+3=-6$, we get

$a^2-6a-27=(a-9)(a+3)$.

25. $w^2+14w+49=w^2+2\cdot w\cdot7+7^2$

$=(w+7)^2$

26. $6m^2n+13mn^2+6n^3=n(6m^2+13mn+6n^2)$

$=n(2m+3n)(3m+2n)$

27. $2x^2 - 16x + 32 = 2(x^2 - 8x + 16)$
$$= 2(x^2 - 2 \cdot x \cdot 4 + 4^2)$$
$$= 2(x - 4)^2$$

28. $x^3 + 5x^2 - 9x - 45 = x^2(x + 5) - 9(x + 5)$
$$= (x^2 - 9)(x + 5)$$
$$= (x - 3)(x + 3)(x + 5)$$

29. $12x^4 + 4x^3 - 40x^2 = 4x^2(3x^2 + x - 10)$
$$= 4x^2(3x - 5)(x + 2)$$

30. $x^3 + 27 = x^3 + 3^3$
$$= (x + 3)(x^2 - x \cdot 3 + 3^2)$$
$$= (x + 3)(x^2 - 3x + 9)$$

31. $-4(x - 2) > 20$
$$-4x + 8 > 20$$
$$-4x > 12$$
$$\frac{-4x}{-4} < \frac{12}{-4}$$
$$x < -3$$
Interval: $(-\infty, -3)$.

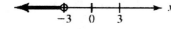

32. $3x - 2y \geq 2$
First, we get y by itself on one side of the inequality:
$3x - 2y \geq 2$
$$-2y \geq -3x + 2$$
$$\frac{-2y}{-2} \leq \frac{-3x + 2}{-2}$$
$$y \leq \frac{3}{2}x - 1$$

The graph of $y \leq \frac{3}{2}x - 1$ is the line $y = \frac{3}{2}x - 1$ and the region below the line. We use a solid line along the boarder to indicate that the points on the line are solutions to $3x - 2y \geq 2$.

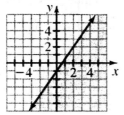

33. $y > \frac{1}{4}x - 3$
$y \geq -2x$

First we sketch the graph of $y > \frac{1}{4}x - 3$ and the graph of $y \geq -2x$. The graph of $y > \frac{1}{4}x - 3$ is the region above the line $y = \frac{1}{4}x - 3$ (graph the line with a dashed line). The graph of $y \geq -2x$ is the line $y = -2x$ and the region above the line (graph the line with a solid line). The graph of the solution set of the system is the intersection of the graphs of the inequalities.

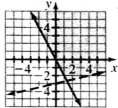

34. $2x + 5 = 7x - 3$
$$2x + 5 - 7x = 7x - 3 - 7x$$
$$-5x + 5 = -3$$
$$-5x + 5 - 5 = -3 - 5$$
$$-5x = -8$$
$$\frac{-5x}{-5} = \frac{-8}{-5}$$
$$x = \frac{8}{5}$$

So, the solution is $\frac{8}{5}$.
This is a linear equation.

35. $6(2x - 3) = 4(3x + 1) - 3(2x - 5)$
$$12x - 18 = 12x + 4 - 6x + 15$$
$$12x - 18 = 6x + 19$$
$$6x = 37$$
$$x = \frac{37}{6}$$

So, the solution is $\frac{37}{6}$.
This is a linear equation.

36.
$$w^2 = 5w + 24$$
$$w^2 - 5w - 24 = 0$$
$$(w-8)(w+3) = 0$$
$$w - 8 = 0 \quad \text{or} \quad w + 3 = 0$$
$$w = 8 \quad \text{or} \quad w = -3$$

So, the solutions are -3 and 8.
This is a quadratic (or second-degree) equation.

37.
$$(2r-1)(3r-2) = 1$$
$$6r^2 - 4r - 3r + 2 = 1$$
$$6r^2 - 7r + 2 = 1$$
$$6r^2 - 7r + 1 = 0$$
$$(6r-1)(r-1) = 0$$
$$6r - 1 = 0 \quad \text{or} \quad r - 1 = 0$$
$$6r = 1 \quad \text{or} \quad r = 1$$
$$r = \frac{1}{6}$$

So, the solutions are $\frac{1}{6}$ and 1.
This is a quadratic (or second-degree) equation.

38.
$$3x^2 - 7 = 13$$
$$3x^2 = 20$$
$$x^2 = \frac{20}{3}$$
$$x = \pm\sqrt{\frac{20}{3}}$$
$$x = \pm\frac{\sqrt{20}}{\sqrt{3}}$$
$$x = \pm\frac{2\sqrt{5}}{\sqrt{3}} \cdot \frac{\sqrt{3}}{\sqrt{3}}$$
$$x = \pm\frac{2\sqrt{15}}{3}$$

So, the solutions are $\pm\frac{2\sqrt{15}}{3}$.
This is a quadratic (or second-degree) equation.

39.
$$(x+4)^2 = 60$$
$$x + 4 = \pm\sqrt{60}$$
$$x + 4 = \pm 2\sqrt{15}$$
$$x = -4 \pm 2\sqrt{15}$$

So, the solutions are $-4 \pm 2\sqrt{15}$.
This is a quadratic (or second-degree) equation.

40. $3x^2 - 5x - 4 = 0$
We solve by using the quadratic formula with $a = 3$, $b = -5$, and $c = -4$.
$$x = \frac{-b \pm \sqrt{b^2 - 4ac}}{2a}$$
$$= \frac{-(-5) \pm \sqrt{(-5)^2 - 4(3)(-4)}}{2(3)}$$
$$= \frac{5 \pm \sqrt{25 + 48}}{6} = \frac{5 \pm \sqrt{73}}{6}$$

So, the solutions are $\frac{5 \pm \sqrt{73}}{6}$.

This is a quadratic (or second-degree) equation.

41.
$$2x(x+3) = -3$$
$$2x^2 + 6x = -3$$
$$2x^2 + 6x + 3 = 0$$
We solve by using the quadratic formula with $a = 2$, $b = 6$, and $c = 3$.
$$x = \frac{-b \pm \sqrt{b^2 - 4ac}}{2a} = \frac{-(6) \pm \sqrt{(6)^2 - 4(2)(3)}}{2(2)}$$
$$= \frac{-6 \pm \sqrt{36 - 24}}{4} = \frac{-6 \pm \sqrt{12}}{4} = \frac{-6 \pm 2\sqrt{3}}{4}$$
$$= \frac{2(-3 \pm \sqrt{3})}{4} = \frac{-3 \pm \sqrt{3}}{2}$$

So, the solutions are $\frac{-3 \pm \sqrt{3}}{2}$.

This is a quadratic (or second-degree) equation.

42. $\dfrac{6}{p-2} = \dfrac{5}{p-3}$

We note that 2 and 3 are excluded values.
$$(p-2)(p-3) \cdot \left(\frac{6}{p-2}\right) = (p-2)(p-3) \cdot \left(\frac{5}{p-3}\right)$$
$$(p-3) \cdot 6 = (p-2) \cdot 5$$
$$6p - 18 = 5p - 10$$
$$p = 8$$

Since 8 is not an excluded value, we conclude that 8 is the solution of the equation. This is a rational equation.

43.
$$\frac{5}{m-3} = \frac{m}{m-2} + \frac{m}{m^2 - 5m + 6}$$

$$\frac{5}{m-3} = \frac{m}{m-2} + \frac{m}{(m-3)(m-2)}$$

We note that 2 and 3 are excluded values.

$$(m-3)(m-2) \cdot \left(\frac{5}{m-3}\right) = (m-3)(m-2) \cdot \left(\frac{m}{m-2} + \frac{m}{(m-3)(m-2)}\right)$$

$$(m-2) \cdot 5 = (m-3)(m-2) \cdot \left(\frac{m}{m-2}\right) + (m-3)(m-2) \cdot \left(\frac{m}{(m-3)(m-2)}\right)$$

$$5m - 10 = (m-3) \cdot m + m$$

$$5m - 10 = m^2 - 3m + m$$

$$0 = m^2 - 7m + 10$$

$$0 = (m-5)(m-2)$$

$$m - 5 = 0 \quad \text{or} \quad m - 2 = 0$$

$$m = 5 \quad \text{or} \quad m = 2$$

Since 2 is an excluded value, it is not a solution. We conclude that 5 is the only solution of the equation. This is a rational equation.

44.
$$5\sqrt{x} - 2 = 1$$

$$5\sqrt{x} = 3$$

$$\sqrt{x} = \frac{3}{5}$$

$$\left(\sqrt{x}\right)^2 = \left(\frac{3}{5}\right)^2$$

$$x = \frac{9}{25}$$

We check that $\frac{9}{25}$ satisfies the original equation:

$$5\sqrt{\frac{9}{25}} - 2 \stackrel{?}{=} 1$$

$$5\left(\frac{3}{5}\right) - 2 \stackrel{?}{=} 1$$

$$3 - 2 \stackrel{?}{=} 1$$

$$1 = 1 \quad \text{True}$$

So, $\frac{9}{25}$ is the solution.

This is a radical equation.

45.
$$\sqrt{x} = x - 2$$

$$\left(\sqrt{x}\right)^2 = \left(x - 2\right)^2$$

$$x = x^2 - 4x + 4$$

$$0 = x^2 - 5x + 4$$

$$0 = (x-4)(x-1)$$

$$x - 4 = 0 \quad \text{or} \quad x - 1 = 0$$

$$x = 4 \quad \text{or} \quad x = 1$$

We check that both 1 and 4 satisfy the original equation:

Check $x = 1$ Check $x = 4$

$\sqrt{1} \stackrel{?}{=} 1 - 2$ $\sqrt{4} \stackrel{?}{=} 4 - 2$

$1 \stackrel{?}{=} -1$ $2 \stackrel{?}{=} 2$

False   True

So, the only solution is 4.

This is a radical equation.

46.
$$a = \frac{v - v_0}{t}$$

$$a \cdot t = \frac{v - v_0}{t} \cdot t$$

$$at = v - v_0$$

$$\frac{at}{a} = \frac{v - v_0}{a}$$

$$t = \frac{v - v_0}{a}$$

47. $3x^2 - 12x = 42$

$$\frac{3x^2 - 12x}{3} = \frac{42}{3}$$

$$x^2 - 4x = 14$$

$$x^2 - 4x + 4 = 14 + 4$$

$$(x-2)^2 = 18$$

$$x - 2 = \pm\sqrt{18}$$

$$x - 2 = \pm 3\sqrt{2}$$

$$x = 2 \pm 3\sqrt{2}$$

The solutions are $2 \pm 3\sqrt{2}$.

48. Answers may vary. Three possibilities follow:
$2x - 6 = 10$; $x - 3 = 5$; $2x = 16$

49. $2x - 7y = 3$ Equation (1)

$-5x + 3y = 7$ Equation (2)

To eliminate the x terms, we multiply both sides of equation (1) by 5 and we multiply both sides of equation (2) by 2, yielding the system

$$10x - 35y = 15$$

$$-10x + 6y = 14$$

The coefficients of the x terms are equal in absolute value and opposite in sign. Add the left and rights sides of the equations and solve for y.

$$10x - 35y = 15$$

$$\underline{-10x + 6y = 14}$$

$$-29y = 29$$

$$y = -1$$

Substitute -1 for y in equation (1) and solve for x:

$$2x - 7(-1) = 3$$

$$2x + 7 = 3$$

$$2x = -4$$

$$x = -2$$

The solution is $(-2, -1)$.

50. $3x - 4y = 35$

$$y = 2x - 5$$

Since the second equation is solved for y, substitute $2x - 5$ for y in the first equation and solve for x:

$$3x - 4(2x - 5) = 35$$

$$3x - 8x + 20 = 35$$

$$-5x + 20 = 35$$

$$-5x = 15$$

$$x = -3$$

Substitute -3 for x in the equation $y = 2x - 5$:

$$y = 2(-3) - 5 = -6 - 5 = -11.$$

The solution is $(-3, -11)$.

51. We solve for y so that the equation will be in slope-intercept form:

$$4x - 3y = 6$$

$$-3y = -4x + 6$$

$$\frac{-3y}{-3} = \frac{-4x + 6}{-3}$$

$$y = \frac{4}{3}x - 2$$

The slope is $m = \frac{4}{3}$ and the y-intercept is $(0, -2)$.

We first plot the y-intercept $(0, -2)$. The slope is $\frac{4}{3}$, so the run is 3 and the rise is 4. From $(0, -2)$, we count 3 units to the right and 4 units up, where we plot the point $(3, 2)$. We then sketch the line that contains these two points.

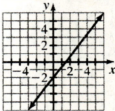

52. $y = 2x^2 - 5$

First, we list some solutions in the table. Then we sketch a curve that contains the points corresponding to the solutions.

x	y
-2	$2(-2)^2 - 5 = 3$
-1	$2(-1)^2 - 5 = -3$
0	$2(0)^2 - 5 = -5$
1	$2(1)^2 - 5 = -3$
2	$2(2)^2 - 5 = 3$

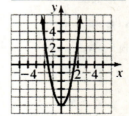

53.
$$2x - 5y = 20$$
$$2x - 5(0) = 20$$
$$2x = 20$$
$$x = 10$$
The solution is 10. Thus, the x-intercept is $(10, 0)$.

54. $y = x^2 + 2x - 48$
$$0 = x^2 + 2x - 48$$
$$0 = (x + 8)(x - 6)$$
$$x + 8 = 0 \quad \text{or} \quad x - 6 = 0$$
$$x = -8 \quad \text{or} \quad x = 6$$
The solutions are -8 and 6. Thus the x-intercepts are $(-8, 0)$ and $(6, 0)$.

55. Slope of ski run $A = \dfrac{130 \text{ yards}}{610 \text{ yards}} \approx 0.21$

Slope of ski run $B = \dfrac{165 \text{ yards}}{700 \text{ yards}} \approx 0.24$

Thus, ski run B is steeper since it has a larger slope.

56. Using the slope formula with $(x_1, y_1) = (-5, -4)$ and $(x_2, y_2) = (-1, -2)$, the slope is
$$m = \frac{y_2 - y_1}{x_2 - x_1} = \frac{-2 - (-4)}{-1 - (-5)} = \frac{-2 + 4}{-1 + 5} = \frac{2}{4} = \frac{1}{2}.$$
The slope is positive, so the line is increasing.

57. We substitute $m = -\dfrac{2}{5}$ and $(x_1, y_1) = (-3, 4)$ into the point slope form and solve for y:
$$y - y_1 = m(x - x_1)$$
$$y - 4 = -\frac{2}{5}(x - (-3))$$
$$y - 4 = -\frac{2}{5}(x + 3)$$
$$y - 4 = -\frac{2}{5}x - \frac{6}{5}$$
$$y = -\frac{2}{5}x - \frac{6}{5} + 4$$
$$y = -\frac{2}{5}x + \frac{14}{5}$$

58. We begin by finding the slope of the line:
$$m = \frac{-3 - 4}{6 - (-2)} = \frac{-7}{8} = -\frac{7}{8}$$
Then we substitute $m = -\dfrac{7}{8}$ and $(x_1, y_1) = (-2, 4)$ into the point slope form and solve for y:
$$y - y_1 = m(x - x_1)$$
$$y - 4 = -\frac{7}{8}(x - (-2))$$
$$y - 4 = -\frac{7}{8}(x + 2)$$
$$y - 4 = -\frac{7}{8}x - \frac{7}{4}$$
$$y = -\frac{7}{8}x - \frac{7}{4} + 4$$
$$y = -\frac{7}{8}x + \frac{9}{4}$$

59. The line has x-intercept $(2, 0)$ and y-intercept $(0, 1)$, so the slope of the line is
$$m = \frac{1 - 0}{0 - 2} = \frac{1}{-2} = -\frac{1}{2}.$$
Since the y-intercept is $(0, 1)$, we have $b = 1$.

So, the equation is $y = -\dfrac{1}{2}x + 1$.

60. The length of the hypotenuse is 10 and the length of one of the legs is 4. We substitute $a = 4$ and $c = 10$ into $a^2 + b^2 = c^2$ and solve for b:
$$4^2 + b^2 = 10^2$$
$$16 + b^2 = 100$$
$$b^2 = 84$$
$$b = \sqrt{84} \text{ (disregard the negative)}$$
$$b = 2\sqrt{21}$$
The length of the second leg is $2\sqrt{21}$.

61. $\dfrac{5}{2x^2 - 7x + 5}$

We set the denominator equal to 0 and solve for x.
$$2x^2 - 7x + 5 = 0$$
$$(2x - 5)(x - 1) = 0$$
$$2x - 5 = 0 \quad \text{or} \quad x - 1 = 0$$
$$x = \frac{5}{2} \quad \text{or} \quad x = 1$$

So, the excluded values are 1 and $\dfrac{5}{2}$.

62. $\dfrac{-7x+14}{x^2-7x+10} = \dfrac{-7(x-2)}{(x-2)(x-5)} = \dfrac{-7}{x-5} = -\dfrac{7}{x-5}$

63. $-3-6=-9$

Over the past two hours, the change in temperature was $-9°F$.

64. Let x be the number of tablespoons of sugar required for 6 servings of bread pudding.

$$\dfrac{x}{6}=\dfrac{3}{8}$$

$$\dfrac{x}{6}\cdot 6 = \dfrac{3}{8}\cdot 6$$

$$x = 2.25$$

So, 2.25 tablespoons of sugar are required for 6 servings of bread pudding.

65. Since p varies directly as w^2, we have $p=kw^2$. We find k by substituting 2 for w and 12 for p:

$$p = kw^2$$
$$12 = k(2)^2$$
$$12 = 4k$$
$$3 = k$$

The equation is $p=3w^2$. We find the value of p when $w=3$ by substituting 3 for w:

$$p = 3(3)^2 = 3(9) = 27$$

So, $p=27$ when $w=3$.

66. Since c varies inversely as r, we have $c=\dfrac{k}{r}$.

We find k by substituting 2 for r and 10 for c:

$$c = \dfrac{k}{r}$$
$$10 = \dfrac{k}{2}$$
$$20 = k$$

The equation is $c=\dfrac{20}{r}$. We find the value of r when $c=4$ by substituting 4 for c and solving for r:

$$4 = \dfrac{20}{r}$$
$$4\cdot r = \dfrac{20}{r}\cdot r$$
$$4r = 20$$
$$r = 5$$

So, $r=5$ when $c=4$.

67. $\dfrac{\dfrac{1}{4x}+\dfrac{3}{x^2}}{\dfrac{1}{2x}-\dfrac{5}{x^2}} = \dfrac{\dfrac{1}{4x}+\dfrac{3}{x^2}}{\dfrac{1}{2x}-\dfrac{5}{x^2}}\cdot\dfrac{4x^2}{4x^2} = \dfrac{\dfrac{1}{4x}\cdot 4x^2+\dfrac{3}{x^2}\cdot 4x^2}{\dfrac{1}{2x}\cdot 4x^2-\dfrac{5}{x^2}\cdot 4x^2}$

$$= \dfrac{x+12}{2x-20} = \dfrac{x+12}{2(x-10)}$$

68. Let x be amount invested in the account paying 4% annual interest, and let y be the amount invested in the account paying 7% annual interest.

The total invested is $12,000, so our first equation is $x+y=12,000$.

The total interest earned in one year is $615. The interest for each account is obtained by multiplying the interest rate (written as a decimal) by the amount invested in that account. Therefore, our second equation is $0.04x+0.07y=615$.

The system is

$$x+y=12,000 \quad \text{Equation (1)}$$
$$0.04x+0.07y=615 \quad \text{Equation (2)}$$

To eliminate the x terms, multiply both sides of equation (1) by -0.04, yielding the system

$$-0.04x-0.04y=-480$$
$$0.04x+0.07y=615$$

Add the left sides and right sides and solve for y.

$$\begin{array}{r} -0.04x-0.04y=-480 \\ \underline{0.04x+0.07y=615} \\ 0.03y=135 \\ y=4500 \end{array}$$

Substitute 4500 for y in equation (1) and solve for x:

$$x+4500=12,000$$
$$x=7500$$

The person should invest $7500 in the account that pays 4% interest and $4500 in the account that pays 7% interest.

69. a. Since the airplane is descending at a rate of 1350 feet per minute, we can model the situation by a linear equation. The slope is -1350 feet per minute. Since the altitude of the plane as 27,500 feet before beginning its descent, the A-intercept is (0, 27500). So, a reasonable model is $A=-1350t+27,500$.

b. We substitute 7 for t in our model:

$$A=-1350(7)+27,500=18,050$$

The plane's altitude 7 minutes after it has begun it descent will be 18,050 feet.

c. We substitute 1200 for A in our model and solve for t:

$$1200 = -1350t + 27,500$$
$$-26,300 = -1350t$$
$$t = \frac{-26,300}{-1350} \approx 19.48$$

The plane's altitude will be 1200 feet about 19.48 minutes after it begins its descent.

d. The slope is $m = -1350$. It means that the airplane descends by 1350 feet per minute.

e. We find the t-intercept by substituting 0 for A and solving for t:

$$0 = -1350t + 27,500$$
$$1350t = 27,500$$
$$t = \frac{27500}{1350} \approx 20.37$$

The t-intercept is approximately (20.37, 0). It means that the airplane will land on the ground (altitude will be 0 feet) about 20.37 minutes after it begins its descent.

70. a.

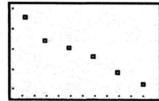

The data can be modeled better by a linear equation than a quadratic equation because the data are close to being in a straight line.

b. Answers may vary depending on the data points selected for creating the model. We use the points (39.5, 25.1) and (59.5, 19.0) to find an equation of the form $p = ma + b$. We begin by finding the slope:

$$m = \frac{19.0 - 25.1}{59.5 - 39.5} = \frac{-6.1}{20.0} \approx -0.31$$

So, the equation of the line is of the form $p = -0.31a + b$. To find b, we substitute the coordinates of the point (59.5, 19.0) into the equation and solve for b:

$$p = -0.31a + b$$
$$19.0 = -0.31(59.5) + b$$
$$19.0 = -18.445 + b$$
$$37.4 \approx b$$

So, the equation is $p = -0.31a + 37.4$.

c. The slope is -0.31. It means that the percentage of American adults who say that they are interested in soccer decreases by 0.31 percentage point per year of age.

d. Answers may vary depending on the model found in part (b). We substitute 25 for a in our model:

$$p = -0.31(25) + 37.4 \approx 29.6$$

We predict that approximately 29.6% of 25-year-old Americans say that they are interested in soccer.

e. Answers may vary depending on the model found in part (b). We substitute 21 for p in our model and solve for a:

$$21 = -0.31a + 37.4$$
$$-16.4 = -0.31a$$
$$53 \approx a$$

We predict that 21% of 53-year-old Americans say that they are interested in soccer.

f. Answers may vary depending on the model found in part (b). We find the a-intercept by substituting 0 for p and solving for a:

$$0 = -0.31a + 37.4$$
$$0.31a = 37.4$$
$$t = \frac{37.4}{0.31} \approx 120.65$$

The a-intercept is approximately (120.65, 0). This means that no 121-year-old Americans are interested in soccer. [Note: the oldest American at the time of publication was 113 years old.]

71. a. $(-0.21t + 6.27) + (0.074t + 1.08)$
$$= -0.21t + 0.074t + 6.27 + 1.08$$
$$= -0.136t + 7.35$$

This expression represents the total average daily oil production in the year that is t years since 1990.

b. We substitute 21 for a:

$$-0.136(21) + 7.35 = -2.856 + 7.35 \approx 4.5$$

It means that, in the year 1990 + 21 = 2011, the total average daily oil production will be about 4.5 million barrels per day.

c. The rate of change is -0.136 million barrels per day per year, or -136 thousands barrels per day per year. In other words, the total average daily oil production declines by 136 thousands barrels each year.

d. From the offshore equation, we substitute $0.074t + 1.08$ for p in the onshore equation:

$$0.074t + 1.08 = -0.21t + 6.27$$

$$0.284t = 5.19$$

$$t \approx 18.3$$

We substitute 18.3 for t in the onshore equation:

$$p = -0.21t + 6.27$$

$$p = -0.21(18.3) + 6.27 \approx 2.43$$

The onshore and offshore average daily oil production will be equal in the year $1990 + 18.3 \approx 2008$. At that time, the average daily oil production will be approximately 2.43 millions barrels per day.

72. First we define c to be the straight line distance (in miles) between the student's home and school. Then we draw a diagram that describes the situation.

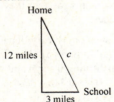

The triangle is a right triangle with legs of length 12 miles and 3 miles.

$$a^2 + b^2 = c^2$$

$$12^2 + 3^2 = c^2$$

$$144 + 9 = c^2$$

$$153 = c^2$$

$$c = \sqrt{153} \approx 12.4 \text{ miles}$$

The length of the trip would be about 12.4 miles.

73. We let p be the number of pens sold when the revenue is \$10,591.80. Our model has the form $R = kp$. We find k by substituting 1275 for p and 8861.25 for R:

$$R = kp$$

$$8861.25 = k(1275)$$

$$6.95 = k$$

The model is $R = 6.95p$. We find the number of pens sold when the revenue is \$10,591 by substituting 10,591 for R:

$$10,591.80 = 6.95p$$

$$1524 = p$$

So, if the revenue is \$10,591.80, then 1524 pens were sold.

74. a.

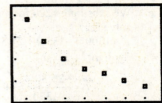

The data can be modeled better by a quadratic equation than a linear equation because the data curve rather than being in a straight line.

b.

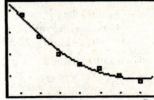

Yes, the model appears to fit the data well.

c. We substitute 23 for a in the model:

$$p = 0.0086(23)^2 - 1.35(23) + 61.3 \approx 34.7994$$

Thus, we estimate that approximately 35% of 23-year-old male drivers were speeding when they became involved in a fatal crash.

d. We substitute 25 for p and solve for a:

$$25 = 0.0086a^2 - 1.35a + 61.3$$

$$0 = 0.0086a^2 - 1.35a + 36.3$$

To solve, we use the quadratic formula with $a = 0.0086$, $b = -1.35$, and $c = 36.3$:

$$a = \frac{-(-1.35) \pm \sqrt{(-1.35)^2 - 4(0.0086)(36.3)}}{2(0.0086)}$$

$$a = \frac{1.35 \pm \sqrt{0.57378}}{0.0172}$$

$$a \approx 34.45 \text{ or } a \approx 122.53$$

We disregard 122.53. We predict that 25% of 34-year-old male drivers were speeding when they became involved in a fatal crash.

75. a.

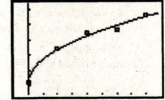

Yes, the model fit the data well.

b. Note that a 30-year-old is represented by $t = 5$. We substitute 5 for a in the model:

$p = 3.4\sqrt{5} + 31 \approx 38.6$

We estimate that 38.6% of 30-year-old Americans go to church in a typical week.

c. We substitute 50% for p in the model and solve for a:

$50 = 3.4\sqrt{a} + 31$

$19 = 3.4\sqrt{a}$

$\dfrac{19}{3.4} = \sqrt{a}$

$a = \left(\dfrac{19}{3.4}\right)^2 \approx 31$

Now, $25 + 31 = 56$. So, we estimate that half of 56-year-old Americans go to church in a typical week.